# Lecture Notes in Physics



Springer
*Berlin*
*Heidelberg*
*New York*
*Barcelona*
*Hong Kong*
*London*
*Milan*
*Paris*
*Tokyo*

Physics and Astronomy ONLINE LIBRARY

http://www.springer.de/phys/

## Editorial Policy

The series *Lecture Notes in Physics* (LNP), founded in 1969, reports new developments in physics research and teaching -- quickly, informally but with a high quality. Manuscripts to be considered for publication are topical volumes consisting of a limited number of contributions, carefully edited and closely related to each other. Each contribution should contain at least partly original and previously unpublished material, be written in a clear, pedagogical style and aimed at a broader readership, especially graduate students and nonspecialist researchers wishing to familiarize themselves with the topic concerned. For this reason, traditional proceedings cannot be considered for this series though volumes to appear in this series are often based on material presented at conferences, workshops and schools (in exceptional cases the original papers and/or those not included in the printed book may be added on an accompanying CD ROM, together with the abstracts of posters and other material suitable for publication, e.g. large tables, colour pictures, program codes, etc.).

## Acceptance

A project can only be accepted tentatively for publication, by both the editorial board and the publisher, following thorough examination of the material submitted. The book proposal sent to the publisher should consist at least of a preliminary table of contents outlining the structure of the book together with abstracts of all contributions to be included.
Final acceptance is issued by the series editor in charge, in consultation with the publisher, only after receiving the complete manuscript. Final acceptance, possibly requiring minor corrections, usually follows the tentative acceptance unless the final manuscript differs significantly from expectations (project outline). In particular, the series editors are entitled to reject individual contributions if they do not meet the high quality standards of this series. The final manuscript must be camera-ready, and should include both an informative introduction and a sufficiently detailed subject index.

## Contractual Aspects

Publication in LNP is free of charge. There is no formal contract, no royalties are paid, and no bulk orders are required, although special discounts are offered in this case. The volume editors receive jointly 30 free copies for their personal use and are entitled, as are the contributing authors, to purchase Springer books at a reduced rate. The publisher secures the copyright for each volume. As a rule, no reprints of individual contributions can be supplied.

## Manuscript Submission

The manuscript in its final and approved version must be submitted in camera-ready form. The corresponding electronic source files are also required for the production process, in particular the online version. Technical assistance in compiling the final manuscript can be provided by the publisher's production editor(s), especially with regard to the publisher's own Latex macro package which has been specially designed for this series.

## Online Version/ LNP Homepage

LNP homepage (list of available titles, aims and scope, editorial contacts etc.):
http://www.springer.de/phys/books/lnpp/

LNP online (abstracts, full-texts, subscriptions etc.):
http://link.springer.de/series/lnpp/

W. Köhler S. Wiegand (Eds.)

# Thermal Nonequilibrium Phenomena in Fluid Mixtures

Springer

**Editors**

Werner Köhler
Universität Bayreuth
Physikalisches Institut
95440 Bayreuth, Germany

Simone Wiegand
Max-Planck-Institut für Polymerforschung
Ackermannweg 10
55128 Mainz, Germany

---

*Cover picture:* see figure 10, page 19, contribution by B. Hafskjold in this volume

---

Library of Congress Cataloging-in-Publication Data.

Die Deutsche Bibliothek - CIP-Einheitsaufnahme

Thermal nonequilibrium phenomena in fluid mixtures / W. Köhler ; S. Wiegand (ed.). - Berlin ; Heidelberg ; New York ; Barcelona ; Hong Kong ; London ; Milan ; Paris ; Tokyo : Springer, 2002
(Lecture notes in physics ; Vol. 584)
(Physics and astronomy online library)
ISBN 3-540-43231-0

ISSN 0075-8450
ISBN 3-540-43231-0 Springer-Verlag Berlin Heidelberg New York



Springer-Verlag Berlin Heidelberg New York
a member of BertelsmannSpringer Science+Business Media GmbH

http://www.springer.de



Typesetting: Camera-data conversion by Steingraeber Satztechnik GmbH Heidelberg
Cover design: *design & production*, Heidelberg

Printed on acid-free paper
SPIN: 10863191 54/3141/du - 5 4 3 2 1 0

# Preface

In September 2000, the University of Bayreuth, Germany, hosted the Fourth International Meeting on Thermodiffusion (IMT4).

The IMT conferences were born from the idea of bringing together researchers in the field of thermodiffusion. Under the auspices of the *European Group of Research in Thermodiffusion* (EGRT) the conference series started in 1994 with IMT1 in Toulouse and has been continued every other year with IMT2 (Pau, 1996), IMT3 (Mons, 1998), and IMT4 (Bayreuth, 2000). The next conference, IMT5, will be held in 2002 in Lyngby, Denmark.

Thermodiffusion, also called thermal diffusion or the Ludwig-Soret effect, describes the coupling between a temperature gradient and a resulting mass flux. Although the effect was already discovered in the 19th century by Ludwig and Soret, it has gained growing interest during the last years due to improved experimental techniques like state-of-the-art thermogravitational columns, modern optical methods, flow channels, and microgravity experiments, to mention only a few. We are still far from a detailed microscopic picture, but analytical theories have been improved and the availability of fast computers and efficient algorithms for nonequilibrium molecular dynamics simulations has provided valuable input from the theoretical side.

The IMT conferences cover all aspects of thermodiffusion from fundamentals to new applications. Traditionally, the focus has been on the fluid state, ranging from mixtures of simple liquids to more complex systems such as critical mixtures, electrolytes, polymers, colloidal dispersions, or magnetic fluids. IMT4 tried to widen the scope by including a plenary lecture about thermodiffusion in ionic solids. Scientific input comes from diverse disciplines such as physics, chemistry, engineering, and geophysics.

Sadly, Leo Kempers passed away while this book was being prepared. Many of us have lost a friend and respected colleague. His manuscript has been brought into its final state by A. Shapiro, whom we want to thank here.

IMT4 would not have been possible without help from many people, ranging from the scientific committee to students, secretaries, and technicians helping with the local organization. We are grateful to the University of Bayreuth for hosting the conference and for financial support from the Deutsche Forschungsgemeinschaft, the Bavarian State Ministry of Education, Culture, Science and Art, the Emil Warburg Foundation, Wyatt Technology Germany, and the Max Planck Institute for Polymer Research.

Bayreuth, Mainz,
August 2001

*Werner Köhler*
*Simone Wiegand*

# Organization of the Book

Since reviews about thermodiffusion research are scarce, it was our intention to provide a comprehensive overview of the current activities in the field in book form. Consequently, the contributions within this volume are not merely the papers presented at IMT4 but were written with the aim to contain both a review-like introduction together with recent research results. Hence, the book should be of value for both the experts and interested scientists working in different areas. It is organized in three Parts, but the classification is not always sharp.

In the First Part, general concepts, theoretical aspects and computer simulations are discussed. **Bjørn Hafskjold** reviews simulation methods to study thermal diffusion. Equilibrium molecular dynamic simulations using Green-Kubo formalism as well as non-equilibrium methods using linear response theory to derive the transport coefficients are discussed. **Jutta Luettmer-Strathmann** provides a summary on the asymptotic and crossover behavior of thermodiffusion and other transport properties close to the critical point. **Konstantin I. Morozov** develops a theory of the Soret effect on surfacted and ionic colloids. He shows that the double layer thickness and the electric potential of the particle surface determine the sign of the Soret coefficient. **Alexander A. Shapiro** and **Erling H. Stenby** discuss to what extent the concept of principle of entropy maximization can be transferred from the framework of equilibrium thermodynamics to non-equilibrium steady states. **Leo J.T.M. Kempers** uses a thermodynamic approach, which includes kinetic contributions to predict the Soret effect in multicomponent mixtures. Comparison with mixtures relevant to the chemical and petroleum industry shows agreement within a factor of two over four decades. **Ryszard Wojnar** derives a kinetic theory for Brownian particles under the influence of a gravity and a temperature field. He applies this theory to the thermodiffusion process in porous media. **Jan V. Sengers** and **José M. Ortiz de Zárate** demonstrate that the Soret effect induces long-range concentration fluctuations in a binary liquid system which is in a stationary thermal non-equilibrium state. **Jürgen Janek**, **Carsten Korte** and **Alan B. Lidiard** summarize the current state of thermodiffusion in ionic solids. Model experiments as well as theoretical approaches are discussed. The last contribution of the first Part is a song which was performed by **Florian Müller-Plathe** during his presentation. It gives a historic overview and presents recent results from computer simulations.

In the Second Part of the book, experimental techniques as well as their application to special systems, such as polymers, are discussed. In the first paper, **Simone Wiegand** and **Werner Köhler** summarize the recent applications and developments in optical grating techniques. The influence of convection and the approach of a critical point are discussed in more detail. **Guy Chavepeyer**, **Jean-François Dutrieux**, **Stéfan Van Vaerenbergh**, and **Jean-Claude Legros** describe experimental effects in thermal diffusion flow cells, which perturb the measurement of small Soret coefficients. They discuss numerical simulation results and compare those to experimental data. **Javier Valencia**, **Mohamed Mounir Bou-Ali**, **Oscar Ecenarro**, **José Antonio Madariaga** and **Carlos M. Santamaría** explain the limitations of the Furry, Jones and Onsager thermogravitational column theory. **Michel Martin**, **Charles Van Batten** and **Mauricio Hoyos** give a background in the thermal field flow fractionation technique, which is a standard liquid chromatography technique for polymer fractionation and characterization. **Martin E. Schimpf** reviews thermodiffusion theories. He compares them with experimental results obtained for various polymers by field flow fractionation and presents a recently developed hydrodynamic model.

The Third Part of the book covers theoretical and experimental aspects of thermodiffusion and convection, including thermodiffusion in porous media. In the first contribution, **Jean-Karl Platten**, **Jean-François Dutrieux** and **Guy Chavepeyer** discuss the combination of free convection and thermodiffusion for the measurement of Soret coefficients in Rayleigh-Bénard cells and thermogravitational columns. **Björn Huke** and **Manfred Lücke** summarize their finding about laterally periodic convection structures in binary mixtures in the Rayleigh–Bénard system for positive Soret effect. **Mark I. Shliomis** focuses on the convective instabilities in ferrofluid layers in a Rayleigh–Bénard cell heated from above or from below in the presence of a magnetic field. **Boris L. Smorodin**, **Bela I. Myznikova** and **Igor O. Keller** present a theoretical study of the influence of transverse vibrations on the formation of instabilities in binary mixtures. They show that, depending on the amplitude and frequency of the modulation, the vibrations can stabilize or destabilize the equilibrium state of the liquid. The last three papers in the third section discuss thermodiffusion in porous media. **Pierre Costesèque**, **Daniel Fargue** and **Philippe Jamet** review the experimental, theoretical and numerical studies performed on thermodiffusion and thermodiffusion-convection transport in porous media. **Mohamed N. Ouarzazi**, **Annabelle Joulin**, **Pierre-Antoine Bois** and **Jean K. Platten** study the pattern formation of a binary mixture in a porous medium heated from below in the presence of a horizontal flow. The book concludes with a paper by **Bruno Lacabanne**, **Serge Blancher**, **René Creff** and **François Montel**, who derive a model and perform numerical simulations for the Soret effect in multicomponent flow through porous media.

An attempt has been made to use a uniform nomenclature, but this has proved to be an almost impossible task with contributions from different scientific disciplines. The most important symbols that most authors could agree on are summarized in a global glossary on page XVII. At the end of every contribution there is a supplementary glossary with the symbols not contained in the global one. Some authors preferred to stay with their own established notation and deviated from the conventions in the global glossary. Hence, the reader is always advised to check the supplementary glossary.

There is no agreement in the literature on how to define the sign of the Soret coefficient. The sign convention used in this book does not depend on the densities of the components and is explained in detail on page XVII.

We would like to express our thanks to the contributors to this volume for their enthusiasm and their ready cooperation in making this book a timely reflection of the progress achieved in various theoretical aspects and applications of thermodiffusion.

# Contents

# List of Contributors

**Serge Blancher**
Laboratoire de Transferts Thermiques
Université de Pau
64000 Pau, France
serge.blancher@univ-pau.fr

**Pierre-Antoine Bois**
Université de Lille
LML, 59655 Villeneuve d'Ascq Cedex, France
pierre-antoine.bois@univ-lille1.fr

**Mohamed Mounir Bou-Ali**
Departamento de Ingeniería Mecánica Energética y de Materiales
Universidad Pública de Navarra
Pamplona, Spain
mounir.bouali@unavarra.es

**Guy Chavepeyer**
Université de Mons-Hainaut
General Chemistry
Avenue du Champ de Mars, 24,
B-7000 Mons, Belgium
guy.chavepeyer@umh.ac.be

**Pierre Costesèque**
Université Paul Sabatier
39, allées Jules Guesde
31000 Toulouse, France
costeseq@cict.fr

**René Creff**
Laboratoire de Transferts Thermiques
Université de Pau
64000 Pau, France
rene.creff@univ-pau.fr

**Jean-François Dutrieux**
Université de Mons-Hainaut
General Chemistry
Avenue du Champ de Mars, 24,
B-7000 Mons, Belgium
jean-francois.dutrieux@umh.ac.be

**Oscar Ecenarro**
Departamento de Física Aplicada II
Universidad del País Vasco, Apdo 644
48080 Bilbao, Spain
wdpecaro@lg.ehu.es

**Daniel Fargue**
Ecole des Mines de Paris
60, bd Saint Michel
75272 Paris Cedex 06, France
fargue@de.ensmp.fr

**Bjørn Hafskjold**
Department of Chemistry
Norwegian University of Science and Technology
N-7491 Trondheim, Norway
bhaf@chembio.ntnu.no

**Mauricio Hoyos**
Ecole Supérieure de Physique et de Chimie Industrielles
Laboratoire de Physique et Mécanique des Milieux Hétérogènes
(UMR CNRS 7636), 10, rue Vauquelin
75231 Paris Cedex 05, France
hoyos@pmmh.espci.fr

**Björn Huke**
Institut für Theoretische Physik
Universität des Saarlandes
Postfach 151150
66041 Saarbrücken, Germany
huke@lusi.uni-sb.de

**Philippe Jamet**
Ecole des Mines de Paris
35, rue Saint-Honoré
77305 Fontainebleau Cedex, France
jamet@isige.ensmp.fr

**Jürgen Janek**
Physikalisch-Chemisches Institut
Justus-Liebig-Universität Gießen
Heinrich-Buff-Ring 58, 35392 Gießen, Germany
juergen.janek@phys.chemie.uni-giessen.de

**Annabelle Joulin**
Université de Lille
LML, 59655 Villeneuve d'Ascq Cedex
France
annabelle.joulin@univ-lille1.fr

**Igor O. Keller**
Intel Israel (74) Ltd.
P.O. Box 1659
Haifa, 31015 Israel
igor.keller@intel.com

**Leo J.T.M. Kempers**
Shell International Exploration & Production
P.O. Box 60, 2280 AB Rijswijk
The Netherlands

**Werner Köhler**
Physikalisches Institut
Universität Bayreuth
95440 Bayreuth, Germany
werner.koehler@uni-bayreuth.de

**Carsten Korte**
Physikalisch-Chemisches Institut
Justus-Liebig-Universität Gießen
Heinrich-Buff-Ring 58, 35392 Gießen, Germany
Carsten.Korte@phys.Chemie.uni-giessen.de

**Bruno Lacabanne**
Laboratoire de Mathématiques Appliquées
Université de Pau
64000 Pau, France
bruno.lacabanne@univ-pau.fr

**Jean-Claude Legros**
Université Libre de Bruxelles
Microgravity Research Center (MRC)
Chemical Physics E.P. Dept. - CP 165
Avenue F.D. Roosevelt, 50
B-1050 Bruxelles, Belgium
jclegros@ulb.ac.be

**Alan B. Lidiard**
The J.J. Thomson Physical Laboratory
University of Reading
Whiteknights, Reading RG6 6AF, UK
A.B.Lidiard@reading.ac.uk

**Manfred Lücke**
Institut für Theoretische Physik
Universität des Saarlandes
Postfach 151150
66041 Saarbrücken, Germany
luecke@lusi.uni-sb.de

**Jutta Luettmer-Strathmann**
Department of Physics
The University of Akron
Akron, OH 44325-4001, USA
jutta@physics.uakron.edu

**José Antonio Madariaga**
Departamento de Física Aplicada II
Universidad del País Vasco, Apdo 644

48080 Bilbao, Spain
wupmazaj@lg.ehu.es

**Michel Martin**
Ecole Supérieure de Physique et de Chimie Industrielles
Laboratoire de Physique et Mécanique des Milieux Hétérogènes
(UMR CNRS 7636), 10, rue Vauquelin
75231 Paris Cedex 05, France
martin@pmmh.espci.fr

**François Montel**
Elf Exploration Production
Centre Scientifique Jean Feger
Pau, Francc
francois.montel@elf-p.fr

**Konstantin I. Morozov**
Institute of Continuous Media Mechanics
UB of Russian Academy of Science
Korolev Str. 1
614013 Perm, Russia
mrk@icmm.ru

**Florian Müller-Plathe**
Max-Planck-Institut für Polymerforschung
Ackermannweg 10
55128 Mainz, Germany
mplathe@mpip-mainz.mpg.de

**Bela I. Myznikova**
Institute of Continuous Media Mechanics
UB of Russian Academy of Science
Korolev Str. 1
614013 Perm, Russia
myz@icmm.ru

**José M. Ortiz de Zárate**
Departamento de Física Aplicada 1
Facultad de Ciencias Físicas
Universidad Complutense
E-28040 Madrid, Spain
fiap102@sis.ucm.es

**Mohamed Najib Ouarzazi**
Université de Lille
LML, 59655 Villeneuve d'Ascq Cedex
France
ouarzazi.najib@univ-lille1.fr

**Jean Karl Platten**
Université de Mons-Hainaut
General Chemistry
Avenue du Champ de Mars, 24,
B-7000 Mons, Belgium
jean.platten@umh.ac.be

**Carlos María Santamaría**
Departamento de Física Aplicada II
Universidad del País Vasco Apdo 644
48080 Bilbao, Spain
wdpsasac@lg.ehu.es

**Martin E. Schimpf**
Department of Chemistry
Boise State University
Boise, ID 83725, USA
mschimpf@boisestate.edu

**Jan V. Sengers**
Institute for Physical Science and Technology
and Department of Chemical Engineering
University of Maryland
College Park, MD 20742, USA
sengers@ipst.umd.edu

**Alexander A. Shapiro**
Engineering Research Center
IVC-SEP
Department of Chemical Engineering
Technical University of Denmark
DK 2800 Lyngby, Denmark
ash@kt.dtu.dk

**Mark I. Shliomis**
Department of Mechanical Engineering
Ben-Gurion University of the Negev
P.O.B. 653, Beer-Sheva 84105, Israel
shliomis@netvision.net.il

**Boris L. Smorodin**
Perm State University
Department of Theoretical Physics
Bukirev str. 15
614600 Perm, Russia
boris.smorodin@psu.ru

**Erling H. Stenby**
Engineering Research Center
IVC-SEP
Department of Chemical Engineering
Technical University of Denmark
DK 2800 Lyngby, Denmark
ehs@kt.dtu.dk

**Javier Valencia**
Departamento de Ingeniería Mecánica
Energética y de Materiales
Universidad Pública de Navarra
Pamplona, Spain
valencia@unavarra.es

**Charles Van Batten**
Ecole Supérieure de Physique
et de Chimie Industrielles
Laboratoire de Physique et Mécanique
des Milieux Hétérogènes
(UMR CNRS 7636), 10, rue Vauquelin
75231 Paris Cedex 05, France
charles.van_batten@yahoo.com

**Stéfan Van Vaerenbergh**
Université Libre de Bruxelles
Microgravity Research Center (MRC)
Chemical Physics E.P. Dept., CP 165
Avenue F.D. Roosevelt, 50
B-1050 Bruxelles, Belgium
svanvaer@ulb.ac.be

**Simone Wiegand**
Max-Planck-Institut
für Polymerforschung
Ackermannweg 10
55128 Mainz, Germany
wiegand@mpip-mainz.mpg.de

**Ryszard Wojnar**
Instytut Podstawowych Problemów
Techniki
Polska Akademia Nauk
ul. Świętokrzyska 21
00-049 Warszawa, Poland
rwojnar@ippt.gov.pl

# Glossary of Symbols

| | |
|---|---|
| $c$ | concentration (weight fraction) |
| $c_{\mathrm{p}}$ | specific heat capacity at constant pressure |
| $c_{\mathrm{v}}$ | specific heat capacity at constant volume |
| $D$ | mutual diffusion coefficient |
| $D_{\mathrm{T}}$ | thermal diffusion coefficient |
| $D_{\mathrm{th}}$ | thermal diffusivity |
| $F$ | Helmholtz free energy |
| $k_{\mathrm{B}}$ | Boltzmann's constant |
| $Le$ | Lewis number |
| $M$ | molar mass |
| $Pr$ | Prandtl number |
| $q$ | wave vector |
| $t$ | time |
| $Ra$ | Rayleigh number |
| $Re$ | Reynolds number |
| $Sc$ | Schmidt number |
| $S_{\mathrm{T}}$ | Soret coefficient (see below for sign convention): $S_{\mathrm{T}} = D_{\mathrm{T}}/D$ |
| $T$ | temperature |
| $x$ | concentration (mole fraction) |
| $u$ | fluid velocity |
| $\alpha$ | cubic expansion coefficient: $\alpha = -\left(\frac{1}{\rho}\frac{\partial \rho}{\partial T}\right)_{p,c}$ |
| $\alpha_{\mathrm{T}}$ | thermal diffusion factor: $\alpha_{\mathrm{T}} = TS_{\mathrm{T}}$ |
| $\beta$ | solutal expansion coefficient: $\beta = \left(\frac{1}{\rho}\frac{\partial \rho}{\partial c}\right)_{p,T}$ |
| $\lambda$ | wavelength of light |
| $\lambda_{\mathrm{T}}$ | thermal conductivity |
| $\mu$ | difference in chemical potential per unit mass between the two species in a binary mixture |
| $\phi$ | concentration (volume fraction) |
| $\psi$ | separation ratio: $\psi = \frac{\beta}{\alpha} S_{\mathrm{T}} c(1-c)$ |
| $\rho$ | mass density |

## Sign convention for $S_{\mathrm{T}}$ and $D_{\mathrm{T}}$

In a binary mixture of A and B, '$S_{\mathrm{T}}$ of A' (and also $D_{\mathrm{T}}$) is positive if A migrates to the cold side. This implies that $S_{\mathrm{T}}$ of B must be negative, since B migrates to the hot side. Usually, one would specify $S_{\mathrm{T}}$ for the component that has been used for the definition of the concentration. Note that this definition does not depend on the densities of the two components.

# Part I

# Concepts, Theory, and Computer Simulations

# Computer Simulations of Thermal Diffusion in Binary Fluid Mixtures

Bjørn Hafskjold

Department of Chemistry, Norwegian University of Science and Technology, N-7491 Trondheim, Norway

**Abstract.** Compared to other transport properties, thermal diffusion has not been much studied by computer simulations. During the last decade, however, a couple of different simulation methods have been applied to a few systems, aiming both at determination of numerical values of the Soret coefficient and at a better understanding of the phenomenon. Based on such simulations, we will present some explanations for why there is a Soret effect in binary fluids.

The computer simulation methods may be put in two main categories; equilibrium molecular dynamics (MD) simulations using Green-Kubo formalism to derive the transport coefficients, and non-equilibrium methods (NEMD) using linear response theory to derive the coefficients. The NEMD methods are in turn of two types, synthetic NEMD where the external thermodynamic force is built into the equations of motion for the system, and boundary-driven NEMD, where the external force stems from perturbations of the systems boundaries. We will discuss possibilities and limitations with these methods, and what we have learned from the simulations.

Some possibilities for future simulation work will be outlined, e.g. coupled transport processes across interfaces and in porous media.

## 1 Introduction

The purpose of computer simulations of thermal diffusion is to describe, understand, and eventually predict the behaviour of mixtures in a temperature gradient. Several industrial processes, such as separation of isotopes, separation of products in a porous catalyst in a chemical reactor, condensing a vapour mixture on a cold surface, and water injection on oil reservoirs, have in common that they involve a mixture subject to a temperature gradient. The phenomenological description of this situation can be made by irreversible thermodynamics [1], shown here for a binary mixture:

$$\boldsymbol{J}_{\mathrm{q}} = -l_{\mathrm{qq}}\frac{\nabla T}{T^2} - l_{\mathrm{q1}}\frac{\nabla_{\mathrm{T}}(\mu_1-\mu_2)}{T} \tag{1a}$$

$$\boldsymbol{J}_1 = -l_{\mathrm{1q}}\frac{\nabla T}{T^2} - l_{11}\frac{\nabla_{\mathrm{T}}(\mu_1-\mu_2)}{T} \tag{1b}$$

where $\boldsymbol{J}_{\mathrm{q}}$ and $\boldsymbol{J}_1$ are the heat flux and diffusive molar flux of component 1, respectively, $T$ is the temperature, $\mu_{\mathrm{i}}$ the chemical potential of component $i$, and $\nabla$ is the grad operator. The subscript $T$ denotes that the temperature variation of the chemical potential is not included. For simplicity, we shall in the following

W. Köhler and S. Wiegand (Eds.): LNP 584, pp. 3–23, 2002.

assume that all thermodynamic forces and fluxes act in the same direction, and use scalar notation instead of vector notation, except when vector notation is required for clarity.

The $l_{\alpha\beta}$ are the phenomenological coefficients, subject to Onsager's reciprocity relations [2],

$$l_{\alpha\beta} = l_{\beta\alpha} \tag{2}$$

The flux of component 2, $J_2$, does not appear in Eq. (1) because it depends on $J_1$ through the reference frame for the velocities. In terms of the average velocities of the two individual species, $v_\mathrm{k}$, each of the two mass fluxes are

$$J_\mathrm{k} = \tilde{c}_\mathrm{k}\left(v_\mathrm{k} - v_0\right) \tag{3}$$

where $\tilde{c}_\mathrm{k}$ is the concentration of component $k$ and $v_0$ is the velocity of the reference frame. In most reported work, the barycentric reference frame is used, *i.e.* $v_0$ is the velocity of the center of mass, given by

$$\left(\sum_{\mathrm{k}=1}^{2} \rho_\mathrm{k}\right) v_0 = \sum_{\mathrm{k}=1}^{2} \rho_\mathrm{k} v_\mathrm{k} \tag{4}$$

In this case, the two *mass* fluxes are related by $J_1 = -J_2$ for a binary mixture, and the only independent chemical potential gradient is $\nabla_\mathrm{T}(\mu_1 - \mu_2)/T$ ($\mu_\mathrm{k}$ is here the partial *specific* Gibbs energy of component $k$).

In chemical applications, it is most convenient to use the mean molar reference frame where $v_0$ is given by

$$\left(\sum_{\mathrm{k}=1}^{2} \tilde{c}_\mathrm{k}\right) v_0 = \sum_{\mathrm{k}=1}^{2} \tilde{c}_\mathrm{k} v_\mathrm{k} \tag{5}$$

and the two *molar* fluxes are connected by $J_1 = -J_2$. Like for the barycentric case, the only independent chemical potential gradient is $\nabla_\mathrm{T}(\mu_1 - \mu_2)/T$ (but $\mu_\mathrm{k}$ is here the partial *molar* Gibbs energy of component $k$).

If we use component 2 as reference frame (the solvent reference frame), only the relative flux,

$$J_\mathrm{r} = \tilde{c}_1\left(v_1 - v_2\right) = J_1 - \frac{x_1}{x_2} J_2 \tag{6}$$

enters into the flux-force relations. The only independent chemical force is in this case $\nabla_\mathrm{T}\mu_1/T$. The solvent reference frame is equivalent to applying the Gibbs-Duhem equation to the chemical potential gradients.

In computer simulations, it is convenient to use the MD cell as frame of reference and represent $v_\mathrm{k}$ by the coordinates used in the simulation. This is equivalent to the choice $v_0 = 0$. For a two-phase system, it is natural to refer the fluxes to the interface, which is then stationary in the MD cell.

For most of the results reported in this work, $J_1 = -J_2$, so that the MD cell and mean molar reference frames are identical.

Heat- and mass transfer are coupled through the Soret- and Dufour effects, expressed by

$$J_1 = \frac{l_{1\mathrm{q}}}{l_{11}} (J_{\mathrm{q}})_{\nabla_{\mathrm{T}}(\mu_1 - \mu_2)=0} \tag{7}$$

and

$$J_{\mathrm{q}} = \frac{l_{\mathrm{q}1}}{l_{11}} (J_1)_{\nabla T=0} \, , \tag{8}$$

respectively.

In terms of operational quantities, the fluxes may alternatively be written

$$J_{\mathrm{q}} = -\lambda_{\mathrm{T}} \nabla T + q^* J_1 \tag{9a}$$

$$J_1 = -D_{\mathrm{T}} \tilde{c} x_1 x_2 \nabla T - D \tilde{c} \nabla x_1 \tag{9b}$$

where $\lambda_{\mathrm{T}}$ is the thermal conductivity at zero mass flux, $q^*$ is the heat of transport, $D_{\mathrm{T}}$ the thermal diffusion coefficient, $D$ the diffusion coefficient, $\tilde{c}$ the overall concentration, and $x_{\mathrm{i}}$ the mole fraction of component $i$. With these quantities,

$$\frac{l_{1\mathrm{q}}}{l_{11}} = RT^2 S_{\mathrm{T}} \left( 1 + \frac{\partial \ln \gamma_1}{\partial \ln x_1} \right) \tag{10a}$$

$$\frac{l_{\mathrm{q}1}}{l_{11}} = q^* \tag{10b}$$

where $R$ is the gas constant and $\gamma_1$ is the activity coefficient of component 1, and $S_{\mathrm{T}}$ the Soret coefficient,

$$S_{\mathrm{T}} = \frac{D_{\mathrm{T}}}{D} \tag{11}$$

The thermal diffusion factor, $\alpha_{\mathrm{T}}$, is related to $D_{\mathrm{T}}$ by

$$\alpha_{\mathrm{T}} = S_{\mathrm{T}} T \tag{12}$$

Determination of $\alpha_{\mathrm{T}}$ and $q^*$ can be made from Eqs. (9-12):

$$\alpha_{\mathrm{T}} = -\frac{T}{x_1 x_2} \left( \frac{\nabla x_1}{\nabla T} \right)_{\mathrm{J}_1=0} \tag{13}$$

and

$$q^* = \left( \frac{J_{\mathrm{q}}}{J_1} \right)_{\nabla \mathrm{T}=0} \tag{14}$$

Incidentally, Eq. (10) also shows why the Soret and Dufour effects become significant in the critical region; $l_{11}$ vanishes whereas $l_{1\mathrm{q}}$ and $l_{\mathrm{q}1}$ do not [3].

## 2 Computer simulation methods

Simulations of coupled heat- and mass transport are made with the molecular dynamics (MD) method. MD algorithms may be put in two main categories; *equilibrium* MD (EMD) using the Green-Kubo formalism (see *e.g.* McQuarrie [4]) to derive the transport coefficients, and *non-equilibrium* MD (NEMD) using linear response theory to derive the coefficients (see *e.g.* Evans and Morriss [5]). In both cases, the fluxes and forces must be expressed in molecular quantities.

The mass flux can be readily obtained in a mechanical approach such as MD. The molar flux is (where we have retained vector notation for clarity)

$$\boldsymbol{J}_{\mathrm{k}} = \frac{1}{V N_{\mathrm{A}}} \left\langle \sum_{\mathrm{i} \in \mathrm{k}} (\boldsymbol{v}_{\mathrm{i}} - \boldsymbol{v}_{0}) \right\rangle \tag{15}$$

where $\boldsymbol{v}_{\mathrm{i}}$ is the velocity of particle $i$, $\boldsymbol{v}_0$ is the velocity of the reference frame, and $N_{\mathrm{A}}$ is Avogadro's number. The $\langle ... \rangle$ in Eq. (15) denotes an ensemble average and the summation includes only particles of type $k$.

The heat flux cannot be computed directly because it is not a mechanical property. It is, however, related to the enthalpy flux, $\boldsymbol{J}_{\mathrm{H}}$, by

$$\boldsymbol{J}_{\mathrm{q}} = \boldsymbol{J}_{\mathrm{H}} - H_1 \boldsymbol{J}_1 - H_2 \boldsymbol{J}_2 \tag{16}$$

where $H_{\mathrm{k}}$ is the partial molar enthalpy of component $k$. To determine the partial molar enthalpy, we need the enthalpy as function of composition, which in principle requires a series of MD simulations. With the chosen mean molar frame of reference, Eq. (16) becomes

$$\boldsymbol{J}_{\mathrm{q}} = \boldsymbol{J}_{\mathrm{H}} - (H_1 - H_2) \boldsymbol{J}_1 \tag{17}$$

The enthalpy flux may be expressed in terms of mechanical quantities [5,6]

$$\boldsymbol{J}_{\mathrm{H}} = \frac{1}{V} \left\langle \sum_{\mathrm{i}} \left( \left[ \frac{1}{2} m_{\mathrm{i}} (\boldsymbol{v}_{\mathrm{i}} - \boldsymbol{v}_0)^2 + \Phi_{\mathrm{i}} \right] (\boldsymbol{v}_{\mathrm{i}} - \boldsymbol{v}_0) - \frac{1}{2} \sum_{\substack{j=1 \\ j \neq i}}^{\mathrm{N}} [(\boldsymbol{v}_{\mathrm{i}} - \boldsymbol{v}_0) \cdot \boldsymbol{F}_{\mathrm{ij}}] \boldsymbol{r}_{\mathrm{ij}} \right) \right\rangle \tag{18}$$

where $m_{\mathrm{i}}$ is the mass of particle $i$, $\Phi_{\mathrm{i}}$ is the potential energy of particle $i$ in the field of all the other particles, $\boldsymbol{F}_{\mathrm{ij}}$ is the force acting on $i$ due to $j$, and $\boldsymbol{r}_{\mathrm{ij}}$ is the vector from the position of $i$ to the position of $j$. $\boldsymbol{J}_{\mathrm{H}}$ is a relative quantity. In Eq. (18) this property is reflected in the reference value for $\Phi_{\mathrm{i}}$. In the present case this reference value is such that $\Phi_{\mathrm{i}} = 0$ when the interparticle distance is infinite.

The heat flux, $\boldsymbol{J}_{\mathrm{q}}$, must be determined from Eq. (16). Three important cases avoid the non-trivial task of determining $H_{\mathrm{k}}$; (1) isotope mixtures with $H_1 = H_2$, giving $\boldsymbol{J}_{\mathrm{q}} = \boldsymbol{J}_{\mathrm{H}}$ in the mean molar reference frame, and (2) $\boldsymbol{J}_1 = \boldsymbol{J}_2 = 0$, again with $\boldsymbol{J}_{\mathrm{q}} = \boldsymbol{J}_{\mathrm{H}}$ as the result, and (3) ideal mixtures with $H_{\mathrm{k}}$ equal to the molar enthalpy of component $k$.

The role of the external force ($\nabla T$ or $\nabla \mu_{\mathrm{k}}$) depends on the chosen MD method, and will be discussed in the context of each method.

### 2.1 Equilibrium MD

The EMD method has been used extensively to compute the diagonal elements of the transport coefficients ($l_{\alpha\alpha}$) for homogeneous systems, and suitable expressions may be found in *e.g.* [4,7,8]. Off-diagonal elements, and in particular the thermal diffusion coefficient, can also be determined from EMD [4,8,9]:

$$D_{\mathrm{T}} = \frac{V}{3RT^2\tilde{c}x_1x_2} \int_0^{\infty} \langle \boldsymbol{J}_1(t) \cdot \boldsymbol{J}_{\mathrm{q}}(0) \rangle \, dt \tag{19}$$

where $V$ is the molar volume and $\boldsymbol{J}_\alpha(t)$ is the instantaneous value of $\boldsymbol{J}_\alpha$. Experience has shown that the thermal conductivity and the self diffusion coefficient can be effectively computed with the Green-Kubo formalism, whereas the cross-coefficients suffer from less efficiency in terms of the amount of compute power for a given level of accuracy in the results, due to the weaker correlation between $\boldsymbol{J}_1(t)$ and $\boldsymbol{J}_{\mathrm{q}}(t)$ and the fact that $\boldsymbol{J}_{\mathrm{q}}$ is not directly accessible in a MD simulation [10].

The main advantage of EMD is that the system is uniform, which in turn means that the properties are well defined. The main disadvantage is that the system is not directly related to an experimental situation with external forces that drive the transport process. Moreover, if the objective is to study coupled heat- and mass transport in heterogeneous systems, such as across an interface or in a porous medium, EMD cannot be applied.

### 2.2 Non-equilibrium MD

NEMD methods are in turn of two types, *synthetic* NEMD, where the external thermodynamic force is built into the Hamiltonian or the equations of motion of the system, and *boundary-driven* NEMD, where the external force stems from perturbations of the systems boundaries. These methods have in recent years become increasingly popular in studies of coupled transport phenomena because they give a better signal-to-noise ratio than EMD, and they have been used to compute transport coefficients as well as studying the transport mechanisms at a molecular scale [9,11,12].

**Synthetic NEMD.** The idea of synthetic NEMD is to create equations of motion that dissipate energy at a certain rate, and then associate the energy dissipation with an external force and a conjugate flux [5]. In some cases, such as diffusion of differently coloured particles (tracer diffusion) and shear flow, one can find a Hamiltonian that completely describes the system, and the simulation is technically speaking essentially the same as an equilibrium simulation. In other cases, like for heat conduction, a corresponding Hamiltonian has not been found, but the equations of motion are still homogeneous and compatible with periodic boundary conditions. The equations of motion used to compute thermal conductivity are of the form [5,11,13]

$$\frac{\partial \boldsymbol{r}_{\mathrm{i}}}{\partial t} = \frac{\boldsymbol{p}_{\mathrm{i}}}{m_{\mathrm{i}}} \tag{20a}$$

$$\frac{\partial \boldsymbol{p}_{\mathrm{i}}}{\partial t} = \boldsymbol{F}_{\mathrm{i}}(t) + \boldsymbol{R}_{\mathrm{i}}(t) \tag{20b}$$

where $\boldsymbol{r}_{\rm i}$ and $\boldsymbol{p}_{\rm i}$ are the position and momentum, respectively, of particle $i$, $m_{\rm i}$ is the mass of particle $i$, $\boldsymbol{F}_{\rm i}$ is the force acting on $i$ due to all the other particles, and $\boldsymbol{R}_{\rm i}$ is the required perturbation that is identified as the external force,

$$\boldsymbol{R}_{\rm i}(t) = (E_{\rm i} - \overline{E})\boldsymbol{F}_{\rm i}(t) - \frac{1}{2}\sum_{\rm j} \boldsymbol{F}_{\rm ij}\left[\boldsymbol{r}_{\rm ij} \cdot \boldsymbol{F}_{\rm i}(t)\right] + \frac{1}{2N}\sum_{\rm j,k} \boldsymbol{F}_{\rm jk}\left[\boldsymbol{r}_{\rm jk} \cdot \boldsymbol{F}_{\rm i}(t)\right] \quad (21)$$

where $E_{\rm i}$ is the kinetic and potential energy of particle $i$ and $\overline{E}$ is the average energy of all the particles at time $t$.

The transport coefficients are derived from the measured linear response in the flux to the force $\boldsymbol{R}_{\rm i}$. Synthetic NEMD is not limited to small perturbations of the equilibrium Hamiltonian, and the signal-to-noise ratio for the transport coefficients is considerably improved compared to equilibrium simulations.

The main advantage of synthetic NEMD is that the equations of motion are compatible with periodic boundary conditions at all spatial and temporal variations of the external force, and that the system is homogeneous. The latter feature is useful in cases where the transport coefficients depend strongly on the thermodynamic state. The main disadvantage of the synthetic NEMD is that the flux-force relationships are indirect, unlike in an experiment. Moreover, it is not clear how the method applies to a heterogeneous system, such as a liquid/vapour interface.

**Boundary-driven NEMD.** The idea in boundary-driven NEMD is to let the external force act through the boundaries of a system, like in an experimental situation [10,14–21]. The external force must be compatible with the boundary conditions, which implies that one has to introduce some sort of walls into the system. The walls can be either directly interacting walls [14], like in a porous system [22], or regions where the temperature, chemical potential, or momentum is set by a local perturbation of the system [15]. In either case, the system is of limited extension, and a careful extrapolation to the thermodynamic limit and to the limit of vanishing force must be made.

A schematic picture of a boundary-driven NEMD cell is shown in Fig. 1. The cell is divided into a number of layers perpendicular to the external force (in this case a force acting in the $x$-direction). The layers form three regions, marked H, M, and L (not all the layers are shown in region M). Regions H and L may be walls formed by particles at fixed positions, a "fuzzy" wall [14], or a layer of particles obeying perturbed equations of motion [18–20].

A thermal force (temperature gradient) is then introduced as a perturbation to the particles' momenta in regions H and L, formally like for synthetic NEMD (Eq. (20)). However, the purpose of the perturbation is to thermostat the regions H and L. $\boldsymbol{R}_{\rm i}(t)$ may be given by a random force like in the Andersen thermostat [23], a Gaussian thermostat [5], or a thermostat based on velocity scaling [24]. It is not trivial to implement a thermostat that works over several time steps, like the Nosé-Hoover thermostat [25], because particles move in and out of the thermostatted regions from one point in time to the next.

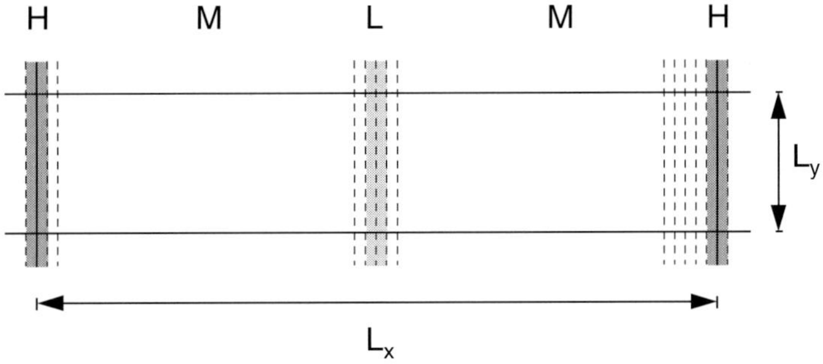


**Fig. 1.** Layout of a MD cell used in boundary-driven NEMD. The cell is usually not cubic.

The enthalpy flux is computed for each layer with Eq. (18), except that the first summation includes only those particles $i$ that are in the given layer. The enthalpy flux is then computed from the amount of energy added and withdrawn.

Instead of imposing a temperature gradient on the system and measure the heat flux, one can impose a heat flux and measure the temperature gradient [18,20]. Technically, this is achieved by adding and withdrawing energy from layers H and L, respectively, which is essentially the way the thermostatting is done. It has, however, the advantage that computation of the heat flux follows directly from the added/withdrawn energy.

A gradient in chemical potential or a concentration gradient is introduced by manipulating the particle types and molecular properties in regions H and L by particle swapping [26–28]. Alternatively, a specified mass flux may be imposed on the system through source- and sink terms at the boundaries [29].

A shear field is introduced by a special choice of $\boldsymbol{R}_{\mathrm{i}}(t)$; an additional velocity in $y$- or $z$- direction in layers H and L will set up a shear, $\partial p_{\mathrm{y}}/\partial x$, that can be used to determine the viscosity of the system [30,31].

The middle regions (marked M) are used for computations of the local temperature, density, pressure, composition, enthalpy, fluxes, and other properties of interest. Two assumptions form the basis for relating the mechanical properties to the thermodynamic properties: (1) each layer in region M is in local equilibrium, and (2) each layer represents a grand canonical ensemble. The first assumption has been examined for medium- to high-density states, where it is found to hold for temperature- and concentration gradients that are much higher than experimental ones [29,32]. The conditions of general validity of the assumptions are not known. In fact, the assumption of local equilibrium must be expected to fail for very dilute gases, unless the external force is extremely weak.

The main advantages of boundary-driven NEMD are that the method is intuitive and conceptually similar to an experimental situation, and it can be applied to heterogeneous systems. The main disadvantage is that suffers from lack of a precise definition of the local quantities because the system is not homogeneous. Moreover, the external force must be compatible with the MD system's periodic boundary conditions, which may severely restrict its spatial and dynamic properties.

Since the system is not uniform, care must be taken in computation of the heat flux. The last term in Eq. (18) is essentially the same as a pressure-tensor calculation. Todd *et al.* [33] have developed an algorithm for heterogeneous systems, the "method of planes", which gives the correct pressure tensor for planar surfaces. This technique was implemented in the heat-flux calculation by Baranyi [19]. Other techniques have been applied to cylindrical [34] and spherical [35] shapes.

# 3 Some simulation results

## 3.1 Bulk fluid

As an illustration of the results from a boundary-driven NEMD simulation, the temperature- and mole fraction profiles for a supercritical state of a binary isotope mixture with $m_2/m_1 = 0.1$ is shown in Fig. 2. The Lennard-Jones/spline potential [36], which is a smoothly truncated Lennard-Jones potential, was used here. The system had $N = 1024$ particles in a fixed volume $V$, the overall number density was $n^* = N\sigma_{11}^3/V = 0.7$ (in reduced LJ units where $\sigma_{11}$ is the diameter of component 1), and the overall composition was $x_1 = 0.5$. The cell was non-cubic with sides $L_\mathrm{x}/4 = L_\mathrm{y} = L_\mathrm{z}$. The mass fluxes were set to $J_1 = J_2 = 0$. A total of $10^6$ time steps of length $dt^* = 0.002$ were generated, the first $0.3 \times 10^6$ were used to establish a stationary state.

The figure shows that component 1 (the heavy component) is enriched in the cold middle of the cell (at $x/L_\mathrm{x} = 0.5$), and component 2 is in surplus at the

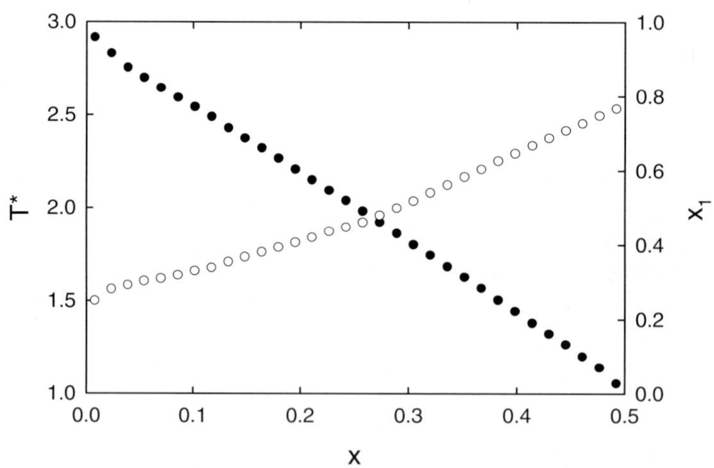


**Fig. 2.** The temperature profile (black circles) and mole fraction of component 1 (white circles) in a binary LJ/spline system at $n^* = 0.7$, $m_2/m_1 = 0.1$, $\sigma_{22}/\sigma_{11} = \varepsilon_{22}/\varepsilon_{11} = 1.0$, and $J_1 = J_2 = 0$.

hot ends (at $x/L_{\mathrm{x}} = 0$). This corresponds to a positive $\alpha_{\mathrm{T}}$. The temperature gradient is almost linear, representing a constant thermal conductivity (the heat flux is constant through the system at stationary state) whereas the mole fraction shows a slight curvature.

The same data are plotted as $x_1$ *versus* $T^*$ ($= k_{\mathrm{B}}T/\varepsilon_{11}$, where $\varepsilon_{11}$ is the potential depth of 11-interactions) in Fig. 3. Differentiating a third-order polynomial fit to the data gives $\alpha_{\mathrm{T}} = 2.24 \pm 0.05$ at a reduced temperature $T^* = 1.5$ in this particular case.

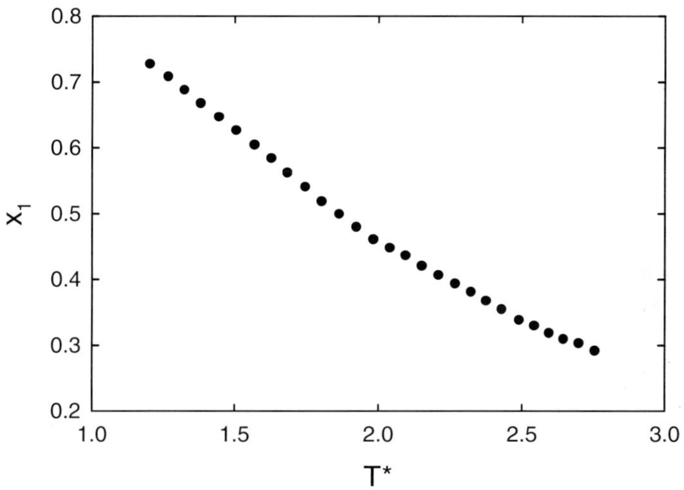


**Fig. 3.** The same data as in Fig. (2), but plotted as $x_1$ *versus* $T^*$.

Results of a systematic set of simulations of LJ/spline binary mixtures are shown in Figs. 4 and 5. In Fig. 4a, the thermal diffusion factor is plotted *versus* mole fraction of component 1, $x_1$, of a mixture with $\sigma_{22}/\sigma_{11} = 1.0$, $\varepsilon_{22}/\varepsilon_{11} = 1.0$, and $m_2/m_1 = 0.1$ at $T^* = 1.5$ (which is approximately 1.6 times the critical temperature for this system) and an overall density of $n^* = 0.4$ (which is approximately the critical density). The purpose of the line is to indicate the trend in the data. We observe that $\alpha_{\mathrm{T}}$ is positive for the whole composition range. The local density corresponding to $T^* = 1.5$ may vary slightly over the composition range, which is not shown in the figure. However, the trend is clearly that $\alpha_{\mathrm{T}}$ decreases with increasing $x_1$, suggesting that the Soret effect decreases with increasing concentration of the species with the higher molecular weight. We are not aware of experimental measurements of the thermal diffusion factor in isotope mixtures, but a study on $\mathrm{He} - \mathrm{C_2H_2}$ mixtures [37] also show that $\alpha_{\mathrm{T}}$ decreases with increasing content of the heavier component.

Fig. 4b shows $\alpha_{\mathrm{T}}$ as function of density, $n^*$, for the same mixture at $x_1 = 0.5$ and $T^* = 1.5$. It is known both from kinetic theory, experiments, and computer simulations [38] that $\alpha_{\mathrm{T}}$ should have this behaviour. At high densities, $\alpha_{\mathrm{T}}$ seems

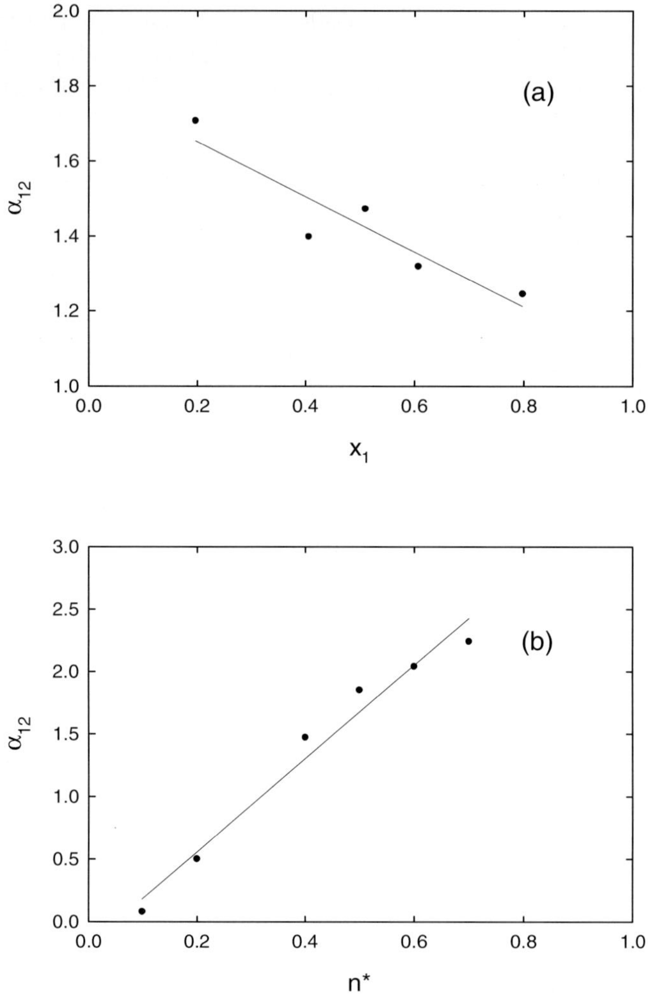


**Fig. 4.** Thermal diffusion factor for a binary LJ/spline mixture with $m_2/m_1 = 0.1$ and $\sigma_{22}/\sigma_{11} = \varepsilon_{22}/\varepsilon_{11} = 1.0$ at $T^* = 1.5$ shown as function of (a) composition at an overall density of $n^* = 0.4$, and (b) density at an overall composition $x_1 = 0.5$.

to flatten out, *i.e.* the Soret effect reaches a saturation point due to reduced mobility of the particles.

Studies of isotope mixtures have revealed that the contribution to the heat flux from the lighter component is predominantly flux of kinetic energy, and this contribution increases from the cold to the hot side [29,38]. The contribution from the heavier component is predominantly energy transfer by molecular collisions, and it increases from the hot to the cold side. Accordingly, the Soret effect can be interpreted as the system's response to the external force by minimising its entropy production at stationary state [39,40]; heat is conducted more effectively

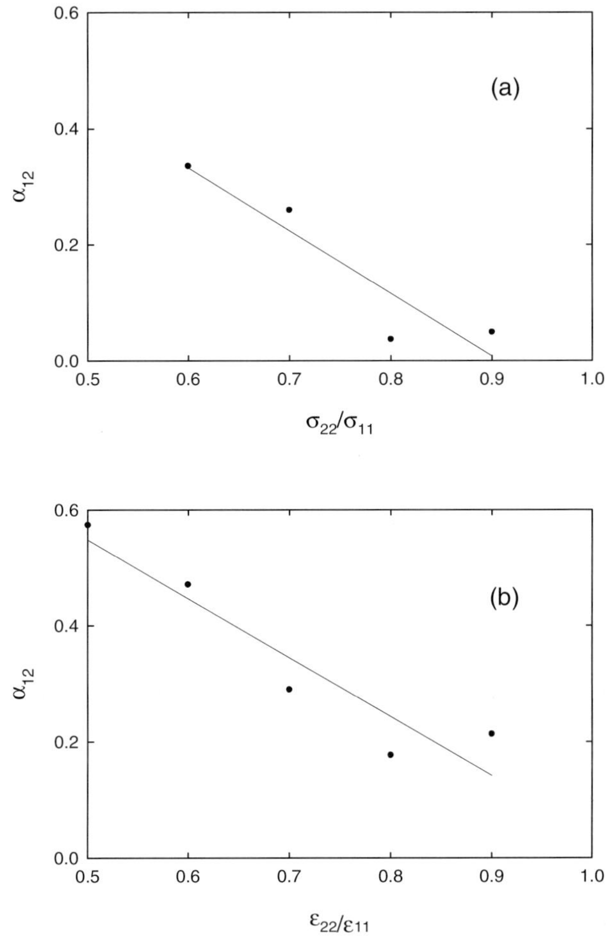


**Fig. 5.** Thermal diffusion factor for a binary LJ/spline mixture with $m_2/m_1 = 0.1$, $x_1 = 0.5$, and $T^* = 1.5$ shown as function of (a) diameter ratio, and (b) potential-depth ratio.

through the system if the lighter component migrates towards the hot side. This is like a "Le Chatelier's principle" for the stationary state.

Let us consider the enthalpy flux, Eq. (18), to consist of three terms:

$$\boldsymbol{J}_{\mathrm{H}} = \left\langle \sum_{\mathrm{i}} e_{\mathrm{kin}} \left(\boldsymbol{v}_{\mathrm{i}} - \boldsymbol{v}_0\right) \right\rangle + \left\langle \sum_{\mathrm{i}} e_{\mathrm{pot}} \left(\boldsymbol{v}_{\mathrm{i}} - \boldsymbol{v}_0\right) \right\rangle + \left\langle \sum_{\mathrm{i}} e_{\mathrm{coll}} \left(\boldsymbol{v}_{\mathrm{i}} - \boldsymbol{v}_0\right) \right\rangle \quad (22)$$

where $e_{\mathrm{kin}}$, $e_{\mathrm{pot}}$, and $e_{\mathrm{coll}}$ are the kinetic energy density, the potential energy density, and the collision energy density, respectively. At low density, the first term in Eq. (22) dominates. In the isotope case, both species have the same kinetic energy density, but the lighter component has a higher velocity. The lighter component is therefore a more effective transporter of energy than the heavier one, especially at the hot side where the difference in velocity is biggest.

In a recent, very nice and systematic study by Reith and Müller-Plathe (RM) [41] on binary, liquid Lennard-Jones mixtures near the triple point, they found that the larger or more strongly interacting particles accumulate in the cold region of the system. They found a linear relationship between $\alpha_T$ and the cohesive energy density ratio, $e_1/e_2$, for $0.5 < e_1/e_2 < 1.5$, where $e_k$ is defined as $\varepsilon_{kk}/\sigma_{kk}^3$. Their simulations were in the liquid state, corresponding to $n^* \approx 1.1$ and $T^* \approx 0.9$ with our definitions of the reduced variables (note that RM used different definitions). The thermal diffusion factor was found to be negative for $e_1/e_2 < 1.0$ (for $\varepsilon_{22}/\varepsilon_{11} = 1.0$ and $\sigma_{22}/\sigma_{11} < 1.0$). They explained that the particles with the higher cohesive energy density would accumulate in the cold region of the system. At high diameter ratios, a different mechanism would dominate; a packing effect would hinder the larger particles to move (shown by a small value of the self diffusion coefficient and a uniform distribution of the number density of the larger particles), and the smaller particles would move to the colder region as an entropic effect to make up the higher density there.

In Fig. 5 we show how $\alpha_T$ varies with the molecular parameters for the LJ/spline system at a fixed temperature, $T^* = 1.5$, and overall composition $x_1 = 0.5$. In both Fig. 5a and 5b, the mass ratio is $m_2/m_1 = 1.0$. Fig. 5a shows the dependency on the molecular-diameter ratio, $\sigma_{22}/\sigma_{11}$, with $\varepsilon_{22}/\varepsilon_{11} = 1.0$, and Fig. 5b the dependency on the potential-well ratio, $\varepsilon_{22}/\varepsilon_{11}$, with $\sigma_{22}/\sigma_{11} = 1.0$. Like RM, we find that the thermal diffusion factor is positive for $\varepsilon_{22}/\varepsilon_{11} < 1.0$, meaning that the component with the deeper potential well accumulates in the cold region. Again, the lines are trend lines rather than a model. For instance, $\alpha_T$ must tend to zero when $\varepsilon_{22}/\varepsilon_{11}$ tends to 1.0.

It is relatively simple to understand the effect of the potential-depth ratio. At the cold side, the density is higher, the interparticle distances are shorter, and the last term of Eq. (22) dominates the enthalpy flux. If the system organises itself so as to transport energy in the most effective way, the more strongly interacting species would migrate towards the cold side. This is the observed effect.

Unlike RM, we find that the thermal diffusion factor is positive also for $\sigma_{22}/\sigma_{11} < 1.0$, meaning that the *larger* particles accumulate in the cold region. A possible explanation is as follows: Our simulations were made at a much lower density, and both species are mobile. However, RM found that $\alpha_T < 0$ also at $n^* \approx 0.8$, and the discrepancy must be caused by something else. The three contributions to the enthalpy flux (which is identical to the heat flux in this case) for the LJ/spline mixture are shown in Fig. 6a for $m_2/m_1 = \varepsilon_{22}/\varepsilon_{11} = 1.0$, $\sigma_{22}/\sigma_{11} = 0.6$, and $x_1 = 0.5$. The kinetic contribution dominates in the hot region, and there is no preference for the two species here since they have the same mass. In the cold region, the collision contribution is larger than the other two. The larger particles, which also have a longer range of the potential, transport energy more effectively here. This situation is different from the LJ potential used by RM; the two components had in their case the same potential range. The contributions to the enthalpy flux from the two species are shown in Fig. 6b. The two species contribute equal amounts to the flux in the hot region, whereas the larger particles contribute more in the cold region.

The difference between the results of RM and us confirms that details of the potential has a significant effect on the thermal diffusion coefficient, as is often stated, and a further investigation of this would be interesting.

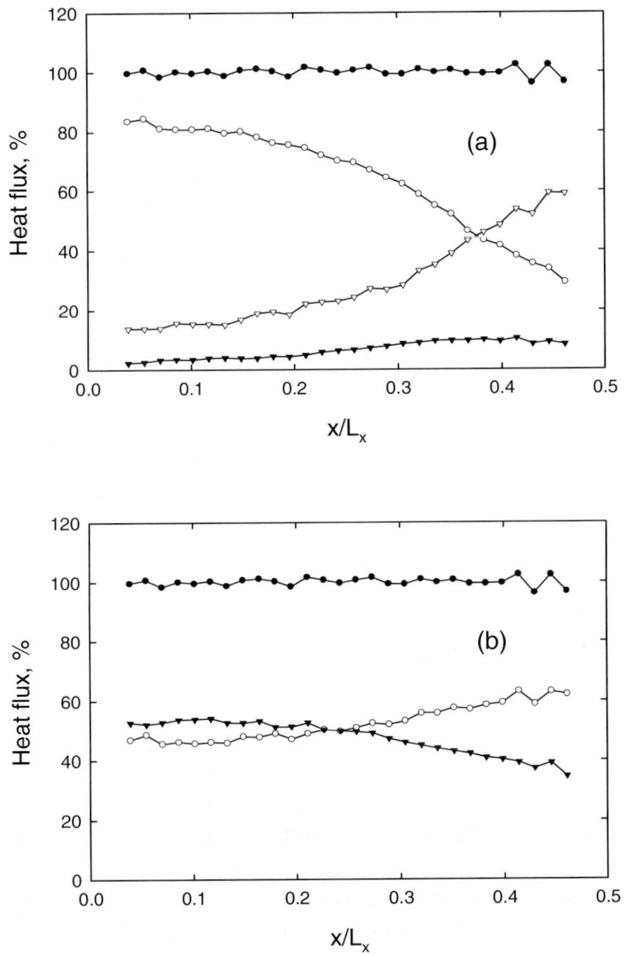


**Fig. 6.** Relative contributions to the heat flux for the LJ/spline mixture at $m_2/m_1 = \varepsilon_{22}/\varepsilon_{11} = 1.0$, $\sigma_{22}/\sigma_{11} = 0.6$ and $x_1 = 0.5$. The full heat flux ($J_{\mathrm{H}}$) is shown with black circles. Fig. (a) shows the kinetic (white circles), potential (black triangles), and collision (white triangles) contributions to $J_{\mathrm{H}}$. Fig. (b) shows the contributions to $J_{\mathrm{H}}$ from component 1 (white circles) and 2 (black triangles).

### 3.2 Coupled heat- and mass transport across a vapour/liquid interface

Heat and mass transfer across an interface occurs frequently in natural and industrial processes, such as in crystallisation and distillation. The phase transition causes dramatic changes in the thermodynamic properties, in the local structure, the transport coefficients, and the molecular mechanisms for the transport processes. For instance, the thermal conductivity changes by 2-3 orders of magnitude from liquid to vapour [42]. If there is a mass flux present, the heat flux across the interface will also change, due to the fact that the heat of vaporisation represents a source term in the heat balance. Here, we shall illustrate how the heat- and mass transport are coupled in a two-phase system with a condensing vapour. We shall also show how the molecular transport mechanisms change from vapour to liquid.

In a series of simulations, Hafskjold and Ratkje studied a liquid/vapour two-phase system [43]. The overall thermodynamic state was chosen to be in the two-phase region, and regions H and L were thermostatted to a high and a low temperature, respectively [43,44]. The liquid/vapour coexistence curve is roughly known for the one-component LJ/s system [45], with a critical temperature and density of $T^* \approx 0.9$ and $n^* \approx 0.4$, respectively, and triple-point temperature and density of $T^* \approx 0.55$ and $n^* \approx 0.7$, respectively. The final configuration of a typical run with $T^*_{\mathrm{H}} = 1.0$, $T^*_{\mathrm{L}} = 0.7$, and overall density $n^* = 0.4$ is shown in Fig. 7 for a mixture with $m_2/m_1 = 1.0$, $\sigma_{22}/\sigma_{11} = 1.0$, and $\varepsilon_{22}/\varepsilon_{11} = 0.8$ [43]. The thickness of the interface was found to be approximately 7 molecular diameters. For each run, $3.6\times 10^6$ time steps of length $dt^* = 0.002$ were generated, the first $0.6 \times 10^6$ time steps were allowed for stabilising the stationary state.

The temperature profile shows a much larger gradient in the vapour than in the liquid due to the smaller thermal conductivity in the vapour. The shoulder in the temperature profile just outside the surface is still somewhat mysterious,

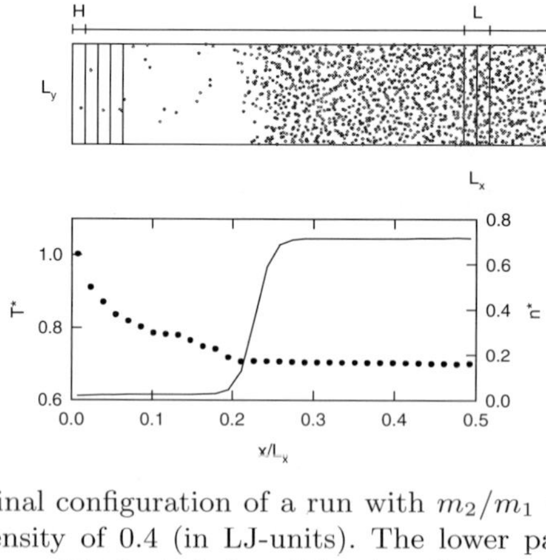


**Fig. 7.** Final configuration of a run with $m_2/m_1 = \sigma_{22}/\sigma_{11} = 1.0$, $\varepsilon_{22}/\varepsilon_{11} = 0.8$ and overall density of 0.4 (in LJ-units). The lower part of the figure shows the average temperature- (black circles) and density- (line) profiles of the system.

but it seems clear that it is not related to the Knudsen layer [46]. This shoulder represents an abnormal variation in the thermal conductivity near the interface; it has a very low value some 10 molecular diameters from the Gibbs dividing surface on the vapour side. The origin of this behaviour is not clear, but trajectories made by Matsumoto [47] for a similar system shows that quite a few particles nearly escape from the surface, but are captured back to the liquid at about this distance. We have earlier shown that this is a feature of the interface; if the heat flux is generated in a one-phase vapour system at the same conditions, the effect disappears [43].

Heat- and mass transfer are coupled through the Soret- and Dufour effects, quantified by the thermal diffusion factor and the heat of transfer, respectively. In bulk phase, these effects are normally small, but when there is a phase transition, the heat of transition makes a strong contribution to the coupling. Under the conditions we have used here, $J_{\mathrm{H}}$, $J_1$, and $J_2$ are all constant across the interface. Eq. (16) then implies that $J_{\mathrm{q}}$ has a source term, $\Delta_{\mathrm{vap}}H_1J_1 + \Delta_{\mathrm{vap}}H_2J_2$ at the interface. This source term is zero if $J_1 = J_2 = 0$ and for an isotope mixture when $J_1 = -J_2$. In general, however, $\Delta_{\mathrm{vap}}H$ represents a significant contribution to $q^*$ in the vapour phase [48].

Fig. 8 shows the enthalpy flux contributions from the two components for the system shown in Fig. 7. An interesting feature here is that the two components contribute in a proportion corresponding to the composition in the gas phase, but component 1 contributes relatively more in the liquid. In fact, in this particular case, component 2 contributes almost zero to the heat flux in the liquid. This is a consequence of the stronger interactions between 1-1 pairs than between 2-2 pairs through the collision energy flux. Again we see that the particles tend to

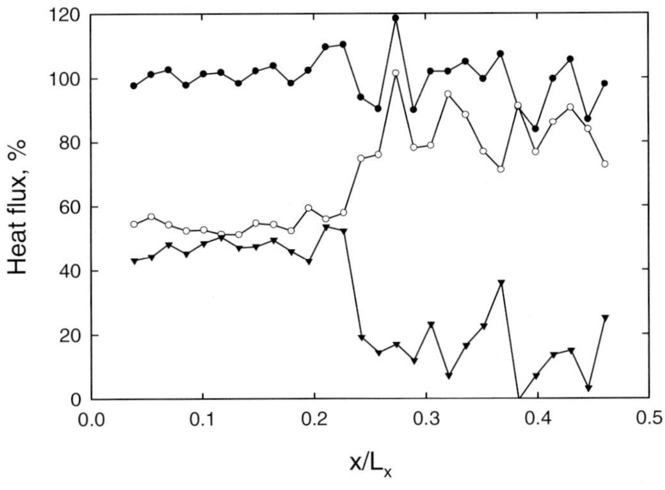


**Fig. 8.** Relative contributions to the heat flux in a vapour/liquid interface for the LJ/spline mixture shown in Fig(7) at $J_1 = J_2 = 0$. The contributions are from component 1 (white circles) and 2 (black triangles). The full flux is shown as black circles.

accumulate where they conduct heat most effectively. At conditions with non-zero and opposite mass fluxes (positive $J_1$), we have found that component 1 carries heat from the bulk phases to the interface, whereas component 2 carries heat in the opposite direction.

### 3.3 Thermal diffusion in porous media

In a porous system, the thermodynamic properties, such as the phase behavior, may deviate very much from the properties of the bulk fluid. Much less is known about the transport properties, and in particular coupled heat- and mass transport. Thermal creep flow may, however, play a role when normal convection is lacking (under zero gravity), when the system has been established over a long time (hydrocarbon reservoirs subject to the geothermal gradient), or when the thermal force is extremely large.

It is very difficult to describe thermal creep, and for that matter almost any transport process, in a porous material at the molecular level due to the complexity of the pore structure and the large span from molecular scale to the scale of the pore structure. Most studies based on computer simulations have therefore been made for a pore model with simple geometry, notably for slit pores [22,49–54]. The studies of transport phenomena include diffusion, which depends only on the fluid particles and the fluid-wall interactions [22]. Thermal conduction depends in addition on the matrix, and is therefore a more difficult problem, both from a phenomenological and a computational point of view.

Recently, two computer simulation studies on thermal creep in pores have appeared [54,55]. Both studies found that there is a flow of matter along the wall from the cold end of the pore to the hot end. A similar result was also found by Mackowski *et al.* [56], solving the continuum equations for mass-, momentum-, energy-, and component conservation. Rosner and Papadopoulos [53] used these results to point out that the no-slip fluid/wall boundary condition cannot be applied in the nonisothermal case because the walls will instead *drive* a convective flow.

The system layout in the work by Wold and Hafskjold [54] is shown in Fig. 9. They used a LJ/spline binary isotope mixture with $m_2/m_1 = 0.1$ and $\sigma_{22}/\sigma_{11} = \varepsilon_{22}/\varepsilon_{11} = 1.0$ in a slit pore. The pore width and the interaction between the fluid particles and the wall were the main variables. The fluid/wall interaction was modelled with the Steele 10-4-3 potential [57]. The simulation cell was closed at each end so that the net mass flux in the $x$-direction was equal to zero. A snapshot of the system with pore width $20\sigma$, temperature varying from 1.0 (in LJ units) at the cold end to 2.0 at the hot end, and fluid density of approximately 0.4 (in LJ units), is shown in Fig. 10. The two-dimensional plot of the flow vectors for this pore in Fig. 11 clearly shows the pattern of the creep flow and the consequential backflow in the middle of the pore. They found that this pattern was typical for strongly attracting fluid-wall potentials (except for a very narrow pore of width $2.5\sigma$). For a purely repulsive fluid/wall interaction, the flow was weak and mostly in the opposite direction. They attributed the creep flow to the gradient in interfacial tension along the wall. The thermal diffusion coefficient

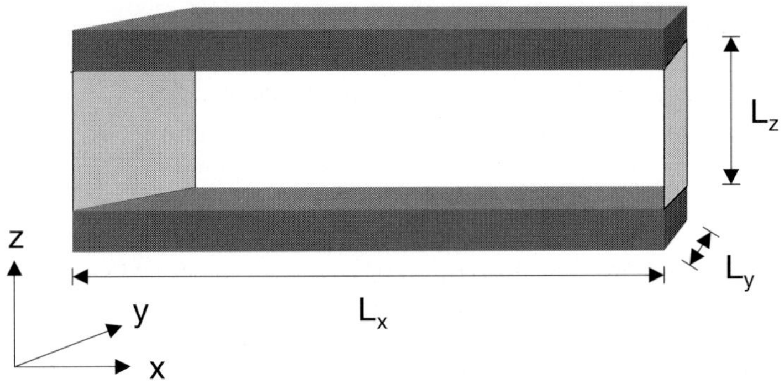


**Fig. 9.** Layout of a slit pore closed at $x = 0$ and $x = L_x$.

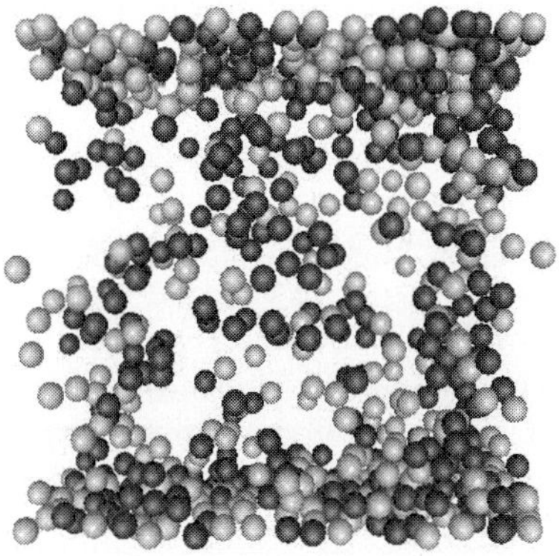

**Fig. 10.** Snapshot of a two-component LJ/spline system in a slit pore of width $20\sigma$. The pore walls are not shown.

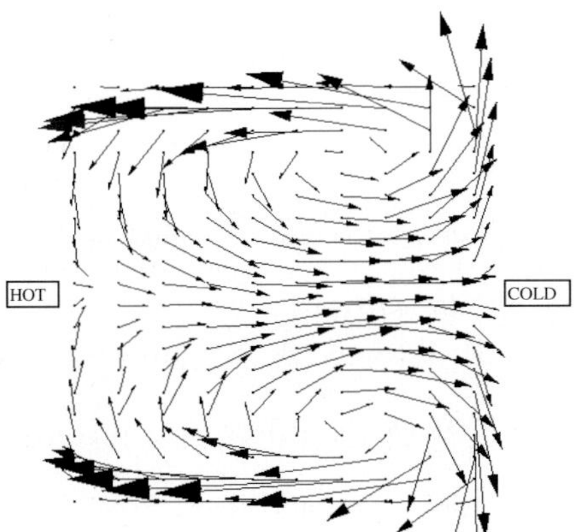


**Fig. 11.** Flow pattern in the system shown in Fig. (10).

was comparable to that in bulk fluid at the same overall density, except when the creep flow was strong, in which case the thermal diffusion was blurred by the convective mixing.

Several simulation studies of porous media have appeared in recent years. The computational methods are still somewhat immature. However, important phenomena like microgravity processes, heterogeneous catalysis with large heats of reaction, and membrane separation call for the more fundamental understanding of mechanisms that computer simulations can provide. More realistic porous geometries and molecular structures must be applied.

## 4 Conclusions

NEMD is an established and well suited method to compute the thermal diffusion factor and heat of transport in binary mixtures. A physical reason for the Soret effect in a mixture is beginning to emerge; the system organises itself in such a way that heat is transported most effectively (with minimum entropy production). In general, the heavier or more strongly interacting species accumulate in the cold region of the system. Simulations of transport properties of two-phase systems are a novel feature of the boundary-driven NEMD. The technique enables us to study in detail how the transport coefficients and transport mechanisms vary across the interface. Thermal diffusion in porous systems is very little understood, but this phenomenon may attract more attention in the future if the computational methodology becomes better established.

## 5 Acknowledgements

Thanks to Professor Signe Kjelstrup and Werner Köhler, and Dr. Øystein Sæther for editorial help with this article.

## References

1. S.R. de Groot, P. Mazur: *Non-equilibrium Thermodynamics.* (Dover Ed., New York 1984)
2. L. Onsager: Phys. Rev. **37**, 405 (1931)
3. R. Haase, K.-H. Drücker, H. Buchner, J. Schwinum: Zeit. Phys. Chem. **186**, 113 (1994)
4. D.A. McQuarrie: *Statistical Mechanics.* (Harper and Row, New York 1976)
5. D.J. Evans, G.P. Morriss: *Statistical Mechanics of Nonequilibrium Liquids.* In: *Theoretical Chemistry*, ed. by D.P. Craig and R. McWeeny (Academic Press, London 1990)
6. J.H. Irving, J.G. Kirkwood: J. Chem. Phys. **18**, 817 (1950)
7. D. Frenkel, B. Smit: *Understanding Molecular Simulation. From Algorithms to Applications.* (Academic Press, San Diego 1996)
8. K.E. Gubbins: In: *Statistical Mechanics*, ed. by K. Singer (The Chemical Society, London 1973)

9. R. Vogelsang, C. Hoheisel, G.V. Paolini, G. Ciccotti: Phys. Rev. A **36**, 3964 (1987)
10. J.-M. Simon, D.K. Dysthe, A.H. Fuchs, B. Rousseau: Fluid Phase Equil. **150-151**, 151 (1998)
11. D. MacGowan, D.J. Evans: Phys. Rev. A **34**, 2133 (1986)
12. G.V. Paolini, G. Ciccotti: Phys. Rev. A **35**, 5156 (1987)
13. D. Evans: Phys. Rev. A **34**, 1449 (1986)
14. W.T Ashurst, W.G. Hoover: Phys. Rev. A **11**, 658 (1975)
15. A. Tenenbaum, G. Ciccotti, R. Gallico: Phys. Rev. A **25**, 2778 (1982)
16. J.M. Kincaid, X. Li, B. Hafskjold: Fluid Phase Equil. **76**, 113 (1992)
17. B. Hafskjold, T. Ikeshoji, S.K. Ratkje: Mol. Phys. **80**, 1389 (1993)
18. T. Ikeshoji, B. Hafskjold: Molec. Phys. **81**, 251 (1994)
19. A. Baranyai: Phys. Rev. E **54**, 6911 (1996)
20. F. Müller-Plathe: J. Chem. Phys. **106**, 6082 (1997)
21. F. Müller-Plathe, D. Reith: Comp. Theor. Polymer Sci. **9**, 203 (1999)
22. R.F. Cracknell, D. Nicholson, N. Quirke: Molec. Sim. **13**, 161 (1994)
23. H.C. Andersen: J. Chem. Phys. **72**, 2384 (1980)
24. L.V. Woodcock, Chem. Phys. Lett. **10**, 257 (1971)
25. S. Nosé: Molec. Phys. **52**, 255 (1984)
26. P. Sindzingre, C. Massobrio, G. Ciccotti, D. Frenkel: Chem. Phys. **129**, 213 (1989)
27. A.P. Thompson, D.M. Ford, G.S. Heffelfinger: J. Chem. Phys. **109**, 6406 (1998)
28. A.P. Thompson, G.S. Heffelfinger: J. Chem. Phys. **110**, 10693 (1999)
29. B. Hafskjold, S.K. Ratkje: J. Stat. Phys. **78**, 463 (1995)
30. S. Hess, W. Loose: Ber. Bunsenges. Phys. Chem. **94**, 216 (1990)
31. F. Müller-Plathe: Phys. Rev. E **59**, 4894 (1999)
32. B. Hafskjold, I. Fujihara, T. Ikeshoji: Molec. Phys. **90**, 999 (1997)
33. B.D. Todd, D.J. Evans, P.J. Daivis: Phys. Rev. E **52**, 1627 (1995)
34. M. Mareschal, M. Baus, R. Lovett: J. Chem. Phys. **106**, 645 (1997)
35. T. Ikeshoji, B. Hafskjold, H. Furuholt (in preparation)
36. B.L. Holian, D.J. Evans: J. Chem. Phys. **78**, 5147 (1983)
37. A.A. Youssef: Gazz. Chim. Ital. **120**, 653 (1990)
38. J.M. Kincaid, B. Hafskjold: Mol. Phys. **82**, 1099 (1994)
39. T.A. Bak: *Contributions to the Theory of Chemical Kinetics.* (Benjamin, New York 1963)
40. I. Prigogine: Bull. classe des Sci. Acad. Roy. Belg. **31**, 600 (1945)
41. D. Reith, F. Müller-Plathe: J. Chem. Phys. **112**, 2436 (2000)
42. R.C. Reid, J.M. Prausnitz, T.K. Sherwood: *The Properties of Gases and Liquids*, 3rd. ed. (McGraw-Hill, New York 1977)
43. B. Hafskjold, S.K. Ratkje: Molecular interpretation of coupled heat- and mass transport across a vapour/liquid interface, Proceedings, ECOS'96, June 25-27, 1996, Stockholm
44. B. Hafskjold, T. Ikeshoji: Molec. Sim. **16**, 139 (1966)
45. R. Halseid: The Critical Point of the Lennard-Jones Spline Fluid, Project Report, Norw. Inst. of Technology (1993)
46. A. Røsjorde, D.W. Fossmo, D. Bedeaux, S. Kjelstrup, B. Hafskjold: Submitted to J. Colloid. Interf. Sci.
47. M. Matsumoto: Molec. Sim. **16**, 209 (1996)
48. S.K. Ratkje, B. Hafskjold: Ind. Eng. Chem. Res. **34**, 3001 (1995)
49. S.-I. Furukawa, T. Shigeta, T. Nitta: J. Chem. Eng. Japan **29**, 725 (1996)
50. S.-I. Furukawa, T. Nitta: J. Chem. Eng. Japan **30**, 116 (1997)
51. D. Nicholson: Carbon **36**, 1511 (1998)

52. L. Xu, T.T. Tsotsis, M. Sahimi: J. Chem. Phys. **111**, 3252 (1999)
53. D.E. Rosner, D.H. Papadopoulos: Ind. Eng. Chem. Res. **35**, 3210 (1996)
54. I. Wold, B. Hafskjold: Int. J. Thermophys. **20**, 847 (1999)
55. D.H. Papadopoulos, D.E. Rosner: Phys. Fluids **7**, 2535 (1995)
56. D.W. Mackowski, V.R. Rao, D.G. Walker, R.W. Knight: J. Crystal Growth **179**, 297 (1997)
57. W.A. Steele: *The Interaction of Gases with Solid Surfaces* (Pergamon, Oxford 1974)

## Supplementary glossary

| | |
|---|---|
| $\tilde{c}$ | Overall concentration (kmol/m$^3$) |
| $\tilde{c}_\mathrm{k}$ | Concentration (component $k$) |
| $dt^*$ | Length of time step |
| $e_\mathrm{k}$ | Element $k$ in cohesive energy density ratio, $e_\mathrm{k} = \varepsilon_\mathrm{kk}/\sigma^3_\mathrm{kk}$ |
| $e_\mathrm{coll}$ | Collision energy density |
| $e_\mathrm{kin}$ | Kinetic energy density |
| $e_\mathrm{pot}$ | Potential energy density |
| $\bar{E}$ | Average energy (all particles) at time $t$ |
| $E_\mathrm{i}$ | Kinetic and potential energy (particle $i$) |
| $\boldsymbol{F}_\mathrm{i}$ | Force upon $i$ due to all other particles |
| $\boldsymbol{F}_\mathrm{ij}$ | Force on $i$ due to $j$ |
| $H_\mathrm{k}$ | Partial molar enthalpy (component $k$) |
| $\boldsymbol{J}_\mathrm{H}$ | Enthalpy flux |
| $J_\mathrm{k}$ | Mass flux (component $k$) |
| $\boldsymbol{J}_\mathrm{k}$ | Diffusive molar flux (component $k$) |
| $\boldsymbol{J}_\mathrm{q}$ | Heat flux |
| $J_\mathrm{r}$ | Relative flux |
| $\boldsymbol{J}_\alpha(t)$ | Instantaneous value of $\boldsymbol{J}_\alpha$ |
| $l_{\alpha\beta}$ | Phenomenological coefficients |
| $L_\mathrm{x}$ | Cell length (x-direction) |
| $m_\mathrm{i}$ | Mass (particle $i$) |
| $n^*$ | Number density |
| $N$ | Number of particles |
| $N_\mathrm{A}$ | Avogadro number |
| $\boldsymbol{P}_\mathrm{i}$ | Momentum of $i$ |
| $q^*$ | Heat of transport |
| $\boldsymbol{r}_\mathrm{i}$ | Position of $i$ |
| $\boldsymbol{r}_\mathrm{ij}$ | Vector from position of $i$ to position of $j$ |
| $R$ | Gas constant |
| $\boldsymbol{R}_\mathrm{i}$ | Required perturbation (external force) |
| $T^*$ | Lennard-Jones temperature |
| $T^*_\mathrm{H}$, $T^*_\mathrm{L}$ | High and low $T^*$ of thermostats |
| $v_\mathrm{k}$ | Average velocities of individual species |
| $\boldsymbol{v}_\mathrm{i}$ | Velocity (particle $i$) |
| $v_0$ | Velocity of reference frame |

| | |
|---|---|
| $\boldsymbol{v}_0$ | Velocity (reference frame) |
| $V$ | Molar volume (also cell volume) |
| $x_{\mathrm{i}}$ | Mole fraction (component $i$) |
| $\gamma_{\mathrm{i}}$ | Activity coefficient (component $i$) |
| $\varepsilon_{11}$ | Potential depth (1-1 interaction) |
| $\Phi_{\mathrm{i}}$ | Potential energy of $i$ in the field of the other particles |
| $\mu_{\mathrm{i}}$ | Chemical potential (component $i$) |
| $\sigma_{11}$ | Diameter of particle (component 1) |

# Thermodiffusion in the Critical Region

Jutta Luettmer-Strathmann

Department of Physics, The University of Akron, Akron, OH 44325-4001, USA

**Abstract.** Long range fluctuations of the order parameter affect thermodynamic as well as transport properties in the critical region. The values of transport coefficients near a critical point are typically enhanced compared to the values in the classical region far away from a critical point. Asymptotically close to a critical point, the critical enhancements obey power laws with universal critical exponents. The asymptotic critical region is surrounded by a large crossover region, where the enhancements are no longer asymptotic but still significant compared to the background values. In binary mixtures, this is also the region where differences are observed in the critical behavior of transport coefficients near liquid-liquid (consolute) and liquid-vapor (plait) critical points. In this contribution, we review the critical dynamics of binary fluid mixtures with a focus on thermodiffusion.

## 1 Introduction

Many transport properties of fluids exhibit singular behavior near a critical point (cf. [1,2]). The transport properties related to heat and mass transport in binary mixtures are no exception. The mutual diffusivity, for example, vanishes at a critical point while the Ludwig-Soret coefficient, describing the transport of matter in response to a temperature gradient, diverges as a critical point is approached. Since transport coefficients can have anomalous values in a large region around a critical point [1,2] a knowledge of their critical behavior is important for many technical and scientific applications. In particular, thermal diffusion in the critical region is important in such fields as supercritical-fluid chemistry, the treatment of hazardous waste, geochemistry of hydrocarbon reservoirs, and the study of the kinetics of phase separation (cf. [3–6] and references therein).

Far away from a critical point, transport coefficients are non-singular and generally slowly varying functions of the temperature and density. When this regular behavior is extrapolated into the critical region, one arrives at a separation of transport coefficients into the so-called background contributions and critical enhancements, see Fig. 1. The critical enhancements are due to long-range fluctuations in the fluid. Asymptotically close to the critical point, the correlation length of the fluctuations diverges leading to universal scaling laws for the critical enhancements of the transport coefficients. The asymptotic critical region, however, is usually much smaller than the region in which critical enhancements are observed experimentally (cf. [1,2]). The behavior of the transport coefficients in this so-called crossover region cannot be described in terms

W. Köhler and S. Wiegand (Eds.): LNP 584, pp. 24–37, 2002.

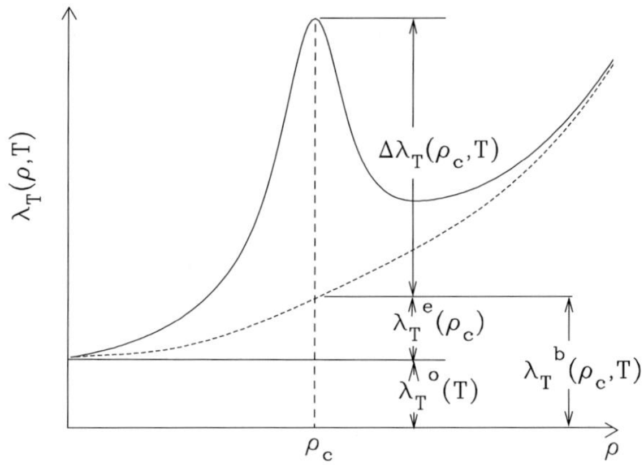


**Fig. 1.** Illustration of the decomposition of a transport coefficient, here the thermal conductivity of a one-component fluid, into critical enhancement $\Delta\lambda_T$ and background contributions $\lambda_T^b = \lambda_T^0(T) + \lambda_T^e(T,\rho)$, where $\lambda_T^0(T)$ is the dilute gas contribution and $\lambda_T^e(T,\rho)$ is the so-called excess term

of simple power laws and depends on details of the physical system under consideration. For binary fluid mixtures, the crossover behavior of the transport coefficients depends qualitatively on the nature of the critical point; this issue will be discussed in Section 4.

Two different theoretical approaches are widely used to investigate dynamic critical phenomena. The dynamic renormalization-group theory formulated by Halperin, Hohenberg and Ma [7] forms the theoretical basis for the concepts of dynamic scaling and universality [8]. It is an extension of the ideas of the static renormalization-group theory of Wilson and Fisher [9,10], which provides the basis for scaling and universality of thermodynamic properties in the critical region. Mode-coupling theories [11–13], on the other hand, originated from the idea of Fixman [14] that the critical enhancements of the transport coefficients are the result of non-linear coupling between the hydrodynamic modes of the system. Dynamic renormalization-group and mode-coupling theories, as well as the related decoupled-mode theory of Perl and Ferrell [15,16], have recently been extended to describe the crossover behavior of transport properties of fluids and their mixtures [17–28].

## 2 Thermodynamic properties in the critical region

Before we address transport properties in the critical region, we will briefly review some aspects of the thermodynamics of fluids near critical points. The critical point of a one-component fluid is characterized by its critical temperature $T_c$, density $\rho_c$ and pressure $P_c$. When a fluid at its critical density is cooled through the critical temperature, it undergoes a continuous phase transition to a state with two coexisting phases that differ in density. The order parameter for the phase transition can be identified with the difference between the actual and the critical density of the fluid, while the chemical potential plays the role of the ordering field [29]. Since the order parameter for the liquid-vapor phase transition is a scalar and since the interactions between molecules in a simple fluid are short ranged, simple fluids belong to the universality class of the three-dimensional Ising model [30]. Asymptotically close to the critical point, the thermodynamic properties of a fluid exhibit singular behavior and satisfy well-known power laws in the reduced temperature $\Delta\tilde{T} = (T - T_c)/T$. Of special interest in this work are the so-called generalized susceptibilities, i.e. derivatives of thermodynamic densities (e.g. the mass density $\rho$ and the specific entropy $s$) with respect to thermodynamic fields (temperature $T$, pressure $P$, etc.) [31]. The specific isochoric heat capacity $c_\mathrm{v} = T(\partial s/\partial T)_V$ and the symmetrized isothermal compressibility $\chi_\mathrm{T} = \rho(\partial\rho/\partial P)_T$ obey the asymptotic power laws

$$c_\mathrm{v} \sim \Delta\tilde{T}^{-\alpha}, \qquad \chi_\mathrm{T} \sim \Delta\tilde{T}^{-\gamma}, \qquad \text{for } \rho = \rho_\mathrm{c} \text{ and } T \geq T_\mathrm{c}. \tag{1}$$

The values of these critical exponents for the universality class of the three-dimensional Ising model are given by [29,32]

$$\alpha = 0.110 \pm 0.003, \qquad \gamma = 1.239 \pm 0.002. \tag{2}$$

Please note that the isochoric heat capacity diverges weakly, i.e. with exponent $\alpha$, while the symmetrized isothermal compressibility diverges strongly, i.e. with exponent $\gamma$. Since the specific isobaric heat capacity $c_\mathrm{p} = T(\partial s/\partial T)_P$ is related to $c_\mathrm{v}$ and $\chi_\mathrm{T}$ through

$$\rho c_\mathrm{p} = \rho c_\mathrm{v} + T\rho^{-2}(\partial P/\partial T)^2_\rho \chi_\mathrm{T}, \tag{3}$$

and since $(\partial P/\partial T)_\rho$ remains finite at the critical point, it follows that $c_\mathrm{p}$ diverges strongly as the critical point is approached.

The critical behavior of the thermodynamic properties is induced by long-range order-parameter fluctuations in the critical region [33]. Denoting the local number density at position $\mathbf{r}$ by $n(\mathbf{r})$ and its average value by $n = \langle n(\mathbf{r})\rangle = \rho/m$, where $\rho$ is the mass density and $m$ is the mass of a molecule, we define the correlation function for the density fluctuations as [34,35]

$$G(|\mathbf{r}_1 - \mathbf{r}_2)|) = \langle n(\mathbf{r}_1)n(\mathbf{r}_2)\rangle - n^2. \tag{4}$$

In general, the correlation function decays exponentially in space

$$G(r) \sim \exp(-r/\xi), \tag{5}$$

where $\xi$ is the correlation length of the fluctuations [34]. The correlation length diverges as the critical point is approached and satisfies the power law [29]

$$\xi = \xi_0(\Delta\tilde{T})^{-\nu} \qquad \text{for } \rho = \rho_c \text{ and } T \geq T_c, \tag{6}$$

where $\xi_0$ is a system-dependent amplitude and $\nu$ a critical exponent. At the critical point, the correlation function decays algebraically, $G(r) \sim r^{-d+2-\eta}$ for

$r \to \infty$ [34], where $d$ is the spatial dimension of the system and $\eta$ is another critical exponent (not to be confused with the viscosity $\tilde{\eta}$). According to the fluctuation dissipation theorem [35], the symmetrized compressibility $\chi_{\mathrm{T}}$ is related to the integral over the correlation function

$$\chi_{\mathrm{T}} = \frac{1}{k_{\mathrm{B}}T} \int d\mathbf{r}\, G(r)\,, \tag{7}$$

where $k_{\mathrm{B}}$ is Boltzmann's constant.

The divergence of the correlation length at the critical point motivates the static scaling hypothesis [33,35] "near the critical point, the correlation length $\xi$ is the only characteristic length of the system in terms of which all other lengths must be measured" [35]. A dimensional analysis [35] then reveals that the singular behavior of the thermodynamic properties of interest can be expressed in terms of the diverging correlation length

$$c_{\mathrm{v}} \sim \xi^{\alpha/\nu}, \quad \chi_{\mathrm{T}} \sim \xi^{\gamma/\nu}, \quad c_{\mathrm{p}} \sim \xi^{\gamma/\nu}, \tag{8}$$

where the critical exponents obey scaling relations

$$2 - \alpha = d\nu, \qquad \gamma = \nu(2 - \eta). \tag{9}$$

This implies that $\nu$ and $\eta$ have the values $\nu = 0.63$ and $\eta = 0.033$. The asymptotic power laws for the thermodynamic properties discussed here are valid only in a very small region around the critical point [29]. To describe fluids and fluid mixtures in the much larger region in which the critical fluctuations affect the thermodynamic properties, so-called crossover equations-of-state have to be employed (cf. [1]).

Thermodynamics of binary fluid mixtures in the asymptotic critical region are most elegantly described by the theory of Griffiths and Wheeler [31]. In this theory, thermodynamic fields are defined from combinations of the temperature, the pressure, and the chemical potentials in such a way, that the first field represents the ordering field, the second the temperature variable, and the third a non-scaling field, which is not present in one-component fluids. By keeping the non-scaling field constant, the critical behavior of binary mixtures is mapped onto that of the three-dimensional Ising model. There is some freedom in the choice of the fields. For binary mixtures near a liquid-liquid critical point (consolute point) one often takes the pressure to be the non-scaling field. This makes the concentration (fluctuating at constant pressure) the order parameter for the phase transition. For binary mixtures near a vapor-liquid critical point (plait point), on the other hand, the non-scaling field is often taken to be the difference between the chemical potentials of the two components [1] making the density (fluctuating at constant chemical potential difference) the order parameter.

According to the theory of Griffiths and Wheeler [31], a derivative of a thermodynamic density with respect to a thermodynamic field diverges strongly, i.e. like $\xi^{\gamma/\nu}$, when two thermodynamic fields are kept constant, weakly, i.e. like $\xi^{\alpha/\nu}$, when one density and one field are kept constant, and not at all, when two densities are kept constant. For the specific heat capacities of binary mixtures, for example, this implies the following asymptotic power laws

$$c_{\mathrm{p},\mu} \sim \xi^{\gamma/\nu}, \quad c_{\mathrm{p,c}} \sim \xi^{\alpha/\nu}, \quad c_{\mathrm{v,c}} \sim \xi^{0}, \tag{10}$$

where $c$ is the mass fraction of component 2 and $v = 1/\rho$, where $\rho = \rho_1 + \rho_2$ is the mass density of the mixture. The chemical potential $\mu = \mu_2/m_2 - \mu_1/m_1$, where $m_i$ is the mass per molecule and $\mu_i$ the chemical potential of component $i \in \{1, 2\}$, is the thermodynamic quantity conjugate to $c$ [36].

The asymptotic power laws introduced above hold near consolute as well as plait points (except at "special points" on the critical line [31]). The asymptotic critical region is (again) surrounded by a large crossover region, where the asymptotic power laws no longer describe the effect of the critical fluctuations on the thermodynamic properties. Crossover phenomena are very much in evidence for binary mixtures near plait points. In those mixtures, the thermodynamic properties suggest pure-fluid like behavior at some distance from the critical point and cross over to the asymptotic, mixture-like behavior only extremely close the critical point (cf. Fig. 16 in [25] and Fig. 15 in [37]).

## 3 Transport coefficients in the asymptotic critical region

According to the concept of dynamic universality [8], the critical dynamics of physical systems are determined by only a few characteristic properties. Systems within a dynamic universality class belong to the same static universality class and have the same conservation laws and Poisson-bracket relations among the order parameter and the conserved densities [8]. In the classification of Hohenberg and Halperin [8] one-component fluids and incompressible mixtures belong to model H, which Siggia *et al.* [38] extended to describe the critical dynamics of compressible fluid mixtures (model H′). The dynamic scaling hypothesis of Ferrell *et al.* [39,40] and Halperin and Hohenberg [41,42] extends scaling with the correlation length to dynamic critical phenomena. If $m(\mathbf{r}, t)$ denotes the value of the order parameter at position $\mathbf{r}$ and time $t$, the correlation function (4) is generalized to

$$G(|\mathbf{r}_1 - \mathbf{r}_2)|) = \langle m(\mathbf{r}_1, t_1) m(\mathbf{r}_2, t_2) \rangle - \langle m(\mathbf{r}_1, t_1) \rangle \langle m(\mathbf{r}_2, t_2) \rangle, \tag{11}$$

where the angular brackets indicate equilibrium averages. The dynamic scaling hypothesis states that the Fourier transform $\tilde{G}(q, \omega)$ of the correlation function can be expressed as

$$\tilde{G}(q, \omega) = \xi^{2-\eta} g(q\xi, \omega\xi^{z}), \tag{12}$$

where $g$ is a universal scaling function, $\eta$ is the static critical exponent, and $z$ is a new dynamic critical exponent. Eq. (12) implies that a system near a critical point does not only have a characteristic length, namely $\xi$, but also a characteristic time $\tau_q$, which has the scaling form $\tau_q = \xi^{z} f(q\xi)$ , where $f$ is another scaling function [33]. For one-component fluids and binary mixtures, the exponent $z$ is related to the critical exponent $z_\eta$ of the viscosity $\tilde{\eta}$

$$\tilde{\eta} \sim \xi^{z_\eta}, \tag{13}$$

through $z = 3 + z_\eta$. Unfortunately, the value for the dynamic critical exponent is not as well known as the values for the static critical exponent (for a recent discussion see [43]). In this work, we will employ the renormalization-group result $z_\eta = 0.063$ by Hao [44].

In order to discuss the asymptotic scaling laws for the transport properties related to heat and mass diffusion, we start from the hydrodynamic equations of Landau and Lifshitz [45]. If the fluid as a whole is at rest then, to first order, the diffusion current $\mathbf{J}_\mathrm{d}$ and the heat current $\mathbf{J}_\mathrm{q}$ are related to the time derivatives of the mass fraction $c$ of component 2 and the specific entropy $s$ through [45]

$$\rho \frac{\partial c}{\partial t} = -\nabla \cdot \mathbf{J}_\mathrm{d}, \tag{14}$$

$$\rho T \frac{\partial s}{\partial t} = -\nabla \cdot (\mathbf{J}_\mathrm{q} - \mu \mathbf{J}_\mathrm{d}), \tag{15}$$

where, as before, $\rho$ is the mass density, $T$ the temperature, and $\mu = \mu_2/m_2 - \mu_1/m_1$ the chemical potential difference. The linear laws relating the currents $\mathbf{J}_\mathrm{d}$ and $\mathbf{J}_\mathrm{q}$ to gradients of the temperature and chemical potential difference define the transport coefficients $\tilde{\alpha}$, $\tilde{\beta}$, $\tilde{\gamma}$, and $\tilde{\delta}$ [45]

$$\mathbf{J}_\mathrm{d} = -\tilde{\alpha}\nabla\mu - \tilde{\beta}\nabla T, \tag{16}$$

$$\mathbf{J}_\mathrm{q} - \mu\mathbf{J}_\mathrm{d} = -\tilde{\delta}\nabla\mu - \tilde{\gamma}\nabla T, \tag{17}$$

where $\tilde{\delta} = T\tilde{\beta}$ due to Onsager reciprocal relations [46].

The transport coefficients $\tilde{\alpha}$, $\tilde{\beta}$, and $\tilde{\gamma}$ can be expressed in terms of experimentally accessible transport properties. The coefficient $\tilde{\alpha}$ plays the role of a mass conductivity and is related to the mutual diffusivity $D$

$$D = \frac{\tilde{\alpha}}{\rho}\left(\frac{\partial \mu}{\partial c}\right)_{T,P}. \tag{18}$$

The thermal diffusion coefficient $D_\mathrm{T} = S_\mathrm{T} D$, where $S_\mathrm{T}$ is the Soret coefficient [46], is given by

$$D_\mathrm{T} = \frac{1}{c(1-c)\rho}\left[\tilde{\alpha}\left(\frac{\partial \mu}{\partial T}\right)_{c,P} + \tilde{\beta}\right]. \tag{19}$$

The thermal conductivity $\lambda_\mathrm{T}$ is defined by

$$\mathbf{J}_\mathrm{q} = -\lambda_\mathrm{T}\nabla T \text{ with } \mathbf{J}_\mathrm{d} = 0\,, \tag{20}$$

and therefore given by

$$\lambda_\mathrm{T} = \tilde{\gamma} - T\tilde{\beta}^2/\tilde{\alpha}. \tag{21}$$

In some experimental situations, the thermal conductivity in the absence of a concentration gradient (rather than in the absence of mass flow) is required [46,47]. This quantity corresponds to the transport coefficient $L_{\mathrm{SS}}$ of [25] and can be expressed as

$$L_{\mathrm{SS}} = \lambda_{\mathrm{T}} + \frac{T}{\tilde{\alpha}}\left(\tilde{\alpha}\left(\frac{\partial\mu}{\partial T}\right)_{c,P} + \tilde{\beta}\right)^2 . \tag{22}$$

As customary in the critical region [1,2], each of the transport coefficients $\tilde{\alpha}$, $\tilde{\beta}$, and $\tilde{\gamma}$ is separated into a regular or background part (indicated by a superscript b) and a critical enhancement (indicated by a prefix $\Delta$)

$$\tilde{\alpha} = \tilde{\alpha}^{\mathrm{b}} + \Delta\tilde{\alpha}, \qquad \tilde{\beta} = \tilde{\beta}^{\mathrm{b}} + \Delta\tilde{\beta}, \qquad \tilde{\gamma} = \tilde{\gamma}^{\mathrm{b}} + \Delta\tilde{\gamma}. \tag{23}$$

This implies a separation of the diffusion coefficient $D$ into background and critical enhancement, $D = D^{\mathrm{b}} + \Delta D$, with

$$D^{\mathrm{b}} = \frac{\tilde{\alpha}^{\mathrm{b}}}{\rho}\left(\frac{\partial\mu}{\partial c}\right)_{T,P} \quad \text{and} \quad \Delta D = \frac{\Delta\tilde{\alpha}}{\rho}\left(\frac{\partial\mu}{\partial c}\right)_{T,P} . \tag{24}$$

Similarly, the thermal diffusion coefficient $D_{\mathrm{T}}$ decomposes as $D_{\mathrm{T}} = D_{\mathrm{T}}^{\mathrm{b}} + \Delta D_{\mathrm{T}}$ with

$$\begin{aligned} D_{\mathrm{T}}^{\mathrm{b}} &= \frac{1}{c(1-c)\rho}\left[\tilde{\alpha}^{\mathrm{b}}\left(\frac{\partial\mu}{\partial T}\right)_{c,P} + \tilde{\beta}^{\mathrm{b}}\right], \\ \Delta D_{\mathrm{T}} &= \frac{1}{c(1-c)\rho}\left[\Delta\tilde{\alpha}\left(\frac{\partial\mu}{\partial T}\right)_{c,P} + \Delta\tilde{\beta}\right]. \end{aligned} \tag{25}$$

For the transport coefficient $L_{\mathrm{SS}} = L_{\mathrm{SS}}^{\mathrm{b}} + \Delta L_{\mathrm{SS}}$ we find

$$\begin{aligned} L_{\mathrm{SS}}^{\mathrm{b}} &= \tilde{\gamma}^{\mathrm{b}} + T\left(\frac{\partial\mu}{\partial T}\right)_{c,P}\left[2\tilde{\beta}^{\mathrm{b}} + \tilde{\alpha}^{\mathrm{b}}\left(\frac{\partial\mu}{\partial T}\right)_{c,P}\right], \\ \Delta L_{\mathrm{SS}} &= \Delta\tilde{\gamma} + T\left(\frac{\partial\mu}{\partial T}\right)_{c,P}\left[2\Delta\tilde{\beta} + \Delta\tilde{\alpha}\left(\frac{\partial\mu}{\partial T}\right)_{c,P}\right]. \end{aligned} \tag{26}$$

The thermal conductivity, finally, is written as $\lambda_{\mathrm{T}} = \lambda_{\mathrm{T}}^{\mathrm{b}} + \Delta\lambda$ with

$$\begin{aligned} \lambda_{\mathrm{T}}^{\mathrm{b}} &= \tilde{\gamma}^{\mathrm{b}} - T(\tilde{\beta}^{\mathrm{b}})^2/\tilde{\alpha}^{\mathrm{b}} , \\ \Delta\lambda_{\mathrm{T}} &= \frac{T}{\tilde{\alpha}}\left[\frac{(\tilde{\beta}^{\mathrm{b}})^2}{\tilde{\alpha}^{\mathrm{b}}}\Delta\tilde{\alpha} - 2\tilde{\beta}^{\mathrm{b}}\Delta\tilde{\beta} - (\Delta\tilde{\beta})^2\right] + \Delta\tilde{\gamma} . \end{aligned} \tag{27}$$

To investigate the asymptotic critical behavior, diffusion coefficients corresponding to the transport coefficients $\tilde{\beta}$ and $\tilde{\gamma}$ are introduced [48] in analogy with equation (18)

$$\tilde{\alpha} = \rho\left(\frac{\partial c}{\partial\mu}\right)_{T,P} D_{\tilde{\alpha}} , \; \tilde{\beta} = \rho\left(\frac{\partial c}{\partial T}\right)_{\mu,P} D_{\tilde{\beta}} , \; \tilde{\gamma} = \rho T\left(\frac{\partial s}{\partial T}\right)_{\mu,P} D_{\tilde{\gamma}} . \tag{28}$$

From mode-coupling calculations, Kawasaki [12] showed that the mutual diffusivity $D$ obeys a Stokes-Einstein diffusion law asymptotically close to the critical point. Mistura [48] predicted that this should be true for the critical enhancements $\Delta D_i$ of all three diffusion coefficients

$$\Delta D_i = R_{\mathrm{D}} \frac{k_{\mathrm{B}} T}{6\pi\tilde{\eta}\xi}, \quad i \in \{\tilde{\alpha}, \tilde{\beta}, \tilde{\gamma}\}, \tag{29}$$

where $\tilde{\eta}$ is the viscosity and where $R_{\mathrm{D}}$ is a universal amplitude. Mode-coupling calculations to first order [12,48] yield $R_{\mathrm{D}} = 1$, while dynamic-renormalization group calculations yield a slightly higher value of $R_{\mathrm{D}} = 1.056$ [19]. We will set $R_{\mathrm{D}} = 1$ in our discussion of the asymptotic scaling laws. For the enhancements of the transport coefficients Eq. (29) yields

$$\Delta\tilde{\alpha} = \frac{k_{\mathrm{B}} T\rho}{6\pi\tilde{\eta}\xi}\left(\frac{\partial c}{\partial \mu}\right)_{T,P}, \quad \Delta\tilde{\beta} = \frac{k_{\mathrm{B}} T\rho}{6\pi\tilde{\eta}\xi}\left(\frac{\partial c}{\partial T}\right)_{\mu,P}, \quad \Delta\tilde{\gamma} = \frac{k_{\mathrm{B}} T^2\rho}{6\pi\tilde{\eta}\xi}\left(\frac{\partial s}{\partial T}\right)_{\mu,P}. \tag{30}$$

In order to determine the asymptotic critical behavior of the experimentally accessible transport properties, we have to take both critical enhancements and background contributions into account [1,2]. The background values $\tilde{\alpha}^{\mathrm{b}}$, $\tilde{\beta}^{\mathrm{b}}$, and $\tilde{\gamma}^{\mathrm{b}}$ are (by definition) unaffected by critical fluctuations and therefore scale like $\xi^0$. Inserting Eqs. (29) and (30) into Eqs. (24) for the mutual diffusivity $D$ we find that $D$ vanishes approximately like the inverse correlation length in the asymptotic critical region:

$$\left.\begin{aligned} \Delta D &= \frac{k_{\mathrm{B}} T}{6\pi\tilde{\eta}\xi} \sim \xi^{1-\eta-z_\eta} \simeq \xi^{-1} \\ D^{\mathrm{b}} &= \tilde{\alpha}^{\mathrm{b}}\rho\left(\frac{\partial \mu}{\partial c}\right)_{T,P} \sim \xi^{\gamma/\nu} = \xi^{-2+\eta} \simeq \xi^{-2} \end{aligned}\right\} \Rightarrow D \sim \xi^{-1-z_\eta} \tag{31}$$

Because of the thermodynamic identity $(\partial c/\partial T)_{\mu,P} = -(\partial \mu/\partial T)_{c,P}(\partial c/\partial \mu)_{T,P}$ the critical enhancements of $\tilde{\alpha}$ and $\tilde{\beta}$ are related in the asymptotic critical region through $\Delta\tilde{\beta} = -(\partial\mu/\partial T)_{c,P}\Delta\tilde{\alpha}$, where the derivative $(\partial\mu/\partial T)_{c,P}$ remains finite at the critical point. Since the combination $\Delta\tilde{\beta}+(\partial\mu/\partial T)_{c,P}\Delta\tilde{\alpha}$ vanishes asymptotically close to the critical point, the critical enhancement $\Delta D_{\mathrm{T}}$ in Eq. (25) vanishes, so that

$$\left.\begin{aligned} \Delta D_{\mathrm{T}} &= 0 \\ D_{\mathrm{T}}^{\mathrm{b}} &= \frac{1}{c(1-c)\rho}\left[\tilde{\alpha}^{\mathrm{b}}\left(\frac{\partial \mu}{\partial T}\right)_{c,P} + \tilde{\beta}^{\mathrm{b}}\right] \sim \xi^0 \end{aligned}\right\} \Rightarrow D_{\mathrm{T}} \sim \xi^0 \tag{32}$$

This implies that the Soret coefficient diverges approximately like the correlation length

$$S_{\mathrm{T}} = D_{\mathrm{T}}/D \sim \xi^{1+z_\eta} \simeq \xi. \tag{33}$$

The critical behavior of the transport coefficient $L_{\mathrm{SS}}$ is deduced with the aid of the thermodynamic identity $c_{P,c} = c_{P,\mu} - T(\partial\mu/\partial T)^2_{c,P}(\partial c/\partial\mu)_{T,P}$. From

Eqs. (26) and (30) we conclude that the critical enhancement $\Delta L_{\mathrm{SS}}$ vanishes in the asymptotic limit

$$\left.\begin{aligned} \Delta L_{\mathrm{SS}} &= \frac{k_{\mathrm{B}}T}{6\pi\tilde{\eta}\xi}\rho c_{P,c} \sim \xi^{-1-z_\eta+\alpha/\nu} \simeq \xi^{-0.9} \\ L^{\mathrm{b}}_{\mathrm{SS}} &= \tilde{\gamma}^{\mathrm{b}} + T\left(\frac{\partial\mu}{\partial T}\right)_{c,P}\left[2\tilde{\beta}^{\mathrm{b}} + \tilde{\alpha}^{\mathrm{b}}\left(\frac{\partial\mu}{\partial T}\right)_{c,P}\right] \sim \xi^0 \end{aligned}\right\} \Rightarrow L_{\mathrm{SS}} \sim \xi^0. \tag{34}$$

For the thermal conductivity, finally, we find from Eqs. (26) and (27) that the critical enhancement $\Delta\lambda_{\mathrm{T}}$ goes to a finite constant asymptotically close to the critical point

$$\left.\begin{aligned} \Delta\lambda_{\mathrm{T}} &= \frac{k_{\mathrm{B}}T}{6\pi\tilde{\eta}\xi}\rho c_{P,c} + T(\rho c(1-c)D_{\mathrm{T}})^2/\alpha^{\mathrm{b}} \sim \xi^0 \\ \lambda^{\mathrm{b}}_{\mathrm{T}} &= \tilde{\gamma}^{\mathrm{b}} - T(\tilde{\beta}^{\mathrm{b}})^2/\tilde{\alpha}^{\mathrm{b}} \sim \xi^0 \end{aligned}\right\} \Rightarrow \lambda_{\mathrm{T}} \sim \xi^0. \tag{35}$$

## 4 Thermodiffusion near consolute and plait points

From the asymptotic analysis presented in the previous section, we expect the mutual diffusion coefficient $D$ to vanish, $D \sim \xi^{-1-z_\eta}$, we expect the thermal diffusion coefficient $D_{\mathrm{T}}$ to approach a constant, $D_{\mathrm{T}} \sim \xi^0$, and we expect the Soret coefficient to diverge, $S_{\mathrm{T}} \sim \xi^{1+z_\eta}$, as the critical point is approached. The thermal diffusion ratio $k_{\mathrm{T}}$ is defined as

$$k_{\mathrm{T}} = c(1-c)TS_{\mathrm{T}}, \tag{36}$$

and behaves like the Soret coefficient $k_{\mathrm{T}} \sim \xi^{1+z_\eta}$. Since the correlation length diverges as $\xi \sim \Delta\tilde{T}^{-\nu}$, where $\Delta\tilde{T} = (T-T_{\mathrm{c}})/T$ is the reduced temperature and $\nu = 0.63$, the asymptotic analysis predicts for the temperature dependence of these transport coefficients

$$D \sim \Delta\tilde{T}^{(1+z_\eta)\nu} \simeq \Delta\tilde{T}^{0.67}, \tag{37}$$

$$k_{\mathrm{T}} \sim \Delta\tilde{T}^{-(1+z_\eta)\nu} \simeq \Delta\tilde{T}^{-0.67} \tag{38}$$

$$D_{\mathrm{T}} \sim k_{\mathrm{T}}D \sim \Delta\tilde{T}^0 \tag{39}$$

Dynamic renormalization-group (RG) calculations by Folk and Moser [21] show that the non-asymptotic critical behavior of $k_{\mathrm{T}}D$ depends qualitatively on the type of critical point. The product $k_{\mathrm{T}}D$ is constant throughout the critical region of a consolute point, whereas $k_{\mathrm{T}}D \sim$ const. is valid only asymptotically close to a plait point.

In 1975, Giglio and Vendramini [49] performed thermal diffusion measurements near the consolute point of an aniline-cyclohexane mixture. In Fig. 2 we compare experimental data of Giglio and Vendramini [49] with the predicted asymptotic behavior. The experimental data for $k_{\mathrm{T}}$ and $D$ agree with the asymptotic power laws close to the critical temperature but show increasing deviations

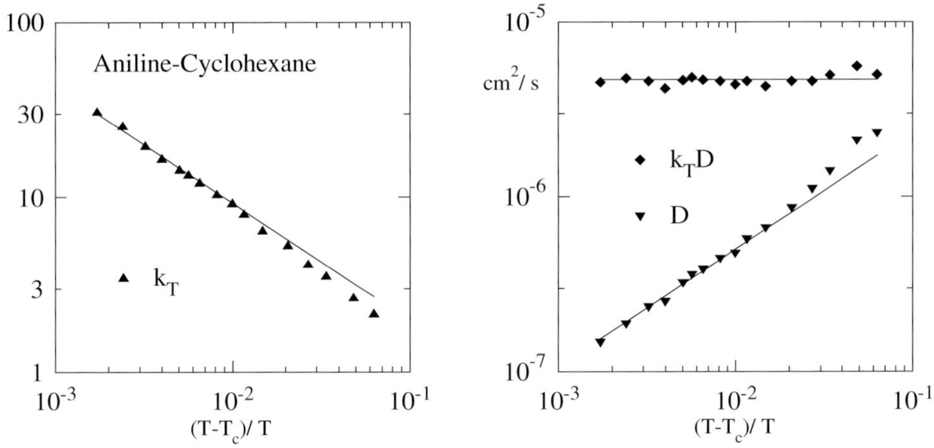


**Fig. 2.** Thermal diffusion data as a function of reduced temperature near the consolute point of an aniline-cyclohexane mixture. The symbols represent experimental data [49], the lines represent the (asymptotic) power laws

with increasing distance from the critical point. This is typical for the crossover behavior of binary mixtures near consolute points. Folk and Moser [21] have applied dynamic RG results to this mixture and find good agreement between their crossover theory and experimental data over the whole temperature range. Note that, as predicted by dynamic RG, the experimental values for $k_T D$ remain constant throughout the critical region.

Thermodiffusion near plait points was investigated for mixtures of $^4$He-$^3$He by Horst Meyer and co-workers [50,51]. In Fig. 3 we present thermal diffusion data [50,51] in the critical region of a $^4$He-$^3$He mixture with mole-fraction $x(^3\text{He}) = 0.8$. Also included in the figure are the asymptotic predictions of Eqs. (37)–(38). In this case, the asymptotic power laws do not describe the experimental data. Instead, the experimental values for the thermal diffusion ratio $k_T$ seem to diverge with an exponent of 1.23 [51], almost twice as high as the asymptotic prediction of 0.67. Until recently, the situation was similar for the thermal conductivity $\lambda_T$ near plait points. Experiments [51,52] suggested a pure-fluid like divergence of $\lambda_T$ near a plait point, in contrast to the asymptotic prediction for mixtures that $\lambda_T$ approach a finite value at the critical point, cf. Eq. (35). It was not until Sakonidou *et al.* [37,53] performed thermal conductivity measurements near the vapor-liquid critical point of a methane-ethane mixture, that the limiting finite behavior of the thermal conductivity near a plait point was observed experimentally. These data also confirmed the predicted crossover [23,25] from the pure-fluid like behavior at some distance from the critical point to the mixture-like behavior, $\lambda_T \sim$ const., very close to the critical point. In Fig. 4 we present mode-coupling predictions [25] for thermal diffusion in a $CO_2$-ethane mixture of mole-fraction $x(\text{ethane}) = 0.75$. Also shown are the asymptotic power laws of Eqs. (37)–(39). Just as seen for the $^4$He-$^3$He mixtures, the

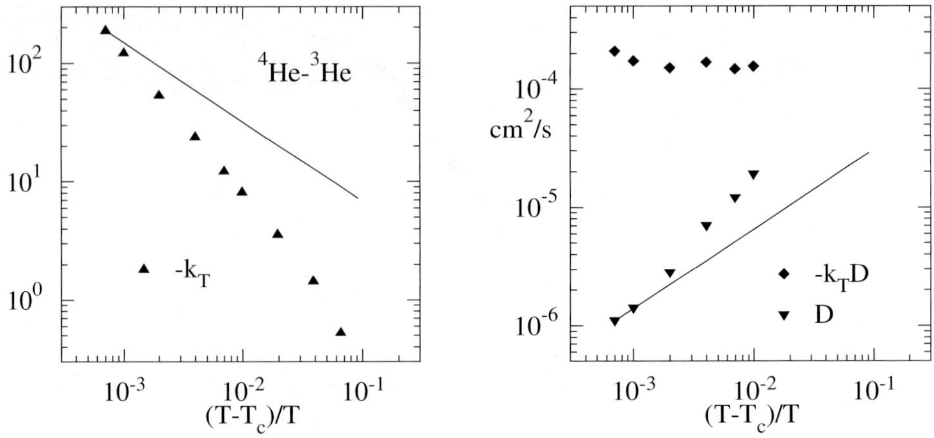


**Fig. 3.** Thermal diffusion data as a function of reduced temperature near the plait point of a $^4$He-$^3$He mixture. The symbols represent experimental data [50,51], the lines represent the asymptotic power laws

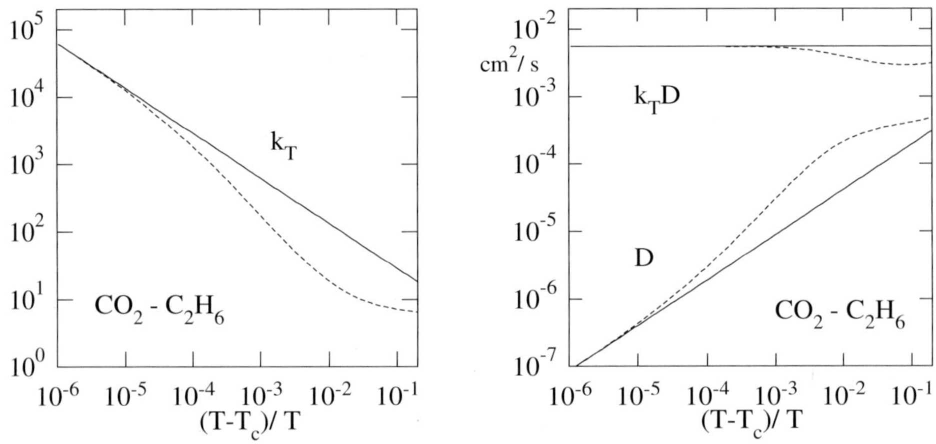


**Fig. 4.** Thermal diffusion near the plait point of a $CO_2$-ethane mixture. The dashed lines represent predictions from a crossover model based on mode-coupling theory [25], the solid lines represent the asymptotic power laws

crossover thermal diffusion ratio increases much more rapidly than predicted from the asymptotic analysis for temperatures not too close to the critical point. For very small reduced temperatures, the crossover prediction [25] approaches the asymptotic power law (38), suggesting that the asymptotic critical region is extremely small and does not cover the temperature range, where experiments were performed on $^4$He-$^3$He mixtures. Folk and Moser [21] recently investigated the critical dynamics (including sound absorption and sound attenuation) of $^4$He-$^3$He mixtures with the aid of dynamic renormalization-group calculations. A comparison of their results with experimental data shows generally good agreement between theory and experiment and confirms the non-asymptotic nature of the experimental data [50,51].

## 5 Discussion

In this contribution, we discussed scaling and universality for both static and dynamic properties of fluids mixtures. We compared experimental thermal diffusion data [49] for an aniline-cyclohexane mixture near its consolute point with the predicted asymptotic power laws and found that the experimental data confirm the asymptotic predictions as the temperature approaches the critical temperature closely. For temperatures farther away from the critical point, the data reflect the crossover [21] from the asymptotic critical to the classical region far away from the critical point. We also compared experimental thermal diffusion data for a $^4$He-$^3$He mixture near a plait point [50,51] with the asymptotic power laws. In this case, experiment and asymptotic prediction differ considerably over the whole temperature range where experiments were performed. By comparing this situation with experimental and theoretical results for other binary mixtures near plait points, we concluded that the asymptotic critical region in binary mixtures near plait points is smaller than in mixtures near consolute points. Furthermore, crossover models [21,25] are required to describe the critical behavior of transport properties near plait points in the whole experimentally accessible region.

## References

1. J. V. Sengers: 'Effects of critical fluctuations on the thermodynamic and transport properties of supercritical fluids.' In: *Supercritical Fluids: Fundamentals for Application*, ed. by E. Kiran, J. M. H. Levelt Sengers (Kluwer Academic, Dordrecht 1994) pp. 231–271
2. J. V. Sengers, J. Luettmer-Strathmann: 'The critical enhancements'. In: *Transport Properties of Fluids*, ed. J. H. Dymond, J. Millat, C. A. Nieto de Castro (Cambridge University Press, Cambridge, UK 1995) Chap. 6, pp 131–159
3. E. Kiran, J. M. H. Levelt Sengers, edts: *Supercritical Fluids – Fundamentals for Application.* NATO ASI Series E: Applied Sciences - Vol. 273. (Kluwer Academic, Dordrecht 1994)
4. T. J. Butenhoff, G. E. Goemans, S. J. Buelow: J. Phys. Chem. **100**, 5982 (1996)
5. A. Firoozabadi, K. Ghorayeb, K. Shukla: AIChE Journal **46**, 892 (2000)
6. J. P. Delville, C. Lalaude, A. Ducasse: Physica A **262**, 40 (1999)
7. B. I. Halperin, P. C. Hohenberg, S. Ma: Phys. Rev. Lett. **29**, 1548 (1972)
8. P. C. Hohenberg, B. I. Halperin: Rev. Mod. Phys. **49**, 435 (1977)
9. K. G. Wilson: Phys. Rev. B **4**, 3174 (1971)
10. M. E. Fisher: Rev. Mod. Phys. **46**, 597 (1974)
11. L. P. Kadanoff, J. Swift: Phys. Rev. **166**, 89 (1968)
12. K. Kawasaki: Ann. Phys. **61**, 1 (1970)
13. K. Kawasaki: 'Mode coupling and critical dynamics'. In: *Phase Transitions and Critical Phenomena Vol.5a.* ed C. Domb, M. S. Green (Academic Press, London, New York 1976) pp 165–403

14. M. Fixman: J. Chem. Phys. **36**, 310 (1962)
15. R. A. Ferrell: Phys. Rev. Lett. **24**, 1169 (1970)
16. R. Perl, R. A. Ferrell: Phys. Rev. A **6**, 2358 (1972)
17. R. Folk, G. Moser: Europhys. Lett. **24**, 533 (1993)
18. R. Folk, G. Moser: J. Low Temp. Phys. **99**, 11 (1995)
19. R. Folk, G. Moser, Phys. Rev. Lett. **75**, 2706 (1995)
20. R. Folk, G. Moser, Int. J. Thermophys. **16**, 1363 (1995)
21. R. Folk, G. Moser, Phys. Rev. E **58**, 6246 (1998)
22. G. A. Olchowy, J. V. Sengers: Phys. Rev. Lett. **61**, 15 (1988)
23. J. Luettmer-Strathmann, J. V. Sengers: Int. J. Thermophys. **15**, 1241 (1994)
24. J. Luettmer-Strathmann, J. V. Sengers, G. A. Olchowy: J. Chem. Phys. **103**, 7482 (1995)
25. J. Luettmer-Strathmann, J. V. Sengers: J. Chem. Phys. **104**, 3026 (1996), J. Chem. Phys. **106**, 438 (1997)
26. S. B. Kiselev, V. D. Kulikov: Int. J. Thermophys **15**, 283 (1994)
27. S. B. Kiselev, V. D. Kulikov: Int. J. Thermophys **18**, 1143 (1997)
28. S. B. Kiselev, M. L. Huber, Fluid Phase Equilib. **142**, 253 (1998)
29. J. V. Sengers, J. M. H. Levelt Sengers: Ann. Rev. Phys. Chem. **37**, 189 (1986)
30. M. Ley-Koo, M. S. Green: Phys. Rev. A **23**, 2650 (1981)
31. R. B. Griffiths, J. C. Wheeler: Phys. Rev. A **2**, 1047 (1970)
32. A. J. Liu, M. E. Fisher, Physica A **156**, 35 (1989)
33. Shang-Keng Ma: *Modern Theory of Critical Phenomena* (Benjamin, Reading, Mass. 1976)
34. M. E. Fisher: J. Math. Phys. **5**, 944 (1964)
35. K. Huang: *Statistical Mechanics*, 2nd edn. (Wiley, New York 1987)
36. L. D. Landau, E. M. Lifshitz: *Statistical Physics.* (Pergamon Press, London 1958)
37. E. P. Sakonidou, H. R. van den Berg, ten Seldam. C. A., J. V. Sengers: J. Chem. Phys. **109**, 717 (1998)
38. E. D. Siggia, B. I. Halperin, P. C. Hohenberg: Phys. Rev. B **13**, 2110 (1976)
39. R. A. Ferrell, N. Menyhárd, H. Schmidt, F. Schwabl, P. Szépfalusy: Phys. Rev. Lett. **18**, 891 (1967)
40. R. A. Ferrell, N. Menyhárd, H. Schmidt, F. Schwabl, P. Szépfalusy: Ann. Phys. **47**, 565 (1968)
41. B. I. Halperin, P. C. Hohenberg: Phys. Rev. Lett. **19**, 700 (1967)
42. B. I. Halperin, P. C. Hohenberg: Phys. Rev **177**, 952 (1969)
43. G. Flossmann, R. Folk, G. Moser: Phys. Rev. E **60**, 779 (1999)
44. H. Hao: Aspects of Dynamic Critical Phenomena in a Single Component Fluid. PhD Thesis, University of Maryland, College Park, Maryland (1991)
45. L. D. Landau, E. M. Lifshitz: *Fluid Mechanics* 2nd edn. (Pergamon Press, Oxford 1987)
46. S. R. de Groot, P. Mazur: *Non-Equilibrium Thermodynamics.* (Dover, New York 1984)
47. R. Haase: *Thermodynamics of Irreversible Processes.* (Addison-Wesley, Reading, MA 1969)
48. L. Mistura: Nuovo Cimento **12 B**, 35 (1972)
49. M. Giglio, A. Vendramini, Phys. Rev. Lett. **34**, 561 (1975)
50. L. H. Cohen, M. L. Dingus, H. Meyer: Phys. Rev. Lett. **50**, 1058 (1983)
51. L. H. Cohen, M. L. Dingus, H. Meyer: J. Low Temp. Phys. **61**, 79 (1985)
52. L. H. Cohen, M. L. Dingus, H. Meyer: J. Low Temp. Phys. **49**, 545 (1982)
53. E. P. Sakonidou, H. R. van den Berg, C. A. ten Seldam, J. V. Sengers: Phys. Rev. E **56**, 4943 (1997)

## Supplementary glossary

| | |
|---|---|
| $T_{\mathrm{c}}$, $\rho_{\mathrm{c}}$, $P_{\mathrm{c}}$ | critical temperature, density, pressure |
| $\Delta\tilde{T} = (T - T_{\mathrm{c}})/T$ | reduced temperature |
| $\chi_{\mathrm{T}} = \rho(\partial\rho/\partial P)_T$ | symmetrized isothermal compressibility |
| $\xi$ | correlation length |
| $\alpha \simeq 0.11$, $\gamma \simeq 1.239$, $\eta \simeq 0.033$, $\nu \simeq 0.63$ | static critical exponents |
| $z_\eta \simeq 0.063$, $z = 3 + z_\eta$ | dynamic critical exponents |
| $\tilde{\eta}$ | viscosity |
| $L_{\mathrm{SS}}$ | thermal conductivity in the absence of a concentration gradient |
| $k_{\mathrm{T}} = c(1 - c)TS_{\mathrm{T}}$ | thermal diffusion ratio |
| $\tilde{\alpha}$, $\tilde{\beta}$, $\tilde{\gamma}$, $\tilde{\delta}$ | Landau transport coefficients |

### Superscript

| | |
|---|---|
| b | indicates a background contribution |

### Prefix

| | |
|---|---|
| $\Delta$ | indicates a critical enhancement |

# On the Theory of the Soret Effect in Colloids

Konstantin I. Morozov

Institute of Continuous Media Mechanics, UB of Russian Academy of Science,
1, Korolev Str., 614013 Perm, Russia

**Abstract.** Movement of colloidal particles can be driven by the inhomogeneous temperature field. In this paper the relation between the phenomenon of the migration of colloidal particles and the characteristics of the protective layer around the grain is established. Two cases of surfacted and ionic colloids are considered. The temperature gradient in the fluid induces the redistribution of surfactant molecules or ions near the particle. The redistribution leads to a loss in the balance of the stress on the particle surface, and that results in the grain movement along the temperature gradient. The values of the Soret coefficient of thermodiffusion motion in colloids turn out to be extremely high – two–four orders larger than the values of the Soret coefficient for molecular systems. The analytical solution of the problem of thermodiffusion motion in surfacted colloids is obtained using an assumption for the solute–particle interaction. The particles of these colloids always migrate towards colder temperature regions of the fluid. In the case of ionic colloids the direction of the particle movement depends on the magnitude of the double layer thickness and the value of the electric potential of the particle surface. The slowing-down of the thermodiffusion motion and the changing of its direction as a result of an increase in the ionic force of the solution are predicted.

## 1 Introduction

The phenomenon of the separation of substances in a binary mixture under the influence of a inhomogeneous temperature field has been known for more than 100 years and is called thermodiffusion or the Soret effect [1]. The thermodiffusion is observed in an enormous number of *molecular* systems, such as gas and liquid mixtures, the solutions of salts, etc. The parameter that characterizes the separation of substances in the mixture is the Soret coefficient $S_{\mathrm{T}}$ (see below). The values of $S_{\mathrm{T}}$ are very small for molecular systems, which makes the practical application of thermodiffusion problematical or even technically impossible. At the same time the role of thermodiffusion in *disperse* systems (colloids and suspensions) can be significant.

Probably the first observation of thermodiffusion in colloids was made by the group of Prof. Bacri [2]. The authors used the method of Forced Rayleigh Scattering and demonstrated that the modulation of the temperature field induces the modulation of the colloidal particle concentration. This result was a qualitative one, however it showed the existence of this new interesting phenomenon. The first quantitative data related to the magnitude of the effect were obtained recently [3–6]. Following these works let us define the Soret coefficient $S_{\mathrm{T}}$ of the colloid. Let $\nabla\phi$ be the stationary gradient of the volume concentration $\phi$

W. Köhler and S. Wiegand (Eds.): LNP 584, pp. 38–60, 2002.

of the colloidal particles subjected to the inhomogeneous temperature field with gradient $\boldsymbol{A}$. The gradients are related to each other according to the following equation

$$\nabla\phi + S_{\mathrm{T}}\phi(1-\phi)\boldsymbol{A} = 0 . \tag{1}$$

Here $S_{\mathrm{T}}$ is the Soret coefficient of colloidal particles.

The experimental values $S_{\mathrm{T}}$ were obtained using the above-mentioned method of Forced Rayleigh Scattering [3,6] as well as by means of a thermodiffusion column [4–6]. The different approaches to the determination of $S_{\mathrm{T}}$ gave well correlated and at the same time unexpected results; these can be summarized as follows:

1. The Soret coefficient of the thermodiffusion motion in colloids turned out to be two–four (!) orders of magnitude greater then the typical $S_{\mathrm{T}}$ value of known molecular systems.

2. The Soret coefficients of ionic and surfacted colloids have opposite signs – the particles of ionic colloids migrate towards the higher-temperature regions of the fluid whereas the grains of surfacted colloids move in the opposite direction.

The latter statement made us recall that it is common to divide disperse systems into two categories, ionic and surfacted, which differ in stability with respect to particle coagulation. The first category is characterized by electrostatic stabilization which is achieved by imparting electric charge to the particles. The second category incorporates disperse systems whose particles are covered by an additional layer of surfactant. The electrostatic repulsions in ionic colloids and steric interactions in surfacted ones prevent the particle from approaching each other and prevent coagulation.

The phenomenon of the thermodiffusion movement of colloidal particles has a concentrational analog – so-called *diffusiophoresis* [7]. This term was introduced by Derjaguin and co-authors to denote the disperse particles migration subjected to the concentration gradient of solute molecules. An elegant approach to describing the diffusiophoresis in non-electrolytes solutions and electrolytes solutions was developed in [8,9]. We use some results of this theory for the description of thermodiffusion motion in colloids.

In this paper the theory of the motion of colloidal particles induced by the temperature field is proposed. Due to the exclusive variety and complexity of colloidal systems we adopt some simplifying assumptions. We ignore the real polydispersedness of colloidal particles and consider them as hard spheres of equal size. The other assumption is connected with the effects of interparticle interactions. The colloids represent three-ingredients systems: colloidal particles, a carrier fluid and solute particles of the protective layer (molecules of surfactants or ions). In general, there are six kinds of interactions between particles of these three substances. The interaction between colloidal particles can be decreased considerably by using low-concentration colloids. This corresponds to the real experimental situation [3–6]. Therefore, we use here an approximation of low colloid concentration and investigate the case of a single colloidal particle. Among the other interactions, we focus attention on the interaction between the colloidal and solute particles. This interaction alone separates colloidal and binary molecular systems. We would expect that the strong temperature dependence of the concentration of solute molecules near colloidal particles is the primary source of the intensive thermodiffusion motion in colloids. We show that taking into account the above-mentioned interaction leads to an adequate description of the phenomenon of thermodiffusion in colloids and predicts experimentally observable values of the Soret coefficient. All other interactions play minor roles. For example, the interaction between colloidal particles and fluid molecules can also be the reason for thermodiffusion motion. In this case, however, the values of the termodiffusion coefficient are expected to be typical of molecular systems, i.e. much smaller than in colloids. We regard this conclusion to be a result of the absence of a noticeable variation of fluid molecules concentration with temperature. Analogously, solute–solute interaction can influence the state of the protective layer around colloidal particles [10] and the thermodiffusion process. Consideration of this interaction is the individual many-body problem and is outside of the scope of this study.

In this paper we take into account only the interaction between colloidal and solute particles and neglect entirely the other types of interactions. We establish that the thermodiffusion motions of the particles in colloids of both types have a common nature and are related to the characteristics of the protective layer of counterions and surfactant molecules in the cases of ionic and surfacted colloids, respectively. In Sect. 2, the simpler case of surfacted colloids is considered. The equations of diffusion of surfactant molecules and fluid motion are obtained using approximations of small Péclet and Reynolds numbers. The effect of the difference in the values of the thermal conductivities of the particle and the fluid on the temperature distribution near the grain is taken into account. The analysis is done with some assumptions on solute–particle interaction, and the analytical solution of the problem is obtained. The peculiarities of the theory of ionic colloids and its main results examined in our papers [11,12] are summarized in Sect. 3. We make some concluding remarks in Sect. 4.

## 2 Surfacted colloids

### 2.1 Why does a colloidal particle move?

The main difference between colloidal and molecular systems is that the rigid colloidal particles are about 100 Å in size, which is much larger than the size of the molecules in a liquid solvent. It allows us to consider the liquid as a continuous medium.

Let us examine a single spherical grain that resides in the fluid with surfactant molecules. The particle is rigid and its radius is $R$. Let $\varphi_s$ be the potential of the interaction between the surfactant molecule and the grain. We put the origin of the coordinate system in the center of the particle. The potential $\varphi_s = \varphi_s(r)$ is assumed to be central. It is assumed that the surfactant molecules do not interact with each other. The concentration of the particles for a constant temperature $T_0$ is described by the Boltzmann distribution

$$n_0(r) = n_\infty \exp(-\Phi) , \tag{2}$$

where $n_\infty$ is the solute concentration far from the particle, $\Phi = \varphi_s/k_B T_0$ is the dimensionless potential, and $k_B$ is the Boltzmann constant.

If the system is in thermodynamic equilibrium the particle and the fluid are at rest. Indeed, the distribution (2) is spherically symmetric and all forces applied to the grain are balanced by each other. Analogously, the body force $-n_0 \nabla \varphi_s$ exerted by the particle on the fluid is balanced by the pressure gradient in the fluid.

For the inhomogeneous temperature field the spherical symmetry of the problem changes to an axial one. The solute concentration $n(\boldsymbol{r})$ is no longer equal to its equilibrium value $n_0(r)$. As a result, the stresses on the opposite sides of the grain do not compensate each other and the particle starts moving. In other words, it is profitable for the particle to shed the surfactant molecule from a hot side of its surface and adsorb another surfactant molecule on a cold side. According to this simple picture the colloidal particles must migrate toward the low temperature region of the fluid.

## 2.2 Equations of solute concentration

Let $\boldsymbol{A}$ be the given gradient of the temperature in the fluid far from the grain. The characteristic time of propagation of thermal perturbation, $\tau_T \sim R^2/D_{th}$, where $D_{th}$ is the thermal diffusivity of the fluid, is much shorter than the time of the rotational Brownian diffusion of the grain, $\tau_B \sim \eta R^3/k_B T_0$ ($\eta$ is the viscosity of the fluid) [13]. Indeed, using the data for an aqueous solution, $\eta \sim$ 1 cP and $D_{th} \sim 0.0015\ \mathrm{cm^2\,s^{-1}}$, we find $\tau_T/\tau_B \sim 10^{-3}$. Hence we may assume the temperature distribution $T(\mathbf{r})$ near the particle is time-independent. This distribution can be written in terms of the thermal conductivities $\lambda_T^{(1)}$ and $\lambda_T^{(2)}$ of the particle and the fluid, respectively, as follows [13]

$$T(\boldsymbol{r}) = T_0 + T_1 = T_0 + \left(1 - \kappa \left(\frac{R}{r}\right)^3\right) \boldsymbol{A}\boldsymbol{r} \,, \tag{3}$$

where $\kappa = (\lambda_T^{(1)} - \lambda_T^{(2)})/(\lambda_T^{(1)} + 2\lambda_T^{(2)})$, $T_0$ is the temperature in the center of the grain, and $T_1$ is the inhomogeneity of the temperature field due to the different values of the thermal conductivities of the particle and fluid as well as the external imposed gradient $\boldsymbol{A}$.

The characteristic scale of the temperature variation is assumed to be macroscopic. Therefore the dimensionless temperature gradient $\varepsilon_T = AR/T_0$ is a small parameter of the problem. In experiments [3–6] $A \sim 10^2$ K/cm, $R \sim 10^{-6}$ cm, $T_0 \sim 10^2$ K and $\varepsilon_T \sim 10^{-6}$.

To determine the grain velocity $\boldsymbol{U}$ let us first find the distribution $n(\boldsymbol{r})$ of the surfactant in the temperature field. We expand it near the equilibrium value (1) upon $\varepsilon_T$

$$n = n_0 + n_\infty \tilde{c} \,, \tag{4}$$

where the quantity $\tilde{c} = \tilde{c}(\boldsymbol{r})$ is linear with respect to $\varepsilon_{\mathrm{T}}$. The diffusion flux of the surfactant molecules near the grain is

$$\boldsymbol{j} = -D\left(\nabla n + \frac{n\nabla\varphi_{\mathrm{s}}}{k_{\mathrm{B}}T}\right) , \tag{5}$$

where $D$ is the solute diffusion coefficient. For simplicity we do not distinguish here and below between the diffusion coefficients near and far from the particle, as is done often [14]. In the partial case of a uniform temperature field ($T = T_0$) (5) reduces to the trivial equilibrium result (2).

Substituting (2) and (4) into (5) and retaining only terms that are linear on $\varepsilon_{\mathrm{T}}$, one can write the diffusion flux of the surfactant in the form

$$\boldsymbol{j} = -Dn_0\left(\nabla Q + \frac{\Phi}{T_0}\nabla T\right) , \tag{6}$$

where we used the notation

$$Q = \tilde{c}\mathrm{e}^{\Phi} - \Phi T_1/T_0 . \tag{7}$$

Let us emphasize that the dimensionless potential $\Phi$ in (6) and (7) is the same as above: it is determined by the temperature $T_0$.

The intensity of the diffusion of the surfactant molecules is characterized by the Péclet number $Pe = UL/D$, where $L$ is the characteristic scale of the solute concentration variation. The value of the Péclet number for this problem is very small (see estimate below). Hence, the convective term $\boldsymbol{u}\nabla n$ ($\boldsymbol{u}$ is the fluid velocity) can be omitted in the equation for the stationary diffusion which takes the simple form $\nabla \cdot \boldsymbol{j} = 0$ [13]. Substituting (6), one obtains

$$\Delta Q - \nabla Q\nabla\Phi + \nabla T\nabla\Phi(1-\Phi)/T_0 = 0 . \tag{8}$$

The boundary conditions for (8) vanish for surfactant concentrations far from the particle, and the conditions for impermeability on its surface are:

$$Q(r \to \infty) \to 0 , \quad j_n(r = R) = 0 . \tag{9}$$

Let the $z$-axis be directed along the external temperature gradient $\boldsymbol{A}$ and $\theta$ be the polar angle. Due to the axisymmetric nature of the problem we can find a solution in the form

$$Q(r,\theta) = q(r)\cos\theta . \tag{10}$$

The resulting equation for $q(r)$ is obtained by the substitution of (3) and (10) into (8)

$$\Delta q - \frac{2q}{r^2} - \frac{\mathrm{d}\Phi}{\mathrm{d}r}\frac{\mathrm{d}q}{\mathrm{d}r} + \frac{A}{T_0}\left[1 + 2\kappa\left(\frac{R}{r}\right)^3\right]\frac{\mathrm{d}\Phi}{\mathrm{d}r}(1-\Phi) = 0 . \tag{11}$$

The supplementary conditions for (11) are

$$q(r \to \infty) \to 0 , \quad \frac{\mathrm{d}q}{\mathrm{d}r}(r = R) = \frac{A\zeta(1+2\kappa)}{T_0} , \tag{12}$$

where $-\zeta = \Phi(R)$ is the value of the solute–particle potential when the surfactant molecule is attached to the surface of the grain.

When the system of equations (11) and (12) is solved, then the solute distribution $n(\boldsymbol{r})$ is known and the hydrodynamic part of the problem can be solved.

## 2.3 A colloidal particle as a "swimmer"

The next step in our consideration is the derivation of an expression for the colloidal particle velocity $\boldsymbol{U}$. Due to the symmetry of the problem it is parallel to the temperature gradient $\boldsymbol{A}$. We assume that the coordinate system is attached to the body of grain, so that the particle velocity $\boldsymbol{U}$ in the laboratory coordinate system is expressed through the fluid velocity $\boldsymbol{u}$ far from the particle as

$$\boldsymbol{U} = -\boldsymbol{u}(r \to \infty) . \tag{13}$$

Therefore, the aim of this section is the determination of the fluid velocity.

We will show below that the Reynolds number of the thermodiffusion motion $Re = \varrho U R/\eta$, where $\varrho$ is the fluid density, is small, as is the Péclet number. Thus, the equations of fluid motion in the Stokes approximation can be used. They are

$$\eta \Delta \boldsymbol{u} - \nabla p + \boldsymbol{f} = 0 , \quad \nabla \cdot \boldsymbol{u} = 0 , \tag{14}$$

where $p$ is the pressure. The term $\boldsymbol{f} = -n(\boldsymbol{r})\nabla\varphi_{\mathrm{s}}$ is the body force exerted on the fluid. Applying (2), (4) and (7), the expression for the body force becomes:

$$\boldsymbol{f} = \nabla(n_0 k_{\mathrm{B}} T_0) - n_0 k_{\mathrm{B}} T_0 (Q + \Phi T_1/T_0)\nabla\Phi . \tag{15}$$

The significance of result (15) is that the body force can not be reduced to the gradient form. Due to the non-potential character of the body force the equilibrium of the fluid is impossible. The colloidal particle behaves like a "swimmer". Forcing the fluid to move, it starts moving too. This is another interpretation of the above-mentioned reasons for why the grain moves under the influence of a temperature field.

Substituting (15) into (14) gives

$$\eta \Delta \boldsymbol{u} - \nabla\tilde{p} - n_0 k_{\mathrm{B}} T_0 (Q + \Phi T_1/T_0)\nabla\Phi = 0 , \tag{16}$$

where $\tilde{p} = p - n_0 k_{\mathrm{B}} T_0$ is the renormalized pressure.

Now let us make the representation which is due to the axisymmetric nature of the problem

$$\boldsymbol{u} = h(r)\cos\theta\, \boldsymbol{i}_r + g(r)\sin\theta\, \boldsymbol{i}_\theta , \tag{17}$$

$$\nabla \times \boldsymbol{u} = G(r)\sin\theta\, \boldsymbol{i}_\phi , \tag{18}$$

$$G(r) = \frac{\mathrm{d}g}{\mathrm{d}r} + \frac{h+g}{r} , \tag{19}$$

where $\boldsymbol{i}_r$, $\boldsymbol{i}_\theta$, $\boldsymbol{i}_\phi$ are the unit vectors of the spherical coordinate system. Using the equation of discontinuity (second relation in (14)) the radial part of the vorticity can be written as

$$G(r) = -\frac{1}{2}h'' r - 2h' . \tag{20}$$

After taking the curl of (16) we obtain the resulting equation for the determination of the fluid velocity

$$\Delta G - \frac{2G}{r^2} - \frac{n_0 k_B T_0}{\eta} \left[ \frac{q}{r} + \frac{A\Phi}{T_0} \left( 1 - \kappa \left( \frac{R}{r} \right)^3 \right) \right] \Phi' = 0 \; . \tag{21}$$

The differential equation (21) is of the fourth order over the variable $h(r)$ (see (20)) and four additional boundary conditions are needed to solve it. Two of them are obtained from the non-slip condition on the surface of the particle, $\boldsymbol{u}(r = R) = 0$. They are

$$h(r = R) = 0 \; , \quad h'(r = R) = 0 \; . \tag{22}$$

The two remaining relations are related to the asymptotic behavior of the velocity field. The value of the fluid velocity $\boldsymbol{u}(r \to \infty)$ far from the grain has to be finite. This value is the objective of our consideration because it defines the thermodiffusion velocity $U$. For this value we obtain from (13) and (17) the following equation

$$U = -h(r \to \infty) \; . \tag{23}$$

The last condition can be obtained from the additional physical constraint [9]. Let us consider a large spherical surface with radius $R_* \gg R$ enclosing the particle and the surrounding fluid. The total force exerted on the body enclosed by this surface must be zero because the solute–particle interactions are negligible far from the particle. This proves to be equivalent to the absence of a Stokeslet in the asymptotic representation $\boldsymbol{u}(r \to \infty)$ or, in other words, the velocity field $\boldsymbol{u}$ decays to $-\boldsymbol{U}$ as $O(r^{-m})$, where $m \geq 2$ [9]. Both supplementary asymptotic conditions can be combined in one relation

$$h(r \to \infty) = -U + O(r^{-m}) \; , \quad m \geq 2 \; . \tag{24}$$

The expressions (2), (11), (12), (20)–(22) and (24) form the complete system of equations to be solved. Solving this system is rather complicated. At first, the solution of the diffusion problem (11), (12) should be found and the obtained value of the radial part $q(r)$ of the perturbed surfactant distribution should be substituted in the last term of (21). Then, the solution of (20)–(22), (24) for the fluid velocity $\boldsymbol{u}$ should be found. Finally, the velocity $U$ of thermodiffusion motion is defined by the asymptotic behavior of the velocity field (23).

### 2.4 The case of step-like potential. Analytical solution of the problem

It is evident that the realization of the above proposed scheme of the determination of the particle velocity is impossible without knowledge of the potential of interaction $\Phi(r)$ between a surfactant molecule and a particle. The quantity of $\Phi(r)$ is difficult to measure and its precise form is unknown. Therefore, there is a large room for various speculations about this characteristic. We make the simplest and reasonable, in our opinion, assumption related to $\Phi(r)$, writing it in the form of a step-like function:

$$\Phi(r) = \begin{cases} -\zeta, & R \le r < R + L\,, \\ 0, & r \ge R + L\,. \end{cases} \tag{25}$$

Here $L$ is the characteristic length of the interaction. The sign of $\zeta$ is assumed to be positive and corresponds to the attractive solute–particle interaction. The values of $L$ and $\zeta$ depend on the nature of the three media, the carrier fluid, the surfactant and the grain material, and are known approximately: $L = 1$–$2$ Å and $\zeta = 20$–$30$ [15].

Let us introduce the Heaviside step function

$$\Theta(x) = \begin{cases} 0, & x < 0\,, \\ 1, & x \ge 0\,. \end{cases} \tag{26}$$

The potential of the interaction between the surfactant molecule and the particle can be rewritten as

$$\Phi(r) = -\zeta\left[1 - \Theta(r - r_0)\right]\,, \tag{27}$$

where $r_0 = R + L$.

The exact analytical solution for this model can be derived. The substitution (27) into the equations of diffusion (11) and fluid motion (21) lead to the singularity at $r = r_0$ because of $\Theta'(x) = \delta(x)$. To achieve further progress let us consider the following differential equation for a function $s(r)$ of spherical coordinate $r$

$$\Delta s - \frac{2s}{r^2} + B\delta(r - r_0) = 0\,. \tag{28}$$

Here $B$ is the parameter. The general solution of (28) is

$$s(r) = b_1 r + \frac{b_2}{r^2} + \frac{B}{3}\theta(r - r_0)\left(-r + \frac{r_0^3}{r^2}\right)\,, \tag{29}$$

where $b_1$ and $b_2$ are the arbitrary constants.

Let us consider also the characteristic terms of (11) and (21): $\Theta(r - r_0)\Phi'$, $\Phi\Phi'$, $e^{-\Phi}\Phi'$ and $e^{-\Phi}\Phi\Phi'$. All these terms contain the indeterminacies that can be covered by the rules

$$\begin{aligned} \Theta(r - r_0)\Phi' &\to \tfrac{1}{2}\zeta\delta(r - r_0)\,, \\ \Phi\Phi' &\to -\tfrac{\zeta^2}{2}\delta(r - r_0)\,, \\ \mathrm{e}^{-\Phi}\Phi' &\to (\mathrm{e}^{\zeta} - 1)\delta(r - r_0)\,, \\ \mathrm{e}^{-\Phi}\Phi\Phi' &\to -\left((\zeta - 1)\mathrm{e}^{\zeta} + 1\right)\delta(r - r_0)\,. \end{aligned} \tag{30}$$

The validity of (30) can be verified by the direct integration of both sides of the relations around the singularity point $r = r_0$. For example, by integrating the first relation in (30) we obtain

$$\int_{r_0-}^{r_0+} \Theta(r - r_0)\Phi' \, \mathrm{d}r = \zeta \int_{r_0-}^{r_0+} \Theta(r - r_0)\mathrm{d}\Theta(r - r_0) = \frac{1}{2}\zeta \, . \tag{31}$$

We utilize the above relations to solve the equations of diffusion (11) and fluid motion (21) which have the form of (28). The final results are rather cumbersome. We present them only for the moderate and high values of the solute–particle interaction ($\zeta > 5$) that correspond to the real experimental situation [15]. Using the boundary conditions (12), (22) and (24) we obtain for this case

$$\begin{aligned} q(r) &= \frac{B}{3}r + \frac{R^3}{2r^2}\left[\frac{B}{3} - \frac{A}{T_0}\zeta(1 + 2\kappa)\right] + \frac{B}{3}\Theta(r - r_0)\left(-r + \frac{r_0^3}{r^2}\right) , \\ B &= -3\frac{A\zeta}{T_0}\frac{\zeta\left(1 - 2\gamma(1 + \kappa)\right) + 2(1 + 2\gamma\kappa)}{\zeta(1 + 2\gamma) - 6} , \quad \gamma = \left(\frac{R}{r_0}\right)^3 , \end{aligned} \tag{32}$$

$$\begin{aligned} h(r) = -\frac{\xi}{15}\Bigg\{ & r^2 - R^2 + 5r_0^3\left(\frac{1}{r} - \frac{1}{R}\right) + \frac{1}{3}R^2(2R^3 - 5r_0^3)\left(\frac{1}{r^3} - \frac{1}{R^3}\right) \\ & -\Theta(r - r_0)\left[r^2 - r_0^2 + 5r_0^3\left(\frac{1}{r} - \frac{1}{r_0}\right) - r_0^5\left(\frac{1}{r^3} - \frac{1}{r_0^3}\right)\right]\Bigg\} , \\ \xi = (1 - \gamma\kappa) & \frac{n_\infty k_\mathrm{B} A \mathrm{e}^\zeta}{\eta} \frac{\zeta^2(2 + \gamma) - \zeta(5 + 4\gamma) + 6}{\zeta(1 + 2\gamma) - 6} . \end{aligned} \tag{33}$$

The concentration of surfactant molecules far from the particle $n_\infty$ is not the measurable physical property. Instead of $n_\infty$, it is common to use the Gibbs excess surface concentration $\Gamma$, defined as

$$\Gamma = n_\infty \int_R^\infty \left(\mathrm{e}^{-\Phi} - 1\right) \, \mathrm{d}r \, . \tag{34}$$

Let us also introduce the dimensionless parameter $\Lambda$, the ratio of the characteristic scale of the solute–particle interaction to the radius of the grain: $\Lambda = L/R$. This parameter in colloids is very small. Therefore, we remain below the main term in the expansion of the velocity using the powers of $\Lambda$, excepting only the factor $1 - \gamma\kappa$ because the experimental values for $\kappa$ are close to unity [3].

Examining (33) for the limit $r \to \infty$ we get finally the surprisingly simple expression for the velocity of thermodiffusion motion of the particle is

$$\boldsymbol{U} = -\frac{\Gamma k_\mathrm{B} L}{\eta}(1 - \kappa\gamma)(\zeta - 1)\boldsymbol{A} \, . \tag{35}$$

Let us analyze the obtained result. Due to the positiveness of the parameters in (35) the velocity of the thermodiffusion motion in surfacted colloids is directed opposite to the temperature gradient $\boldsymbol{A}$, i.e. the particle migrates toward the low-temperature region, which is in agreement with the experimental data [2–6] and our intuitive arguments presented at the beginning of Sect. 2.

The physical sense of (35) is clear. Indeed, if the right part of (35) is multiplied and divided by $T_0$, then the block $\sigma = \Gamma k_\mathrm{B} T_0(\zeta - 1)$ is the total energy of solute–particle interactions per unit of area of particle surface. It may be considered as the effective surface tension at the fluid–grain boundary. The factor $(1 - \kappa\gamma)A$ is the tangential temperature gradient at the upper boundary $r_0 = R + L$ of the protective layer (see (3)) and the multiplier $p_\tau = \sigma(1 - \kappa\gamma)A/T_0$ plays the role of the tangential stresses at $r_0$. Because of the small thickness of the protective layer the flow inside it is a Couette one. Thus, the combination $p_\tau L/\eta$ represents the tangential fluid velocity at the upper boundary of the protective layer. Therefore, we can think of the motion of the fluid and particle as the result of the tangential viscous stresses acting along the boundary of the protective layer and the outer fluid.

The author of [3] interpreted the obtained experimental data on thermodiffusion in colloids by analogy with the well-known problem of the thermocapillary movement of a liquid drop due to the Marangoni effect [16]. At first sight, our consideration above corresponds completely to this interpretation. However, it does not. The point is that the thermodiffusion velocity $U$ in (35) is proportional to the solute–particle interaction length $L$. Thus, if we intend to avoid the consideration of the problem inside the protective layer and present its properties in the form of effective surface tension, we have to carry out the limiting transition $L \to 0$. This procedure gives the trivial result for the particle velocity, which is not striking because of the no-slip conditions (22) at the grain surface. Therefore, the attempt to explain the thermodiffusion motion of colloidal particles by the Marangoni-like effect [3] fails and the similarity of the migration of rigid and liquid particles driven by an inhomogeneous temperature field proves to be at least superficial.

Let us estimate the order of magnitude of the thermodiffusion velocity. The substitution of the experimental data $R = 50$ Å, $\eta = 10^{-2}$ P, $A \sim 10^2$ K/cm, $\kappa = 0.95$ [3] and typical data for adsorption of surfactant molecules on the grain surface $\Gamma = 10^{14}\ \mathrm{cm}^{-2}$, $L = 1$ Å and $\zeta = 20$ [15] gives the estimate $U \sim 10^{-6}$ cm/s. Now we can estimate the values of the Reynolds $Re = \varrho UR/\eta$ and the Péclet $Pe = UL/D$ numbers. Using $\varrho \sim 1\ \mathrm{g/cm}^3$ for the fluid density and $D \sim 10^{-5}\ \mathrm{cm}^2/\mathrm{c}$ for the diffusion coefficient of the surfactant molecules we obtain $Re \sim 10^{-10}$ and $Pe \sim 10^{-9}$. As we assumed above the Reynolds and the Péclet numbers proved to be small.

### 2.5 Calculation of the Soret coefficient. The problem of "hidden parameters"

As soon as the thermodiffusion velocity is known, one can calculate the Soret coefficient $S_\mathrm{T}$ of the colloid. The thermodiffusion flux of grains $\boldsymbol{j}_\mathrm{T} = n_\mathrm{p}\boldsymbol{U}$, with $n_\mathrm{p}$ being the colloidal particle concentration, should be balanced by the particle diffusion flux $\boldsymbol{j}_d$, that for the low-concentration ($\Phi \ll 1$) colloid equals [17]

$$\boldsymbol{j}_{\mathrm{d}} = -\frac{k_{\mathrm{B}} T_0}{6\pi\eta R_{\mathrm{H}}} \nabla n_{\mathrm{p}} \ , \tag{36}$$

where $R_{\mathrm{H}}$ is the so-called hydrodynamic radius of the particle. We recall that the particle moves in fluid surrounded by the cloud of the surfactant molecules which effectively increases the particle radius from $R$ to the hydrodynamic radius $R_{\mathrm{H}}$. The Soret coefficient of the surfacted colloid is found from the comparison of the equation of the balance these two fluxes and (1).

$$S_{\mathrm{T}} = 6\pi \frac{\Gamma R_{\mathrm{H}} L}{T_0} (1 - \kappa\gamma)(\zeta - 1) \ . \tag{37}$$

It is seen from (37) that the values of the Soret coefficient of the surfacted colloids are positive. Let us compare our result with the experimental data. In [3–6] it was found that the Soret coefficient of colloids is extremely large. It has values 0.1–0.2 $\mathrm{K}^{-1}$ which are two–four orders larger than the values of the Soret coefficient for the molecular systems. Our formula (37) gives the right estimation $S_{\mathrm{T}}$. Indeed, using the mentioned data for $\Gamma$, $L$, $\kappa$, $\zeta$, $R$ and the experimental parameters $R_{\mathrm{H}} = 150$ Å and $T_0 = 323$ K in formula (37) we find $S_{\mathrm{T}} \sim 0.17\ \mathrm{K}^{-1}$. However, a more precise comparison with the experimental data is impossible because such quantities as the Gibbs adsorption $\Gamma$, the length $L$ and the energy $\zeta$ of the solute–particle interaction are not well-measurable properties and are known only approximately. The problem is that the Soret coefficients of colloids are determined with much better accuracy than the values $\Gamma$, $L$ and $\zeta$. Following the terminology of the well-known topic of quantum mechanics these quantities can be considered as the "hidden parameters". Let us emphasize that all listed parameters are responsible for the formation of a protective layer around the particle. Thus, we can state that at present our knowledge about the protective layer is very restricted.

The uncertainty with "hidden parameters" does not allow an experimenter to describe exhaustingly the phenomenon of thermodiffusion in colloids. Thus, the theory must provide an experimenter with the prescription for how to treat his/her data without information about the solute–particle interactions and the state of the protective layer. As we showed above the reduction of this problem to the determination of the effective value of the surface tension at the grain–fluid boundary is incorrect. Fortunately, all the "hidden parameters" can be combined in a single block in (35), (37) and their separate definitions are unnecessary.

Let us introduce the dimensionless parameter $P$, given by the equality

$$P = 6\pi \Gamma R_{\mathrm{H}} L (1 - \kappa\gamma)(\zeta - 1) \ . \tag{38}$$

It is seen from the definition (38) that the quantity $P$ characterizes the protective layer of the surfactant molecules surrounding the single colloidal particle. Therefore, we shall call it the *protective layer state (PLS) parameter*. As far as a PLS parameter is related to the properties of the single grain it is the *microscopic* characteristic of a colloid. In terms of $P$ the equation (35) takes the form of a Stokes force balance equation

$$-6\pi\eta R_{\mathrm{H}}\boldsymbol{U} + \boldsymbol{F}_{\mathrm{T}} = 0\,, \tag{39}$$

where the second term $\boldsymbol{F}_{\mathrm{T}} = -Pk_{\mathrm{B}}\boldsymbol{A}$ has the sense of "thermodiffusion force" acting on the colloidal particle subjected to the inhomogeneous temperature field. It does not mean that there is an external regular force which induces the particle movement. It is evident that there are no regular forces in the system and the particle drift is caused only by internal forces inside the protective layer. Equation (39) signifies that the particle under temperature gradient $\boldsymbol{A}$ moves with the same velocity as the rigid spherical particle in a viscous fluid does, driven by the regular force $\boldsymbol{F}_{\mathrm{T}}$.

The expression of the Soret coefficient can be rewritten in term of $P$.

$$S_{\mathrm{T}} = \frac{P}{T_0}\,. \tag{40}$$

This relation connects the microscopic quantity, protective layer state parameter $P$, with the macroscopic one, the Soret coefficient. It is usual to determine the quantity $\alpha_{\mathrm{T}} = S_{\mathrm{T}}T_0$ as the *thermal diffusion factor* [13]. We obtain the result that the protective layer state parameter and the thermal diffusion factor coincide: $P = \alpha_{\mathrm{T}}$. Hence, we make clear the nature of the thermal diffusion factor. When we determine the Soret coefficient as well as the thermal diffusion factor experimentally, we obtain the integral information about the state of the protective layer of the colloidal particles by the same token. In conclusion in this section devoted to the problem of the thermodiffusion motion in surfacted colloids we point out that the PLS parameter in such systems is positive, as we saw already, and has typical values $P \sim 20$–$50$ [3,4,6].

# 3 Ionic colloids

## 3.1 The protective layer of ionic colloids

Let us go to the investigation of the thermodiffusion motion in the ionic colloids. The consideration of these systems has a number of common features with the case of surfacted colloids. Thus, below we use the same notation for the one-type quantities as in Sect. 2.

The ionic colloids differ from the surfacted ones by the different method of stabilization against the particle coagulation. The stability of the ionic colloids is achieved by the mechanism of the selective adsorption of ions of one sign on the colloidal particle surface. As a result, each particle carries the superficial electric charge. The sign of the superficial electric charge may be positive as well as negative depending on the nature of the carrier fluid. The ions which do not take part in the process of adsorption are situated in a solution (usually a water-based one) and provide the electrostatic neutrality of the colloid. The ions with the same sign of charge as the colloidal particle are said to be coions, while the ions of opposite sign of charge are named counterions. The co- and counterions form the Debye protective double layer around the grain that prevents the coagulation of colloidal particles.

Let us examine the case of a single charged spherical grain which resides in the solution of electrolyte. As earlier we denote the radius of the grain as $R$ and the grain charge is assumed to be positive for definiteness. We consider a symmetric electrolyte, i.e. the charges of co- and counterions are equal to $e$ and $-e$, respectively, and the concentration of ions far from the particle is $n_\infty$. The distribution $T(\boldsymbol{r})$ of the temperature around the grain is stationary and defined by (3). We assume that the electrolyte ions do not interact with each other. Hence, for their concentrations at isothermal conditions ($\boldsymbol{A} = 0$) we have (cf. with (2))

$$n_+^0(r) = n_\infty \exp(-\Phi) \ , \quad n_-^0(r) = n_\infty \exp(\Phi) \ , \tag{41}$$

where $\Phi = e\varphi_0/k_\mathrm{B}T_0$ is the dimensionless energy of the ion–grain interaction, and $\varphi_0 = \varphi_0(r)$ is the electrostatic potential of the grain. The potential at the particle surface, $\zeta = \Phi(R)$, is assumed to be a parameter of the problem.[1] The quantity $\Phi$ satisfies the Poisson equation

$$\Delta\Phi = \sinh\Phi/L^2 \ , \tag{42}$$

where $L = (\varepsilon k_\mathrm{B}T_0/8\pi n_\infty e^2)^{\frac{1}{2}}$ is the Debye length, and $\varepsilon$ is the dielectric constant of the fluid. In contrast to the surfacted colloids the protective layer thickness $L$ in the case of ionic colloids is a well-controlling parameter of the problem (the increase in the number of ions in the solution by adding salt leads to its decrease).

Note that as before the label "0" in (41) designates the state of thermodynamic equilibrium, where the temperature of the system is constant. In this case the electric field $\boldsymbol{E}_0 = -\nabla\varphi_0$ around the grain is spherically symmetric, so that the net force acting on the particle is zero. Similarly, the electric body force on the electrolyte is balanced by the pressure gradient in the fluid.

In a nonuniform temperature field, the ion concentrations $n_+(\boldsymbol{r})$ and $n_-(\boldsymbol{r})$, and the potential $\varphi(\boldsymbol{r})$ differ from the equilibrium values $n_+^0$, $n_-^0$ and $\varphi_0$. The electric stresses at the opposite sides of the particle do not balance each other and the particle begins to move. To determine the grain velocity $\boldsymbol{U}$ we must first find the ion distributions $n_+(\boldsymbol{r})$ and $n_-(\boldsymbol{r})$ in the temperature field and then investigate the hydrodynamic part of the problem as previously.

### 3.2 Reduction of number of variables

There are three temperature-dependent variables $n_+(\boldsymbol{r})$, $n_-(\boldsymbol{r})$ and $\varphi(\boldsymbol{r})$ in our present problem instead of the single one (the surfactant molecules concentration $n(\boldsymbol{r})$) in the case of surfacted colloids. We expand them near equilibrium values (cf. (4))

$$n_+ = n_+^0 + n_\infty\tilde{c}_+ \ , \quad n_- = n_-^0 + n_\infty\tilde{c}_- \ , \quad \varphi = \varphi_0 + \varphi_1 \ . \tag{43}$$

[1] In contrast to the case of surfacted colloids (see comment to (12)) we defined here $\zeta$ with a different sign.

Here the quantities $\tilde{c}_+$, $\tilde{c}_-$, $\varphi_1$ are linear with respect to the dimensionless temperature gradient $\varepsilon_{\mathrm{T}}$.

Let us examine the diffusion problem of coions as an example. The diffusion flux of coions in an electrostatic field of the charged grain is (cf. (5))

$$\boldsymbol{j}_+ = -D_+ \left( \nabla n_+ + \frac{n_+ e \nabla \varphi}{k_{\mathrm{B}} T} \right) , \tag{44}$$

where $D_+$ is the coion diffusion coefficient. Substituting (3), (41), (43) into (44) and retaining only terms linear on $\varepsilon_{\mathrm{T}}$ one can write the diffusion flux of coions in the form

$$\boldsymbol{j}_+ = -D_+ n_+^0 \left( \nabla Q_+ + \frac{\Phi}{T_0} \nabla T \right) , \tag{45}$$

where we used the notation

$$Q_+ = \tilde{c}_+ e^{\Phi} + \Phi_1 - \Phi T_1 / T_0 \, , \quad \Phi_1 = e\varphi_1 / k_{\mathrm{B}} T_0 \, . \tag{46}$$

The flux of counterions is obtained from (45) and (46) by using a simple replacement index "+"→ index "–", $\Phi \to -\Phi$, $\Phi_1 \to -\Phi_1$.

The forms of the coion flux (45) and that for surfactant molecules (6) coincide outwardly. The only difference between them is the appearence of the additional term $\Phi_1$ in the expression for the quantity $Q_+$. In principle, the equation for $\Phi_1$ can be derived by the linearization on $\varepsilon_{\mathrm{T}}$ of the Poisson equation. However, it is unnecessary to solve the separate problem for the electric field $\varphi(\boldsymbol{r})$ around the grain in an inhomogeneous temperature field. As we show below, the diffusion and hydrodynamic parts of our problem contain only the variables $Q_+$ and $Q_-$ and the information about the $\Phi_1$-terms becomes redundant. Therefore, it is sufficient for our purpose to use the isothermal value $\varphi_0$ which is the solution of (42). Finally, the number of unknowns is reduced: instead of initial quantities $\tilde{c}_+$, $\tilde{c}_-$ and $\varphi_1$ we use $Q_+$ and $Q_-$. The analogous reduction of variables takes place in the problem of diffusiophoresis as well [9].

Because the fluxes of coions (45) and surfactant molecules (6) have the same form, the final equations for the variable $Q_+(r,\theta)$ and its radial part $q_+(r) = Q_+/\cos\theta$ in the approximation of the small Péclet number have the forms (8) and (11), respectively. The boundary conditions for $q_+$ are (see (12) and the footnote on the previous page)

$$q_+(r \to \infty) \to 0 \, , \quad \frac{\mathrm{d}q_+}{\mathrm{d}r}(r = R) = -\frac{A\zeta(1 + 2\kappa)}{T_0} \, . \tag{47}$$

The equation and the supplementary conditions for the counterion variable $q_-(r)$ can be found from those for $q_+$ by using the above-mentioned replacement.

### 3.3 Equations of the ionic colloid dynamics

Let us examine the hydrodynamic part of the problem of thermodiffusion motion in ionic colloids. The equations for fluid movement have the form shown in (14) with a different body force $\boldsymbol{f}$. In this case the force is of electrical origin and given by the formula [18]

$$\boldsymbol{f} = -e(n_+ - n_-)\nabla\varphi - \frac{(\nabla\varphi)^2}{8\pi}\,\frac{\mathrm{d}\varepsilon}{\mathrm{d}T}\,\nabla T\ . \tag{48}$$

Clearly, the value of the body force acting on the fluid is determined by two terms: the Coulomb term, related to the presence of free charges (ions) dependence of the dielectric constant $\varepsilon$ of the fluid. Since in the fluid, and the dielectrophoretic term, reflecting the temperature $\mathrm{d}\varepsilon/\mathrm{d}T$ is negative, the dielectrophoretic term forces the fluid to move along temperature gradient $\boldsymbol{A}$. This is equivalent to the particle moving in the opposite direction, i.e. into colder layers of the fluid. In contrast to the second term, the projection of the Coulomb term on the direction of $\boldsymbol{A}$ has no definite sign, i.e. it can be either positive or negative. As a result the particle can migrate either up or down the temperature gradient (see below).

Substituting (41) and (43) into (48) and retaining only terms linear on $\varepsilon_\mathrm{T}$ one can write the body force in the form

$$\begin{aligned} \boldsymbol{f} = \nabla(...) - n_\infty k_\mathrm{B}T_0 \Bigg\{ \Bigg( \mathrm{e}^{-\Phi}\left[Q_+ + \frac{T_1}{T_0}\Phi\right] - \mathrm{e}^{\Phi}\left[Q_- - \frac{T_1}{T_0}\Phi\right]\Bigg) \nabla\Phi + \\ + \frac{\nabla T}{T_0}\,\nu L^2 \left(\frac{\mathrm{d}\Phi}{\mathrm{d}r}\right)^2 \Bigg\}\ , \end{aligned} \tag{49}$$

where $\nu = \mathrm{d}\ln\varepsilon/\mathrm{d}\ln T$ is the logarithmic derivative of the dielectric constant with respect to temperature. The gradient term renormalizing only the pressure of a fluid is not given in (49). As we mentioned above, the electrical body force is determined totally by the variables $Q_+$ and $Q_-$ and there is no need to define the perturbed value $\Phi_1$ of the electric potential separately. Thus, the electric properties of the particles are not important for the description of the phenomenon of thermodiffusion in ionic colloids. Our consideration of the problem is equally applicable for the cases of dielectric and conductive colloidal particles. A comparison between (48) and (49) shows that the quantities $n_\infty \exp(-\Phi)(Q_+ + T_1\Phi/T_0)$ and $n_\infty \exp(\Phi)(Q_- - T_1\Phi/T_0)$ correspond to effective concentrations of co- and counterions, respectively.

After taking the curl of the Stokes equations (14) and (49) we have the resulting expression for the radial part of the vorticity (18)

$$\begin{aligned} \Delta G - \frac{2G}{r^2} - \frac{n_\infty k_\mathrm{B}T_0}{\eta}\Bigg[\frac{\mathrm{e}^{-\Phi}q_+ - \mathrm{e}^{\Phi}q_-}{r} \\ + 2\frac{A}{T_0}\left(1 - \kappa\left(\frac{R}{r}\right)^3\right)\left(\Phi\cosh\Phi - \nu L^2\Phi''\right)\Bigg]\Phi' = 0\ . \end{aligned} \tag{50}$$

Expression (50) with the additional conditions of (22) and (24) together with the Poisson equation (42) for the isothermal electrostatic potential $\Phi$ and the diffusion equations for the radial parts $q_+$ and $q_-$ of the effective ions concentrations constitute the complete set of equations for determining the migration velocity $U$ of a colloidal grain.

### 3.4 Calculation of the particle velocity

The obtained system of equation is very complicated and can not be solved analytically. In general, its solution depends on four dimensionless parameters: the potential $\zeta$ at the particle surface, the ratio $\Lambda = L/R$ of the Debye length to the particle radius, $\kappa$ (see the commentary to (3)) and $\nu$. We write the velocity of the grain motion in the temperature field $\boldsymbol{A}$ as $\boldsymbol{U} = \boldsymbol{A} n_\infty k_\mathrm{B} L^2 \tilde{u}/\eta$, specifying the dimensionless migration velocity $\tilde{u}$ explicitly. Before starting the numerical investigation of the problem, let us consider the simplest approximation of an *infinitely small double layer* ($\Lambda \to 0$). Using the method of matched asymptotic expansions [19] in a small parameter $\Lambda$ we find analytically the value of the migration velocity of the grain [11]

$$\tilde{u} = -\frac{2}{3}\,(1-\kappa)\left[\zeta^2 - 8(3+\nu)\ln\cosh\frac{\zeta}{4}\right] . \tag{51}$$

For arbitrary values of $\Lambda$ the system can be solved only numerically. Hence we will fix the values of the parameters $\kappa = 0.75$ and $\nu = -1.5$. Both values correspond to the experiment [3,6] and are characteristic of aqueous colloidal systems, for which the thermal conductivity coefficient of the dispersed phase exceeds that of the solvent by a factor of 10.

Figure 1 presents the result of the numerical calculation for the dimensionless particle velocity $\tilde{u}$ as a function of $\zeta$, the surface potential, for several values of $\Lambda$ (the respective values are placed at the curves). For an infinitely thin double layer (see the lowest curve in Fig. 1 and (51)), the particle migration velocity is always directed opposite the temperature gradient, as in the case of surfacted colloids. The situation changes dramatically when the values of $\Lambda$ become finite. All curves with $\Lambda \neq 0$ have a minimum. As the surface potential increases further, the particle velocity grows, going through zero at $\zeta = \zeta_0$. The latter means that the direction of particle motion reverses. Now the grains move toward the higher temperature layer of the fluid.

Figure 2 shows the "neutral" curve $\zeta_0 = \zeta_0(\Lambda)$ corresponding to the disappearance of the thermodiffusion movement of the particle [12]. This curve divides the plane of parameters $\zeta$ and $\Lambda$ into two parts. Region 1 corresponds to the particle migration opposite to the temperature gradient. In region 2 the grain moves along $\boldsymbol{A}$. Why the grain ceases to move on the "neutral" curve becomes clear if we analyze Fig. 3, which depicts the streamlines and the lines of constant counterion concentration near the particle. A stream function $\Upsilon(r,\theta)$ is defined as [13]

$$\boldsymbol{u} = \frac{\nabla \Upsilon \times \boldsymbol{i}_\phi}{r\sin\theta} , \tag{52}$$

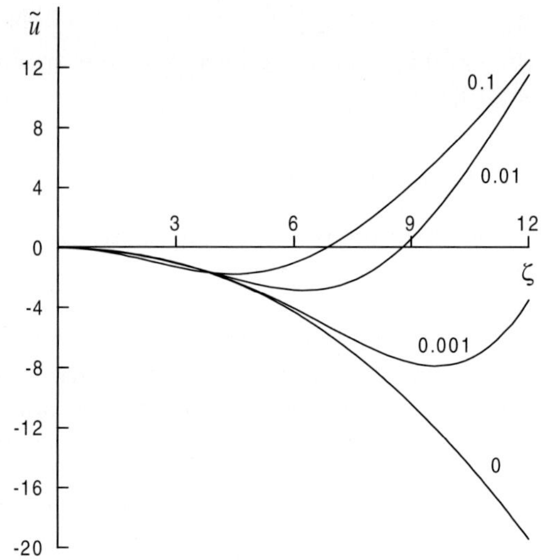


**Fig. 1.** Particle migration velocity in units $An_\infty k_B L^2/\eta$ as a function of the particle surface potential for several values of $\Lambda$

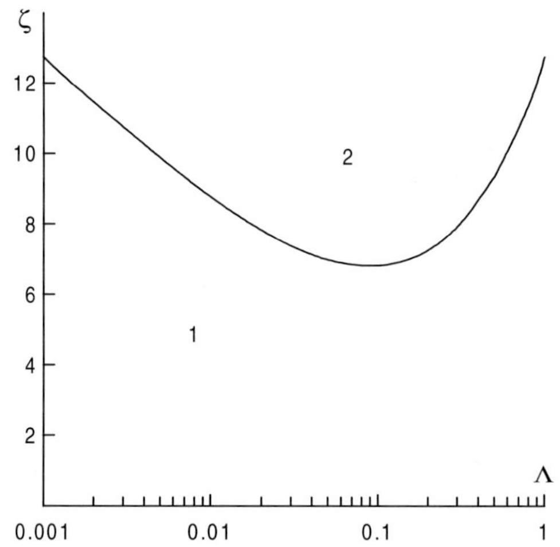


**Fig. 2.** "Phase" diagram of thermodiffusion in terms of the dimensionless surface potential and the Debye length. Particle moves opposite to the temperature gradient in parameter region 1 and along the temperature gradient in region 2

and can be expressed in term of function $h(r)$ (see (17))

$$\Upsilon(r,\theta) = \frac{1}{2} h(r) r^2 \sin^2\theta \; . \qquad (53)$$

The streamlines in Fig. 3 are presented for several values of $\Upsilon$ labeling the corresponding curves. The effective values of counterion concentration $n_\infty \exp(\Phi)(Q_- - T_1\Phi/T_0)$ are shown in Fig. 3 without the large multiplier $\exp(\Phi)$.

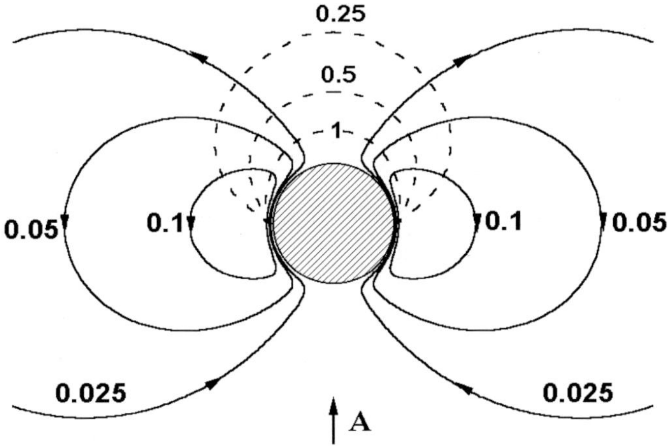


**Fig. 3.** Stream functions (solid curves) in units $An_\infty k_B L^2 R^2/\eta$ and contours of constant excess in the concentration of counterions (dashed curves) in units $AnR/T_0$ in the absence of particle migration ($\Lambda = 0.1$ and $\zeta_0 = 6.82$)

The dielectrophoretic body force (the last term in (48) and (49)) forms the fluid flow mainly along the temperature gradient $\boldsymbol{A}$ and forces the particle to move in the opposite direction, i.e. against $\boldsymbol{A}$. However, the Coulomb force opposes this motion due to an excess of negatively charged ions near the "hot" pole of the grain. For the values of $\Lambda$ and $\zeta$ depicted in Fig. 3, these tendencies are balanced. If $\zeta < \zeta_0$ holds, the first tendency is predominant, i.e. the particle migrates into the low-temperature region. In this case the stream functions become unclosed and turn into a "spider" (see Fig. 4).

If $\zeta > \zeta_0$ holds, the Coulomb term is predominant and the grain moves into the hot layers of the fluid. This situation is shown in Fig. 5. The closed stream functions in Fig. 3 now become more compact and the lateral stream functions of opposite flow arise.

### 3.5 Calculation of the Soret coefficient

Since the thermodiffusion velocity of the particle is known we can calculate the Soret coefficient $S_T$ of the ionic colloid in terms of the dimensionless velocity $\tilde{u}$ shown in Fig. 1. Carrying out the calculation following the approach described in Sect. 2 we obtained finally the relation for the Soret coefficient [12]

$$S_T = -\frac{3}{4} \frac{R_H}{L_B} \frac{\tilde{u}}{T_0} , \qquad (54)$$

where $R_H$ is the hydrodynamic radius of the particle, as before, and $L_B = e^2/\varepsilon k_B T_0$ is the Bjerrum length, interpreted as the distance at which the energy of the ion interaction is equal to the thermal energy. For aqueous electrolytes, $L_B \approx 7$ Å.

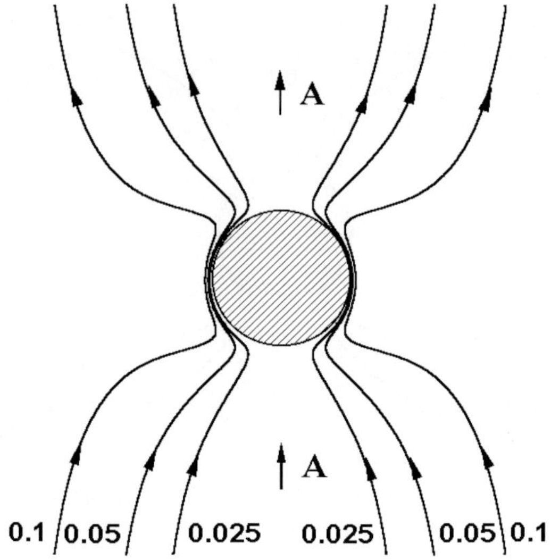


**Fig. 4.** Stream functions in units $An_\infty k_B L^2 R^2/\eta$ in the case of particle migration against the temperature gradient $\boldsymbol{A}$; $\Lambda = 0.1$, $\zeta$ potential is slightly lower than the value $\zeta_0$ ($\zeta = 6.81$)

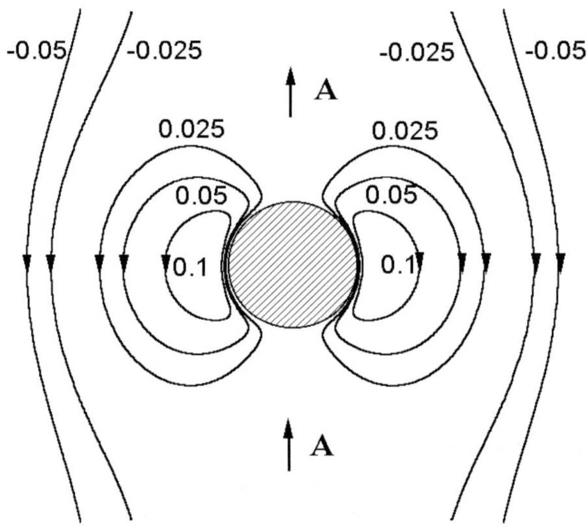


**Fig. 5.** Stream functions in units $An_\infty k_B L^2 R^2/\eta$ in the case of particle movement along the temperature gradient $\boldsymbol{A}$; $\Lambda = 0.1$, $\zeta$ potential is slightly higher than the value $\zeta_0$ ($\zeta = 6.83$)

Let us estimate the magnitude of the Soret coefficient. Substituting into (54) the values $R_H \sim 100$ Å, $T_0 \sim 300$ K, $\tilde{u} \sim 1$ we find $S_T \sim 0.1\ \mathrm{K}^{-1}$. Thus the Soret coefficient of ionic colloids has the same order as that of surfacted colloids. Equation (54) and Fig. 1 suggest that the Soret coefficient of the ionic colloid, in contrast to surfacted ones, may be either positive and negative, depending on the magnitudes of the electrostatic surface potential $\zeta$ and the dimensionless Debye length $\Lambda$. In our opinion the negative sign of the Soret coefficient in experiment [3,6] is due to the high value of the $\zeta$ potential of the samples used in the experiment. The substitution of the experimental values $R_{\mathrm{H}} = 100$ Å, $T_0 = 300$ K, $S_{\mathrm{T}} = -0.11\ \mathrm{K}^{-1}$, $\Lambda = 0.4$ [6] in (54) gives the estimate for the particle surface potential $\zeta = 9.5$. This value is in comparatively good agreement with the value $\zeta = 7.5$ which can be calculated from (42) and the independent data for the particle radius $R = 64$ Å and the surface charge density $\sigma = 0.2\ \mathrm{C/m^2}$. A better agreement could be achieved if a number of factors that were ignored here, such as an existence of the Stern layer near the particle surface [14], the ion–ion correlations [20], and the finite size of the ions, were taken into account. However, at present only a qualitative test of the proposed theory is feasible. Indeed, the increase of the ionic force of the solution leads to a decrease in the double layer thickness and the surface potential. According to the behavior of the dimensionless velocity $\tilde{u} = \tilde{u}(\zeta)$ presented in Fig. 1 and the expression for the Soret coefficient (54) we predict that these alterations must be accompanied by a slowing down of the thermodiffusion motion and a change in its direction and the sign of $S_{\mathrm{T}}$.

## 4 Conclusion

The phenomena of thermodiffusion in surfacted and ionic colloids have similar natures. The interactions between the particle and surfactant molecules or electrolyte ions is the source of the particle migration under the influence of the temperature field. The magnitude and the sign of the Soret coefficient depends on the kind of solute–grain potential – its characteristic depth and spatial scale. If the thickness of the protective layer is negligible, the particle always moves into the colder region of the fluid irrespective of the type of colloid, and the sign of the Soret coefficient is positive. The Soret coefficient of ionic dispersions may have either sign depending on the magnitudes of two parameters – the ratio of the double layer thickness to the particle radius and the electric potential of the particle surface. The control of these parameters gives the experimenter the unique possibility to change the direction of the thermodiffusion movement.

The values of the Soret coefficients of colloidal particles are very high. The effect is due to the proportionality of the Soret coefficient to the particle size $R_{\mathrm{H}}$ or $R$ (see (37) and (54)). If the particle radius is comparable to the molecular size the correct estimation $S_{\mathrm{T}} \approx 0.001\ \mathrm{K}^{-1}$ for the molecular system follows from (37) and (54). The colloidal particle radius is 100 times greater than the molecular one and the Soret coefficient for colloids is $S_{\mathrm{T}} \approx 0.1\ \mathrm{K}^{-1}$ [3–6]. The size of the suspension particles is two orders higher compared to colloids. Therefore, extremely large values of the Soret coefficients of suspension, $S_{\mathrm{T}} \approx 10\ \mathrm{K}^{-1}$, are expected in experiments. Another way of increasing the Soret coefficient of disperse systems is synthesis of colloids and suspensions with similar thermal properties of the carrier fluids and dispersed particles decreasing the distortions of the temperature field around the particles. The high values of the Soret coefficient make disperse systems attractive for possible practical employment.

## Acknowledgments

This work was carried out with the financial support of the Russian Fund for Fundamental Research (Project 01-02-17839). The author is grateful to A.F. Pshenichnikov for many stimulating discussions on the problem.

## References

1. S.R. de Groot, P. Mazur: *Non-equilibrium Thermodynamics* (North-Holland, Amsterdam 1972)
2. J.-C. Bacri, A. Cebers, A. Bourdon, G. Demouchy, B.M. Heegard, R. Perzynski: Phys. Rev. Lett. **74**, 5032 (1995)
3. J. Lenglet: Generation de Second Harmonique et Diffusion Rayleigh Forcee dans les Colloides Magnetiques. PhD Thesis, L'Universite Paris 7, Paris (1996)
4. E. Blums, A. Mezulis, M. Maiorov, G. Kronkalns: JMMM **169**, 220 (1997)
5. E. Blums, S. Odenbach, A. Mezulis, M. Maiorov: Phys. Fluids **10**, 2155 (1998)
6. A. Mezulis: Mass Transport Phenomenon in Non-isothermic Magnetic Colloids. PhD Thesis, L'Universite Paris 7, Paris (1999)
7. B.V. Derjaguin, S.S. Dukhin, A.A. Korotkova: Kolloidn. Zh. **23**, 53 (1961)
8. J.L. Anderson, M.E. Lowell, D.C. Prieve: J. Fluid. Mech. **117**, 107 (1982)
9. D.C. Prieve, J.L. Anderson, J.P. Ebel, M.E. Lowell: J. Fluid Mech. **148**, 247 (1984)
10. D.H. Napper: *Polymeric Stabilization of Colloidal Dispersions* (Academic Press, London 1983)
11. K.I. Morozov: JMMM **201**, 248 (1999)
12. K.I. Morozov: JETP **88**, 944 (1999)
13. L.D. Landau, E.M. Lifshitz: *Fluid Mechanics*, 2nd edn. (Pergamon, New York 1987)
14. V.G. Levich: *Physicochemical Hydrodynamics* (Prentice-Hall, Englewood Cliffs, 1962)
15. R.E. Rosensweig: *Ferrohydrodynamics* (Cambridge University Press, Cambridge 1985)
16. N.O. Young, J.S. Goldstein, M.J. Block: J. Fluid. Mech. **6**, 350 (1959)
17. K.I. Morozov: Phys. Rev. E **53**, 3841 (1996)
18. L.D. Landau, E.M. Lifshitz: *Electrodynamics of Continuous Media*, 2nd edn. (Pergamon, New York 1984)
19. M. Van Dyke: *Perturbation Methods in Fluid Mechanics* (Academic, New York 1964)
20. R. Kjellander, S. Marčelja: J. Chem. Phys. **82**, 2122 (1985)

## Supplementary glossary

| | |
|---|---|
| $\boldsymbol{A}$ | external temperature gradient |
| $\tilde{c}$ | dimensionless perturbation of surfactant concentration |
| $\tilde{c}_+$ | dimensionless perturbation of coion concentration |
| $\tilde{c}_-$ | dimensionless perturbation of counterion concentration |
| $e$ | elementary charge |
| $G$ | radial part of fluid vorticity |
| $\boldsymbol{j}$ | diffusion flux of surfactant molecules |
| $\boldsymbol{j}_{\mathrm{d}}$ | diffusion flux of particles |
| $\boldsymbol{j}_{\mathrm{T}}$ | thermodiffusion flux of particles |
| $L$ | length of surfactant (ion) concentration variation |
| $L_{\mathrm{B}}$ | Bjerrum length |
| $n$ | surfactant molecule concentration |
| $n_\infty$ | surfactant (ion) concentration far from particle |
| $n_0$ | equilibrium surfactant concentration |
| $n_{\mathrm{p}}$ | colloidal particle concentration |
| $n_+$ | coion concentration |
| $n_-$ | counterion concentration |
| $p$ | pressure in fluid |
| $Pe = UL/D$ | Péclet number |
| $q$ | radial part of perturbed surfactant distribution |
| $q_+$ | radial part of perturbed coion distribution |
| $q_-$ | radial part of perturbed counterion distribution |
| $Q$ | perturbation of surfactant distribution |
| $Q_+$ | perturbation of ion distribution |
| $Q_-$ | perturbation of counterion distribution |
| $r_0 = R + L$ | upper boundary of protective layer |
| $R$ | particle radius |
| $R_{\mathrm{H}}$ | hydrodynamic radius of particle |
| $T_0$ | temperature in center of particle |
| $\tilde{u}$ | dimensionless migration velocity |
| $U$ | particle velocity |
| $\gamma = (R/r_0)^3$ | dimensionless parameter |
| $\Gamma$ | Gibbs excess surface concentration |
| $\varepsilon_{\mathrm{T}} = AR/T_0$ | dimensionless temperature gradient |
| $\varepsilon$ | dielectric constant of fluid |
| $\zeta$ | dimensionless solute–particle potential |
| $\eta$ | fluid viscosity |
| $\theta$ | polar angle |

| | |
|---|---|
| $\kappa = (\lambda_{\mathrm{T}}^{(1)} - \lambda_{\mathrm{T}}^{(2)})/(\lambda_{\mathrm{T}}^{(1)} + 2\lambda_{\mathrm{T}}^{(2)})$ | thermal conductivities parameter |
| $\lambda_{\mathrm{T}}^{(1)}$ | thermal conductivity of particle |
| $\lambda_{\mathrm{T}}^{(2)}$ | thermal conductivity of fluid |
| $\Lambda = L/R$ | dimensionless protective layer thickness |
| $\nu = \mathrm{d}\ln\varepsilon/\mathrm{d}\ln T$ | logarithmic derivative of dielectric constant |
| $\tau_{\mathrm{T}}$ | thermal time |
| $\tau_{\mathrm{B}}$ | Brownian diffusion time |
| $\varphi_{\mathrm{s}}$ | potential of surfactant–particle interaction |
| $\varphi$ | electrostatic potential |
| $\varphi_0$ | equilibrium electrostatic potential |
| $\Phi = \varphi_{\mathrm{s}}/k_{\mathrm{B}}T_0$ | dimensionless solute–particle potential |
| $\varrho$ | fluid density |

# Principle of Entropy Maximization for Nonequilibrium Steady States

Alexander A. Shapiro and Erling H. Stenby

Engineering Research Center IVC-SEP, Department of Chemical Engineering, Technical University of Denmark, DK 2800 Lyngby, Denmark

**Abstract.** The goal of this contribution is to find out to what extent the principle of entropy maximization, which serves as a basis for the equilibrium thermodynamics, may be generalized onto non-equilibrium steady states. We prove a theorem that, in the system of thermodynamic coordinates, where entropy has a maximum in a steady state with regard to some thermodynamic variables, the matrix of the Onsager phenomenological coefficients becomes diagonal. The theorem requires consistent rules of the coordinate transformations in the non-equilibrium thermodynamics. Such rules are formulated. The results make it possible, in some cases, to reduce the number of unknown transport coefficients in thermodynamic description of the transport processes.

## 1 Introduction

Linear non-equilibrium thermodynamics has become a well grounded and comprehensive tool for modeling transport and relaxation processes reasonably close to equilibrium [1,2]. However, in spite of the advanced development of the theory [3], its solid statistical grounds [1,4,5] and almost perfect logical structure [6], some questions of theoretical and practical importance still remain unanswered.

An important general problem concerns the relations between the equilibrium and the non-equilibrium thermodynamics [7–9]. In particular, this problem is connected with the problem of modeling the non-convective heats of transport and thermodiffusion ratios, since the quasi-equilibrium approach has been widely used for this purpose [9–14] (see [10] for a review of the earlier works). Among other approaches to modeling the transport properties, the approach based on the gas kinetic theory is usually applied to gases, including their flows in porous media [15–19], and the approach based on the idea of activation energy is applied to liquids [20–24]. In the present work, we will concentrate on the quasi-equilibrium approach.

A particular question addressed here is to what extent the principle of entropy maximization may be generalized onto non-equilibrium steady states. This is the basic principle of the equilibrium thermodynamics, however, it cannot be spread directly onto non-equilibrium processes without generalization of the concept of entropy [9]. In order to characterize the steady states, a non-equivalent Prigogine principle of the minimum entropy production is used [25] . Meanwhile, the entropy maximization (at least, with regard to some variables) is equivalent to the quasi-equilibrium considerations, which have been used previously for prediction of the transport properties.

W. Köhler and S. Wiegand (Eds.): LNP 584, pp. 61–73, 2002.

We will show that the principle of entropy maximization is valid for a specific choice of the thermodynamic variables for which the matrix of phenomenological coefficients becomes diagonal. This result allows us to address another question, about reduction of the number of unknown dependencies characterizing transport processes. The Onsager theory expresses the relations between thermodynamic forces and fluxes in terms of the phenomenological coefficients, whose amount increases quadratically with the number of independent fluxes. Reciprocal relations make it possible to reduce the number of forces and fluxes approximately twice, but their number still remains quadratic. Apart from this, the Onsager theory says nothing (or almost nothing) about phenomenological coefficients. As a result, empirical or semi-empirical models of transport coefficients are used in practice for multicomponent mixtures under more or less realistic thermodynamic conditions [26] (some other approaches have been mentioned above). Assumption about the existence of the diagonal systems of coordinates, together with some restrictions on the allowed coordinate transformations, makes it possible to reduce the number of unknown coefficients to $2N$, where $N$ is the number of independent thermodynamic forces. This opens a new way of modeling the transport processes (other different attempts of such a reduction are presented in [27,28]).

It should be stressed that in a non-equilibrium state maximization of the entropy is not performed with regard to all the possible variables, since this would lead to the situation of complete thermodynamic equilibrium. The extensive thermodynamic variables should be subdivided onto the two sets, of the exchange and the free variables. The first type of variables participate in the flow through the system, and with regard to them the Prigogine principle of minimum of the entropy production is obeyed. The rest of the extensive variables are conserved within the system, and with regard to them maximization may be (or may not be) performed.

Unlike most of the theoretical works mentioned above, we do not try to develop the theory in a possible general form. Instead, we select a proper physical object corresponding to a typical imaginary experiment to measure the transport coefficients or to the schematic representation of a transport process used in chemical engineering [29]. In spite of its simplicity, the object selected for study exhibits all the necessary non-equilibrium thermodynamic properties. On the other hand, the simplicity of the object allows us to develop a mathematical representation of the non-equilibrium thermodynamic relationships, with particular emphasis on the transformations of thermodynamic coordinates. We try to combine physical and mathematical way of arguing. More rigorous mathematical considerations may be found elsewhere [30,31].

This paper is organized as follows. Section 2 provides the mathematical description of the steady states and introduces the transformations of the thermodynamic fluxes and forces, which are in agreement with the common practice of non-equilibrium thermodynamics. The main result is proven and discussed in Section 3.

# 2 Description of a steady state

## 2.1 The physical system

In order to introduce designations and to prepare for further comparison, let us first consider a classical object of the equilibrium thermodynamics (Fig. 1a). It consists of a thermodynamic system to be studied, and of a reservoir (for example thermostat) which serves as an infinite source of the extensive values (mass, energy...) and determines the conditions of equilibrium. Let us denote by $\mathbf{x}$ the vector of extensive variables and by $\mathbf{X}$ the vector of the conjugate intensive values:

$$X^{\mathrm{i}} = \partial S / \partial x_{\mathrm{i}} \tag{1}$$

We split these vectors into two parts, exchange ($E$) and free ($F$) variables: $\mathbf{X} = (\mathbf{X}_{\mathrm{E}}, \mathbf{X}_{\mathrm{F}})$ and $\mathbf{x} = (\mathbf{x}_{\mathrm{E}}, \mathbf{x}_{\mathrm{F}})$. The parameters $\mathbf{x}_{\mathrm{E}}$ are subject to exchange between the reservoir and the system, and this is the way the values $\mathbf{X}_{\mathrm{E}}$ of the system are determined by the values of the reservoir. More precisely, using the subscripts $R$, $S$ to denote the reservoir and the system, we may write

$$dS_{\mathrm{R}} = \mathbf{X}_{\mathrm{ER}} d\mathbf{x}_{\mathrm{ER}}, \quad dS_{\mathrm{S}} = \mathbf{X}_{\mathrm{ES}} d\mathbf{x}_{\mathrm{ES}} \quad (d\mathbf{x}_{\mathrm{FR}} = d\mathbf{x}_{\mathrm{FS}} = 0)$$

and maximization of the entropy under the conditions

$$d\mathbf{x}_{\mathrm{ER}} = -d\mathbf{x}_{\mathrm{ES}} = d\mathbf{x}_{\mathrm{E}}$$

gives

$$(\mathbf{X}_{\mathrm{ER}} - \mathbf{X}_{\mathrm{ES}}) d\mathbf{x}_{\mathrm{E}} = 0, \qquad \text{or} \qquad \mathbf{X}_{\mathrm{ER}} = \mathbf{X}_{\mathrm{ES}}. \tag{2}$$

Such a formulation of the equilibrium thermodynamics allows for straightforward generalization onto non-equilibrium steady states. Instead of one reservoir and one system, we consider the two reservoirs $R1$ and $R2$ in contact with the two systems $S1$ and $S2$ as shown in Fig. 1b. The systems are connected by a conducting element (or, for brevity, *a conductor*), which may be thought of as a capillary or a porous membrane [1]. The reservoirs $Ri$ are in equilibrium with the systems $Si$, but not with each other. This creates the fluxes $\dot{\mathbf{x}}_{\mathrm{E}}$ of the exchange

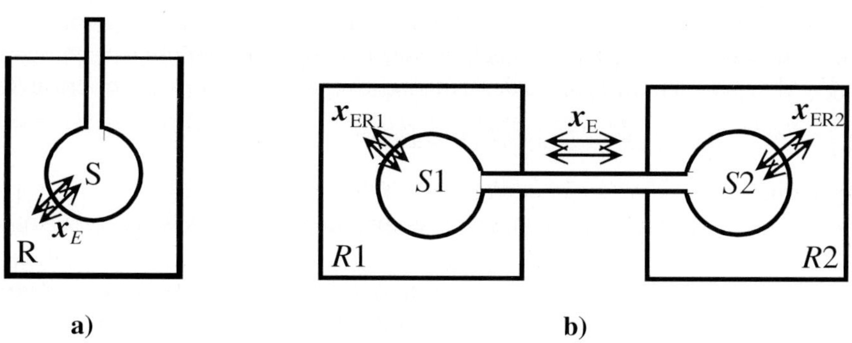


**Fig. 1.** Schematic picture of a simple discontinuous system.

parameters through the whole system. Since the reservoirs are infinite, the flow of $\mathbf{x}_\mathrm{E}$ continues over an infinite time and reaches a steady state. The free variables $\mathbf{x}_\mathrm{F}$ are conserved within $S1+S2$, but may be redistributed between the two systems. For small deviations from equilibrium, this steady state is determined by phenomenological relations of the linear non-equilibrium thermodynamics [25]:

$$\dot{\mathbf{x}} = \mathbf{L}(\mathbf{X})\delta\mathbf{X} \tag{3}$$

or, in the coordinate form,

$$\dot{x}_\mathrm{i} = L_\mathrm{ij}(\mathbf{X})(X_1^\mathrm{j} - X_2^\mathrm{j}) \tag{4}$$

Here and further on we use a conventional agreement about summation by a repeating subscript and superscript. Phenomenological coefficients $L_\mathrm{ij}$ are supposed to be functions of the average intensive values $\mathbf{X}$, which are well defined as the differences $\delta X^\mathrm{j}$ become infinitesimal. In this case the fluxes $\dot{x}_\mathrm{i}$ are also infinitely small, and the systems (3), (4) may be considered as relations between first order differential values. The coefficients $L_\mathrm{ij}$ are supposed to obey the common conditions of non-degenerativeness and the Onsager reciprocal relations [1]. For simplicity, we assume that these relations are symmetric:

$$L_\mathrm{ij} = L_\mathrm{ji}$$

Among the values $\mathbf{x}$, only $\mathbf{x}_\mathrm{E}$ may flow while $\mathbf{x}_\mathrm{F}$ remain conserved within the two systems. Correspondingly, the systems (3) or (4) should be completed by the condition

$$\dot{\mathbf{x}}_\mathrm{F} = 0 \tag{5}$$

While the exchange between the two systems proceeds according to the phenomenological relations and never reaches the equilibrium (provided that the reservoirs are infinite), the exchange between a reservoir $Ri$ and an adjacent system $Si$ is supposed to be instantaneous and bring them in equilibrium. According to Eq. (2):

$$\mathbf{X}_\mathrm{ER1} = \mathbf{X}_\mathrm{ES1} = \mathbf{X}_\mathrm{E1}, \quad \mathbf{X}_\mathrm{ER2} = \mathbf{X}_\mathrm{ES2} = \mathbf{X}_\mathrm{E2} \tag{6}$$

Under known $\mathbf{X}_\mathrm{E,i}$ the system (3), (5) is sufficient to determine the fluxes $\dot{x}_\mathrm{E,i}$, as well as the differences $\delta X_\mathrm{F}^\mathrm{j}$. Then the values $X_\mathrm{F1}^\mathrm{j}$, $X_\mathrm{F2}^\mathrm{j}$ may be determined, for example, by use of the known amounts $x_\mathrm{F1,i} + x_\mathrm{F2,i}$ in the two systems.

A standard mathematical object, which is usually considered in connection with the thermodynamic system described above, is the entropy production form:

$$\dot{S} = L_\mathrm{ij}\delta X^\mathrm{i}\delta X^\mathrm{j} = \dot{x}_\mathrm{i}\delta X^\mathrm{i} = \dot{x}_\mathrm{E,i}\delta X_\mathrm{E}^\mathrm{i} \tag{7}$$

It is assumed here that no entropy is produced due to exchange between the systems and their reservoirs. The last equality in Eq. (7) takes place because of Eq. (5).

Another mathematical object, which may be introduced with the thermodynamic system being considered, is the differential of the total entropy:

$$dS = d(S_\mathrm{R1} + S_\mathrm{S1} + S_\mathrm{S2} + S_\mathrm{R2})$$

$$= \mathbf{X}_{\mathrm{ER1}} d\mathbf{x}_{\mathrm{ER1}} + \mathbf{X}_{\mathrm{ER2}} d\mathbf{x}_{\mathrm{ER2}} + \mathbf{X}_{\mathrm{ES1}} d\mathbf{x}_{\mathrm{ES1}} + \mathbf{X}_{\mathrm{ES2}} d\mathbf{x}_{\mathrm{ES2}} + \mathbf{X}_{\mathrm{FS1}} d\mathbf{x}_{\mathrm{FS1}} + \mathbf{X}_{\mathrm{FS2}} d\mathbf{x}_{\mathrm{FS2}} \quad (8)$$

This expression may be simplified by the following considerations. First, the value of $d\mathbf{x}_{\mathrm{E}}$ is introduced in order to satisfy the equalities:

$$d\mathbf{x}_{\mathrm{ES1}} = -d\mathbf{x}_{\mathrm{ER1}} + d\mathbf{x}_{\mathrm{E}}, \quad d\mathbf{x}_{\mathrm{ES2}} = -d\mathbf{x}_{\mathrm{ER2}} - d\mathbf{x}_{\mathrm{E}} \quad (9)$$

The differential $d\mathbf{x}_{\mathrm{E}}$ expresses the exchange of $\mathbf{x}_{\mathrm{E}}$ between the two systems, as separated from the exchanges $d\mathbf{x}_{\mathrm{ER1}}$, $d\mathbf{x}_{\mathrm{ER2}}$ of the systems with the reservoirs. Since the values $\mathbf{x}_{\mathrm{F}}$ are exchanged only between the two systems, we may also write

$$d\mathbf{x}_{\mathrm{FS1}} = -d\mathbf{x}_{\mathrm{FS2}} = d\mathbf{x}_{\mathrm{F}} \quad (10)$$

Substitution of Eqs. (6), (9), (10) into expression (8) for the entropy differential makes it possible to cancel the terms related to exchange with the reservoirs and to reduce Eq. (8) to the form of

$$dS = \delta\mathbf{X}_{\mathrm{E}} d\mathbf{x}_{\mathrm{E}} + \delta\mathbf{X}_{\mathrm{F}} d\mathbf{x}_{\mathrm{F}} \quad (11)$$

This equation is similar, to some extent, to expression (7) for the entropy production, where the time variations $\dot{\mathbf{x}}$ are substituted by the corresponding differentials $d\mathbf{x}$. However, unlike in the expression for entropy production, the differentials $d\mathbf{x}_{\mathrm{F}}$ cannot be neglected, since virtual exchange of these values between the two systems remains possible.

As an example, let us consider an imaginary experiment by means of which the thermodiffusion factors or thermal heats of transport are commonly determined. Let the system $S1 + S2$ contain an N-component mixture, whose components do not react with each other. The molar amounts of the components are denoted by $N_1,...,N_{\mathrm{N}}$. For such a system, it is natural to select the thermodynamic fluxes as molar fluxes of each component and the molar flux of energy. The list of fluxes and of the conjugate thermodynamic forces for such a system is shown in Table 1 (where $\mu^{\mathrm{i}}$ are chemical potentials of the components). This selection of forces and fluxes will further on be referred to as "thermodynamic" set of variables.

In order to obtain thermodiffusional effects, the systems $S1$, $S2$ are kept at different temperatures $T_1$, $T_2$. This is achieved by the energy exchange between the systems and the adjacent reservoirs, which, in this case, are thermostats. Thus, the internal energy $U$ is the only exchange value which may flow through

**Table 1.** Thermodynamic fluxes and forces for a multicomponent mixture in standard thermodynamic coordinates

| | Substance | Energy |
|---|---|---|
| Fluxes | $\dot{N}_{\mathrm{i}}$ | $\dot{U}$ |
| Forces | $\delta(\mu^{\mathrm{i}}/T)$ | $\delta(-1/T)$ |

the system. The differences $\delta N_i$ are established according to the Onsager reciprocal relations (3) and the conservation conditions (5). The entropy production form is expressed by the only term

$$\dot{S} = \dot{U}\delta\left(-\frac{1}{T}\right)$$

Here the sign "−" is due to the fact that $\delta A$ is determined as $A_1 - A_2$ for any intensive value $A$. The entropy differential for the considered system is

$$dS = dU\delta\left(-\frac{1}{T}\right) + dN_i\delta\left(\frac{\mu^i}{T}\right)$$

## 2.2 Transformations of thermodynamic variables

Instead of the "thermodynamic" set of forces and fluxes presented in Table 1, the transport processes are usually described in terms of the "hydrodynamic" set of variables containing diffusion and convective mass fluxes, as well as the non-convective heat flux. A main advantage of such a set of variables is that the convective flux is uncoupled from other fluxes, and the corresponding non-diagonal phenomenological coefficients are zero. This is a far consequence of the Galilean principle of invariance and of the Curie principle of fluxes on the micro-level inside the conductor. Thermodynamic fluxes and forces for the last system of thermodynamic coordinates are shown in Table 2 [1,29].

An important distinction of the "hydrodynamic" set of variables from the "thermodynamic" set is that the fluxes $\dot{x}_c$, $\dot{x}_{D,i}$, $\dot{x}_q$ are not the time variations of the extensive values, but the linear combinations of such variations with the coefficients depending on the intensive parameters of the system. Meanwhile, thermodynamic forces are still introduced as $\delta$-differences of intensive variables.

Availability of different systems of thermodynamic variables together with the fact that some of the variables may not be full derivatives (or differentials) requires consistent rules for the choice of the admissible systems and for the transformations between them. These rules serve as a generalization of the linear transformations with constant coefficients, which have been considered in classical books [1,13], and of the more general linear transformations presented in [32]. In the following, we denote by $z$-set the set of thermodynamic variables with the fluxes $\dot{z}_i$ and the forces $\delta Z^i$.

Let us consider the transition between the $x$- and the $y$-set. First, we postulate that for any set of variables the thermodynamic forces should be in the form

**Table 2.** Thermodynamic fluxes and forces for a multicomponent mixture in "hydrodynamic" set of variables

| | Mass diffusion | Convection | Non-convective heat transfer |
|---|---|---|---|
| Fluxes | $\dot{x}_{D,i} = M_i\dot{N}_i - \rho T c_i \dot{x}_c$ | $\dot{x}_c = \sum M_i\dot{N}_i/(\rho T)$ | $\dot{x}_q = \dot{U} - (Th/v)\dot{x}_c$ |
| Forces | $\delta(\mu^i/M_iT - \mu^N/M_NT)$ | $\delta P$ | $\delta(-1/T)$ |

$\delta X^{\mathrm{i}}$, where $X^{\mathrm{i}}$ are real thermodynamic variables (usually, fields [33]). Hence, transformations of the thermodynamic forces are given by a simple rule

$$\delta Y^{\mathrm{i}} = \frac{\partial Y^{\mathrm{i}}}{\partial X^{\mathrm{j}}} \delta X^{\mathrm{j}} \tag{12}$$

Further, any transformation of the thermodynamic coordinates should preserve the entropy production form. Correspondingly,

$$\dot{S} = \delta X^{\mathrm{i}} \dot{x}_{\mathrm{i}} = \delta Y^{\mathrm{j}} \dot{y}_{\mathrm{j}} = \delta X^{\mathrm{i}} \frac{\partial Y^{\mathrm{j}}}{\partial X^{\mathrm{i}}} \dot{y}_{\mathrm{j}}$$

Since the differentials $\delta X^{\mathrm{i}}$ are independent, the rule of transformation for thermodynamic fluxes is

$$\dot{x}_{\mathrm{i}} = \frac{\partial Y^{\mathrm{j}}}{\partial X^{\mathrm{i}}} \dot{y}_{\mathrm{j}} \tag{13}$$

or, in the vector form,

$$\left( \frac{\partial \mathbf{x}}{\partial \mathbf{y}} \right)^{\mathrm{T}} = \frac{\partial \mathbf{Y}}{\partial \mathbf{X}} \tag{14}$$

The values $\dot{x}_{\mathrm{i}}$ given by Eq. (13) do not necessarily have to be the time derivatives of the thermodynamic variables. However, it is easy to verify that the transformations between $x$- and $y$-systems given by Eqs. (13), (14) retain the form (13) of the thermodynamic fluxes. It can also be shown that these transformations do not only preserve the entropy production form, but also the entropy differential (11). This differential always has the form $\delta Y^{\mathrm{i}} dy_{\mathrm{i}}$, where $dy_{\mathrm{i}}$ must not be the differentials of the extensive variables, but may be first order differential forms of the type $(\partial Z^{\mathrm{j}} / \partial Y^{\mathrm{i}}) dz_{\mathrm{j}}$, similar to Eq. (13). This confirms the mathematical correctness of the introduced transformations.

Finally, let us consider the transformation of the matrix of the phenomenological coefficients. We denote by $L_{\mathrm{ij}}(\mathbf{X})$, $L_{\mathrm{kl}}(\mathbf{Y})$ etc. matrices of the kinetic coefficients in the variables $X^{\mathrm{i}}$, $Y^{\mathrm{i}}$ etc. According to Eq. (7),

$$\dot{S} = L_{\mathrm{ij}}(\mathbf{X}) \delta X^{\mathrm{i}} \delta X^{\mathrm{j}} = L_{\mathrm{ij}}(\mathbf{X}) \frac{\partial X^{\mathrm{i}}}{\partial Y^{\mathrm{k}}} \frac{\partial X^{\mathrm{j}}}{\partial Y^{\mathrm{l}}} \delta Y^{\mathrm{k}} \delta Y^{\mathrm{l}}$$

Thus,

$$L_{\mathrm{kl}}(\mathbf{Y}) = L_{\mathrm{ij}}(\mathbf{X}) \frac{\partial X^{\mathrm{i}}}{\partial Y^{\mathrm{k}}} \frac{\partial X^{\mathrm{j}}}{\partial Y^{\mathrm{l}}} \tag{15}$$

In other words, $L_{\mathrm{ij}}$ are transformed as coefficients of the Riemannian metric $\dot{S}$ in the space of intensive variables. It is possible (although not necessary for the purposes of the present work) to interpret the whole theory in terms of the appropriate Riemannian geometry. That is why we will refer to different sets of variables (forces and fluxes) as to the thermodynamic systems of coordinates.

Another choice of the "hydrodynamic" system of coordinates, that may be found in the literature (cf. [25], Section 4.1) is represented in Table 3. In this

**Table 3.** Thermodynamic fluxes and forces for a multicomponent mixture in the second "hydrodynamic" set of variables

| | Mass diffusion | Convection | Non-convective heat transfer |
|---|---|---|---|
| Fluxes | $\dot{x}_{D,i} = M_i \dot{N}_i - \rho T c_i \dot{x}_c$ | $\dot{x}_c = \sum M_i \dot{N}_i/(\rho T)$ | $\dot{x}_Q = \dot{U} - h^i \dot{N}_i$ |
| Forces | $\delta_T(\mu^i/M_i T - \mu^N/M_N T)$ | $\delta P$ | $\delta(-1/T)$ |

table, designation $\delta_T$ means that the difference is taken at constant temperature, but variable composition and pressure:

$$\delta_T(\mu^i/T) = \delta(\mu^i/T) + h^i \delta(-1/T)$$

In this case both $dy_i$ and $\delta Y^i$ are not differentials of variables, but first order differential forms. If they are allowed to be *arbitrary* differential forms, then one may introduce an arbitrary non-degenerate transformation matrix $\mathbf{A}(\mathbf{X})$ and perform the transition from $x$- to $y$-variables by the rules

$$\delta \mathbf{Y} = \mathbf{A} \delta \mathbf{X}, \quad \dot{\mathbf{y}} = (\mathbf{A}^T)^{-1} \dot{\mathbf{x}}, \quad \mathbf{L}(\mathbf{Y}) = \mathbf{A}^T \mathbf{L}(\mathbf{X}) \mathbf{A} \tag{16}$$

This is the most general possible transformation, retaining the physical structure of the theory. It should be mentioned, however, that mathematically there are other possibilities. For example, it is possible to simply perform a transformation $\mathbf{y} = \mathbf{y}(\mathbf{x})$ and, then, to calculate the conjugate parameters $Y^i$ according to Eq. (1). However, on this way the new variables $\mathbf{y}$ may loose an important physical property, to be conserved quantities. For example, while the equality $\delta \mathbf{x}_{ES} = -\delta \mathbf{x}_{ER}$ is valid *exactly* for the exchange between a system and a reservoir, a similar equality with regard to the $y$-variables may only be valid within the second-order terms. This leads to more complicated rules for transformation of the second order differentials, like the entropy production form. It can be shown that the only transformations $\mathbf{y}(\mathbf{x})$ that preserve the conservation property are linear transformations with constant coefficients. This is a too narrow class of transformations to be considered. The transformations (13), (16) considered here retain the conservation property, however, they have to be applied to the differential forms, rather than to the true variables.

# 3 Entropy maximization

## 3.1 Main theorem

An equilibrium state of a closed system corresponds to the maximum of the entropy, in particular, $dS = 0$. The goal of this section is to find out to what extent the maximization of the entropy may be applied to the non-equilibrium steady states.

Let us consider the non-equilibrium system introduced in Section 2.1 and the corresponding differential of the entropy (11). Condition $dS = 0$ means that all the intensive parameters $X^{\mathrm{i}}$ of the two systems, and therefore of the two reservoirs, are equal and that all the systems and reservoirs are in equilibrium. This equilibrium will be assumed to be stable.

In order to avoid this situation of full equilibrium and to allow for non-equilibrium steady states, maximization with regard to some of the differentials $dx_{\mathrm{i}}$ should be prohibited. It is necessary to prohibit the maximization with regard to the exchange variables $dx_{\mathrm{E,i}}$, since inequality of the parameters $X^{\mathrm{i}}_{\mathrm{E}}$ in the reservoirs creates a non-equilibrium steady state. The physical significance of the parameters $X^{\mathrm{i}}_{\mathrm{E}}$ is that they are controlled and may assume different values in the two reservoirs, independently of each other. This control is lost if the maximization with regard to $dx_{\mathrm{E,i}}$ is performed.

Maximization with regard to the rest of the variables results in the following condition:

$$dS|_{\mathbf{x}_{\mathrm{E}}=\mathrm{const}} = \delta\mathbf{X}_{\mathrm{F}} d\mathbf{x}_{\mathrm{F}} = 0 \Leftrightarrow \delta\mathbf{X}_{\mathrm{F}} = 0 \tag{17}$$

We call this condition *the Kelvin maximization*, since Lord Kelvin was the first to apply it (or similar considerations about the quasi-equilibrium character of transport) in his studies of thermoelectricity [34]. It should be noted that condition (17) is not invariant with regard to the choice of a thermodynamic system of coordinates, since this choice determines the values $X^{\mathrm{i}}_{\mathrm{F}}$, which should be equated for the two systems. This fact was not taken into account in works [9,10,12–14] applying Kelvin maximization, where common systems of thermodynamic coordinates were used (like the systems presented in the Tables of this paper).

Assume that in a given system of thermodynamic coordinates $\mathbf{x}$, $\mathbf{X}$ the condition of Kelvin maximization is true for any choice of free and exchange variables. According to Eq. (5), free parameters $\mathbf{x}_{\mathrm{F}}$ may be determined as parameters whose fluxes vanish independently of the values $\delta\mathbf{X}$. On the other hand, condition (17) of Kelvin maximization means that intensive free parameters $\mathbf{X}_{\mathrm{F}}$ of the two systems must be equal. Hence, the following rule is obeyed:

$$\dot{x}_{\mathrm{i}} = 0 \Leftrightarrow \delta X^{\mathrm{i}} = 0. \tag{18}$$

Comparing this rule with reciprocal relations (4), we see that the rule may be valid if and only if the matrix of the kinetic coefficients is diagonal:

$$\dot{x}_{\mathrm{i}} = L^{\mathrm{i}}(\mathbf{X})\delta X^{\mathrm{i}} \tag{19}$$

Vice versa, if the matrix of kinetic coefficients is diagonal, rule (18) is obeyed. This means that $\delta\mathbf{X}_{\mathrm{F}} = 0$, as the fluxes $\dot{\mathbf{x}}_{\mathrm{F}}$ of the free variables are zeros. According to Eq. (17), differential $dS|_{\mathbf{x}_{\mathrm{E}}=\mathrm{const}}$ vanishes and the steady state considered is a stationary point of the entropy with regard to free variables. For the states sufficiently close to the equilibrium, this stationary point should be a maximum, since the set of maxima is open in the set of the stationary points.

We have proven the following central theorem of the present work:

**Theorem 1.** *Let us consider sufficiently small deviations from equilibrium. Entropy obeys the principle of Kelvin maximization in some system of thermody-*

*namic coordinates, for any choice of the free and exchange variables, if and only if reciprocal relations in that system of coordinates are diagonal.*

### 3.2 Discussion

The obtained results show to what extent the principle of entropy maximization, playing the central role in the equilibrium thermodynamics, may be generalized onto non-equilibrium steady states. Such a generalization is incomplete since the maximization is not absolute. It is performed with regard to the free variables, under constant exchange variables. The exchange variables obey another fundamental principle, of the minimum of the entropy production. As a consequence, the maximization depends on a choice of a system of thermodynamic coordinates. It is valid only for such systems where reciprocal relations become diagonal.

As a simple illustration, let us consider the flows in a binary mixture (described as above) and assume that the "hydrodynamic" set of variables (Table 2) *already* leads to diagonal reciprocal relations. This results in simple relations between forces and fluxes:

$$\dot{x}_{\mathrm{c}} = L_{\mathrm{c}}\delta P, \quad \dot{x}_{\mathrm{D}} = L_{\mathrm{D}}\delta\left(\frac{\mu^1}{M^1 T} - \frac{\mu^2}{M^2 T}\right), \quad \dot{x}_{\mathrm{q}} = L_{\mathrm{q}}\delta\left(-\frac{1}{T}\right)$$

In particular, we obtain the (correct) result that the convective flux is independent of other fluxes [29], and the Denbigh–Haase formula for the non-convective heat of transport [10,13]:

$$Q = \frac{h^1}{M_1} - \frac{h^2}{M_2}$$

Unfortunately, this example cannot be easily generalized onto multicomponent mixtures, in view of the asymmetry of the thermodynamic forces with regard to the $N$th component.

If, instead of the system of coordinates represented in Table 2, the system from Table 3 is considered, application of the entropy maximization results in the heats of transport equal to zero. For the coordinates from Table 1 this application leads to complex and hardly interpretable expressions, which, to the best of our knowledge, have not been presented in the literature. Additionaly, the principle that the convective flux should be independent from other fluxes is violated for this case, unless all the phenomenological coefficients are set to be equal.

Comparison of these examples shows that the results of the entropy maximization is strongly dependent on the choice of a system of the thermodynamic coordinates, even if this choice is made among the commonly used systems. A choice of the "right" system should be done on the basis of additional physical considerations.

A question about "right" system may be reverted and put in the following way: whether there always exists such a (Kelvin) system of thermodynamic coordinates, in which the matrix of reciprocal relations becomes diagonal, and, correspondingly, the principle of entropy maximization is valid. With the transformations (16), existence of a Kelvin system of coordinates becomes, simply, a well known theorem of the linear algebra, about diagonalization of a positive definite quadratic form. However, if a narrower class of transformations, (12) to (15) is considered, the corresponding existence theorem, to the best of our knowledge, cannot be proven, except for the systems with 2 or 3 independent thermodynamic forces. In this case assumption about existence of the Kelvin system of coordinates is a new physical principle. This assumption makes it possible to simplify the problem of predicting the transport coefficients and modeling the transport processes. For a system with $N$ independent forces, there are $N(N+1)/2$ independent phenomenological coefficients, each being the function of all the thermodynamic characteristics of the system. For example, the flow of the $N$-component system considered above is determined by $N(N-1)/2$ diffusion coefficients, $N-1$ thermodiffusion ratios, and the viscosity coefficient. The transformation to the diagonal system of coordinates makes it possible to reduce the number of unknown parameters to $2N$: $N$ thermodynamic variables, corresponding to the diagonal matrix of the reciprocal coefficients, and $N$ diagonal coefficients in this matrix.

Future development of the diagonal non-equilibrium thermodynamics assumes physical interpretation and creation of particular models for the diagonal coordinates – "the flow potentials" – and for the corresponding kinetic coefficients – "conductivities".

**Acknowledgements**

The authors would like to devote the paper to the memory of Dr. Leo J.T.M. Kempers. Without his fundamental and innovative work in the area of thermodiffusion and without intensive discussions with him the present work would not become possible.

## References

1. S.R. de Groot, P. Mazur: Non-Equilibrium Thermodynamics, North-Holland, Amsterdam, 1962; Dover, New York, 1984.
2. S. Sieniutycz, J.S. Shiner: J. Non-Equilib. Thermodyn. **19**, 303 (1994)
3. B.J. Edwards, H.C. Öttinger, R.J.J. Jongschaap: J. Non-Equilib. Thermodyn. **22**, 356 (1997)
4. P. Mazur: Physica A **261**, 451 (1998)
5. R. Luzzi, A.R. Vasconellos, J. Casas-Vazques, D. Jou: Physica A **248**, 111 (1998)
6. B.H. Lavenda: Physical Review A **9**, 929 (1974)
7. S. Sieniutycz: Open Sys. & Information Dyn. **4**, 185 (1997)
8. A.A. Bhalekar: J. Non-Equilib. Thermodyn. **21**, 330 (1996)
9. J. Keizer: J. Chem. Phys. **69**, 2609 (1978)
10. K.G. Denbigh: The Thermodynamics of the Steady State, Methuen & Co. Ltd., London, 1951.
11. W.M. Rutherford: A.I.Ch.E. Journal **9**, 841 (1963)

12. R. Haase: Z. Phys. **127**, 1 (1950)
13. R. Haase: Thermodynamics of Irreversible Processes, Addison–Wesley, Reading, 1969.
14. L.J.T.M. Kempers: J. Chem. Phys. **90**, 6541 (1989)
15. J.M. Kincaid, E.G.D. Cohen, M. Lopez de Haro: J. Chem. Phys. **86**, 963 (1987)
16. A. Ern, V. Giovangigli: Physica A **214**, 526 (1995)
17. J.D. Ramshaw: J. Non-Equilib. Thermodyn. **21**, 99 (1996)
18. E.A. Mason, A.P. Malinauskas: Gas Transport in Porous Media: The Dusty-Gas Model. Elsevier, Amsterdam, Oxford, 1983.
19. V.M. Zhdanov: Advances in Colloid and Interface Science **66**, 1 (1996)
20. W.M. Rutherford, H.G. Drickamer: J. of Chem. Phys. **22**, 1157 (1954)
21. E.L. Dougherty, H.G. Drickamer: J. of Chem. Phys. **23**, 295 (1955)
22. L.J. Tichacek, W.S. Kmak, H.G. Drickamer: J. of Phys. Chem. **60**, 660 (1956)
23. K. Shukla, A. Firoozabadi: Ind. Eng. Chem. Res. **37**, 3331 (1998)
24. A. Firoozabadi, K. Ghorayeb, K. Shukla: A.I.Ch.E. Journal **46**, 892 (2000)
25. I. Prigogine: Thermodynamics of Irreversible Processes, John Wiley & Sons, Interscience Publishers, 1967.
26. R.C. Reid, J.M. Prausnitz, B.E. Poling: The Properties of Gases and Liquids, McGraw-Hill Book Company, New York, 1987.
27. L.B. Benano-Melly, J-P. Caltagirone, B. Faissat, F. Montel: Entropie **217**, 41 (1999)
28. S. Subramaniam: J. Non-Equilib. Thermodyn. **24**, 1 (1999)
29. A.A. Shapiro, E.H. Stenby: Transport in Porous Media **41**, 305 (2000)
30. Z. Banach, S. Piekarski: Arch. Mech. **44**, 191 (1992)
31. H.W. Haslach: Reports on Mathematical Physics **39**, 147 (1997)
32. E. Piña: Physica **98A**, 613 (1979)
33. R.B. Griffits, J.C. Wheeler: Physical Review A **2**, 1047 (1970)
34. W. Thomson (Lord Kelvin): Proc. Roy. Soc. Edinburgh **3**, 225 (1854)

## Supplementary glossary

| | |
|---|---|
| $h$ | molar enthalpy |
| $L$ | phenomenological coefficient |
| $M$ | molar mass |
| $N$ | molar amount |
| $\mathrm{N}$ | number of components |
| $P$ | pressure |
| $S$ | entropy |
| $v$ | molar volume |
| $X, Y, Z$ | (generalized) intensive value |
| $x, y, z$ | (generalized) extensive value |
| $U$ | internal energy |

### Superscripts

| | |
|---|---|
| $\mathrm{i, j, k, l}$ | component number |
| $\mathrm{T}$ | transposed (matrix) |

**Subscripts**

| | |
|---|---|
| c | convective |
| D | diffusive |
| E | exchange |
| F | free |
| i, j, k, l | component number |
| q | heat |
| R | reservoir |
| S | system |
| T | under constant temperature |

# A Comprehensive Theory of the Soret Effect in a Multicomponent Mixture

Leo J.T.M. Kempers

Shell International Exploration & Production,
P.O. Box 60, 2280 AB Rijswijk, the Netherlands

**Abstract.** A comprehensive theory for the Soret effect (also called: Ludwig-Soret effect and thermal diffusion) is presented which incorporates both the thermodynamic contribution and the kinetic contribution. The new theory is an extension of a theory presented in our previous paper (J. Chem. Phys. (1989) 90, 6541) in which the thermodynamic contribution only was modelled. The single assumption of the theory is that the Soret effect in the steady state is the macroscopic state accomplished by a maximum number of microstates with respect to the ideal gas state. The methodology of the new theory, which can be used to model other cross-effects in irreversible thermodynamics, is further validated by application to the textbook case of gravity equilibrium in an isothermal fluid column. As a result, the Soret effect in a multicomponent mixture can be calculated by using input from an equation-of-state of the mixture and kinetic gas theory.

However, it is crucial to critically examine the input data derived from an equation-of-state, particularly if calculations are compared to measurements to draw conclusions about the validity of the theory. The present equations-of-state, even those that are calibrated for use in the chemical and petroleum industry, require modification for the calculation of the Soret effect, because of a higher demand in accuracy. Further research to improve the accuracy of the equation-of-state is recommended, particularly for near-critical conditions. In addition, it is also important to examine which frame of reference (center of volume or center of mass) applies to a particular measurement or practical application, because the frame of reference determines which mathematical expression for the thermal diffusion factor must be used.

## 1 Introduction

The Soret effect is the tendency of a mixture of two or more components to separate as a result of a spatial temperature difference. It is one of the well known cross-effects in irreversible thermodynamics. This general theory assumes the component and heat fluxes to be a linear function of the concentration differences and the temperature difference. In the steady state, in which the component fluxes have vanished, and only a heat flux remains, the Soret effect is quantified by the equation:

$$\Delta x_{\mathrm{i}} = -x_{\mathrm{i}}(1 - x_{\mathrm{i}})\alpha_{\mathrm{Ti}}\frac{\Delta T}{T} \qquad (i = 1, \ldots, N) \tag{1}$$

where $x_{\mathrm{i}}$ is the mole fraction of component $i$, $T$ the absolute temperature, $N$ the number of components and $\alpha_{\mathrm{Ti}}$ the thermal diffusion factor of component

W. Köhler and S. Wiegand (Eds.): LNP 584, pp. 74–92, 2002.

$i$ and the subject of this paper. For dilute gases, kinetic gas theory enables calculation of $\alpha_{\mathrm{Ti}}$ [1]. It appears that in dilute gases $\alpha_{\mathrm{Ti}}$ is nearly independent of the component mole fractions.

In 1989 we published a thermodynamic theory of the Soret effect for multicomponent liquids [2], in which a common physical principle from statistical thermodynamics, i.e. "a macroscopic state is accomplished by a maximum number of possible microstates", is applied to the steady non-equilibrium state. No other assumptions were made, except for mechanical equilibrium. The result was a set of equations that provides a numeric prediction of the thermal diffusion factors of a multicomponent mixture. This set of equations allows for prediction of the thermodiffusion ratios based on the equilibrium thermodynamic properties of a muxture. It does not contain additional matching parameters. For a binary mixture, the explicit expression for the thermal diffusion factor according to that theory is:

$$\alpha_{\mathrm{T1}} = \frac{v_1 v_2}{v_1 x_1 + v_2 x_2} \frac{\frac{h_2}{v_2} - \frac{h_1}{v_1}}{x_1 \frac{\partial \mu_2}{\partial x_1}} \tag{2}$$

where $v_{\mathrm{i}}$, $h_{\mathrm{i}}$, $\mu_{\mathrm{i}}$ are respectively the molar volume, molar enthalpy and chemical potential of component $i$. The above expression was validated with experimental data of three mixtures in our 1989 paper. The qualitative observations of Kramers and Broeder [3] (i.e. in a mixture of hydrocarbons, the sequence from the hot to the cold region is: light normal paraffins, heavy normal paraffins, branched paraffins, naphtenes, monocyclic aromatics, bicyclic aromatics) are all predicted by our theory. Our theory was also confirmed by the qualitative measurements of Costesque and Riviére [4].

To our knowledge there are three other models/expressions in the literature for the thermal diffusion factor, apart from kinetic gas theory. There is the expression of Haase [5], which is equal to (2) but with the molar volumes replaced by molar masses with an extra term for the so-called kinetic contribution. We will discuss this expression again in this paper (see expression (16)). The second model is the kinetic model of Dougherty and Drickamer [1], which predictive power is lowered by the fact that it contains a fitting parameter [2]. The third model was developed recently by Shukla and Firoozabadi [6] along the kinetic approach of Dougherty and Drickamer. They introduce two new parameters, $\tau_1$ and $\tau_2$, with a definite physical meaning, but they use them as matching parameters, which strongly reduces predictivity of the model. An alternative way to predict the thermodiffusion ratio would be to apply molecular dynamic simulations based on the Green-Kubo formalizm. This approach has a deep physical background, however, presently it may be applied to a limited number of systems. We do not consider it in the paper (see, for example, the recent review [7] and the papers of the present volume).

Despite its success in comparing measurements in various mixtures, there are still three shortcomings in our purely thermodynamic theory and expression (2):

1. The theory ignores the kinetic contribution (due to selective friction between the components) to the Soret effect (it quantifies the thermodynamic

contribution only). As a result, the above expression cannot be used for dilute gases, in which the kinetic contribution dominates the thermodynamic contribution. Therefore expression (2) is valid only for non-ideal mixtures, such as near-critical mixtures, as was shown in the previous paper [2] by comparison to measured data.
2. The *total* molar enthalpy is required as input, while generally no more than the deviation of the enthalpy from a reference state is available. (The reference state is normally taken as the ideal gas state of the pure component.)
3. Haase's expression compares generally better to the measured data of the mixture of methane and n-butane than expression (2), probably due to a special measurement procedure, as was shown recently by Høier [8].
4. Another shortcoming lies in the test data that are used for validation of expression (2) and of other models. The calculated thermal diffusion factor is very sensitive to input data from an equation of state, while the accuracy of these data is insufficient.

There is thus a need for a comprehensive theory that 1) takes the kinetic contribution into account, 2) does not need input of the total molar enthalpy and 3) combines Haase's expression and our previous theory in one. In Section 2 of this paper, we will present such a comprehensive approach, which is a subtle variation of the old theory. Several special cases of the comprehensive theory are given in Section 3. The new theory is tested against experimental data in Section 4. Finally, in Section 5 proper selection of test data and extension of the comprehensive theory to other cross-effects will be discussed.

## 2 A comprehensive thermodynamic theory

As stated in Section 1, the Soret effect consists of two contributions, a kinetic and a thermodynamic. The kinetic contribution is caused by selective friction: the magnitude of the interaction in molecular collisions is generally different for each pair of mixture components. For dilute gases this contribution is quantified by kinetic gas theory. This theory predicts a non-zero Soret effect for an ideal gas. We consider the kinetic contribution, denoted by the thermal diffusion factor of the ideal gas state $\alpha^0_{\mathrm{T}i}$, as input to our thermodynamic theory.

The thermodynamic contribution to the Soret effect is caused by selective attraction/repulsion between the components. In mixtures in which a non-uniform temperature is maintained, it is thermodynamically advantageous for some components to concentrate in the region where the temperature is high and for other components to concentrate in the region where the temperature is low. This contribution, which is zero for an ideal gas, is modelled here.

For the calculation of the magnitude of the Soret effect we consider the following set-up (Fig. 1). Two bulbs, A and B, with equal and constant volume $V$ are joined by an insulated rigid tube of small diameter and filled with an $N$-component mixture ($N \geq 2$). The diameter of the tube is small enough to eliminate convection currents. The two-bulb system is in no way intended as a restriction; it helps to simplify the mathematics.

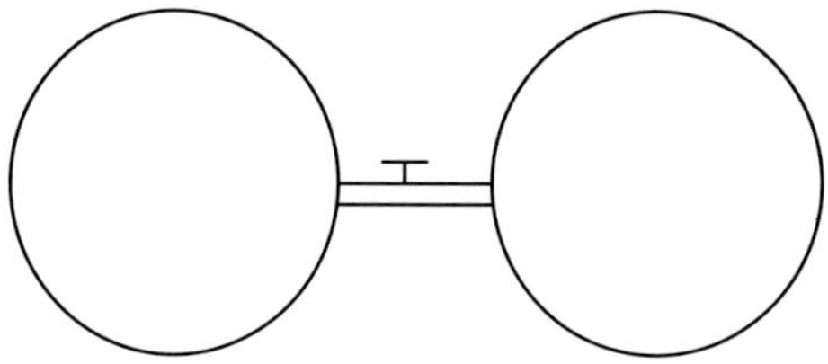


**Fig. 1.** Set-up with the two connected bulbs

In the *initial state*, the whole set-up is kept at a uniform and constant temperature and the composition of the mixture is uniform everywhere. After closing the valve in the tube the temperature $T$ in bulb A, is increased by a small amount $\Delta T/2$ to $T^{\mathrm{A}}$ and the temperature of bulb B is lowered by the same amount to $T^{\mathrm{B}}$. After this *intermediate state*, the valve is reopened. Measures have been taken such that $T^{\mathrm{A}}$ and $T^{\mathrm{B}}$ remain constant. Because the bulbs are in communication after the valve is opened, there is some interchange of mass between the bulbs by thermal expansion and diffusion. After some time a *final, stationary state* is reached, in which there is still a constant flux of heat from bulb A to bulb B but the mass flux of each component has vanished; in each of the two bulbs the thermodynamic properties such as the mole fractions $x_{\mathrm{i}}^{\mathrm{A}}$ and $x_{\mathrm{i}}^{\mathrm{B}}$ $(i = 1, \ldots, N)$ are uniform and constant. Due to the Soret effect there is a difference in mole fraction $\Delta x_{\mathrm{i}}$ $(i = 1, \ldots, N)$ between the bulbs.

To calculate the magnitude of the concentration differences $\Delta x_{\mathrm{i}}$ in the final, steady state due to the Soret effect, we make use of the canonical partition function $Z$ of the two-bulb system. If it is assumed that the steady state is the macroscopic state accomplished by a maximum number of microstates, the steady state can be calculated from the maximum of the partition function. So far, everything resembles the approach presented in our previous paper.

The new element is the following: we determine the maximum number of microstates with respect to the ideal gas state, so we use the *deviation* of the canonical partition function $Z$ from the partition function $Z^0$ in the ideal gas state. In this way we separate the Soret effect in the ideal gas state (which is due to the kinetic contribution) from the thermodynamic contribution. (Later in the mathematical procedure, the Soret effect in the ideal gas state will be included.) Hence instead of determining $\max\{Z\}$ as in our previous paper, we determine:

$$\max\left\{\frac{Z}{Z^0}\right\} \tag{3}$$

The canonical partition function $z^{\mathrm{A}}$ of bulb (or subsystem) A is a function of the variables $T^{\mathrm{A}}$, $V^{\mathrm{A}}(=V)$, $n_1^{\mathrm{A}}, \ldots, n_{\mathrm{N}}^{\mathrm{A}}$. Similarly, the partition function $z^{\mathrm{B}}$ of bulb B is a function of the variables $T^{\mathrm{B}}$, $V^{\mathrm{B}}(=V)$, $n_1^{\mathrm{B}}, \ldots, n_{\mathrm{N}}^{\mathrm{B}}$. The canonical partition function $z^{\mathrm{A}^0}$ of bulb A in the ideal gas state is a function of the variables $T^{\mathrm{A}}, V, n_1^{\mathrm{A}^0}, \ldots, n_{\mathrm{N}}^{\mathrm{A}^0}$. Finally, the canonical partition function $z^{\mathrm{B}^0}$ of bulb B in the ideal gas state is a function of the variables $T^{\mathrm{B}}$, $V$, $n_1^{\mathrm{B}^0}, \ldots, n_{\mathrm{N}}^{\mathrm{B}^0}$. If we ignore for a moment the interdependence of the number of moles of a particular component in a bulb and the number of moles of the same type in the other bulb, then we may treat the bulbs as independent subsystems of the two-bulb system. If the contribution of the connecting tube is ignored, the partition function $Z(z^{\mathrm{A}}, z^{\mathrm{B}})$ of the total system is then equal to the product of the partition functions of the bulbs:

$$Z(z^{\mathrm{A}}, z^{\mathrm{B}}) = z^{\mathrm{A}} \times z^{\mathrm{B}} \tag{4}$$

The value of $\Delta n_{\mathrm{i}} = n_{\mathrm{i}}^{\mathrm{A}} - n_{\mathrm{i}}^{\mathrm{B}}$ in the steady state can be calculated from the maximum statement (3) with the following constraints. The first constraint is material conservation of the components, expressed by

$$n_{\mathrm{i}}^{\mathrm{A}} + n_{\mathrm{i}}^{\mathrm{B}} = n_{\mathrm{i}}^{\mathrm{t}} \quad (i = 1, \ldots, N), \tag{5}$$

where $n_{\mathrm{i}}^{\mathrm{t}}$ are given constants. The second constraint is material conservation of the components in the ideal gas state:

$$n_{\mathrm{i}}^{\mathrm{A}} + n_{\mathrm{i}}^{\mathrm{B}} = n_{\mathrm{i}}^{\mathrm{A}^0} + n_{\mathrm{i}}^{\mathrm{B}^0} \quad (i = 1, \ldots, N) \tag{6}$$

The third constraint deals with the reference frame in which the diffusion process takes place. In a system without convection currents, the fluid motion is one-dimensional and the movement of the fluid parcels is constrained by the walls of the system. In the set-up of Fig. 1, the diffusion process takes place in a two-bulb system that is attached to a laboratory coordinate system. Since the interchange of mass between the bulbs, which occurs after the valve has been reopened, usually causes a translation of the center of mass of the fluid mixture, an external force must be exerted on the two-bulb system to keep it fixed in the laboratory coordinate system. The result of this force is that the center of volume of the fluid mixture does not translate. Because of the absence of convection, the zero translation of the center of volume applies to every fluid parcel in the system. Since, initially, the volume of the fluid inside bulb A is equal to the volume of the fluid inside bulb B, the zero translation of the center of volume means that these volumes remain equal:

$$\sum_{i=1}^{N} n_{\mathrm{i}}^{\mathrm{A}} v_{\mathrm{i}}^{\mathrm{A}} = \sum_{i=1}^{N} n_{\mathrm{i}}^{\mathrm{B}} v_{\mathrm{i}}^{\mathrm{B}} \tag{7}$$

Note that the constraint of mechanical equilibrium is not applied in contrast with our previous derivation. Instead, the constraint of zero translation of the center of volume, eq. (7) is applied. We consider this last constraint as more obvious than the constraint of mechanical equilibrium. In fact, mechanical equilibrium, which is expressed by $p^{\mathrm{A}} - p^{\mathrm{B}}$, is a consequence of the constraint of zero translation of the center of volume: if we apply $p^{\mathrm{A}} = p^{\mathrm{B}}$ as a fourth constraint, the result appears to be the same.

An alternative frame of reference is the center of mass. In that case it is assumed that the center of mass does not translate. In that case the constraint

(7) must be replaced by

$$\sum_{i=1}^{N} n_i^A m_i = \sum_{i=1}^{N} n_i^B m_i \tag{8}$$

With (8) instead of (7), the molar volume $v_i$ in the following mathematical expressions must be replaced by the molar mass $m_i$. In which cases the center of mass is a better frame of reference than the center of volume is discussed at the end of this Section.

By substituting Eq. (4) into (3) and applying the thermodynamic relationship $Z = \exp(-F/kT)$, where $F(T, V, n_1, ..., n_N)$ is the Helmholtz free energy of a bulb, and including the constraints (5)–(7) into the extremum determination, the mathematical problem is to find $n_i^A$ and $n_i^B$ ($i = 1, \ldots, N$) from

$$\min \left\{ \frac{F^A - F^{A^0}}{T^A} + \frac{F^B - F^{B^0}}{T^B} \right.$$

$$\left. - \sum_{i=1}^{N} [\lambda_i (n_i^A + n_i^B) + \kappa_i (n_i^A + n_i^B - n_i^{A^0} - n_i^{B^0})] - \nu \sum_{i=1}^{N} (n_i^A v_i^A - n_i^B v_i^B) \right\}, \tag{9}$$

where $\lambda_1, \ldots, \lambda_N$, $\kappa_1, \ldots, \kappa_N$ and $\nu$ are Lagrange multipliers.

The mathematical treatment of (9) is not given here but straightforward. The first step in the treatment is differentiation of the expression between brackets in equation (9) with respect to the independent variables $n_1^A, \ldots, n_N^A$, $n_1^B, \ldots, n_N^B$ and $n_1^{A^0}, \ldots, n_N^{A^0}$, $n_1^{B^0}, \ldots, n_N^{B^0}$. This results in $4N$ equations from which the Lagrange multiplie rs are eliminated. Furthermore, another step in the mathematics is the recognition of the term $x_i^{A^0} - x_i^{B^0}$ as the result of the kinetic contribution, which depends on the thermal diffusion factor $\alpha_{Ti}^0$ of the ideal gas state. This brings the kinetic contribution back into the equation. The solution is the following set equations in the unknown thermal diffusion factors:

$$\sum_{j=1}^{N-1} \left( \frac{1}{v_i} \frac{\partial \mu_i}{\partial x_j} - \frac{1}{v_N} \frac{\partial \mu_N}{\partial x_j} \right) x_j (1 - x_j) \alpha_{Tj}$$

$$= \frac{h_N - h_N^0}{v_N} - \frac{h_i - h_i^0}{v_i} + RT \left( \frac{\alpha_{Ti}^0 (1 - x_i)}{v_i} - \frac{\alpha_{TN}^0 (1 - x_N)}{v_N} \right) \tag{10}$$

$$(i = 1, \ldots, N - 1)$$

where the partial derivatives refer to the variables $T$, $p$, $n$, $x_1, \ldots, x_{N-1}$. The set of equations (10) is to be implemented in a phase behaviour computer package, together with the relation (which follows from definition (1)):

$$\sum_{j=1}^{N} x_j (1 - x_j) \alpha_{Tj} = 0 \tag{11}$$

The result is a numeric solution for the thermal diffusion factors $\alpha_{\mathrm{Ti}}$. It is evident that the above set of equations does not suffer from the shortcomings (1) and (2) mentioned in Section 1.

A value for $\alpha_{\mathrm{T}}^{0}$, the thermal diffusion factor of the same mixture at the same temperature in the ideal gas state can be calculated with kinetic gas theory. Hirschfelder, Curtiss and Bird [9] provide a practical guide for its computation. Calculated values of $\alpha_{\mathrm{T}}^{0}$ agree with measurements within 20%. A treatment for binary isotope mixtures is given by Jones and Furry [10]. Experimental values of $\alpha_{\mathrm{T}}^{0}$ can also be found in Vargaftik [11].

Finally, an obvious demand is that the derivation presented in this Section, when applied to a case of thermodynamic equilibrium, gives a result that is confirmed by standard equilibrium thermodynamics. This can be demonstrated for the case of gravity equilibrium in a vertical fluid column at isothermal condition.

### Frame of reference

The question which frame of reference applies, and hence whether constraint (7) or constraint (8) is to be used, does not have a straightforward answer in every case. It is evident that in experimental cells that are fixed to the laboratory, and in which the material fluxes are one-dimensional (no convection), the center of volume is the frame of reference to be used (and hence constraint (7)). This is confirmed for the case of gravity equilibrium in a system at rest. Also large-scale convection, such as in a Clausius–Dückel column, is determined by the system walls, and requires the center of volume as a frame of reference. However, if small-scale convection is present, the movement of a fluid parcel is constrained less by the walls of the experimental set-up. Then the thermal diffusion between adjacently moving fluid parcels is better described with the center of mass of the adjacent fluid parcels as a frame of reference, and hence with equation (8). An example of such a situation might be the application of thermal diffusion to the compositional variation in underground oil reservoirs in which convection currents are most likely to be present. Therefore it is recommended to consider which frame of reference applies in every new case.

## 3 Special cases

### Alternative expression for multicomponent mixture (center of volume is frame of reference)

One can show that an alternative expression for (10) and (11) is the following system with $N-1$ independent equations):

$$\sum_{j=1}^{N-1} \frac{\partial \mu_{\mathrm{i}}}{\partial x_{\mathrm{j}}} x_{\mathrm{j}}(1-x_{\mathrm{j}})\alpha_{\mathrm{Tj}} = \frac{v_{\mathrm{i}}}{v}(h-h_0) - (h_{\mathrm{i}} - h_{\mathrm{i}}^{0}) + RT\alpha_{\mathrm{Ti}}^{0}(1-x_{\mathrm{i}}) \tag{12}$$

$$(i = 1, \ldots, N)$$

where $h$ and $v$ are the molar quantities of the mixture. Although this set is valid for $N$ equations, only $N-1$ are independent.

**Binary mixture (center of volume)**

Substitution of (11) into (10) and the thermodynamic relationship:

$$x_1 \frac{\partial \mu_1}{\partial x_1} = x_2 \frac{\partial \mu_2}{\partial x_2}$$

yields the following explicit expression for the thermal diffusion factor:

$$\alpha_{\mathrm{T1}} = \frac{v_1 v_2}{v_1 x_1 + v_2 x_2} \frac{\frac{h_2 - h_2^0}{v_2} - \frac{h_1 - h_1^0}{v_1}}{x_1 \frac{\partial \mu_1}{\partial x_1}} + \frac{RT\alpha_{\mathrm{T1}}^0}{x_1 \frac{\partial \mu_1}{\partial x_1}} \tag{13}$$

where the partial derivatives refer to the variables $T$, $p$, $n$, $x_1$. An alternative expression follows from (12):

$$\alpha_{\mathrm{T1}} = \frac{\frac{v_1}{v}(h - h^0) - (h_1 - h_1^0)}{x_1 x_2 \frac{\partial \mu_1}{\partial x_1}} + \frac{RT\alpha_{\mathrm{T1}}^0}{x_1 \frac{\partial \mu_1}{\partial x_1}} \tag{14}$$

in which $h = h_1 x_1 + h_2 x_2$ and $v = v_1 x_1 + v_2 x_2$. Expression (13) or (14) replaces expression (2) of the old theory. Expression (14) yields the following qualitative observations. The positive sign of $\alpha_{\mathrm{T}}$ is determined by which species has the smallest molar enthalpy per molar volume. The species with the largest molar volume has a tendency for $\alpha_{\mathrm{T}} > 0$, hence tends to concentrate in the cold region. Finally, the Soret effect increases with increasing non-ideal behaviour.

**Multicomponent mixture (center of mass is frame of reference)**

In some cases, which are discussed at the end of Section 2, the center of mass is a better frame of reference than the center of volume. Hence constraint (8) must then be used instead of constraint (7). The set of equations (12) has then to be replaced by:

$$\sum_{j=1}^{N-1} \frac{\partial \mu_{\mathrm{i}}}{\partial x_{\mathrm{j}}} x_{\mathrm{j}}(1 - x_{\mathrm{j}})\alpha_{\mathrm{Tj}} = \frac{m_{\mathrm{i}}}{m}(h - h_0) - (h_{\mathrm{i}} - h_{\mathrm{i}}^0) + RT\alpha_{\mathrm{Ti}}^0(1 - x_{\mathrm{i}}) \tag{15}$$

$$(i = 1, \ldots, N)$$

where $m$ is the molar mass of the mixture.

**Binary mixture (center of mass)**

Applying (15) to a binary mixture yields:

$$\alpha_{\mathrm{T1}} = \frac{m_1 m_2}{m_1 x_1 + m_2 x_2} \frac{\frac{h_2 - h_2^0}{m_2} - \frac{h_1 - h_1^0}{m_1}}{x_1 \frac{\partial \mu_1}{\partial x_1}} + \frac{RT\alpha_{\mathrm{T1}}^0}{x_1 \frac{\partial \mu_1}{\partial x_1}} \tag{16}$$

This corresponds to Haase's expression [5], which lacked a derivation but was an educated guess [2]. Hence our methodology provides a basis for his expression. The conditions under which (16) can be applied are discussed at the end of Section 2.

### Binary mixture of which molar volume of a component is small (center of volume)

In some near-critical mixtures, the molar volume of one of the components (say component labeled 1) may be very small compared to the molar volume of the other component. In such a case, we have the following approximation for (13):

$$\alpha_{\mathrm{T1}} = -\frac{h_1 - h_1^0}{x_1 x_2 \frac{\partial \mu_1}{\partial x_1}} + \frac{RT\alpha_{\mathrm{T1}}^0}{x_1 \frac{\partial \mu_1}{\partial x_1}} \tag{17}$$

Hence, the magnitude of the Soret effect is determined by the enthalpy of the lower concentrated species.

### Tracer solutes in a liquid solvent (center of volume)

In a solvent of species $N$, tracer solutes are characterised by the same relation as for an ideal gas: $\partial \mu_{\mathrm{i}}/\partial x_{\mathrm{j}} = RT/x_{\mathrm{i}}\delta_{\mathrm{ij}}$. Equation (12) then gives:

$$\alpha_{\mathrm{Ti}} = \frac{\frac{v_{\mathrm{i}}}{v_{\mathrm{N}}}(h_{\mathrm{N}} - h_{\mathrm{N}}^0) - (h_{\mathrm{i}} - h_{\mathrm{i}}^0)}{RT} + \alpha_{\mathrm{Ti}}^0 \tag{18}$$

### No input from kinetic gas theory available (both frames)

If the thermal diffusion factor of the ideal gas state is not available, one may still calculate a value for the thermal diffusion factor if one measured value for $\alpha_{\mathrm{T}}$ of the same mixture at the same temperature but at another pressure and/or composition is available. The reason is that $\alpha_{\mathrm{T}}^0$ is independent of pressure and nearly independent of the mole fractions.

### Dilute gas: negligible thermodynamic contribution (both frames)

In the ideal gas limit, we have $h_{\mathrm{i}} = h_{\mathrm{i}}^0$ and $\partial \mu_{\mathrm{i}}/\partial x_{\mathrm{j}} = RT/x_{\mathrm{i}}\delta_{\mathrm{ij}}$, hence: $\alpha_{\mathrm{Ti}} = \alpha_{\mathrm{Ti}}^0$.

### Near-critical mixtures: negligible kinetic contribution (both frames)

The thermal diffusion factor as a result of the kinetic contribution, $\alpha_{\mathrm{T}}^0$, is generally $< 0.3$, while $\alpha_{\mathrm{T}}$ is above 3 for many non-ideal mixtures. Hence in highly non-ideal liquids and dense gases, the kinetic contribution can be neglected. The thermodynamic contribution is particularly dominant for near-critical mixtures (in which $\partial \mu_1/\partial x_1$ is close to 0 for a binary mixture).

**System with large temperature difference and/or large composition difference (both frames)**

In the second part of the derivation it is assumed that the differences in temperature and composition between the bulbs are small. If this condition is not satisfied, one may use the following equation:

$$\frac{\frac{\mu_i^A - \mu_i^{A^0}}{T^A} - \frac{\mu_i^B - \mu_i^{B^0}}{T^B}}{v_i^A + v_i^B} = \frac{\frac{\mu_N^A - \mu_N^{A^0}}{T^A} - \frac{\mu_N^B - \mu_N^{B^0}}{T^B}}{v_N^A + v_N^B} \tag{19}$$

and solve this set for the molefractions. (For the center of mass as a frame of reference: replace in (19) numerator $v_i^A + v_i^B$ by $m_i$.) There is thus no restriction to systems with a small composition or temperature difference.

# 4 Comparison with experimental data

The new theory has been compared with measured data for both frames of reference: center of volume and center of mass. The comparison has been done for all measured Soret data, known to us, of mixtures that consisting of mainly hydrocarbons and simple gases, for which thermodynamic models are available and well known. The equation of state (EOS) is a modified version of the Soave EOS; the modification is the inclusion of an interaction parameter in the mixing rule for the a-parameter. Interaction parameters are smaller than 0.1 for mixtures with carbondioxide and smaller than 0.02 for all other mixtures. The mixtures are listed in Table 1.

The $\alpha_T$ values calculated with expression (13) (frame of reference: center of volume) are compared to data in Fig. 2. The $\alpha_T$ values calculated with expression (16) (frame of reference: center of mass) are compared to data in Fig. 3. The $\alpha_T$ values refer to component 2; for all mixtures the measured value is positive. The graphs have logarithmic scales to represent $\alpha_T$ values over 4 decades. The $\alpha_T$ signs of all mixtures, except a few datapoints of mixtures 4–7, 9 and 10, correspond to data. Because of the logarithmic scale, these exceptions are not taken up in Figs. 2 and 3.

When the Soret data of all mixtures are taken together, both graphs show that there is at least a qualitative agreement between measured and calculated data over 4 decades. Generally, the agreement is better for the center of mass as a frame of reference than for the center of volume.

In the case of some mixtures, i.e. the gaseous mixtures 13, 15–17, there is good to fair agreement for both frames of reference, and slightly better for the center of volume. In the case of the near-critical mixture 2, the agreement is fair for the center of mass; for the near-critical mixture 1 the agreement is poor for both frames. In the case of liquid hydrocarbon mixtures 4–12 the agreement varies between fair and poor for both frames.

For some mixtures the calculated $\alpha_T$-value is very sensitive to the value of the molar volume or the specific molar enthalpy. This means that calibration of the

**Table 1.** List of mixtures that are used as test cases for expressions (13) and (16)

| | component 1 | component 2 | $p$ (bar) | $T$ (K) | molefraction comp.1 | # | reference |
|---|---|---|---|---|---|---|---|
| 1 | methane | propane | 40–75 | 346.08 | 0.23–0.58 | 17 | [12] |
| 2 | methane | n-butane | 95–204 | 319–394 | 0.40, 0.49 | 96 | [13] |
| 3 | n-hexane | n-octane | 1 | 320 | 0.55 | 1 | [14] |
| 4 | cyclo-hexane | benzene | 1 | 313 | 0.2–0.8 | 3 | [15] |
| 5 | n-hexane | benzene | 1 | 309 | 0.1–0.9 | 5 | [16] |
| 6 | n-heptane | benzene | 1 | 309 | 0.1–0.9 | 9 | [16] |
| 7 | n-octane | benzene | 1 | 309 | 0.1–0.9 | 9 | [16] |
| 8 | n-heptane | n-dodecane | 1 | 296 | 0.5 | 5 | [17] |
| 9 | n-heptane | n-tetradecane | 1 | 306–336 | 0.5 | 3 | [17] |
| 10 | n-heptane | n-pentadecane | 1 | 306 | 0.5 | 1 | [17] |
| 11 | n-heptane | n-hexadecane | 1 | 308 | 0.10–0.90 | 8 | [18] |
| 12 | n-heptane | n-octadecane | 1 | 306 | 0.5 | 1 | [17] |
| 13 | methane | carbondioxide | 3–81 | 357 | 0.48 | 4 | [19] |
| 14 | methane | nitrogen | 4–80 | 357 | 0.50 | 4 | [19] |
| 15 | nitrogen | carbondioxide | 3–81 | 357 | 0.52 | 4 | [19] |
| 16 | hydrogen | nitrogen | 3–78 | 357 | 0.50 | 4 | [19] |
| 17 | hydrogen | carbondioxide | 3–81 | 223–363 | 0.24–0.53 | 12 | [19,20] |

molar volume or molar enthalpy of the mixture to measured density data may not be sufficient for an accurately calculated value for $\alpha_{\mathrm{T}}$. The sensitivity ratio $s_{\mathrm{v}}$ for the specific molar volume is defined as the ratio between the sensitivity of $\alpha_{\mathrm{T}}$ to the specific molar volume by the sensitivity of the molar volume of the mixture to the specific molar volume:

$$s_{\mathrm{v}} = \max(|s_{\mathrm{v1}}|, |s_{\mathrm{v2}}|) \qquad \text{with} \quad s_{\mathrm{vi}} = \frac{v_{\mathrm{i}}}{\alpha_{\mathrm{T}}} \frac{\partial \alpha_{\mathrm{T}}}{\partial v_{\mathrm{i}}} \Big/ \frac{v_{\mathrm{i}}}{v} \frac{\partial v}{\partial v_{\mathrm{i}}} \tag{20}$$

For example, an $s_{\mathrm{v}}$-value of 5 means that a 10% uncertainty in the molar mixture volume due to one of the components corresponds to a 50% uncertainty in $\alpha_{\mathrm{T}}$. A similar ratio sh for the specific molar enthalpy is defined. Table 2, which lists the values of these ratios for each mixture, shows that the uncertainty in $\alpha_{\mathrm{T}}$ of many datapoints is more than 100% if the uncertainty in enthalpy or volume is a few percent.

Table 2 shows that the hydrocarbon mixtures 3–12 are very sensitive to specific molar volume end specific molar enthalpy. For some test data of these mixtures, a change in the molar specific enthalpy or molar specific molar volume has a 100 times greater effect on the thermal diffusion factor than on the molar enthalpy or molar volume of the mixture.

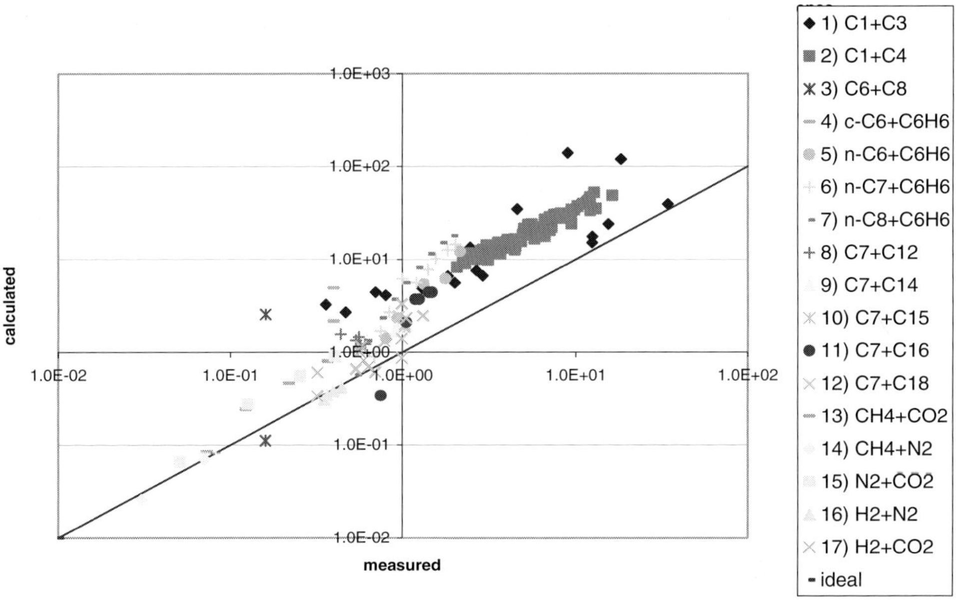


**Fig. 2.** Comparison to measurements for 17 mixtures if center-of-volume is frame-of-reference

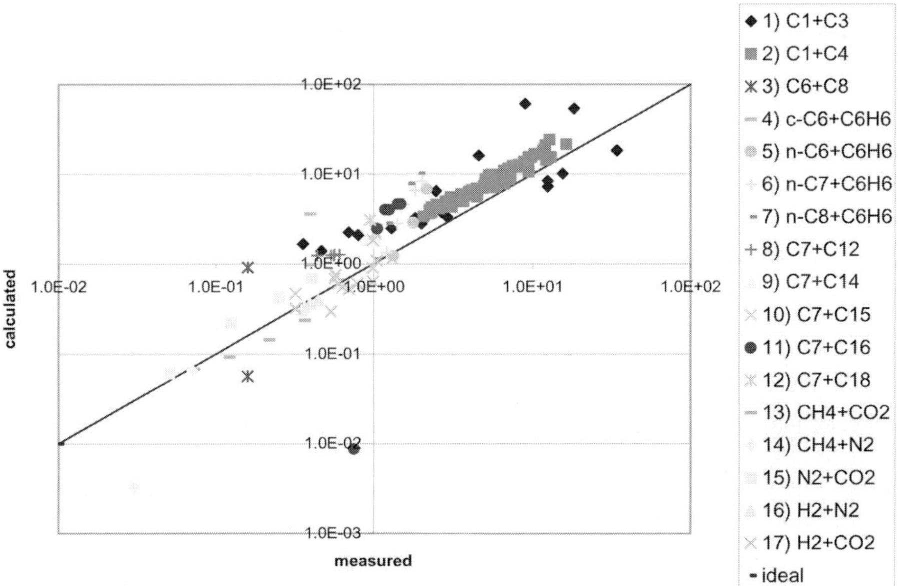


**Fig. 3.** Comparison to measurements if center-of-mass is frame-of-reference

An alternative test for mixture no. 2 is to use one of the 96 Soret measurements on this mixture as a calibration, e.g. the smallest measured $\alpha_T$ value (2.1)

**Table 2.** List of mixtures with type of agreement to data, some experimental conditions (temperature difference between chambers, presence of large-scale convection) and sensitivity of $\alpha_T$ to specific molar and specific molar enthalpy.
[a]: of the 96 datapoints 92 have $s_h$ below 3

| | type mixture | agreement for c.o.m., expr. (13) | agreement for c.o.v., expr. (16) | $\Delta T$, K | presence convection | sensitiv. ratio $s_v$ | sensitiv. ratio $s_h$ |
|---|---|---|---|---|---|---|---|
| 1 | near-critical | poor | poor | 8 | stirring | 0.6–1.0 | 0.6–1.0 |
| 2 | near-critical | poor | fair | 10–20 | stirring | 0.1–2.2 | 0.1–26[a] |
| 3 | liquid at rm. cond. | poor | poor | 40 | yes | 24 | 24 |
| 4 | liquid at rm. cond. | poor | poor | 10 | stirring | 12–390 | 12–390 |
| 5 | liquid at rm. cond. | poor | poor | not given | no stirring | 7–93 | 7–93 |
| 6 | liquid at rm. cond. | poor | poor | not given | no stirring | 4–168 | 4–168 |
| 7 | liquid at rm. cond. | poor | poor | not given | no stirring | 4–297 | 4–297 |
| 8 | liquid at rm. cond. | poor | poor | 5 | yes | 23–41 | 3–21 |
| 9 | liquid at rm. cond. | fair | fair | 5–10 | yes | 24–32 | 1–36 |
| 10 | liquid at rm. cond. | fair | poor | 5 | yes | 31 | 197 |
| 11 | liquid at rm. cond. | poor | poor | 15 | stirring | 15–70 | 6–122 |
| 12 | liquid at rm. cond. | poor | poor | 5 | yes | 30 | 17 |
| 13 | gaseous at high $p$ | fair | good | 140 | no stirring | 1.5–2.1 | 2.1–2.7 |
| 14 | gaseous at high $p$ | fair | poor | 140 | no stirring | 1.4–1.6 | 0.9–1.9 |
| 15 | gaseous at high $p$ | fair | fair | 140 | no stirring | 0.8–1.0 | 0.5 |
| 16 | gaseous at high $p$ | good | good | 140 | no stirring | 0.2–0.3 | 0.2–0.3 |
| 17 | gaseous at high $p$ | good | good | 140, 100 | no stirring | 0.5–1.3 | 0.0–1.6 |

at the lowest temperature, by using the $\alpha_{\mathrm{T}}^0$ value calculated from this calibration point as input in all other 95 measurements. The result of this exercise is shown for a selection of the 96 datapoints in Fig. 4 for both frames. This figure shows that a better match is obtained for both frames, and that the match for the center of mass is best.

The exercise is also done for the 17 datapoints of mixture no. 1 (with the smallest measured $\alpha_{\mathrm{T}}$-value (0.8) at the lowest pressure as a calibration point). Fig. 5, in which some results of the calibration are shown, shows an improved match for both frames, particularly for the center of volume. There is, however, no good match near the critical point.

## 5 Discussion

### Equation of state

In all papers in which our and other's expression for the thermal diffusion factor are compared to measured data [2,6,8] it is implicitly assumed that the EOS that

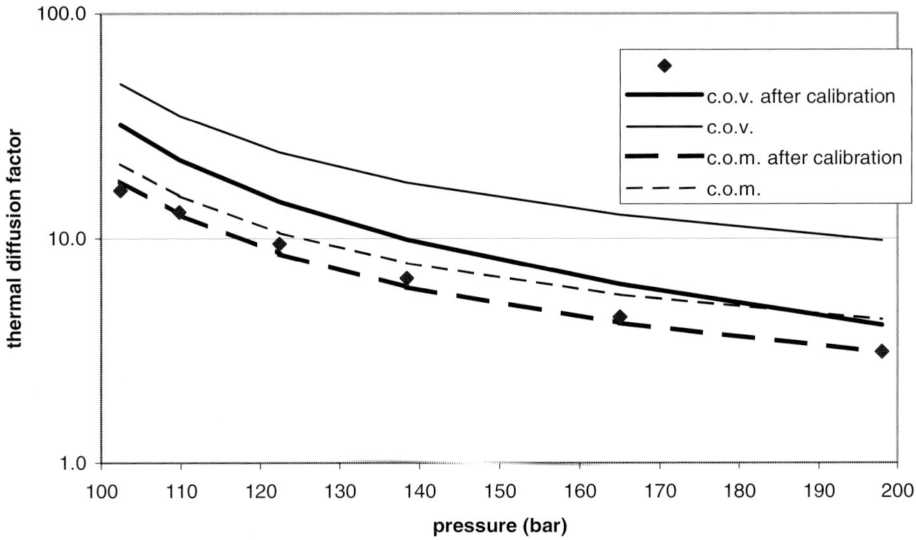


**Fig. 4.** Comparison to measurements for methane (molefraction 0.40) + n-butane at 395 K

is used for the calculation of molar enthalpy and molar volume yields accurate values. This assumption is doubtful. The effect of the EOS on the value of the thermal diffusion factor can be illustrated very well by comparing our calculation of the thermal diffusion factor of the methane/propane mixture in our previous paper to the calculation by Shukla and Firoozabadi [6]. Their calculated values for methane/propane at 346 K and 55 bar are 2–3 times larger than the values presented in our previous paper, although the same mathematical expression for the thermal diffusion factor is used.

Another illustration of the insufficient accuracy of the EOS, is the prediction of the critical pressure of the methane/propane system: 62 bar according to the measurements of Haase and 65.8 bar according to the Soave-Redlich-Kwong or to the Peng-Robinson EOS (and 0.28 methane mole fraction at the temperature of the experiment). Since the numerator in the expression for $\alpha_T$ becomes zero at the critical point, $\alpha_T$ is very sensitive to the location of the critical point. The deviation by an order of magnitude between measurement and theory of $\alpha_T$ in the near-critical regime of the methane/propane mixture must therefore be attributed in the first place to an inadequate EOS. Furthermore, the EOS in the supercritical region is an extrapolation of the EOS in the subcritical region and is not based on data relating to the physical properties in the supercritical region. This means that small errors in the EOS of the supercritical liquid region are likely.

The present EOS's are generally not intended for predicting the Soret effect, which depends on the difference between component properties. They are calibrated only for predicting quantities that depend on the addition of component properties. The prediction of the magnitude of the Soret effect with quantities derived from the EOS is therefore generally more sensitive to an error in a component property (specific molar volume and specific molar enthalpy) than is, for example, the fluid density. But even the density poses a problem. For example, near-critical gas/condensate mixtures, are notorious for a bad prediction of density and molar volume; deviations of more than 10% are not uncommon. Although the volume translation method has been developed to match the measured density of the mixture, this method has never been tested on the prediction of the molar volume of each component in the mixture. Hence the accuracy of the prediction of the magnitude of the Soret effect, that depends on molar volumes, is worse than 10%. According to Table 2, a few extreme examples are the mixtures no. 4–7: the sensitivity of the thermal diffusion factor to the molar volume of one of the components is about 10–several 100's times the sensitivity of the mixture density to the molar volume of the same component.

There is thus a need for calibrating the EOS to new demands like the prediction of the Soret effect. Further research in this area is recommended. One may even reverse the problem: measurements of the Soret effect may reveal more about the phase behaviour and thermodynamic properties of the components in the mixture. With the present EOS accuracy, it seems from Figs. 4 and 5 that an improved prediction of the Soret effect can be obtained by using one measured datapoint as a calibration.

### Linearization

In most tested mixtures the temperature difference between the chambers of the cell was small compared to the average absolute temperature. However, in mixture 3 and the gaseous mixtures 13–17 this is not the case (see Table 2). For these mixtures $\alpha_T$ values are not symmetric over the temperature range around the average temperature. As a consequence, linearization in the derivation of (10) is not allowed and one must solve the implicit equation (19). For the comparison to measurements one then needs the composition of each chamber, which is not available from the relevant paper [19]. (For mixture 3 these compositions are available, but the measurement is done in the presence of large-scale convection for which the mathematical description with eq. (19) is invalid.) Hence only a qualitative comparison to data is possible for these mixtures.

### Suitability of measurements to test theory

It appears from Table 2 that there are no test data available that fulfill the following demands:

- Sufficiently reliable EOS, hence small sensitivity to EOS;
- Small temperature difference or all relevant data available in paper in case of large temperature difference;
- No large-scale convection;
- No stirring (see also next paragraph).

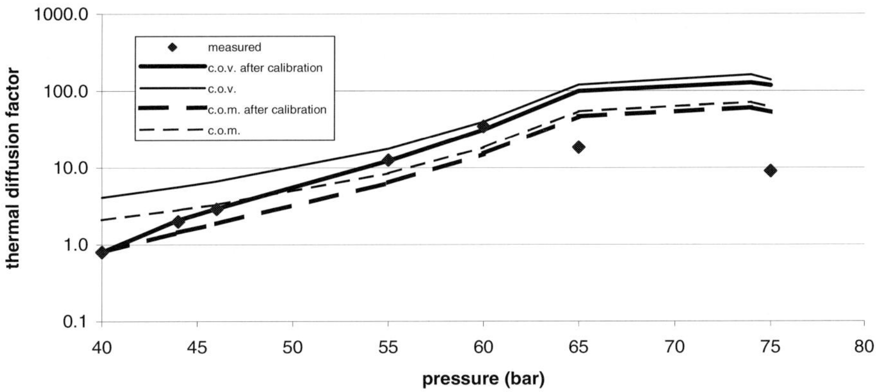


**Fig. 5.** Comparison to measurements for methane (molefraction 0.34) + n-propane at 346 K

If one drops the demand of no stirring, then two mixtures remain: those data of mixture 1 and 2 that are far from critical condition. For future measurements it is recommended that these demands be fulfilled.

## Frame of reference

We have shown in Section 2 that the frame of reference is the center of volume if the experimental set-up is attached to the laboratory and if the fluid motion is one-dimensional. We have shown that applying the methodology to gravity equilibrium at isothermal condition falsifies the center of mass. However, the comparison with measurements in Section 5, in particular those for mixture 2 of methane and n-butane, shows that the center of mass is generally more appropriate than the center of volume. Since the experimental set-up is attached to the laboratory, a possible conclusion is that the fluid motion was not one-dimensional in the measurements. Measurements that were designed to eliminate convection, might have had convection because of the stirring in the chambers of the experimental cell. Stirring, which promotes a uniform mixture composition in each chamber and helps to attain the steady state quicker, introduces two- or three-dimensional fluid movement. Tichachek et al. [15] found a relation between separation and stirring speed.

Experiments without stirring in a microgravity environment are the most reliable way to eliminate convection. Therefore, the Microgravity Research Center in Bruxelles carries out a "Soret Coefficients in Crude Oil" project [21] for the European Space Agency. The experiment will take place in the Space Shuttle. It may result in a reliable test of our theory.

## Application to cross-effect in isothermal diffusion ("Hertz effect")

The methodology of the theory presented in Section 2 is not restricted to application to the Soret effect. In principle, every cross-effect from irreversible thermodynamics can be modelled with this methodology. An example is the cross-effect between diffusion fluxes in an isothermal mixture of three or more components with a spatial concentration difference. This cross-effect is the cause of diffusion of a component against its own concentration gradient. Gustav Hertz was the first to use this cross-effect for the separation of isotopes. The calculation of the magnitude of the cross-effect, which may be called the Hertz effect, is possible if all coefficients of the diffusion tensor of the mixture are known. For dilute gases, kinetic gas theory enables the calculation of the diffusion tensor, but calculation methods for dense gases, liquids and solids require input of measured binary or tracer diffusion coefficients. Using the methodology presented in this paper, we have developed for the first time a calculation method for the magnitude of the Hertz effect without using input of any measured diffusion coefficient [22].

## 6 Conclusions and recommendations

1. The present comprehensive theory of the Soret effect, which yields a set of equations (10) for a multicomponent mixture and expression (12) for a binary mixture, incorporates both the thermodynamic and the kinetic contribution to the Soret effect. It does not require absolute values for the thermodynamic quantities, but only the difference with respect to the ideal gas state.
2. The new description applies to any multicomponent mixture, gaseous, liquid or solid. The description is particularly useful for non-ideal mixtures, such as concentrated solutes and near-critical mixtures.
3. The frame of reference for thermal diffusion is normally the center of volume. When small-scale convection is present, the frame of reference for the thermal diffusion between moving fluid parcels can be better described by the center of mass. The frame of reference determines which mathematical expression for the thermal diffusion factor must be used ((12) or (15) for a multicomponent mixture and (14) or (16) for a binary mixture).
4. The thermodynamic approach of the theory passes the falsification test of application to the textbook case of gravity equilibrium. In this way it is shown that the center of volume (and not the center of mass) is the appropriate frame of reference for a system at rest.
5. The theory with the center of mass as a frame of reference matches measured data better than the same theory with the center of volume as a frame of reference. This can be explained by the presence of two- or three-dimensional fluid movement in the experimental cells. A possible cause is the stirring that was applied to reach the steady state quicker.
6. The present equations of state, even those that are calibrated for use in the chemical and petroleum industry, require improvement for the calculation of the Soret effect, because of a higher demand in accuracy. Further research to improve the accuracy of the EOS is recommended, particularly for near-critical conditions.
7. It is recommended to carry out measurements of the Soret effect in which a small temperature difference is applied (or to mention the composition

of each chamber if a large temperature difference is applied), in which no stirring is applied, and to select mixtures of which the thermal diffusion factor is not very sensitive to input data from the EOS.

## References


1. E.L. Dougherty, H.G. Drickamer: J. Phys. Chem. **59**, 443 (1955)
2. L.J.T.M. Kempers: J. Chem. Phys. **90**, 6541 (1989)
3. H. Kramers: J.J. Broeder, Anal. Chim. Acta **2**, 687 (1948)
4. P. Costesque, E. Rivière: Comptes Rendues Acad. Sci. Paris **315** Série II, 179 (1992)
5. R. Haase: Z. Phys. **127**, 1 (1949)
6. K. Shukla, A. Firoozabadi: Ind. Eng. Chem. Res. **37**, 3331 (1998)
7. S.S. Sarman, D.J. Evans, P.T. Cummings: Physics Reports **305**, 1 (1998)
8. L. Høier: Miscibility variations in compositionally grading petroleum reservoirs. PhD Thesis, University of Trondheim (1997)
9. J.O. Hirschfelder, C.F. Curtiss, R.B. Bird: "Molecular theory of gases and liquids", Wiley, New York (1954).
10. R.C. Jones, W.H. Furry: Reviews of Modern Physics, **18**, 151 (1946)
11. N.B. Vargaftik: "Tables on the thermophysical properties of liquids and gases", Chapter 12 (1975), Hemisphere, Washington.
12. R. Haase, H.-W. Borgmann, K.-H. Dücker, W.-P. Lee: Z. Naturforsch. **26A**, 1224 (1971)
13. W.M. Rutherford, J.G. Roof: J. Phys. Chem. **63**, 1506 (1959)
14. H. Korsching, K. Wirtz: Ber. Dtsch. Chem. Ges. **73**, 249 (1940)
15. L.J. Tichachek, W.S. Kmak, H.G. Drickamer: J. Phys. Chem. **60**, 660 (1956)
16. H. Korsching: Z. Naturforsch. **24A**, 444 (1969)
17. D.J. Trevoy, H.G. Drickamer: J. Chem. Phys. **17**, 1120 (1949)
18. J.J. Shieh: J. Phys. Chem. **73**, 1508 (1969)
19. V.E.W. Becker: Z. Naturforsch. **5A**, 457 (1950)
20. T.K.S. Narayanan, G. Dickel: Z. Naturforsch. **28A**, 290 (1973)
21. S. Van Vaerenbergh, J.-C. Legros, J.-C. Dupin: Advances In Space Research 16 (1995) 869. S. Van Vaerenbergh: "Microgravity experiment for the measurement of Soret coefficients in crude oil (SCCO)", Microgravity Research Center, Université Libre de Bruxelles (1999).
22. L.J.T.M. Kempers: "A comprehensive thermodynamic theory of the cross-effect in isothermal diffusion ("Hertz effect") in a multicomponent mixture", paper submitted.


## Supplementary glossary

| | |
|---|---|
| c.o.m. | center of mass |
| c.o.v. | center of volume |
| EOS | equation-of-state |
| $g$ | gravitational acceleration |
| $H$ | height of center of mass of bulb |
| $h$ | molar enthalpy |

| | |
|---|---|
| $k$ | Boltmann's constant |
| $m$ | molar mass |
| $N$ | number of components in mixture |
| $n$ | mole number |
| $p$ | pressure |
| $R$ | gas constant |
| $V$ | volume of bulb |
| $v$ | molar volume |
| $Z$ | canonical partition function of two-bulb system |
| $z$ | canonical partition function of a bulb |
| $\Delta$ | operator for the difference between bulb A and bulb B |

## Superscripts

| | |
|---|---|
| $A$, $B$ | bulb label |
| 0 | ideal gas state |

## Subscripts

$i$, $j$, $k$, $N$ component label

# Thermodiffusion and Nonlinear Heat Equation

Ryszard Wojnar

Instytut Podstawowych Problemów Techniki, Polska Akademia Nauk,
ul. Świętokrzyska 21, 00-049 Warszawa, Poland

**Abstract.** After recalling the main ideas of kinetic theory of gases and of various scales of description of a material medium, the diffusion of the Brownian particles in a continuum subject to gravitation forces and a temperature field is discussed. The nonlinear energy conservation law for a thermodiffusion process with the Streater effect in which the potential energy of the Brownian particle is converted to a heat due to friction, is obtained, cf. [1]. The second law is satisfied if a nonlinear relation between the probability density function and temperature is postulated. The examples of the Brownian motion corresponding to particular gravitational fields and temperature differences are included. Also, a homogenisation method is used to describe a thermodiffusion process in porous media.

## 1 Introduction

Penrose and Fife [2] giving a general framework for the phenomenological kinetic theory of non-equilibrium processes in which the temperature may vary in space and time, formulated the requirement of a thermodynamic consistency that the mathematical description of a system should obey the first and the second laws of thermodynamics.

Streater [1] was the first who gave a complete thermodynamic description of the Brownian motion in a fluid that complies with the postulate of Penrose and Fife. He combines the Smoluchowski diffusion equation with drift with a nonlinear heat equation in which a heat supply term, as an effect of conversion of the Brownian particle potential energy into heat due to friction, is introduced; and proves that such a system is thermodynamically admissible. In the Streater system of equations so called thermal diffusion forces are absent.

In the present paper the Streater theory is generalized to include the thermal diffusion forces which are important in describing the diffusive processes in presence of temperature gradients. It is believed that such generalization can be used in a research on Brownian motors developed recently [3–6].

### 1.1 Research on thermodiffusion

A phenomenon of thermodiffusion in liquids was discovered apparently for the first time by Ludwig [7] in 1856 and independently by Soret in 1879 [8, 9]. They observed a direct thermodiffusion in which a difference of a solute concentration (and thus the matter flow) is caused by a temperature gradient. In particular, they found that the change of concentration occurs in a tube which is filled

W. Köhler and S. Wiegand (Eds.): LNP 584, pp. 93–120, 2002.

with an initially homogeneous salt solution and has both ends kept at different temperatures. For example, they showed that for a solution of sodium chloride NaCl the salt concentration at the heated end of tube decreases, while at the cold end it increases.

A reciprocal effect, in which a difference of temperatures is caused by the gradient of the particle concentration was observed by Dufour in 1872 [10, 11]. He investigated a diffusion of different pairs of gases (among others the hydrogen and air) through a porous dividing wall. The experiment was the following: the porous wall was filled by the air until a state of thermal equilibrium was reached. Next the hydrogen $H_2$ was allowed to diffuse and it was found that the temperature increases on the wall side emitting the hydrogen and it decreases on the opposite wall side. This means that the thermal diffusion coefficient $D_{\mathrm{T}}$ of $H_2$ in the air is positive as $H_2$ migrates to the cold side. Similar experiments were carried out and described in detail by Feddersen who apparently for the first time introduced the concept of a thermodiffusion [12]. The direct and reciprocal thermodiffusive phenomena are called sometimes Soret-Dufour effects, cf. [13].

At the end of his paper [12] Feddersen expressed the following opinion on the results obtained: *Although the forces evoked by thermodiffusion are for the most part not too great, one can not assume without reserve that the phenomenon plays only a secondary role in the household of Nature because the conditions to its manifestation appear to be rather widely overspread* (translated from German).

The coefficient of thermodiffusion was not introduced by Meyer in his famous dissertation [14], where a free path theory was developed. Enskog (1912) and Chapman (1916) derived the coefficient of thermodiffusion for a gas described by the Boltzmann equation. An experimental evidence of the phenomenon was gained by Chapman and Dootson in 1917, [15], and by Clusius and Waldmann in 1943 [16, 17].

Chapman and Cowling in their monography on the gas theory indicated that an error of the first approximation for $S_{\mathrm{T}}$ is greater than for any other gas coefficients, [15], p. 144. The story described in [15] is characteristic in the following sense. When Chapman's 1917 paper was communicated to the Royal Society, J.H. Jeans who was famous in the kinetic theory of gases as he had improved the elementary methods by considering the persistence of velocities after collision, and who examined the paper for the Society wrote to the author suggesting that operation of the thermal diffusion term in the equation of diffusion would be almost infinitely slow. However, Jeans did not notice that the rate of thermodiffusion was comparable with that of ordinary diffusion, as was shown by the equation and was soon confirmed by the Chapman-Dootson experiments [15], p. 401.

In 1917 Chapman suggested that a thermal diffusion process can be used to separate gases and isotopes. A practical method of such separation was found by Clusius and Dickel in 1938 [18].

In 1939 Furry, Jones and Onsager [19] introduced the thermal diffusion factor $\alpha_{\mathrm{T}}$ defined as $\alpha_{\mathrm{T}} = (D_{\mathrm{T}}/D)\,T$, where $D$ is the diffusion coefficient and $T$ is the temperature. The experiment and theory show that $\alpha_{\mathrm{T}}$ is only slightly dependent on concentration $c$ and in experiments on separation of isotopes it is almost independent on $c$ [20]. Jones and Furry discussed the separation of isotopes by thermal diffusion in 1946 [21]. The Jones and Furry paper was submitted to redaction in 1940, but was voluntarily withheld from publication during the war.

The Enskog-Chapman method of analysis of the Boltzmann equation is somehow awkward, and trials to give more immediate and intuitive explanation of the thermodiffusive effect were proposed by Frankel in 1940 [22] on the basis of a Stefan method [23] and Fürth in 1941 on the basis of the free path theory [24]. These elementary treatments were discussed by Furry in 1948 [25]. However, a simple elementary explanation of the reciprocal thermodiffusive effect is apparently missing in the literature. Such an explanation based on the Pohl approach [26] is proposed in Appendix A.

## 1.2 Phenomenological theory of thermodiffusion

The macroscopic description of a thermodiffusion phenomenon in frame of the Onsager theory of irreversible processes, is to be found in several renowned textbooks, as for example de Groot and Mazur [13], Grew and Ibbs [27], Waldmann [28] and Landau and Lifshits [29].

The rate of change of total entropy $S$ of a viscous fluid occupying a region $\Omega$ in which mass and heat transport processes take place is given by, see [29], p. 222,

$$\dot{S} = -\int_{\Omega} \frac{\mathbf{j}\cdot\nabla\mu}{T} d\Omega - \int_{\Omega} \frac{\mathbf{q}\cdot\nabla T}{T^2} d\Omega \tag{1.1}$$

where the positive viscosity terms are omitted. Above $\mu$ is an appropriately defined chemical potential of the mixture and $T$ is the absolute temperature, while two vectors, $\mathbf{j}$ and $\mathbf{q}$, denote the diffusion and heat fluxes; and $\nabla$ stands for the gradient operator. We see that the associated to $\mathbf{j}$ and $\mathbf{q}$ thermodynamic forces are

$$\frac{\nabla\mu}{T} \quad \text{and} \quad \frac{\nabla T}{T^2}$$

Thus the corresponding transport relations read

$$\begin{aligned} \mathbf{j} &= -L_{\mathrm{D}}T\frac{\nabla\mu}{T} - L_{\mathrm{DT}}T^2\frac{\nabla T}{T^2} \\ \mathbf{q} &= -L_{\mathrm{TD}}T\frac{\nabla\mu}{T} - L_{\mathrm{T}}T^2\frac{\nabla T}{T^2} \end{aligned} \tag{1.2}$$

where $L_{\mathrm{D}}, L_{\mathrm{DT}}, L_{\mathrm{TD}}$ and $L_{\mathrm{T}}$ are certain kinetic coefficients. By the Onsager symmetry we have

$$L_{\mathrm{TD}}T = L_{\mathrm{DT}}T^2 \qquad \text{or} \qquad L_{\mathrm{TD}} = TL_{\mathrm{DT}}$$

Hence the equations for diffusion and heat fluxes are as follows

$$\begin{aligned} \mathbf{j} &= -L_{\mathrm{D}}\nabla\mu - L_{\mathrm{DT}}\nabla T \\ \mathbf{q} &= -L_{\mathrm{DT}}T\nabla\mu - L_{\mathrm{T}}\nabla T \end{aligned} \tag{1.3}$$

and the flux of total energy is given by

$$\mathbf{j}^{\mathrm{E}} = \mathbf{q} + \mu \mathbf{j} \tag{1.4}$$

The gradient $\nabla\mu$ is taken at constant temperature, see [13], p. 273. If we replace the chemical potential $\mu$ by more usual variable of concentration $c$ we obtain

$$\nabla\mu = \left(\frac{\partial\mu}{\partial c}\right)_{T,p} \nabla c \tag{1.5}$$

and substituting this in (1.2) we get

$$\begin{aligned} \mathbf{j} &= -D\nabla c - L_{\mathrm{DT}}\nabla T \\ s\mathbf{q} &= -\lambda_{\mathrm{D}}\nabla c - \lambda_{\mathrm{T}}\nabla T \end{aligned} \tag{1.6}$$

where

$$D \equiv L_{\mathrm{D}}\left(\frac{\partial\mu}{\partial c}\right)_{T,p} \qquad \lambda_{\mathrm{D}} \equiv L_{\mathrm{DT}} T\left(\frac{\partial\mu}{\partial c}\right)_{T,p} \qquad \text{and} \quad \lambda_{\mathrm{T}} \equiv L_{\mathrm{T}} \tag{1.7}$$

The condition that entropy must increase places certain restrictions on the coefficients, for example the diffusion coefficient $D$ and the thermal conductivity $\lambda_{\mathrm{T}}$ must be positive. The quantity $L_{\mathrm{DT}}$ proportional to the thermodiffusion coefficient $D_{\mathrm{T}}$

$$L_{\mathrm{DT}} \equiv c(1-c)D_{\mathrm{T}} \tag{1.7a}$$

may be either positive or negative.

The chemical potential of dissolved substance with low concentration $c$ equals to

$$\mu = k_{\mathrm{B}}T \ln c + f(p,T) \tag{1.8}$$

where $k_{\mathrm{B}}$ is the Boltzmann constant and $f(p,T)$ is a certain function. Then

$$\lambda_{\mathrm{D}} = L_{\mathrm{DT}} k_{\mathrm{B}} T^2 \frac{1}{c} \tag{1.9}$$

and

$$\frac{L_{\mathrm{DT}}}{\lambda_{\mathrm{D}}} = \frac{c}{k_{\mathrm{B}}T^2} \tag{1.10}$$

or

$$\frac{L_{\mathrm{DT}}}{\lambda_{\mathrm{D}}} = \frac{c}{(k_{\mathrm{B}}T)^2} k_{\mathrm{B}} \tag{1.10a}$$

Similar relation appears below, cf. Eq. (4.5).

## 1.3 Elementary theory of thermodiffusion

The different transport phenomena, such as transport of mass, momentum and energy (this means the diffusion, viscosity and heat conduction) are mutually related. The elementary kinetic theory of gases (the free path length theory) gives the following relations for the diffusion, thermal conductivity and viscosity coefficients (see [14, 20, 26])

$$D = \frac{\ell u}{3} \qquad \lambda_{\mathrm{T}} = s k_{\mathrm{B}} n D \qquad \text{and} \qquad \eta = \rho D \tag{1.11}$$

respectively. Here $\ell$ denotes the free path length, $u$ is a mean thermal velocity of gas particles, $s$ is the number of degree of freedom of the diffunding particle, $n$ represents the mean number of particles *per* unit volume (the number density), and $\rho$ is the mass density

$$\rho = nm$$

where $m$ is a mass of the diffunding particle.

This theory is too simple in some sense. If in a binary mixture the particles of type A (or B) are rare, the diffusion coefficients are $D^{\mathrm{A}} = \ell^{\mathrm{A}} u^{\mathrm{A}}/3$ (or $D^{\mathrm{B}} = \ell^{\mathrm{B}} u^{\mathrm{B}}/3$). One finds that

$$\frac{D^{\mathrm{A}}}{D^{\mathrm{B}}} = \frac{m^{\mathrm{B}}}{m^{\mathrm{A}}}$$

It follows from this formula that the diffusion of a small admixture of hydrogen $H_2$ through carbon dioxide $CO_2$ should be 22 times as large as its value for the case of small admixture of $CO_2$ diffusing through $H_2$. This prediction is contradicted by experiment in which the total change of the diffusion coefficient with concentration, in this rather extreme, case is less then 15 percent [20, 25].

The trial to give a theory of thermodiffusion in terms of the free path length approach were undertaken by Fürth [24]. His free path theory has of course the same minuses as the theory of the ordinary diffusion. Therefore Furry [25] regards that the proper way of consideration of thermodiffusion is not the mean free path method of Meyer, but of the momentum transfer method of Stefan [23] applied by Frankel [22] to the thermodiffusion in mixture of two gases, of light and heavy particles. Frankel indicated that the dependence of the cross-section $\sigma$ on the velocity $v$ of light particles can be found by simple dimensional argument. If the particles interact by an inverse $k$-th power repulsion, $F = -K/r^k$, where $K$ and $k$ are constants and $r$ is distance between particles and $\mu$ is the reduced mass $\mu$ of particles, then from the dimensions of this quantities the following proportionalities are inferred

$$\sigma \propto \left(\frac{K}{\mu v^2}\right)^{\frac{2}{k-1}} \propto v^{\frac{-4}{k-1}} \qquad \text{and} \qquad \sigma v \propto v^{\frac{k-5}{k-1}}$$

This means that the momentum transfer of light particles to heavy ones which is proportional to $\sigma v$ varies as $v^{(k-5)/(k-1)}$. Since $v$ is greater for particles coming from the hot side that from those in the reverse direction, there is observed an excess of light component at the hot side for particles harder then Maxwellian particles ($k = 5$) or at the cold side for $k < 5$, and no thermal diffusion for Maxwellian particles $k = 5$.

However, the mean free path method may reflect some properties of thermodiffusion, cf. Appendix A.

### 1.4 Brownian motion

The phenomenon of chaotic motion of a suspension particle anticipated by the ancient philosophers (Democritus, Lucretius) was observed apparently for the first time in modern times by Ingenhousz who was looking at powdered charcoal on an alcohol surface. Also, in 1826–28 Brown published his investigations on movement of fine particles in various organic and inorganic suspensions [30]. The Brownian motion in gases was observed in an experiment by Bodaszewski [31, 32] in 1881 who used different types of smoke containing the Brownian particles of order of $10^{-4}$mm in diameter.

Today Brownian motion means a random motion of heavy particle of suspension (heavy in comparison with weight of particles of surrounding medium). We have the relation

$$D = \frac{\overline{x^2}}{2t} \tag{1.12}$$

where $x$ is displacement of the particle at time $t$ and $\overline{\cdots}$ means an average taken over particle paths. The Brownian diffusion reveals a thermal motion of liquid molecules not on macroscopic (as the temperature) but rather on mezoscopic scale (observed in a microscope). Instead of the third relation in (1.11) we have for the Brownian particle of radius $a$

$$D = \frac{k_{\mathrm{B}}T}{6\pi\eta a} \tag{1.13}$$

The two last equalities are known as the Einstein relations for a Brownian motion, 1905 [33].

The third Einstein relation for a Brownian motion links the force $\mathbf{F}$ acting on a Brownian particle with its velocity $\mathbf{v}$

$$\mathbf{F} = \beta\mathbf{v} \tag{1.14}$$

where $\beta$ is known as the friction constant [34]. In Einstein's approximation

$$\beta = 6\pi\eta a = \frac{k_{\mathrm{B}}T}{D}$$

The presence of a constant gravitational field causes an additional motion with constant velocity $v$ superposed on the random motion. Then, as was shown by Smoluchowski [35, 36]

$$\overline{x^2} = 2Dt + (vt)^2 \tag{1.15}$$

For sufficiently short times the second term is very small compared to the first one and the mean square displacement satisfies (1.12). Smoluchowski indicated also that under the influence of external force $\mathcal{F}(\mathbf{x})$ the Brownian particles acquire velocity $\mathcal{F}/(k_{\mathrm{B}}T)$. Therefore the stream of particles which flow accross a unit surface area situated at point $\mathbf{x}$ is the result of superposition of this motion and the stream of diffusion

$$\mathbf{j} = -D\nabla c + \frac{\mathcal{F}}{k_{\mathrm{B}}T} \tag{1.16}$$

and under these condition the diffusion equation has the form

$$\frac{\partial c}{\partial t} = \mathrm{div}\left( D\nabla c - \frac{\mathcal{F}}{k_{\mathrm{B}}T} \right) \tag{1.17}$$

which is known as the diffusion equation with a drift.

Einstein's formula (1.13) containing a macroscopical constant $\eta$ – the viscosity of the surrounding medium, needs to be justified on a microscopic level.

Note that the first research on a one-dimensional Brownian diffusion in which the heavy spherical particle interacts with the light molecules of gas was made by Lord Rayleigh in 1891 [37]. The problem was investigated by a simplified method by Smoluchowski [38,39]. Despite of approximations Smoluchowski's formula for friction constant was correct in what concerns its dependence on the temperature of gas, the mass of the Brownian particle and the mass and concentration of molecules but has inproper numerical constant. The correct formula was obtained by Lorentz by modification of the Rayleigh method [40,41].

The arise of modern kinetic theory, begining from the methods of BBGKY hierarchy (Bogoliubov-Born-(H.S.) Green-Kirkwood-Yvon) [42] has lead to evaluation of the friction constant $\beta$. According to M.S. Green[41]

$$\beta = \frac{1}{3Mk_{\mathrm{B}}T}\int_{-t}^{0} \langle \mathbf{F}(0)\mathbf{F}(\tau)\rangle d\tau$$

where $M$ is the mass of the Brownian particle, $\mathbf{F}(0)$ is the force on the Brownian particle at particular time chosen as 0, $\mathbf{F}(\tau)$ is the force on the particle at time $\tau$ later, $-t$ is sufficiently long before the collision starts ($t$ is more than the time of collision but small compared to the times between collisions.) The force on the Brownian particle is given by

$$\mathbf{F}(\tau) = \sum_{i}^{N} \mathbf{F}_i(\tau) \qquad i = 1, \cdots, N$$

where $\mathbf{F}_i(\tau)$ is the force on Brownian particle due to molecule number $i$ at time $\tau$. The symbol $\langle \cdots \rangle$ denotes averaging over all the particular initial phases of the surrounding molecules which are assumed to be subdued to the equilibrium distribution.

In the frame of this theory the response of the system to an external small stimulus is a linear functional of the external forces. The transport coefficients are particular examples of the response functions.

A general kinetic equation for the time evolution of a marked particle in a fluid has been derived in several ways. The derivations lead to equations which are non-local in time according to the discussion given by Kubo [43]. Equation of this form reduces to the Fokker-Planck type equation as a particular case cf. Stecki and Narbutowicz [44] and Narbutowicz [45]. Their calculations were performed under assumption of the linear response theory of Kubo [42, 43, 46]. In this approximation the marked particle moves between the equilibrized particles of medium, the theory is isothermal and thermal diffusion phenomenon is omitted.

### 1.5 Brownian motion and nonlinear heat equation

It was pointed out by Streater [1, 47, 48] that the friction work performed by Brownian particles falling in the gravity field is converted to an irreversible heat source energy, so the diffusion and thermal processes are coupled. Therefore, the motion of Brownian particle in an external force field when thermal effects are taken into account can be influenced by the two processes: (*i*) a conversion of the potential energy of the particle into heat, and (*ii*) a thermodiffusion process. In the process (*i*) the heat source appears which introduces a nonlinearity into the heat equation.

Also, Streater [49] performed an analysis of Soret-Dufour effects for a special statistical model of gas of molecules with repulsive cores. In this model only one phenomenological (macroscopic) coefficient occurs, namely the diffusion coefficient which has interpretation of the microscopic hopping rate, and the heat conductivity and cross-coefficients are expressed by the diffusion coefficient, density and a maximum density of the diffusing Brownian particles.

In the present paper a Brownian motion in which both the thermodiffusion effect and an external potential force exist, is studied, cf. [50–53]. We do not enter into a micromechanics of the thermodiffusion and only phenomenological coefficients, namely those of diffusion, thermodiffusion and heat conductivity are taken into account.

Below the temperature $T$ is measured in the energetic scale. Its relation to the temperature $T^{\mathrm{Kelv}}$ given in the degrees of Kelvin and used above is

$$T = k_{\mathrm{B}} T^{\mathrm{Kelv}} \tag{1.18}$$

where $k_{\mathrm{B}} = 1.38 \cdot 10^{-23} \frac{J}{K}$.

## 2 The First Law for Brownian particle

Let $\Omega$ be a bounded, sufficiently regular domain with boundary $\partial\Omega$, cf. Fig. 1. Let the function $f = f(\mathbf{x}, t)$ represents the probability density of finding a tagged Brownian particle at $\mathbf{x}$ at time $t$. Consider the motion of the Brownian particle in a fluid at temperature $T = T(\mathbf{x})$, under the influence of a conservative field of force $\mathbf{F} = -\nabla V$, where $V$ denotes a potential.

Let $\mathbf{j}$ and $\mathbf{q}$ denote the flux of Brownian particle diffusion and heat flow, respectively, considered in that domain. The corresponding continuity equations read

$$\frac{\partial f}{\partial t} + j_{i,i} = 0 \qquad c_{\mathrm{p}} \frac{\partial T}{\partial t} + q_{i,i} = r \tag{2.1}$$

where $c_{\mathrm{p}} = c_{\mathrm{p}}(T)$ stands for the specific heat at constant pressure and $r$ denotes a heat source. Moreover, $j_i$ denotes the $i$th component of the vector $\mathbf{j}$ and the comma denotes differentiation with respect to the spatial variable; thus $j_{i,i} \equiv \mathrm{div}\mathbf{j}$, $(i = 1, 2, 3)$. Hence, alternatively

$$\frac{\partial f}{\partial t} + \mathrm{div}\mathbf{j} = 0 \qquad c_{\mathrm{p}} \frac{\partial T}{\partial t} + \mathrm{div}\mathbf{q} = r \tag{2.1a}$$

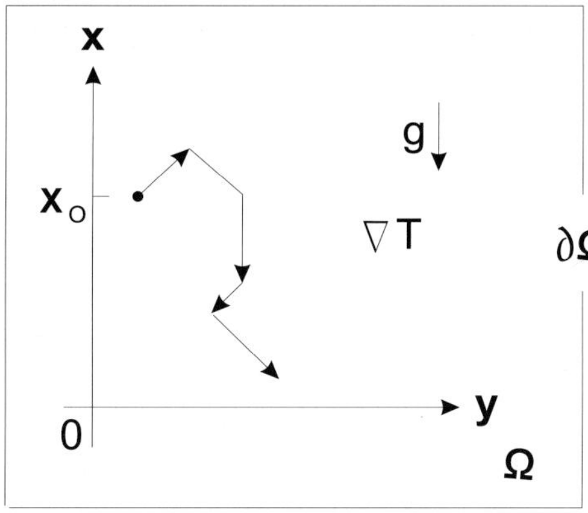


**Fig. 1.** Motion of a Brownian particle in domain $\Omega$ subject to a temperature gradient and a gravitational field.

The term $r$ represents the rate at which the Brownian particle does work converted subsequently into heat

$$r = \mathbf{j} \cdot \mathbf{F} \qquad \text{or} \qquad r = -\mathbf{j} \cdot \nabla V \tag{2.2}$$

We make the following boundary assumptions

$$\mathbf{j} \cdot \mathbf{n} = j_\Omega \qquad \text{and} \qquad \mathbf{q} \cdot \mathbf{n} = q_\Omega \qquad \text{on } \partial\Omega \tag{2.3}$$

where $\mathbf{n}$ denotes the exterior normal to the boundary surface $\partial\Omega$ while $j_\Omega$ and $q_\Omega$ are given scalar functions of $\mathbf{x}$ and $t$. Combining equations (2.1) with the second equation of (2.2) we obtain the following continuity relation

$$\frac{\partial e}{\partial t} + \mathrm{div}\mathbf{j}^{\mathrm{E}} = 0 \tag{2.4}$$

where $e$ denotes the energy density and $\mathbf{j}^{\mathrm{E}}$ is the current of total energy

$$e \equiv Vf + \int_{T_0(\mathbf{x})}^{T(\mathbf{x},t)} c_{\mathrm{p}}(\Theta)d\Theta \qquad \text{and} \qquad \mathbf{j}^{\mathrm{E}} \equiv V\mathbf{j} + \mathbf{q} \tag{2.5}$$

where $T_0 = T_0(\mathbf{x})$ denotes a temperature of the initial state. Here the relation of differentiation of an integral depending on parameter was used

$$c_{\mathrm{p}}(T)\frac{\partial T}{\partial t} = \frac{\partial}{\partial t}\int_{T_0(\mathbf{x})}^{T(\mathbf{x},t)} c_{\mathrm{p}}(\Theta)d\Theta$$

According to conditions (2.3)

$$\mathbf{j}^{\mathrm{E}} \cdot \mathbf{n} = Vj_\Omega + q_\Omega \qquad \text{on}\partial\Omega \tag{2.6}$$

The energy of the system is

$$E \equiv \int_\Omega e d\mathbf{x} = \int_\Omega \left[V(\mathbf{x})f(\mathbf{x},t) + \int_{T_0(\mathbf{x})}^{T(\mathbf{x},t)} c_{\mathrm{p}}(\Theta)d\Theta\right] d\mathbf{x} \tag{2.7}$$

and by integration of (2.4) we find

$$\frac{\partial E}{\partial t} = -\int_{\Omega} \mathrm{div}\mathbf{j}^{\mathrm{E}} d\mathbf{x} \tag{2.8}$$

or by virtue of the divergence theorem and boundary condition (2.6) we get

$$\frac{\partial E}{\partial t} = -\int_{\partial\Omega} (V j_{\Omega} + q_{\Omega})\, dS \tag{2.9}$$

If we assume the vanishing boundary conditions

$$\mathbf{j} \cdot \mathbf{n} = 0 \qquad \text{and} \qquad \mathbf{q} \cdot \mathbf{n} = 0 \qquad \text{on } \partial\Omega \tag{2.10}$$

we get

$$\mathbf{j}^{\mathrm{E}} \cdot \mathbf{n} = 0 \qquad \text{on } \partial\Omega \tag{2.11}$$

Hence for the isolated system

$$\frac{\partial E}{\partial t} = 0 \tag{2.12}$$

Thus the first law of thermodynamics holds true.

## 3 The transport relations

Let $D$ and $\lambda_{\mathrm{T}}$ be the diffusion and heat conductivity coefficients, respectively; and let $M$ and $N$ denote the thermal diffusion direct and reciprocal coefficients. Therefore, the following replacement is made in comparison with Eq. (1.6)

$$L_{\mathrm{DT}} \to M \qquad \text{and} \qquad \lambda_{\mathrm{D}} \to N$$

We admit that $D$, $\lambda_{\mathrm{T}}$, $M$ and $N$ are functions of position $\mathbf{x}$, time $t$, temperature $T$ and probability density $f$. The equations for diffusion and heat fluxes are as follows, cf. [33, 34]

$$\begin{aligned} \mathbf{j} &= -D(\nabla f + \frac{f}{T}\nabla V) - M\nabla T \\ \mathbf{q} &= -N(\nabla f + \frac{f}{T}\nabla V) - \lambda_{\mathrm{T}}\nabla T \end{aligned} \tag{3.1}$$

Substituting the expressions (3.1) into the set of continuity equations (2.1) we get

$$\begin{aligned} \frac{\partial f}{\partial t} &= \mathrm{div}\left[D(\nabla f + \frac{f}{T}\nabla V) + M\nabla T\right] \\ c_{\mathrm{P}}\frac{\partial T}{\partial t} &= \mathrm{div}\left[N(\nabla f + \frac{f}{T}\nabla V) + \lambda_{\mathrm{T}}\nabla T\right] + r \end{aligned} \tag{3.2}$$

The first equation of the last set for $M = 0$ becomes the Smoluchowski equation (1.17).

**Einstein's formula**

The mean velocity of the Brownian particle is, by the first equation of (2.1) and the divergence theorem,

$$v_i = \frac{\partial}{\partial t}\int_\Omega x_i f(\mathbf{x},t)d\mathbf{x} = \int_\Omega x_i \frac{\partial f(\mathbf{x},t)}{\partial t}d\mathbf{x} = -\int_\Omega x_i j_{k,k}\, d\mathbf{x}$$

$$= -\int_{\partial\Omega} x_i j_k n_k dA + \int_\Omega j_i d\mathbf{x}.$$

So, similarly as in [1], $\mathbf{v} = \int_\Omega \mathbf{j}d\mathbf{x}$, where the vanishing boundary condition (2.10) for $\mathbf{j}\cdot\mathbf{n}$ is used, or, by the first equation of the set (3.1),

$$\mathbf{v} = -\int_\Omega \left[D(\nabla f + \frac{f}{T}\nabla V) + M\nabla T\right] d\mathbf{x}. \tag{3.3}$$

If $D$, $T$ and $\nabla V$ are constants, and $f = 0 \quad \text{on} \partial\Omega$, the relation (3.3) passes into Einstein's formula (1.14) $\mathbf{v} = \frac{1}{T}D\mathbf{F}$ where the relation $\int_\Omega f(\mathbf{x})d\mathbf{x} = 1$ was used.

## 4 The Second Law of thermodynamics

In analogy with the entropy of ideal solution we have the following expression for the entropy of our model

$$S = \int_\Omega \left[-f(\mathbf{x},t)\ln f(\mathbf{x},t) + \int_{T_0(\mathbf{x})}^{T(\mathbf{x},t)} \frac{c_{\mathrm{p}}(\Theta)}{\Theta}d\Theta\right] d\mathbf{x} \tag{4.1}$$

where $T_0$ denotes a temperature of the initial state. Hence

$$\frac{\partial S}{\partial t} = \int_\Omega \left[-\frac{\partial f}{\partial t}\ln f(\mathbf{x},t) - \frac{\partial f}{\partial t} + c_{\mathrm{p}}\frac{1}{T}\frac{\partial T}{\partial t}\right] d\mathbf{x} \tag{4.2}$$

Here the relation of differentiation of an integral depending on parameter was exploited

$$\frac{\partial}{\partial t}\int_{T_0(\mathbf{x})}^{T(\mathbf{x},t)} \frac{c_{\mathrm{p}}(\Theta)}{\Theta}d\Theta = \frac{c_{\mathrm{p}}(T)}{T}\frac{\partial T}{\partial t}$$

We substitute the continuity relations (2.1) into (4.2) and use the divergence theorem to obtain

$$\frac{\partial S}{\partial t} = \int_{\partial\Omega} \left(j_i \ln f + j_i - \frac{1}{T}\, q_i\right) n_i dS$$
$$+ \int_\Omega \left[-j_i \frac{1}{f} f_{,i} - \frac{1}{T^2}T_{,i}q_i + \frac{1}{T}r\right] d\mathbf{x} \tag{4.3}$$

If we deal with a closed system, the boundary conditions (2.3) reduce to (2.10), and after substituting $r$ from (2.2) we receive

$$\frac{\partial S}{\partial t} = \int_\Omega \left[-j_i\frac{1}{f}\left(f_{,i} + \frac{f}{T}V_{,i}\right) - q_i\frac{1}{T^2}T_{,i}\right] d\mathbf{x} \tag{4.4}$$

*Remark*

From (4.4) we see that the thermodynamic forces for the diffusion and heat fluxes $\mathbf{j}$ and $\mathbf{q}$ are

$$\frac{1}{f}\left(\nabla f + \frac{f}{T}\nabla V\right) \qquad \text{and} \qquad \frac{1}{T^2}\nabla T \qquad \text{respectively.}$$

Thus the proper form of equations for diffusion and heat fluxes (3.1) is as follows

$$\begin{aligned} \mathbf{j} &= -Df\frac{1}{f}\left(\nabla f + \frac{f}{T}\nabla V\right) - MT^2\frac{1}{T^2}\nabla T \\ \mathbf{q} &= -Nf\frac{1}{f}\left(\nabla f + \frac{f}{T}\nabla V\right) - \lambda_{\mathrm{T}}T^2\frac{1}{T^2}\nabla T \end{aligned} \tag{4.5}$$

and the symmetry of kinetic coefficients gives

$$MT^2 = Nf \tag{4.6}$$

Thus, in this condition not only coefficients $M$ and $N$ are involved but also the unknown fields $f$ and $T$, cf. Eq. (1.10) and Appendices.

Next, from Eq. $(3.1)_1$ we find

$$\nabla f + \frac{f}{T}\nabla V = -\frac{1}{D}\left(\mathbf{j} + M\nabla T\right) \tag{4.7}$$

and after Eq. $(3.1)_2$ we can write

$$\mathbf{q} = \frac{N}{D}\mathbf{j} - \left(\lambda_{\mathrm{T}} - \frac{N}{D}M\right)\nabla T \tag{4.8}$$

Using (4.6) and (4.7) we find

$$\begin{aligned} \mathbf{j}\left(\nabla f + \frac{f}{T}\nabla V\right) &= -\frac{1}{D}\mathbf{jj} - \frac{1}{D}M\mathbf{j}\nabla T \\ \mathbf{q}\nabla T &= \frac{N}{D}\mathbf{j}\nabla T - \left(\lambda_{\mathrm{T}} - \frac{N}{D}M\right)\nabla T\nabla T \end{aligned}$$

Substituting these relations into (4.4) we obtain

$$\frac{\partial S}{\partial t} = \int_{\Omega}\left\{\frac{1}{fD}\mathbf{jj} + \left(\frac{1}{f}\frac{M}{D} - \frac{1}{T^2}\frac{N}{D}\right)\mathbf{j}\nabla T + \left(\lambda_{\mathrm{T}} - \frac{MN}{D}\right)\frac{1}{T^2}\nabla T\nabla T\right\}d\mathbf{x}. \tag{4.9}$$

Therefore, if (4.5) is satisfied and

$$\lambda_{\mathrm{T}} - \frac{N}{D}M \geq 0 \tag{4.10}$$

or the matrix

$$\begin{bmatrix} D & M \\ N & \lambda_{\mathrm{T}} \end{bmatrix}$$

is positive definite, the second law of thermodynamics

$$\frac{\partial S}{\partial t} \geq 0 \tag{4.11}$$

holds true in the system.

## 5 A driven system in a slab

Again, treading in footsteps of [1], we look for a stationary solution far from equilibrium. Consider one-dimensional isotropic model for which a Brownian particle is confined to $0 \leq x \leq 1$.

We impose the condition that the particle current is zero

$$j = 0 \tag{5.1}$$

and that time derivatives vanish. From (2.2) and (5.1), we get

$$r = 0. \tag{5.2}$$

Also we assume that there is a nonzero flow of heat driven by maintaining the ends of the interval $[0, 1]$ at different temperatures, $T_0$ and $T_1$. Eqs. (2.1) for our case read

$$\begin{aligned} D\left(f_{,x} + \frac{f}{T}V_{,x}\right) + MT_{,x} &= 0 \\ \left[N\left(f_{,x} + \frac{f}{T}V_{,x}\right) + \lambda_{\mathrm{T}}T_{,x}\right]_{,x} &= 0. \end{aligned} \tag{5.3}$$

After use of (4.5) in the first equation of the last set we have

$$D\left(f_{,x} + \frac{f}{T}V_{,x}\right) + N\frac{f}{T^2}T_{,x} = 0. \tag{5.4}$$

Hence after integration we find

$$\ln\frac{f}{f_0} = -\int_0^x \left[\frac{1}{T(\xi)}V_{,\xi} + \frac{N}{D}\frac{T_{,\xi}}{T^2}\right] d\xi \tag{5.5}$$

where $f_0$ is a normalizing factor. Using (5.4) we write the second equation of the set (5.3) in the form

$$\left[\left(\lambda_{\mathrm{T}} - \frac{N^2}{D}\frac{f}{T^2}\right)T_{,x}\right]_{,x} = 0 \tag{5.6}$$

After integration we get

$$\left(\lambda_{\mathrm{T}} - \frac{N^2}{D}\frac{f}{T^2}\right)T_{,x} = a \tag{5.7}$$

where $a$ = const and $f$ is given by (5.5). Thus (5.7) is complicated nonlinear integro-differential equation for $T$ which can be solved by approximative methods.

We can write an alternative form of (5.7)

$$T(x) = T_0 + a \int_0^x \frac{d\xi}{\lambda_{\mathrm{T}} - \frac{N^2}{D}\frac{f}{T^2}} \tag{5.8}$$

For $N \to 0$ and $\lambda_{\mathrm{T}}$ = const we have an expression from [1].

We can estimate the influence of the thermodiffusion on distribution $f$ considering the simplified situation in which $D$, $N$ and $\lambda_{\mathrm{T}}$ are constants. As an example of $V$ we take the potential of a homogeneous gravity field in the form

$$V = gx \tag{5.9}$$

where $g$ = constant. Then Eq.(5.5) can be rewritten as

$$f = f_0 \exp\left[\frac{N}{D}\left(\frac{1}{T} - \frac{1}{T_0}\right) - g\int_0^x \frac{1}{T(\xi)}d\xi\right] \tag{5.10}$$

In the first approximation when the second term in the denominator under the integral in (5.8) is neglected, the temperature is a linear function of $x$

$$T(x) = T_0 + (T_1 - T_0)x \tag{5.11}$$

Substituting this expression into (5.5) we get the first approximation of $f$

$$f = f_0 \left(\frac{T}{T_0}\right)^{[-g/(T_1-T_0)]} \exp\left[\frac{N}{D}\left(\frac{1}{T} - \frac{1}{T_0}\right)\right] \tag{5.12}$$

Hence the distribution $f$ is no more Gibbsian. The examples of the function $f$ given by (5.12) for different parameters $g$ and $N/D$ are given in Fig. 2. The plots for $N/D = 1$ are enclosed to show that even for such high unrealistic value the qualitative dependence of the distribution function on temperature is preserved.

We can, however, pursue how this essential change of character of the distribution is accomplished. Let us assume for a moment (until the end of this section) that $N = 0$. Then (5.10) takes the quasi-Gibbsian form

$$f = f_0 \ \exp\left[-g\int_0^x \frac{1}{T(\xi)}d\xi\right] \tag{5.13}$$

and for $T$ given by (5.10) we get

$$f = f_0 \exp\left[-\frac{g}{T_1 - T_0}\ln\left(1 - \frac{T_1 - T_0}{T_0}x\right)\right] \tag{5.14}$$

and in approximation in which

$$\ln\left(1 - \frac{T_1 - T_0}{T_0}x\right) \approx \frac{T_1 - T_0}{T_0}x \tag{5.15}$$

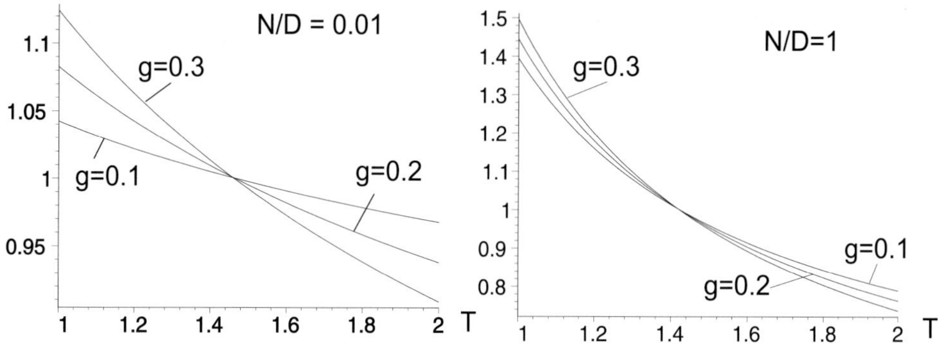


**Fig. 2.** The non-Gibbsian distribution function *versus* $T$ ($T_0 = 1, T_1 = 2$) for different values $g = 0.1, 0.2, 0.3$ and for two values of $N/D$.

we get the barometric formula being a special realization of Gibbsian distribution

$$f = f_0 \mathrm{e}^{-gx/T_0} \tag{5.16}$$

Thus the Gibbsian distribution rules not only for $T = T_0 =$ constant, as is seen immediately from (5.13), but also for temperatures slightly declining from the constant, when approximation (5.15) is possible.

## 6 System in a slab with given temperature

We consider a one-dimensional isotropic model for which a Brownian particle at $t = 0$ is at $x = x_0$, and the particle current $j = 0$ at the bottom of the slab $x = 0$ for $t \geq 0$, see Fig. 1. We assume that the temperature distribution is known $T = T(x)$.

Evolution of the distribution function is described by the equation $(2.3)_1$ which for (1-dim, isotropic) case reads

$$\frac{\partial f}{\partial t} = \left[ D \left( f_{,x} + \frac{f}{T}\, V_{,x} \right) + M T_{,x} \right]_{,x} \tag{6.1}$$

or after use of (4.5) we have

$$\frac{\partial f}{\partial t} = \left[ D \left( f_{,x} + \frac{f}{T}\, \mathcal{V}_{,x} \right) \right]_{,x} \tag{6.2}$$

where

$$\mathcal{V}_{,x} = V_{,x} + \frac{N}{D} \frac{T_{,x}}{T} \tag{6.3}$$

is a given function. If $N$ and $D$ are constants, we have

$$\mathcal{V} = V + \frac{N}{D} \ln \frac{T}{T_0} \tag{6.4}$$

which means that a thermodiffusion leads to a modification of the external potential $V$.

Solution of equations of type (6.2) was discussed by Smoluchowski [35, 36], cf. also Cherkasov [54] and Ricciardi [55].

Eq. (6.2) can be used as a mathematical model for the considered system. We take into account (5.9) as a potential $V$ and (5.10) as a temperature distribution $T(x)$. Then $T_{,x} = T_1 - T_0$ and we make the approximation

$$\frac{T_{,x}}{T} \approx 2\frac{T_1 - T_0}{T_1 + T_0}$$

From (6.3) we have

$$D\mathcal{V}_{,x} = Dg + 2\frac{T_1 - T_0}{T_1 + T_0}N \tag{6.5}$$

In the equation

$$\frac{\partial f}{\partial t} = D\frac{\partial^2 f}{\partial x^2} + D\mathcal{V}_{,x}\frac{\partial f}{\partial x} \tag{6.6}$$

we change the scale of length and time as follows

$$\xi = \frac{x}{x_0} \qquad \tau = \frac{D}{x_0^2}t \tag{6.7}$$

and obtain

$$\frac{\partial f}{\partial \tau} = \frac{\partial^2 f}{\partial \xi^2} + c\frac{\partial f}{\partial \xi} \tag{6.8}$$

where

$$c \equiv \frac{\partial \mathcal{V}}{\partial \xi} = \text{const} \tag{6.9}$$

A solution of Eq. (6.8) for $(\xi, \tau) \in (0, \infty) \times (0, \infty)$ subject to the conditions

$$\begin{aligned} f(\xi, 0) &= \delta(\xi - 1) && \xi \in (0, \infty) \\ \frac{\partial f}{\partial \xi}(0, \tau) + cf(0, \tau) &= 0 && \tau > 0 \\ f(\xi, \tau) &\to 0 && \xi \to \infty, \tau > 0 \end{aligned} \tag{6.10}$$

takes the form

$$\begin{aligned} f(\xi, \tau) = & \frac{1}{2\sqrt{\pi\tau}}\left[\mathrm{e}^{\frac{(\xi-1)^2}{4\tau}} - \mathrm{e}^{-\frac{(\xi+1)^2}{4\tau}}\right]\mathrm{e}^{-c\frac{\xi-1}{2} - c^2\frac{\tau}{4}} \\ & + \frac{c}{\sqrt{\pi}}\mathrm{e}^{-c\xi}\int_{\frac{\xi+1-c\tau}{2\sqrt{\tau}}}^{\infty} \mathrm{e}^{-z^2}dz \end{aligned} \tag{6.11}$$

In Fig. 3 the function $f(\xi, \tau)$ and the corresponding heat source $r = r(\xi, \tau)$ as functions of $\xi$ for a given moment of time are plotted.

An interpretation of the heat source function $r = r(\xi)$ shown in Fig.3 is the following. For $\xi < 1.1$, $(x < 1.1x_0)$ a Brownian particle that falls in the gravity

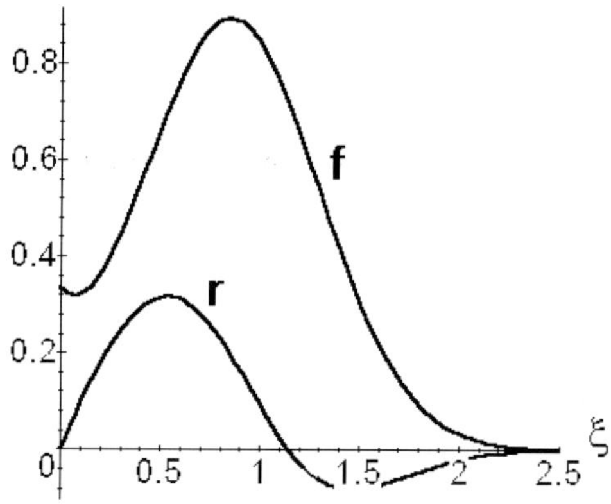


**Fig. 3.** The distribution $f$ and the heat source $r$ as function of $\xi$ for $\tau = 1.5$ and $c = 1.5$. The quantity $r$ is calculated as $r = jc = (cf + df/d\xi)c$ and in the figure the quantity $r/10$ is displayed. The quantities $\breve{x} \equiv x_0$ and $\breve{t} \equiv x_0^2/D$ are the new units of length and time, cf. Eq. (6.7)

field produces a positive amount of heat ($r > 0$). For For $1.1 < \xi < \infty$ the particle goes up due to absorption of an environmental heat, which corresponds to $r < 0$ over that interval.

In Figs. 4, 5 and 6 the solution $f(\xi, \tau)$ as a function of $\xi$ for a set of values of $c$ and for fixed values of time $\tau$ is presented.

It follows from these figures that if

$$g + 2\frac{T_1 - T_0}{T_1 + T_0}\frac{N}{D}$$

changes sign from positive to negative ($T_0 \gg T_1$) then $c$ becomes negative, cf. Eqs. (6.5) and (6.9), and the Brown particles move towards a colder part of the region, *i.e.* most of them do not fall down in the gravitation field. This means that in case $c < 0$ the thermodiffusive forces are greater than the gravitational ones.

## 7 Thermal diffusion in porous medium

Now we shall consider the thermodiffusion of a Brownian particle in liquid filling a periodic porous medium (a skeleton). The skeleton is made of an impermeable rigid but heat conducting material. The domain occupied by the system liquid–skeleton is denoted by $\Omega$. The subdomains occupied by the liquid and skeleton are denoted by $\Omega_{\mathrm{L}}^{\varepsilon}$ and $\Omega_{\mathrm{S}}^{\varepsilon}$ respectively, *i.e.* $\Omega_{\mathrm{L}}^{\varepsilon} \cup \Omega_{\mathrm{S}}^{\varepsilon} = \Omega$. Thus, we assume that the porous medium exhibits a microperiodic structure characterized by elementary $\varepsilon Y$-cells, cf. Fig. 7.

Here $Y$ is an open cube of $R^3$, the so called elementary or basic cell, and $\varepsilon > 0$ is a small parameter

$$\varepsilon = \frac{l}{\mathcal{L}}$$

where $l$ and $\mathcal{L}$ are typical length scales associated with dimensions of pores and of the region $\Omega$, respectively. Let $Y_{\mathrm{L}}$ represents the part of $Y$ occupied by the liquid and $Y_{\mathrm{S}}$ the part occupied by the solid: $Y_{\mathrm{L}} \cup Y_{\mathrm{S}} = Y$.

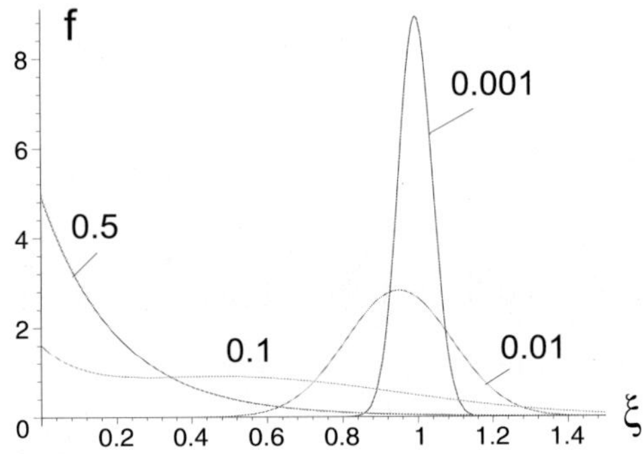


**Fig. 4.** The distribution $f$ *versus* $\xi$ for $c = 5$ and times $\tau = 0.001, 0.01, 0.1, 0.5$.

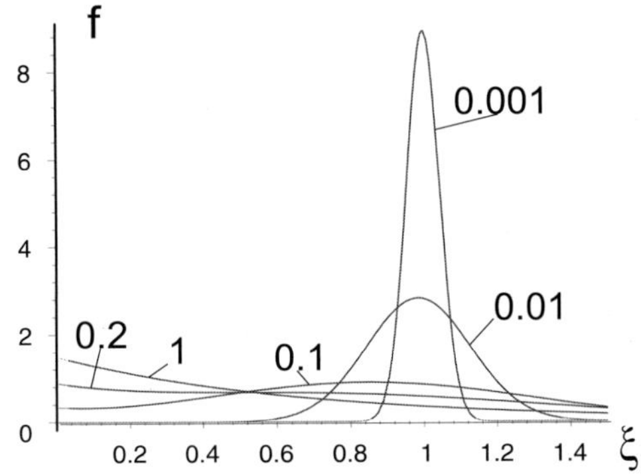


**Fig. 5.** The distribution $f$ *versus* $\xi$ for $c = 1.5$ and times $\tau = 0.001, 0.01, 0.1, 0.2, 1$.

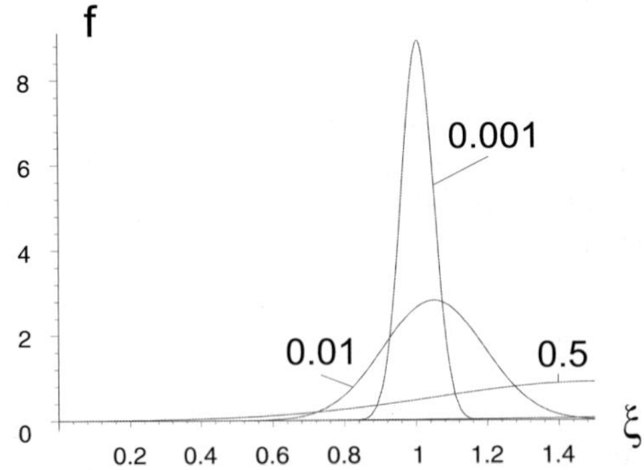


**Fig. 6.** The distribution $f$ *versus* $\xi$ for $c = -5$ and times $\tau = 0.001, 0.01, 0.1, 0.5$.

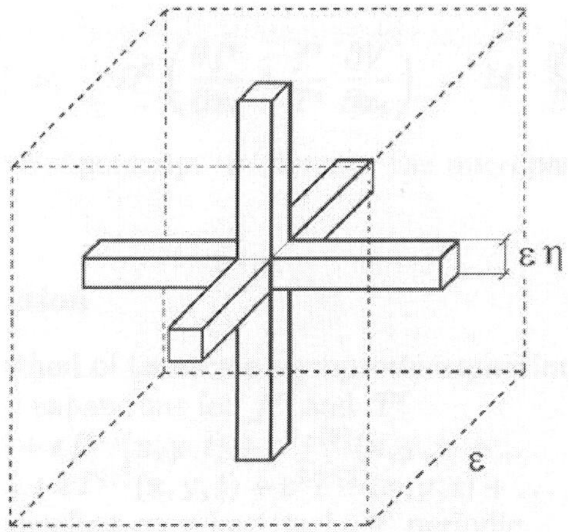


**Fig. 7.** Example of a skeleton of porous medium. Parameter $\varepsilon > 0$ indicates smallness of the basic cell and parameter $\eta$ indicates smallness of the obstacle, courtesy of Dr. A. Gałka

The two spatial variables appear in the method: $\mathbf{x}$ and $\mathbf{y} = \mathbf{x}/\varepsilon$. The variable $\mathbf{x} \in \Omega$ is called the macroscopic (or slow) variable while $\mathbf{y} \in Y$ is called the microscopic (or quick) one.

For a function $f$ of $\mathbf{y} \in Y$ we define the average

$$\langle f \rangle = \frac{1}{|Y|} \int f(\mathbf{y}) d\mathbf{y}$$

The material coefficients of the skeleton are denoted as $c_{\mathrm{p}}^{\mathrm{S}}$, $D^{\mathrm{S}}$, $M^{\mathrm{S}}$, $N^{\mathrm{S}}$ and $\lambda_{\mathrm{T}}^{\mathrm{S}}$, while those of the liquid by $c_{\mathrm{p}}^{\mathrm{L}}$, $D^{\mathrm{L}}$, $M^{\mathrm{L}}$, $N^{\mathrm{L}}$ and $\lambda_{\mathrm{T}}^{\mathrm{L}}$.

In what follows we consider the case when the spatial dependence of potential is mild, $V = V(\mathbf{x})$ and does not depend on $\mathbf{y}$ what means that it changes at the distances great in comparison with the cell dimensions.

## 7.1 Basic system of equations

For a fixed $\varepsilon > 0$ and we formulate the boundary value problem:

$$\begin{aligned} \dot{f}^\varepsilon &= \frac{\partial}{\partial x_i} \left[ D^\varepsilon \left( \frac{\partial f^\varepsilon}{\partial x_i} + \frac{f^\varepsilon}{T^\varepsilon} \frac{\partial V}{\partial x_i} \right) + M^\varepsilon \frac{\partial T^\varepsilon}{\partial x_i} \right] \\ c_{\mathrm{p}}^\varepsilon \dot{T}^\varepsilon &= \frac{\partial}{\partial x_i} \left[ N^\varepsilon \left( \frac{\partial f^\varepsilon}{\partial x_i} + \frac{f^\varepsilon}{T^\varepsilon} \frac{\partial V}{\partial x_i} \right) + \lambda_{\mathrm{T}}^\varepsilon \frac{\partial T^\varepsilon}{\partial x_j} \right] + r^\varepsilon \end{aligned} \tag{7.1}$$

where

$$r^\varepsilon = -j_i^\varepsilon \frac{\partial V}{\partial x_i} \tag{7.2}$$

and

$$j_i^\varepsilon = -\, D^\varepsilon \left( \frac{\partial f^\varepsilon}{\partial x_i} + \frac{f^\varepsilon}{T^\varepsilon} \frac{\partial V}{\partial x_i} \right) \; - \; M^\varepsilon \frac{\partial T^\varepsilon}{\partial x_i} \tag{7.3}$$

Here $\dot{f} = \partial f/\partial t$ and superscript $\varepsilon$ denotes the microperiodic behaviour of a variable.

## 7.2 Homogenisation

According to the method of two-scale asymptotic expansions, cf. *e.g.* [56–58], we assume the following expansions for $f^\varepsilon$ and $T^\varepsilon$

$$
\begin{aligned}
f^\varepsilon &= f^{(0)}(\mathbf{x},\mathbf{y},t) + \varepsilon f^{(1)}(\mathbf{x},\mathbf{y},t) + \varepsilon^2 f^{(2)}(\mathbf{x},\mathbf{y},t) + \ldots \\
T^\varepsilon &= T^{(0)}(\mathbf{x},\mathbf{y},t) + \varepsilon T^{(1)}(\mathbf{x},\mathbf{y},t) + \varepsilon^2 T^{(2)}(\mathbf{x},\mathbf{y},t) + \ldots
\end{aligned}
$$

All the functions depending on $\mathbf{y}$ have to be $Y$-periodic.

Replacing differentiation $\partial/\partial x_i$ by $(\partial/\partial x_i + \varepsilon^{-1}\partial/\partial y_i)$ and comparing terms associated with the same power of $\varepsilon$ we find by analyzing terms at $\varepsilon^{(-2)}$ that

$$
f^{(0)}(\mathbf{x},\mathbf{y},t) = f^{(0)}(\mathbf{x},t) \qquad \text{and} \qquad T^{(0)}(\mathbf{x},\mathbf{y},t) = T^{(0)}(\mathbf{x},t)
$$

Next, one can show that terms at $\varepsilon^{(-1)}$ are

$$
\begin{aligned}
\frac{\partial}{\partial y_i}\left[D\left(\frac{\partial f^{(0)}}{\partial x_i} + \frac{\partial f^{(1)}}{\partial y_i}\right) + M\left(\frac{\partial T^{(0)}}{\partial x_i} + \frac{\partial T^{(1)}}{\partial y_i}\right)\right] &= 0 \\
\frac{\partial}{\partial y_i}\left[N\left(\frac{\partial f^{(0)}}{\partial x_i} + \frac{\partial f^{(1)}}{\partial y_i}\right) + \lambda_{\mathrm{T}}\left(\frac{\partial T^{(0)}}{\partial x_i} + \frac{\partial T^{(1)}}{\partial y_i}\right)\right] &= 0
\end{aligned}
$$

These two relations can be satisfied if

$$
f^{(1)}(\mathbf{x},\mathbf{y},t) = \varphi_m(\mathbf{y},t)\frac{\partial f^{(0)}}{\partial x_m} + \psi_m(\mathbf{y},t)\frac{\partial T^{(0)}}{\partial x_m}
$$

and

$$
T^{(1)}(\mathbf{x},\mathbf{y},t) = \chi_m(\mathbf{y},t)\frac{\partial f^{(0)}}{\partial x_m} + \vartheta_m(\mathbf{y},t)\frac{\partial T^{(0)}}{\partial x_m}
$$

where the following local problems for the functions $\varphi_m, \psi_m, \chi_m$ and $\vartheta_m$ appear in $Y$

$$
\begin{aligned}
\frac{\partial}{\partial y_i}\left[D\left(\delta_{im} + \frac{\partial \varphi_m}{\partial y_i}\right) + M\frac{\partial \chi_m}{\partial y_i}\right] &= 0 \\
\frac{\partial}{\partial y_i}\left[D\frac{\partial \psi_m}{\partial y_i} + M\left(\delta_{im} + \frac{\partial \vartheta_m}{\partial y_i}\right)\right] &= 0 \\
\frac{\partial}{\partial y_i}\left[N\left(\delta_{im} + \frac{\partial \varphi_m}{\partial y_i}\right) + \lambda_{\mathrm{T}}\frac{\partial \chi_m}{\partial y_i}\right] &= 0
\end{aligned}
$$

and

$$
\frac{\partial}{\partial y_i}\left[N\frac{\partial \psi_m}{\partial y_i} + \lambda_{\mathrm{T}}\left(\delta_{im} + \frac{\partial \vartheta_m}{\partial y_i}\right)\right] = 0
$$

## 7.3 Macroscopic equations of thermodiffusion in porous medium

By analyzing terms at $\varepsilon^0$ and averaging we arrive at the homogenised set of equations

$$\dot{f}^{(0)} = \frac{\partial}{\partial x_i}\left(D^{\mathrm{H}}_{im}\frac{\partial f^{(0)}}{\partial x_m} + \langle D\rangle \frac{f^{(0)}}{T^{(0)}}\frac{\partial V}{\partial x_i}\right) + M^{\mathrm{H}}_{im}\frac{\partial^2 T^{(0)}}{\partial x_i \partial x_m}$$

$$\langle c_{\mathrm{p}}\rangle \dot{T}^{(0)} = \frac{\partial}{\partial x_i}\left(N^{\mathrm{H}}_{im}\frac{\partial f^{(0)}}{\partial x_m} + \langle D\rangle \frac{f^{(0)}}{T^{(0)}}\frac{\partial V}{\partial x_i}\right) + \lambda^{\mathrm{H}}_{\mathrm{T}im}\frac{\partial^2 T^{(0)}}{\partial x_j \partial x_m} + \langle r^{(0)}\rangle \qquad (7.4)$$

where

$$D^{\mathrm{H}}_{im} = \langle D\left(\delta_{im} + \frac{\partial \varphi_m}{\partial y_i}\right) + M\frac{\partial \chi_m}{\partial y_i}\rangle$$

$$M^{\mathrm{H}}_{im} = \langle D\frac{\partial \psi_m}{\partial y_i} + M\left(\delta_{im} + \frac{\partial \vartheta_m}{\partial y_i}\right)\rangle$$

$$N^{\mathrm{H}}_{im} = \langle D\left(\delta_{im} + \frac{\partial \varphi_m}{\partial y_i}\right) + M\frac{\partial \chi_m}{\partial y_i}\rangle$$

$$\lambda^{\mathrm{H}}_{\mathrm{T}im} = \langle D\frac{\partial \psi_m}{\partial y_i} + M\left(\delta_{im} + \frac{\partial \vartheta_m}{\partial y_i}\right)\rangle \qquad (7.5)$$

and

$$\langle r^{(0)}\rangle = -\langle j_i^{(0)}\rangle \frac{\partial V}{\partial x_j} \qquad (7.6)$$

where

$$\langle j_i^{(0)}\rangle = -D^{\mathrm{H}}_{im}\frac{\partial f^{(0)}}{\partial x_m} - \langle D\rangle \frac{f^{(0)}}{T^{(0)}}\frac{\partial V}{\partial x_j} - M^{\mathrm{H}}_{im}\frac{\partial T^{(0)}}{\partial x_m} \qquad (7.7)$$

These equations describe the macroscopic scale diffusion of Brownian particles in a porous medium subject to thermal field and smooth external forces. Clearly, the drift term is present in the macroscopic particle flux (7.7). We observe that the homogenized coefficients $D^{\mathrm{H}}_{ij}$, $M^{\mathrm{H}}_{ij} N^{\mathrm{H}}_{ij}$ and $\lambda^{\mathrm{H}}_{\mathrm{T}ij}$ may acquire anisotropic character.

The energy balance is satisfied by virtue of the following equations

$$\frac{\partial E}{\partial t} = -\int_\Omega V(\mathbf{x})\langle j_i^{(0)}\rangle_{,i} - \int_\Omega \left(\langle q_i^{(0)}\rangle_{,i} - \langle r^{(0)}\rangle\right) d\mathbf{x}$$

or

$$\frac{\partial E}{\partial t} = \int_\Omega \Big\{ -\frac{\partial}{\partial x_i}\left[V(\mathbf{x})\langle j_i^{(0)}\rangle\right] + \frac{\partial V(\mathbf{x})}{\partial x_i}\langle j_i^{(0)}\rangle - \frac{\partial \langle q_i^{(0)}\rangle}{\partial x_i} + \langle r^{(0)}\rangle\Big) d\mathbf{x} = 0$$

if the fluxes $\langle j_i^{(0)}\rangle$ and $\langle q_i^{(0)}\rangle$ vanish over $\partial\Omega$.

To obtain the numerical values of the homogenized tensors for the skeleton of thin obstacles, ($\eta \to 0$, cf. Fig. 7), one can use a method due to Cioranescu and Saint Jean Paulin [59].

If the pore configuration is not arranged into a periodic lattice but has a random distribution then the effective coefficients are scalars. The method of averaging of (7.5) in this case was given by Burridge and Keller [60].

## 8 Prigogine conjecture

After substitution of (3.1) into (4.4) we get

$$\frac{\partial S}{\partial t} = \int_{\Omega} \left\{ \left[ D(f_{,i} + \frac{f}{T} V_{,i}) + M T_{,i} \right] \frac{1}{f} \left[ f_{,i} + \frac{f}{T}\, V_{,i} \right] \right.$$
$$\left. + \left[ N(f_{,i} + \frac{f}{T} V_{,i}) + \lambda_{\mathrm{T}} T_{,i} \right] \frac{1}{T^2} T_{,i} \right\} d\mathbf{x}$$

For a special case $D = M = N = 0$ we get

$$\frac{\partial S}{\partial t} = \int_{\Omega} \frac{1}{T^2} \lambda_{\mathrm{T}} T_{,i} T_{,i} d\mathbf{x} \tag{8.1}$$

A function $T = T(\mathbf{x})$ that minimizes $\dot{S}$ is to be found from the variational equation

$$\left[ \frac{d}{d\varepsilon} \int_{\Omega} \frac{1}{(T + \varepsilon \tilde{T})^2} \lambda_{\mathrm{T}}\ (T_{,i} + \varepsilon \tilde{T}_{,i})(T_{,i} + \varepsilon \tilde{T}_{,i}) d\mathbf{x} \right]_{\varepsilon=0} = 0 \tag{8.2}$$

with the condition $\tilde{T} = 0$ on the boundary $\partial\Omega$. This leads to the nonlinear equation

$$\frac{1}{T^3} \lambda_{\mathrm{T}} \frac{\partial T}{\partial x_i} \frac{\partial T}{\partial x_i} = - \frac{\partial}{\partial x_i} \left( \frac{1}{T^2} \lambda_{\mathrm{T}} \frac{\partial T}{\partial x_i} \right) \tag{8.3}$$

On the another hand from the second equation of (3.2) we have

$$c_{\mathrm{P}} \frac{\partial T}{\partial t} = \frac{\partial}{\partial x_i} \left( \lambda_{\mathrm{T}} \frac{\partial T}{\partial x_i} \right) \tag{8.4}$$

and for $\dot{T} = 0$ we get the quasi-linear equation

$$\frac{\partial}{\partial x_i} \left( \lambda_{\mathrm{T}} \frac{\partial T}{\partial x_i} \right) = 0 \tag{8.5}$$

For 1-dim case, (8.3) and (8.5) imply

$$\frac{1}{T^3} \lambda_{\mathrm{T}} \left( \frac{dT}{dx} \right)^2 = - \frac{d}{dx} \left( \frac{1}{T^2} \lambda_{\mathrm{T}} \frac{dT}{dx} \right) \tag{8.6}$$

and

$$\frac{d}{dx} \left( \lambda_{\mathrm{T}} \frac{dT}{dx} \right) = 0 \tag{8.7}$$

If we adjoin the boundary conditions

$$T(0) = T_0 \qquad \text{and} \qquad T(l) = T_1$$

we obtain the solution of (8.6)

$$T = T_0 \mathrm{e}^{AI(x)} \equiv T_{\mathrm{A}}(x) \tag{8.8}$$

and of (8.7)

$$T = T_0 + BI(x) \equiv T_{\rm B}(x) \tag{8.9}$$

where

$$A = \frac{1}{I(l)} \ln \frac{T_1}{T_0} \qquad B = \frac{T_1 - T_0}{I(l)} \qquad \text{and} \qquad I(x) = \int_0^x \frac{1}{\lambda_{\rm T}(\xi)} d\xi \tag{8.10}$$

The entropy production (8.1) for 1-dim case reads

$$\dot{S} = \int_0^l \frac{1}{T^2} \lambda_{\rm T} \left( \frac{dT}{dx} \right)^2 dx \tag{8.11}$$

and for the cases (8.8) and (8.9) we have respectively

$$\dot{S}_{\rm A} = \frac{1}{I(l)} \ln^2 \frac{T(l)}{T(0)} \qquad \text{and} \qquad \dot{S}_{\rm B} = \left[ \frac{T_1 - T_0}{I(l)} \right]^2 \frac{T_1 - T_0}{T_0 T_1} \tag{8.12}$$

For the special case $\lambda_{\rm T} = \text{const} = 1$, $T(0) = 1$, $T(l) = 2$ and $l = 1$ this solution reduces to the example given by Streater in [1]

$$T_{\rm A} = {\rm e}^{x \ln 2} \qquad T_{\rm B} = 1 + x \qquad \dot{S}_{\rm A} = (\ln 2)^2 \qquad \dot{S}_{\rm B} = \frac{1}{2} \tag{8.13}$$

Therefore $\dot{S}_{\rm A} < \dot{S}_B$ what means that the condition of minimum entropy production differs from that describing a stationary distribution of temperature. The system moves according to the heat equation to a state of higher entropy production. However, if the linearizations in (8.8) are postulated

$$T_{\rm A} \approx T_0[1 + AI(x)] \qquad \text{and}$$

$$A = \frac{1}{I(l)} \ln \frac{T_0 + (T_1 - T_0)}{T_0} \approx \frac{1}{I(l)} \frac{T_1 - T_0}{T_0} \tag{8.14}$$

then $T_{\rm A} \approx T_{\rm B}$ and $\dot{S}_{\rm A} \approx \dot{S}_{\rm B}$ and a Prigogine conjecture [61, 62] that for a driven system the rate of entropy production $\dot{S}$ decreases with time towards a minimum is valid, as was already discussed in [61] and [62]. Comparing this result with that of the end of the section 5, we find that the Prigogine conjecture is valid if the distribution function is Gibbsian.

## Conclusions

1. A consistent thermodynamic description of the motion of Brownian particles subject to both thermodiffusive forces and the gravity forces, is given by following the Einstein-Smoluchowski and Streater approaches to the Brownian particle motion problem, cf [2].
2. In particular, it is shown that the Brownian particles in a gravity field move upward provided the particles absorb the heat from the surrounding, and they fall down when generating a heat in a friction process.

3. A relation (4.5) between the coefficients of direct and reciprocal thermodiffusion is found, and exemplified using the models of the elementary gas theory (Appendix A) and the rarified gas theory (Appendix B).
4. A homogenisation of the thermodiffusion equations for the process in porous medium is carried out.
5. A relationship between the Streater result and the Prigogine conjecture is discussed.

## Appendix A. Onsager Relations in an elementary thermodiffusion theory

Let $n, u, \ell$ and $\varepsilon$ be the number density of gas particles, the mean velocity of gas particles, the length of mean free path and the mean energy of a particle, respectively. Then a one-dimensional stream of particles $j$ and of heat $q$ is respectively

$$j = -\frac{1}{6}\frac{d(nu)}{dx}2\ell \qquad \text{and} \qquad q = -\frac{1}{6}\frac{d(\varepsilon u)}{dx}2\ell \tag{A.1}$$

where $\varepsilon = n\frac{3}{2}T$ (we admit that the number of degree of freedom of the diffunding particle is 3). Since

$$\frac{1}{2}m\,u^2 \approx \frac{3}{2}\,T \tag{A.2}$$

where $m$ is a mass of particle, we find that

$$u = \sqrt{\frac{3T}{m}} \qquad \text{and} \qquad \frac{du}{dx} = u\frac{1}{2T}\frac{dT}{dx} \tag{A.3}$$

Hence

$$j = -\frac{1}{3}\ell\left(u\frac{dn}{dx} + nu\frac{1}{2T}\frac{dT}{dx}\right) \quad \text{and} \quad q = -\frac{1}{3}\ell\frac{3}{2}\left(Tu\frac{dn}{dx} + \frac{3}{2}nu\frac{dT}{dx}\right) \tag{A.4}$$

Therefore, the thermodiffusion coefficients

$$M = \frac{1}{3}\,\ell\,n\,u\,\frac{1}{2T} \quad \text{and} \quad N = \frac{1}{3}\frac{3}{2}\ell T u \tag{A.5}$$

and

$$\frac{M}{N} = \frac{n}{3T^2} \tag{A.6}$$

This is an expression of the type Eq. (4.5).

The goal of this calculation was to investigate on a simple example existence of the relations between different kinetic coefficients. Equations (A.4) describe the diffusion of the admixture of a light gas in a heavy one. As the diffusion coefficient in this example is proportional to the mean velocity ($D = \frac{1}{3}\ell u$) the description also applies to a gas flow in porous medium in a near-Knudsen regime [63].

If we deal with the mutual diffusion of $A$ and $B$ gases of comparable properties, a Fürth method [24] can be used. Then, the elementary model loses its transparency. To obtain a better theory, the Boltzmann kinetic equation should be taken into account.

# Appendix B. Onsager Relations in an Enskog-Chapman thermodiffusion theory

The set of thermodiffusion equations as derived from the Boltzmann equation for the gas mixture in which the partial density distribution of one constituent is $n_1 = n_1(\mathbf{r}, t)$ and the density distribution of the second constituent is $n_2 = n_2(\mathbf{r}, t)$, and the conditions of the constant pressure $p$ prevail, reads

$$\begin{aligned}\mathbf{V}_1 - \mathbf{V}_2 &= -\frac{n^2}{n_1 n_2}\left(D_{12}\mathbf{d}_{12} + D_{\mathrm{T}}\frac{1}{T}\frac{\partial T}{\partial \mathbf{r}}\right)\\ \mathbf{q} &= -\lambda_{\mathrm{T}}\frac{\partial T}{\partial \mathbf{r}} + nTS_{\mathrm{T}}\left(\mathbf{V}_1 - \mathbf{V}_2\right)\end{aligned} \tag{B.1}$$

where

$$\mathbf{d}_{12} \equiv \frac{\partial n_{10}}{\partial \mathbf{r}} - \frac{\rho_1\rho_2}{p\rho}(\mathbf{F}_1 - \mathbf{F}_2) \qquad n_{10} \equiv \frac{n_1}{n} \quad \text{and} \quad p = nT \tag{B.2}$$

cf. Chapman and Cowling [7], Chapter 8. Here, $D_{12}$, $D_{\mathrm{T}}$ and $\lambda_{\mathrm{T}}$ are the coefficients of diffusion, thermal diffusion and thermal conduction, respectively, while $S_{\mathrm{T}}$ is the thermodiffusion ratio (Soret coefficient)

$$S_{\mathrm{T}} = \frac{D_{\mathrm{T}}}{D_{12}} \tag{B.3}$$

moreover, $\mathbf{V}_1$ and $\mathbf{V}_2$ are the mean peculiar velocities (measured relative to the mean mass velocity of the gas) of the particle of the constituent number 1 and of number 2, respectively; the number density of the $s$th constituent is $n_s (s = 1, 2)$, the mass of the particle is $m_s$, the partial density of this constituent is $\rho_s = n_s m_s$ and the density $\rho$ of the whole gas is $\rho = \rho_1 + \rho_2$. The total number density is denoted by $n = n_1 + n_2$.

We combine equations of (B.1) and obtain

$$\begin{aligned}\mathbf{V}_1 - \mathbf{V}_2 &= -\frac{n^2}{n_1 n_2}\left(D_{12}\mathbf{d}_{12} + D_{\mathrm{T}}\frac{1}{T}\frac{\partial T}{\partial \mathbf{r}}\right)\\ \mathbf{q} &= -TD_{\mathrm{T}}\frac{n^3}{n_1 n_2}\mathbf{d}_{12} - \left(\lambda_{\mathrm{T}} + S_{\mathrm{T}}\frac{n^3}{n_1 n_2}D_{\mathrm{T}}\right)\frac{\partial T}{\partial \mathbf{r}}\end{aligned} \tag{B.4}$$

For the case of Brownian particle

$$n_1 \ll n_2 \qquad \text{thus} \quad n_2 \approx n$$

Then

$$\mathbf{d}_{12} = \frac{1}{n}\frac{\partial n_1}{\partial \mathbf{r}} - \frac{\rho_1}{nT}(\mathbf{F}_1 - \mathbf{F}_2)$$

and instead of (B.4) we have the following system

$$\begin{aligned}n_1(\mathbf{V}_1 - \mathbf{V}_2) &= -\left[\frac{\partial n_1}{\partial \mathbf{r}} - \frac{\rho_1}{T}(\mathbf{F}_1 - \mathbf{F}_2)\right] - nD_{\mathrm{T}}\frac{1}{T}\frac{\partial T}{\partial \mathbf{r}}\\ \mathbf{q} &= -TD_{\mathrm{T}}\frac{n}{n_1}\left[\frac{\partial n_1}{\partial \mathbf{r}} - \frac{\rho_1}{T}(\mathbf{F}_1 - \mathbf{F}_2)\right] - \left(\lambda_{\mathrm{T}} + S_{\mathrm{T}}\frac{n^2}{n_1}D_{\mathrm{T}}\right)\frac{\partial T}{\partial \mathbf{r}}\end{aligned} \tag{B.5}$$

Comparing the system (B.5) with Eq. (2.1) we find

$$M = \frac{n}{T} D_{\mathrm{T}} \qquad N = T D_{\mathrm{T}} \frac{n}{n_1} \tag{B.6}$$

Hence

$$\frac{M}{N} = \frac{n_1}{T^2} \tag{B.7}$$

If the distribution function $f$ is identified with the number density $n_1$ the Eq. (B.7) is identic with Eq. (4.5).

**Acknowledgement**

The author is indebted to Professor Ignaczak of IPPT for a discussion on the subject. He is also obliged to a Referee whose remarks essentially improved the paper.

## References

1. R. F. Streater: Rep. Math. Phys. **40**, 557 (1997)
2. O. Penrose, P.C. Fife: Physica D **43**, 44 (1990)
3. J. Łuczka, T. Czernik, P. Hänggi: Phys. Rev. E **56**, 3968 (1997)
4. J.L. Mateos: Phys. Rev. Lett. **84**, 258 (2000)
5. K.L. Sebastian: Phys. Rev. E **61**, 937 (2000)
6. P. Reimann: Brownian motors: noisy transport far from equilibrium, Institut für Physik, Universität Augsburg, Augsburg 2000 (submitted to Physics Reports)
7. C. Ludwig: Sitzungsber. Kaiser. Akad. Wiss. (Mathem.-Naturwiss. Cl.), Wien **20**, 539 (1856)
8. Ch. Soret: Arch. Sci. phys. natur., Genéve **2**, 48 (1879)
9. Ch. Soret: C.R. Acad. Sci., Paris **91**, 289 (1880)
10. L. Dufour: Arch. Sci. phys. natur., Genéve **45**, 9 (1872)
11. L. Dufour: Ann. Physik [5] (hgb. J.C. Poggendorff, Leipzig) **28**, 490 (1873)
12. W. Feddersen: Ann. Physik [5] (hgb. J.C. Poggendorff, Leipzig) **28**, 302 (1873)
13. S.R. de Groot, P. Mazur: Non-equilibrium thermodynamics, North-Holland Publ. Co., Amsterdam 1962
14. O.Ae. Meyer: De gasorum theoria, Dissertatio inauguralis math.-phys., Maelzer, Vratislaviae 1866
15. S. Chapman, T.G. Cowling: The mathematical theory of non-uniform gases, At the University Press, Cambridge 1960
16. K. Clusius, L. Waldmann: Naturwiss. **30**, 711 (1942)
17. L. Waldmann: Naturwiss. **31**, 204 (1943)
18. K. Clusius u. G. Dickel: Naturwiss. **26**, 546 (1938)
19. W.H. Furry, R. Clark Jones, L. Onsager: Phys. Rev. **55**, 1083 (1939)
20. J. Jeans: Dynamical theory of gases, 4th edition, Dover Publications, New York 1954
21. R. Clark Jones, W.H. Furry: Rev. Mod. Phys. **18**, 151 (1946)
22. S.P. Frankel: Phys. Rev. **57**, 661 (1940)
23. J. Stefan: Sitzungsber. Kaiser. Akad. Wiss. (Mathem.- Naturwiss. Cl.), Wien **65**, 323 (1872)

24. R. Fürth: Proc. R. Soc. A **179**, 461 (1942)
25. W.H. Furry: Amer. J. Phys. **16**, 63 (1948)
26. R.W. Pohl: Mechanik, Akustik und Wärmelehre, Springer Verlag, Berlin-Göttingen-Heidelberg 1955.
27. K.E. Grew, T.L. Ibbs: Thermal diffusion in gases, Cambridge University Press 1952
28. L. Waldmann: Transporterscheinungen in Gasen von mittleren Druck, in: Handbuch der Physik, hgb. S. Flügge, Bd. XII, Springer Verlag, Berlin-Göttingen-Heidelberg 1958
29. L.D. Landau, M. Lifshits: Fluid mechanics, Pergamon Press, Oxford 1963
30. J. Klafter, M.F. Schlesinger, G. Zumofen: Physics Today, February 1996, p. 33
31. L.J. Bodaszewsky: Chem. Zentralblatt (Dritte Folge) **12**, 209 (1881)
32. Ł.J. Bodaszewski: Kosmos **7**, 177 (1882)
33. A. Einstein: Ann. Phys. **17**, 549 (1905)
34. A. Einstein: Zs. Elektrochemie **14**, 235 (1908)
35. M. v. Smoluchowski: Phys. Zeitschr. **17**, 557-571, 585 (1916)
36. S. Chandrasekhar, M. Kac, R. Smoluchowski: Marian Smoluchowski - his life and scientific work, ed. by R.S. Ingarden, PWN, Warszawa 1986
37. Lord Rayleigh: Phil. Mag. **32**, 424 (1891)
38. M. Smoluchowski: Essai d'une théorie cinétique du mouvement Brownien et des milieux trubles, Imprimerie de l'Université, Cracovie 1906 (extrait du Bull. de l'Acad. des Sciences de Cracovie, Cl. des Sc. Math. et Naturelles, Juillet 1906)
39. M. v. Smoluchowski: Ann. Phys. **21**, 756 (1906)
40. H.A. Lorentz: Les théories statistiques en thermodynamique, Conférences faites au Collége de France en novembre 1912. Redigées en 1913 par L. Dunoyer, B.G. Teubner, Leipzig 1916
41. M.S. Green: J. Chem. Phys. **19**, 1036 (1951)
42. R. Balescu: Equilibrium and nonequilibrium statistical mechanics, New York-London-Sydney-Toronto 1975
43. R. Kubo: J. Phys. Soc. Japan **12**, 570 (1957)
44. J. Stecki, M.A. Narbutowicz: Chem. Phys. Lett. **5**, 345 (1970)
45. M.A. Narbutowicz: Rep. Math. Phys. **8**, 1 (1975)
46. R. Kubo: Rap. Prog. Phys. **29**, 255 (1966)
47. R. F. Streater: J. Math. Phys. **38**, 4570 (1997)
48. R. F. Streater: J. Stat. Phys. **88**, 447 (1997)
49. R.F. Streater: The Soret and Dufour effects in statistical dynamics, Preprint KCL-MTH-98-32, Department of Mathematics, King's College, London 1998
50. R. Wojnar: Entropie **35**, 67 (1999)
51. R. Wojnar: Rep. Math. Phys. **44**, 291 (1999)
52. R. Wojnar: Rep. Math. Phys. **46**, 295 (2000)
53. R. Wojnar: Acta Phys. Polon. B **32**, 333 (2001)
54. I.D. Cherkasov: Theory Probab. Its Applic. **2**, 373 (1957)
55. L.M. Ricciardi: J. Math. Analysis and Applications **54**, 185 (1976)
56. E. Sanchez-Palencia: Non-homogeneous media and vibration theory, Springer, Berlin 1980
57. A. Gałka, J.J. Telega, R. Wojnar: Arch. Mech. **46** , 267 (1994)
58. R. Wojnar: Rep. Math. Phys. **40**, 585 (1997)
59. D. Cioranescu, J. Saint Jean Paulin: J. math. pures et appl. **65**, 403 (1986)
60. R. Burridge, J.B. Keller: J. Acoust. Soc. Am. **70**, 1140 (1981)
61. I. Prigogine: Etude thermodynamique des phénomènes irréversibles, Desoer, Liège 1947

62. P. Glandsdorff, I. Prigogine: Physica **31**, 1242 (1965)
63. E.A. Mason, A.P. Malinauskas: Gas transport in porous media: the dusty-gas model, Elsevier, Amsterdam-Oxford-NewYork 1983

## Supplementary glossary

| | |
|---|---|
| $A, B$ | indices of the mixture components in Sect. 1; constants of integration in Sect. 8 |
| $c$ | in Eqs. (6.8)–(6.11) – gradient of modified potential; otherwise – the concentration |
| $e$ | density of energy |
| $E$ | energy of the system |
| $f$ | the probability density function (the distribution function) for a Brownian particle |
| $f_{,i}$ | $\frac{\partial f}{\partial x_{,i}}$ $(i = 1, 2, 3)$ where $f = f(x_1, x_2, x_3)$ |
| $\mathbf{F}$ | external force acting on Brownian particle |
| $\mathbf{j}$ | flux of diffusion |
| $\mathbf{j}^E$ | flux of total energy |
| $\ell$ | free path length |
| $L_{\mathrm{D}}, L_{\mathrm{DT}}, L_{\mathrm{TD}}, L_{\mathrm{T}}$ | kinetic coefficients introduced in Eqs.(1.2) |
| $L_{\mathrm{T}} \equiv \lambda_{\mathrm{T}}$ | thermal conductivity, cf. Eq. (1.7) |
| $M \equiv L_{\mathrm{DT}}$ | cf. Eqs. (3.1) |
| $N \equiv \lambda_{\mathrm{D}}$ | cf. Eqs. (3.1) |
| $n$ | number density (mean number of particles *per* unit volume) |
| $p$ | pressure |
| $\mathbf{q}$ | flux of heat |
| $r$ | heat source |
| $s$ | number of degree of freedom |
| $S$ | total entropy of the system |
| $T^{\mathrm{Kelv}}$ | temperature in degrees of Kelvin |
| $T$ | temperature in energetic scale, Eq.(1.18) |
| $u$ | mean thermal velocity |
| $\mathbf{v}$ | velocity of a Brownian particle |
| $V$ | potential |
| $\mathcal{V}$ | modified potential, cf. Eq.(6.4) |
| $\mathbf{x}$ | position vector of a Brownian particle |
| $\beta$ | constant of friction |
| $\varepsilon$ | parameter of smallness (in Sect. 7) |
| $\eta$ | coefficient of viscosity; in Sect. 7 – parameter of smallness |
| $\lambda_D$ | cf. Eqs. (1.6) and (1.7) |
| $\sigma$ | cross-section for collision |
| $\theta_i, \phi_i, \chi_i, \psi_i$ | local vectorial functions, cf. Sect. 7 |
| $\Omega$ | space domain occupied by the system |

# Nonequilibrium Concentration Fluctuations in Binary Liquid Systems Induced by the Soret Effect

Jan V. Sengers[1] and José M. Ortiz de Zárate[2]

[1] Institute for Physical Science and Technology and Department of Chemical Engineering, University of Maryland, College Park, MD 20742, USA
[2] Departamento de Física Aplicada 1, Facultad de Ciencias Físicas, Universidad Complutense, E-28040 Madrid, Spain

**Abstract.** When a binary liquid system is brought into a stationary thermal nonequilibrium state by the imposition of a temperature gradient, the Soret effect induces long-range concentration fluctuations even in the absence of any convective instability. The physical origin of the nonequilibrium concentration fluctuations is elucidated and it is shown how the intensity of these concentration fluctuations can be derived from the linearized random Boussinesq equations. Relevant experimental information is also discussed.

## 1 Introduction

In this chapter we consider a binary liquid system located between two horizontal plates with two different but constant temperatures. A temperature gradient in a liquid mixture induces a concentration gradient through the Soret effect [1–5]. Many studies reported in the literature have been concerned with the effect of the Soret-induced concentration gradients on the convective Rayleigh-Bénard instability [6–11]. However, it turns out that there are also interesting fluctuation phenomena present, when the liquid mixture is in a hydrodynamically quiescent stable state, far away from any convective instability. It is the purpose of the present chapter to review the origin and the nature of concentration fluctuations in a binary liquid in stable convection-free thermal nonequilibrium states.

About 50 years ago a procedure for dealing with nonequilibrium phenomena in fluids was proposed by Bogoliubov [12,13]. It was based on a postulate that a fluid in a nonequilibrium state would proceed to thermodynamic equilibrium in two distinct stages: first a microscopic kinetic stage with a time scale of the order of the interval between successive molecular collisions, which for liquids is of the same order as the duration of the molecular collisions, after which local equilibrium is established; second a macroscopic stage during which the fluid evolves in accordance with the hydrodynamic equations. Implicit in this postulate is the idea that no long-range dynamic correlations would be present in a fluid of molecules with short-range forces, unless the system would be near an incipient thermodynamic or hydrodynamic instability.

The subsequent 25 years in the history of nonequilibrium statistical mechanics have revealed a basic flaw in this picture. In evaluating the randomizing

W. Köhler and S. Wiegand (Eds.): LNP 584, pp. 121–145, 2002.

nature of molecular collisions one must make a distinction between quantities like mass, momentum and energy that are conserved in a binary collision on the one hand and nonconserved quantities on the other hand. The fast modes associated with nonconserved quantities do indeed decay during a short kinetic stage, but the slow modes associated with the conserved quantities persist well beyond the time between successive collisions and a coupling between these modes can result in long-range (mesoscopic) dynamic correlations [14]. The classical picture of short-range dynamic correlations first appeared to be inadequate in fluids near the critical point when experiments [15] could no longer be explained by the Van Hove theory of the critical slowing down of the fluctuations which was based on strictly thermodynamic considerations; this observation led to the development of the mode-coupling theory of critical dynamics [16–19]. Some time later it turned out that the same mode-coupling theory could also account for the presence of long-time tails in the Green-Kubo correlation functions that were originally noticed from computer simulations of molecular dynamics [20–22].

Around 1980 it became evident that the mode-coupling theory would also predict the existence of long-range temperature and viscous fluctuations in fluids that are kept in stationary nonequilibrium states [23]. Specifically, Kirkpatrick et al. were the first to derive the appropriate expressions for the temperature and the viscous fluctuations in a one-component fluid subjected to a stationary temperature gradient [24]. Subsequently, others have recovered the same expressions on the basis of Landau's fluctuating hydrodynamics [25–31]. Physically, the presence of a temperature gradient $\nabla T$ leads to nonequilibrium fluctuations because the stationary temperature gradient causes a coupling between the component of the velocity fluctuations parallel to the gradient and the temperature fluctuations. Velocity fluctuations parallel to the gradient are probing regions with different local equilibrium temperatures, thus causing a nonequilibrium contribution to the temperature fluctuations. Since both the temperature and the viscous (transverse velocity) fluctuations do not propagate, but decay diffusively, the nonequilibrium contributions cannot depend on whether the temperature gradient is in the positive or negative direction and, hence, they depend on the square of the temperature gradient. Moreover, each mode contributes a factor $q^{-2}$, so that the intensity of the nonequilibrium fluctuations becomes inversely proportional to the fourth power of the wave number $q$ of the fluctuations. The dependence of the intensity of the nonequilibrium fluctuations on $(\nabla T)^2$ and on $q^{-4}$ has been confirmed by light-scattering experiments [32–35]. An algebraic divergence of the nonequilibrium fluctuations as a function of the wave number $q$ is now believed to be a general feature of fluctuations in fluids in stationary nonequilibrium states [14,36,37].

The ultimate divergence of the intensity of nonequilibrium fluctuations for small wave numbers, i.e., for large wavelengths, will be prevented by gravity and finite-size effects. Specifically, for very small wave numbers the dependence of the nonequilibrium fluctuations on $q$ is suppressed by the presence of gravity, as predicted theoretically by Segrè et al. [38] and confirmed experimentally by

Vailati and Giglio [39,40], and also modified by finite-size effects, as evaluated by Ortiz de Zárate et al. [41,42].

In the case of a liquid mixture, a temperature gradient $\nabla T$ is accompanied by a concentration gradient $\nabla c$ through the Soret effect as mentioned earlier. Just as a temperature gradient causes a coupling between the temperature fluctuations with wave vector $\boldsymbol{q}$ perpendicular to the temperature gradient and the transverse-velocity fluctuations in the direction of the temperature gradient, so will a concentration gradient cause a coupling between the concentration fluctuations with wave vector $\boldsymbol{q}$ perpendicular to the concentration gradient and the transverse-velocity fluctuations in the direction of the concentration gradient. Hence, in a liquid mixture subjected to a temperature gradient not only nonequilibrium temperature and viscous fluctuations will be present, but also nonequilibrium concentration fluctuations as first pointed out by Law and Nieuwoudt [43,44], and as also analyzed by Velasco and García Colín [45] and by Segrè and Sengers [46]. The intensity of the nonequilibrium concentration fluctuations should be proportional to the square of the concentration gradient $\nabla c$ and again inversely proportional to the fourth power of the wave number $q$. At very small $q$, the nonequilibrium concentration fluctuations will again be modified by gravity effects [39,46] and by finite-size effects [47].

The case of nonequilibrium concentration fluctuations in a colloidal suspension in the presence of a concentration gradient has been considered by Schmitz [48]. Vailati and Giglio [49] have extended the theory to treating the effect of time-dependent concentration profiles on the structure factor of a binary liquid.

The complete expressions for the contributions from the nonequilibrium viscous fluctuations, temperature fluctuations and concentration fluctuations to the dynamic structure factor of binary fluids are rather complicated [43,44,46]. However, in the case of (nearly) incompressible liquid mixtures a number of simplifying approximations can be made. First of all, as documented elsewhere [34,46,50], to a good approximation the decay rates of the viscous fluctuations, the temperature fluctuations and the concentration fluctuations can be directly identified with $\nu q^2$, $D_{\mathrm{th}} q^2$, and $Dq^2$, respectively, where $\nu$ is the kinematic viscosity, $D_{\mathrm{th}}$ the thermal diffusivity and $D$ the (mass) diffusion coefficient. Furthermore, the diffusion coefficient $D$ is much smaller than the kinematic viscosity $\nu$ and the thermal diffusivity $D_{\mathrm{th}}$. Thus, the viscous and thermal fluctuations have in practice decayed at the times that the concentration fluctuations are still important. Moreover, because of the dependence of the refractive index on the concentration, the intensity of light scattered by the concentration fluctuations is usually appreciably larger than that from the viscous or the temperature fluctuations. Hence, in many cases one can readily determine the concentration fluctuations with Rayleigh scattering independently from the much faster decaying viscous and temperature fluctuations [34,35,50,51].

In this chapter we shall present a simple derivation of the intensity of the nonequilibrium concentration fluctuations and their contribution to Rayleigh scattering. Instead of considering the full (complicated) structure factor [43,46] and then make a series of simplifying approximations, we shall show how the intensity of the nonequilibrium concentration fluctuations can be readily derived from the same linearized random Boussinesq equations commonly used for studying the convective instability in binary fluids [6–11], including the bouyancy term and, thus, the effects of gravity. Specifically, we shall adopt an approximation to the Boussinesq equations proposed by Velarde and Schechter to simplify the study of the linear stability problem [52]. Therefore, instead of using the full hydrodynamic equations, making a complicated derivation of the structure factor and, then, simplify it by retaining only the experimentally important terms, we shall make the simplification *a priori*, by adopting a set of approximate Boussinesq equations supplemented with random noise terms in accordance with fluctuating hydrodynamics [53,54]. The advantage of this approach is that the calculation of the structure factor associated with the nonequilibrium concentration fluctuations becomes simple and straightforward, while elucidating the physical origin of the nonequilibrium fluctuations. This simplified approach facilitates not only the inclusion of gravity effects to be discussed in this chapter, but also the incorporation of finite-size effects on the nonequilibrium concentration fluctuations to be considered in a future publication [47].

## 2 Theory

We consider a horizontal layer of a binary liquid subjected to a stationary temperature gradient to be designated as $\nabla T_0$, which as a result of the Soret effect is accompanied by a stationary concentration gradient to be designated as $\nabla c_0$. The temperature gradient is assumed to act in the vertical $Z$-direction, while being uniform in the $X$ and $Y$ directions. The thermal conductivity is assumed to vary weakly with temperature so that $\nabla T_0 = dT_0/dz$ and $\nabla c_0 = dc_0/dz$ are (scalar) constants independent of the position in the liquid layer. The relationship between the temperature gradient $\nabla T_0$ and the resulting concentration gradient $\nabla c_0$ is given by:

$$\nabla c_0 = -c\,(1-c)\;S_T\;\nabla T_0, \tag{1}$$

where $c$ is the (average) concentration expressed as weight fraction (w/w) of component 1 of the mixture. Equation (1) defines the Soret coefficient $S_T$ of component 1 in component 2. Hence, component 1 is considered the solute and component 2 the solvent. For isotropic mixtures $S_T$ is a scalar quantity and the induced concentration gradient is parallel to the imposed temperature gradient, but depending on the sign of $S_T$ can have the same or the opposite direction. When $S_T$ is positive, concentration and temperature gradients have opposite directions, with component 1 migrating to the colder region. When $S_T$ is negative, concentration and temperature gradients have the same direction, with component 1 migrating to the warmer region. If $c$ were to represent the concentration of component 2, then (1) would define the Soret coefficient $S_T$ of component 2 in component 1. For a given mixture mass conservation implies that when $S_T$ of component 1 in component 2 is positive, the $S_T$ of component 2 in component

1 must be negative and *vice versa*. In this chapter we adopt the convention that $c$ designates the concentration of component 1.

Let $\rho$ be the density of the liquid layer, $\alpha = -\rho^{-1}\ (\partial\rho/\partial T)_c$ the thermal expansion coefficient and $\beta = \rho^{-1}\ (\partial\rho/\partial c)_T$ the concentration expansion coefficient. Since the liquid mixture is assumed to be (nearly) incompressible, the pressure is not a relevant physical variable. We note that $\beta$ is positive when component 1 is the heavier component and $\beta$ is negative when component 1 is the lighter component. A dimensionless separation ratio $\psi$ may be defined by:

$$\psi = c\ (1 - c)\ S_T\ \frac{\beta}{\alpha}. \tag{2}$$

The parameter $\psi$ represents the ratio of the density gradient $\beta\ \nabla c_0$ produced by the concentration gradient and the density gradient $-\alpha\ \nabla T_0$ produced by the temperature gradient. Note that, in contrast to $S_T$, the sign of $\psi$ is independent of whether the lighter or the heavier component is chosen as component 1 of the mixture, so that it is independent of which component is actually chosen to represent the concentration $c$ of the mixture.

Finally, the Rayleigh number $Ra$ is defined as:

$$Ra = -\frac{\alpha g L^4\ \nabla T_0}{\nu D_{\text{th}}}, \tag{3}$$

where $L$ is the height of the liquid layer and $g$ the gravitational acceleration constant. Note that the Rayleigh number is negative when $\nabla T_0 = dT_0/dz$ is positive, i.e., when the temperature gradient is pointing in the positive $Z$-direction, opposite to the direction of the gravitational force $\boldsymbol{g}$.

## 2.1 Linearized Random Boussinesq Equations for a binary liquid

We want to study small fluctuations around the thermally conducting state, i.e., fluctuations around the quiescent and stable state of the fluid. Following a notation adopted by some previous investigators [6,7,55], we denote the local fluctuation in the temperature $T$ by $\theta$ ($= \delta T$), the local fluctuation in the $z$-component of the fluid velocity by $w$ ($= \delta u_z$) and the local fluctuation in the concentration $c$ of component 1 of the mixture by $\Gamma$ ($= \delta c$). The fluctuations $\theta(\boldsymbol{r},t)$, $w(\boldsymbol{r},t)$ and $\Gamma(\boldsymbol{r},t)$ depend on the position $\boldsymbol{r}(x,y,z)$ and on the time $t$. In the spirit of Landau's fluctuating hydrodynamics [53], we may describe the space and time dependence of these fluctuations by the deterministic linearized random Boussinesq equations for a binary fluid supplemented with random noise terms [6–8,10]:

$$\frac{\partial}{\partial t}\left(\nabla^2 w\right) = \nu\ \nabla^2\left(\nabla^2 w\right) + g\ \left(\frac{\partial^2}{\partial x^2} + \frac{\partial^2}{\partial y^2}\right)\left[\alpha\,\theta - \beta\,\Gamma\right] + \\ + \frac{1}{\rho}\left\{\nabla \times \left[\nabla \times \left(\nabla \cdot \delta\mathsf{T}\right)\right]\right\}_z, \tag{4a}$$

$$\frac{\partial \theta}{\partial t} = D_{\mathrm{th}}\, \nabla^2 \theta - w\, \nabla T_0 - \frac{D_{\mathrm{th}}}{\lambda_{\mathrm{T}}} \nabla(\delta \boldsymbol{Q}), \tag{4b}$$

$$\frac{\partial \Gamma}{\partial t} = D \left[ \nabla^2 \Gamma + \frac{\alpha}{\beta} \psi\, \nabla^2 \theta \right] - w\, \nabla c_0 + \nabla(\delta \boldsymbol{J}), \tag{4c}$$

where $\lambda_{\mathrm{T}}$ is the thermal conductivity of the mixture. The notations $\nabla T_0$ and $\nabla c_0$ for the stationary gradients of temperature and concentration remind us that both $\nabla T_0$ and $\nabla c_0$ in (4) are constants and not gradients of locally fluctuating temperature and concentration. As in the standard fluctuating-hydrodynamics approach [53,54], we have included in (4) random noise terms to represent the contributions from rapidly varying short-range fluctuations [43,46,56–58]. Thus, we have introduced $\delta\mathsf{T}(\boldsymbol{r}, t)$ as the random stress tensor and $\delta\boldsymbol{Q}(\boldsymbol{r}, t)$ as the random heat flux following [56,57]. Additionally, we have introduced $\delta\boldsymbol{J}(\boldsymbol{r}, t)$ as the random solute flux following [43,46,58]. Note that the subscript $z$ in (4a) indicates that the random noise term is to be identified with the $z$-component of the vector between curly brackets. This procedure of supplementing the random Boussinesq equations for a binary liquid with random noise terms was previously used by Schöpf and Zimmermann to study the influence of noise close to the convective instability [59]. We do not consider any pressure fluctuations in (4), since pressure fluctuations cause Brillouin scattering, but they do not contribute to the Rayleigh spectrum of the scattered light.

In applying the linearized random Boussinesq equations, as given by (4), we have assumed that all thermophysical properties of the binary fluid depend only weakly on temperature or concentration so that the variation of these properties as a function of $z$ is negligible small; in practice this is a very good approximation [33]. For the same reason, in (2) the concentration $c$ of component 1 represents an average value through the fluid layer; the substitution of the local concentration by an average concentration is only questionable for extremely dilute solutions [7]. Moreover, the Boussinesq approximation assumes the liquid to be incompressible [55], so the coefficient $D_{\mathrm{th}}$ in (4b) can indeed be identified with the thermal diffusivity $\lambda_{\mathrm{T}}/\rho c_{\mathrm{P}}$, where $c_{\mathrm{P}}$ is the specific isobaric heat capacity. As usual, the Dufour effect has been neglected in (4), since it is relevant only in binary gas mixtures and in liquids near the vapor-liquid critical point [11,60,61]. Following Chandrasekhar [55], to eliminate the hydrostatic pressure gradient we find it convenient to consider (4a) for $\nabla^2 w$, by taking a double curl in the equation for the fluctuating fluid velocity $\delta\boldsymbol{u}$.

## 2.2 Stability of the solution of the Linearized Boussinesq Equations

There is a considerable literature applying the linearized Boussinesq equations to binary systems, mainly focused on the analysis of the stability of the conductive solution (see, for instance, [6–11,62] and references therein). An excellent updated review, including nonlinear-stability analysis, has recently appeared in this Springer series of Lecture Notes in Physics [63]. The stability of the conductive solution depends on whether the temperature and concentration gradients are stabilizing or not. Thus, stability will depend on the sign of the Rayleigh

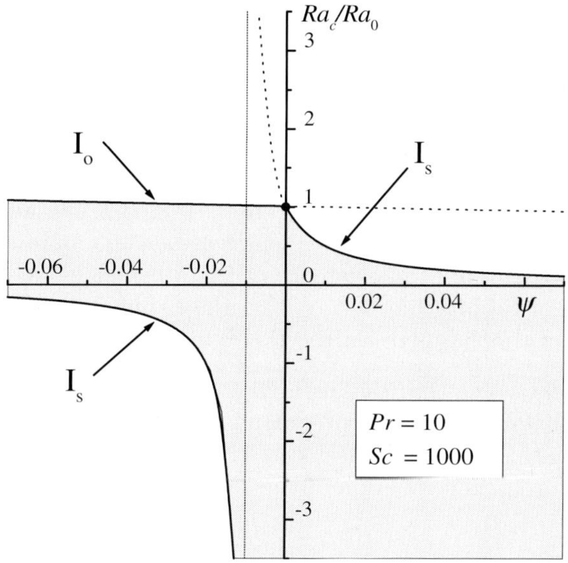


**Fig. 1.** Linear stability diagram deduced from the Boussinesq equations with free-slip-permeable boundary conditions for a binary liquid with typical values of the Schmidt and Prandtl numbers, according to [7]. The region of stability corresponds to the shadowed area.

number $Ra$ and on the sign of the separation ratio $\psi$. In Fig. 1 we present the typical linear stability diagram in the $\{Ra, \psi\}$ plane, plotted after Schechter et al. [7]. The stability diagram depends on the values of the Schmidt and Prandtl numbers. The plot in Fig. 1 was obtained, as indicated, for $Sc$=1000 and $Pr$=10, which are typical values for an ordinary liquid mixture. Depending on the signs of $Ra$ and $\psi$ one can identify four regions in the diagram of Fig. 1:

1. If $Ra > 0$ and $\psi < 0$, the temperature gradient is destabilizing whereas the Soret-induced concentration gradient is weakly stabilizing. Note that if $\beta > 0$, $S_T$ has to be negative to have an overall $\psi < 0$, see (2), and the denser component 1 migrates to the warm lower plate. If $\beta < 0$, the lighter component 1 migrates to the cold upper plate. In both cases the Soret effect is weakly stabilizing. In this situation, the instability appears at a nonzero horizontal wave number and at a nonzero frequency. The instability mechanism is thus an oscillatory periodic Hopf bifurcation or, in the language of Cross and Hohenberg [10], an $I_o$-type instability. Above the threshold, the instability leads to a traveling-waves pattern [59].
2. When $Ra > 0$ and $\psi > 0$, both the temperature and the concentration gradient are destabilizing. For given $\psi$ the instability appears at a lower $Ra$ than in the case of a pure fluid ($\psi = 0$). In this situation the instability appears at a finite wave number but at zero frequency, as in a pure liquid: it is an $I_s$-type instability. Above the threshold, the instability leads to a square pattern.

3. When $Ra < 0$ and $\psi < 0$, the temperature gradient is stabilizing, but the Soret coupling is destabilizing. For given $Ra$, the separation ratio $\psi$ could be negative enough for an instability to appear. Note that this instability develops while heating the fluid from above, a configuration in which a pure liquid is always stable. As above, it is an $\mathrm{I_s}$-type instability. It was recently experimentally observed by La Porta and Surko [64].
4. Finally, when $Ra < 0$ and $\psi > 0$, both temperature and concentration gradients are stabilizing and the system is always in a quiescent stable state.
5. It is interesting to note the existence of a codimension-two instability point where the $\mathrm{I_o}$-instability line meets with the $\mathrm{I_s}$-instability line. In the linear approximation this happens on the $\psi = 0$ axis [65].

It is worth mentioning that the instability diagram depends on the boundary conditions. The diagram in Fig. 1 corresponds to free and permeable boundaries, which is mathematically the simplest case. For this particular case, the critical Rayleigh number $Ra_\mathrm{c}$ as a function of the separation ratio for the $\mathrm{I_o}$-instability line is given by [7]:

$$\left(\frac{Ra_\mathrm{c}}{Ra_0}\right)_\mathrm{o} = \frac{(1+Pr)(Sc+Pr)(1+Sc)}{Sc^2(\psi\, Pr + Pr + 1)}, \tag{5}$$

while the critical Rayleigh for the $\mathrm{I_s}$-instability line is given by [7]:

$$\left(\frac{Ra_\mathrm{c}}{Ra_0}\right)_\mathrm{s} = \frac{Pr}{Pr + \psi\,(Sc+Pr)} \tag{6}$$

with $Ra_0 = 27\pi^4/4$, the critical Rayleigh number for a pure fluid with free-slip boundary conditions [55]. These were the expressions employed to construct Fig. 1. The more realistic case of fixed and impermeable boundaries has also been analyzed [9,66]. But, as in the case of a pure fluid, different boundary conditions cause only quantitative differences while qualitatively the diagram is more or less the same. In addition, nonlinear terms that have been neglected in (4) cause also minor changes in the stability diagram, particularly shifting the position of the codimension-two instability point [65]. In any case, the preceding discussion of the relationship between stability and the signs of $Ra$ and $\psi$ still holds. The only situation for which stability is always assured is when $Ra < 0$ and $\psi > 0$. In this chapter we consider small fluctuations around the conductive state. Therefore, in the following, we shall assume $Ra < 0$ and $\psi > 0$, the situation in which fluctuations do not drive the system outside of a stable quiescent configuration.

### 2.3 Linearized Random Boussinesq Equations in the small-Lewis-number approximation

In principle, it is possible to calculate the structure factor of the binary liquid from the system of coupled differential equations given by (4), as was done in the simpler case of a pure liquid [41]. However, there are several complications which make the calculation here more involved. A first complication arises from the fact that, since they have the same tensorial character, the random heat flux $\delta\boldsymbol{Q}$ and the random solute flux $\delta\boldsymbol{J}$ are coupled and their cross-correlations are not zero: $\langle \delta Q_i \cdot \delta J_j \rangle \neq 0$ [43,58]. (But note that under some approximations these cross correlations may be neglected, as done by Schöpf and Zimmermann [59]). A second complication is that density or refractive-index fluctuations come from both temperature *and* concentration fluctuations; hence, these two types of fluctuations have to be calculated, contrary to the case of a pure liquid, where there are no concentration fluctuations. As mentioned in the introduction, in the case of binary liquid systems, the diffusion coefficient $D$ is much smaller than the thermal diffusivity $D_{\mathrm{th}}$ and the kinematic viscosity $\nu$. Hence, to obtain the structure factor of a binary liquid system we can consider to a very good approximation the limit of vanishing Lewis number $Le = D/D_{\mathrm{th}}$. Since $\nu > D_{\mathrm{th}}$, a small value of the Lewis number $Le$ also implies a small value of the inverse Schmidt number $Sc^{-1} = D/\nu$.

Velarde and Schechter [52] have considered the Boussinesq equations of a liquid mixture for small values of the Lewis number. By supplementing the hydrodynamic equations obtained in [52] with the corresponding random noise terms, one deduces from (4) in the limit $Le \to 0$:

$$0 = \nu\, \nabla^2 \left(\nabla^2 w\right) - \beta g \left(\frac{\partial^2 \Gamma}{\partial x^2} + \frac{\partial^2 \Gamma}{\partial y^2}\right) + \frac{1}{\rho} \left\{\nabla \times \left[\nabla \times (\nabla \cdot \delta \mathsf{T})\right]\right\}_z , \tag{7a}$$

$$\frac{\partial \Gamma}{\partial t} = D\, \nabla^2 \Gamma - w\, \nabla c_0 + \nabla \delta \boldsymbol{J} . \tag{7b}$$

In the limit $Le \to 0$, any coupling between the concentration fluctuations and the temperature fluctuations vanishes and one only needs to consider coupling between the concentration fluctuations and the transverse-velocity fluctuations, as mentioned in the introduction. Moreover, since the transverse-velocity fluctuations decay much faster than the concentration fluctuations, one does not need to include the viscous mode directly in the dynamic structure factor, but only its indirect effect on the concentration fluctuations. That is, in the limit $Sc^{-1} \to 0$ the time derivative $\partial/\partial t(\nabla^2 w)$ on the left-hand side of (7a) can be neglected.

We may mention that the small-Lewis-number approximation has been amply used in the literature to simplify the study of the stability of binary liquid mixtures [7,52,65,67]. As discussed in the literature [7,52], the $Le \to 0$ approximation has several shortcomings; in particular, it does not describe correctly the situation with $Ra > 0$ and $\psi < 0$, missing the interesting Hopf bifurcation and codimension-two instability point. But in this chapter we are only concerned with the case $Ra < 0$ and $\psi > 0$, that is, the case of the quiescent stable thermally conducting state, for wich the $Le \to 0$ approximation is adequate to describe the concentration fluctuations, as will be demonstrated in the subsequent sections.

### 2.4 Calculation of the structure factor

Having introduced the hydrodynamic simplifications described in Sect. 2.3, we can now readily calculate the structure factor of the binary liquid, following a procedure similar to the one adopted in a previous publication [41] in the case of a one-component liquid. In the absence of any boundary conditions we apply a temporal and spatial Fourier transformation to (7) so as to obtain:

$$\begin{pmatrix} -\nu\, q^4 & -\beta\, g\, q_{\parallel}^2 \\ \nabla c_0 & i\omega + D\, q^2 \end{pmatrix} \begin{pmatrix} w(\omega,\boldsymbol{q}) \\ \Gamma(\omega,\boldsymbol{q}) \end{pmatrix} = \begin{pmatrix} F_1(\omega,\boldsymbol{q}) \\ F_2(\omega,\boldsymbol{q}) \end{pmatrix}, \tag{8}$$

with $q_{\parallel}$ representing the magnitude of the wave vector $\boldsymbol{q}$ in the XY plane, i.e., $q_{\parallel}^2 = q_x^2 + q_y^2$. To simplify the notation, we have introduced in (8) Langevin random noise terms $F_1$ and $F_2$, which are related to the Fourier transforms of the random stress tensor $\delta\mathsf{T}(\omega,\boldsymbol{q})$ and of the random solute flux $\delta\boldsymbol{J}(\omega,\boldsymbol{q})$ by:

$$\begin{pmatrix} F_1(\omega,\boldsymbol{q}) \\ F_2(\omega,\boldsymbol{q}) \end{pmatrix} = \begin{pmatrix} i\,\rho^{-1}\,\left\{q_z\,\boldsymbol{q}\cdot[\boldsymbol{q}\cdot\delta\mathsf{T}(\omega,\boldsymbol{q})] - q^2\,[\boldsymbol{q}\cdot\delta\mathsf{T}(\omega,\boldsymbol{q})]_z\right\} \\ -i\,\boldsymbol{q}\cdot\delta\boldsymbol{J}(\omega,\boldsymbol{q}) \end{pmatrix}. \tag{9}$$

Since we have adopted the small-Lewis-number approximation, the temperature fluctuations can be neglected and the Rayleigh component $S(\omega,\boldsymbol{q})$ of the structure factor can be attributed to refractive-index fluctuations caused by the concentration fluctuations only. Hence, the relationship between $S(\omega,\boldsymbol{q})$ and the autocorrelation function $\langle\Gamma^*(\omega,\boldsymbol{q})\cdot\Gamma(\omega',\boldsymbol{q}')\rangle$ of the concentration fluctuations is simply given by:

$$\langle\Gamma^*(\omega,\boldsymbol{q})\cdot\Gamma(\omega',\boldsymbol{q}')\rangle = \left(\frac{\partial c}{\partial n}\right)_T^2 S(\omega,\boldsymbol{q})\,(2\pi)^4\,\delta(\omega-\omega')\,\delta(\boldsymbol{q}-\boldsymbol{q}'). \tag{10}$$

To deduce the autocorrelation $\langle\Gamma^*(\omega,\boldsymbol{q})\cdot\Gamma(\omega',\boldsymbol{q}')\rangle$ of the concentration fluctuations from (8), we need the correlation functions for the Langevin noise terms $F_1$ and $F_2$. They can be calculated from the known correlation functions between the different components of the random stress and the random solute flux. In nonequilibrium fluctuating hydrodynamics it is assumed that the correlation functions of the random stress tensor and the random solute flux retain their local equilibrium values [43]. From the expressions for the correlation functions of the random stress tensor $\delta\mathsf{T}$ as, for instance, given by (3.12) in [56] and for the correlation functions of the random solute flux $\delta\boldsymbol{J}$, as given by [43,58], and using the definition (9) of the Langevin noise terms, we obtain:

$$\begin{aligned} \langle F_1^*(\omega,\boldsymbol{q})\cdot F_1(\omega',\boldsymbol{q}')\rangle &= 2k_{\mathrm{B}}T\frac{\nu}{\rho}\,q_{\parallel}^2\,q^4\,(2\pi)^4\,\delta(\omega-\omega')\,\delta(\boldsymbol{q}-\boldsymbol{q}'), \\ \langle F_2^*(\omega,\boldsymbol{q})\cdot F_2(\omega',\boldsymbol{q}')\rangle &= 2k_{\mathrm{B}}T\frac{D}{\rho}\left(\frac{\partial c}{\partial\mu}\right)_T q^2\,(2\pi)^4\,\delta(\omega-\omega')\,\delta(\boldsymbol{q}-\boldsymbol{q}'), \\ \langle F_1^*(\omega,\boldsymbol{q})\cdot F_2(\omega',\boldsymbol{q}')\rangle &= \langle F_2^*(\omega,\boldsymbol{q})\cdot F_1(\omega',\boldsymbol{q}')\rangle = 0, \end{aligned} \tag{11}$$

where $\mu = \mu_1 - \mu_2$ is the difference between the chemical potential $\mu_1$ per unit mass of component 1 and the chemical potential $\mu_2$ per unit mass of component

2, while $k_\mathrm{B}$ is Boltzmann's constant. Note that the correlations between the random noise terms $F_1$ and $F_2$ are short ranged both in space and in time so that they can be represented by delta functions. Now, inverting (8) and using (10) and (11) one obtains for the dynamic structure factor $S(\omega, \boldsymbol{q})$ for the Rayleigh scattering in the binary liquid:

$$S(\omega, \boldsymbol{q}) = \left(\frac{\partial n}{\partial c}\right)_T^2 \frac{2k_\mathrm{B}T}{\rho} \frac{Dq^2 \left[\left(\frac{\partial c}{\partial \mu}\right)_T + \frac{(\nabla c_0)^2 \, q_\parallel^2}{\nu D \, q^6}\right]}{\omega^2 + D^2 q^4 \left(1 - \psi \frac{Ra}{Le} \frac{\tilde{q}_\parallel^2}{\tilde{q}^6}\right)^2} . \tag{12}$$

In the denominator of (12) we have introduced a dimensionless wave number $\tilde{q} = qL$, so as to express the effect of gravity $g$ through the Rayleigh number defined by (3), thus facilitating a comparison with previous work [46,49]. The quantity $\psi Ra/Le$ in (12) is given by:

$$\psi \frac{Ra}{Le} = -\psi \frac{D_\mathrm{th}}{D} \frac{\alpha g L^4 \; \nabla T_0}{\nu \; D_\mathrm{th}} = \frac{\beta g L^4 \; \nabla c_0}{\nu \; D} . \tag{13}$$

A convective instability will appear when the expression (12) for $S(\omega, \boldsymbol{q})$ diverges for some finite value of $q$. For such a divergence to appear it is required that $\beta \; \nabla c_0 > 0$, i.e., the concentration of the heavier component has to be higher above. This is the so-called salt-fountain, double-diffusive, or salt-fingers instability [7,62,65]. In our case it appears driven by the Soret effect, but it can also be studied in a more general context considering a concentration gradient independent of the temperature gradient, in which case it is called the thermohaline problem [7,68]. However, we are here considering the situation with $Ra < 0$ (thus $\alpha \; \nabla T_0 > 0$) and $\psi > 0$, so that in our case analyticity of (12) is always assured.

Note that, although we have assumed that the stationary concentration gradient appears through the Soret effect, upon substitution of (13) in (12), our final result for the dynamic structure factor $S(\omega, \boldsymbol{q})$ can be written in a form in which $\nabla T_0$ and $\psi$ do not appear explicitly. This fact suggests that our result will be valid whenever there exists a concentration gradient in the system, independent of whether it is induced by the Soret effect or by any other cause. Hence, we may compare our present results with those of Vailati and Giglio [49], who considered nonequilibrium concentration fluctuations in an isothermal free-diffusion process. We find agreement between our present result (12), and equation (24) of Vailati and Giglio [49] for the contribution of the nonequilibrium concentration fluctuations to the Rayleigh spectrum of scattered light. Vailati and Giglio also obtained a second Lorentzian contribution due to the viscous fluctuation. We have neglected this contribution here, since the hydrodynamic approximation employed by us means in practice a small value of the inverse Schmidt number $Sc^{-1} = D/\nu$. If we would have retained $\partial/\partial t(\nabla^2 w)$ on the left-hand side of (7a) we would have recovered this second Lorentzian contribution as well. A minor difference with the result of Vailati and Giglio is that the latter have assumed small scattering angles, so that $q_{\|} \simeq q$. Equation (12) for $S(\omega, \boldsymbol{q})$ is the generalization to arbitrary scattering angles. The fact that our result for the nonequilibrium Rayleigh spectrum in a binary liquid in the presence of a temperature gradient appears to be identical to that for the isothermal case with a concentration gradient is a direct consequence of the $Le \to 0$ approximation, which neglects the temperature fluctuations and its coupling with concentration fluctuations.

By integrating (12) over all frequencies $\omega$, we obtain for the static structure factor of a nonequilibrium binary liquid due to concentration fluctuations:

$$S(\boldsymbol{q}) = S_{\mathrm{E}} \left\{ 1 + \tilde{S}_{NE}(\boldsymbol{q}) \right\}, \tag{14}$$

with

$$\tilde{S}_{\mathrm{NE}}(\boldsymbol{q}) = \left[ \left( \frac{\partial \mu}{\partial c} \right)_T \frac{(\nabla c_0)^2}{\nu\, D} - \frac{\beta\, g\, \nabla c_0}{\nu\, D} \right] \frac{1}{1 - \psi \dfrac{Ra}{Le} \dfrac{\tilde{q}_{\|}^2}{\tilde{q}^6}} \frac{q_{\|}^2}{q^6}. \tag{15}$$

In (14) and (15), $S_{\mathrm{E}}$ is the equilibrium structure factor which is obtained when $\nabla c_0 = 0$, and $\tilde{S}_{\mathrm{NE}}$ represents the normalized nonequilibrium contribution to the structure factor, which is a dimensionless quantity referred to as nonequilibrium enhancement in several publications [34,51]. As in the case of the dynamic structure factor (12), our result (15) for the static structure factor will be valid only in the $\{Ra < 0, \psi > 0\}$ quadrant of the stability diagram in Fig. 1. In any other case a Soret-driven double diffusive instability will appear as a divergence in the structure factor for some finite value of $q_{\|}$. A problem is that (15) has also no divergence for $Ra > 0$ and $\psi < 0$, where a instability in the form of a Hopf bifurcation is expected [7,10,63,65]. The fact that we do not find this instability here is a consequence of the $Le \to 0$ approximation [7,52]. Also in deriving (15) we did not take into account finite-size effects. Linear instability theory of the binary Boussinesq equations suggests that, when boundary conditions are considered together with (7), a modified structure factor could be calculated which would be valid in some bounded region outside the $\{Ra < 0, \psi > 0\}$ quadrant (the region where the conductive solution is stable). Again as before, the static structure factor given by (15) can be written in such a way that it depends explicitly only on the value of the stationary concentration gradient $\nabla c_0$ and not on $\psi$ or $\nabla T_0$. Consequently (15) can be also employed to analyze nonequilibrium concentration fluctuations in isothermal diffusion processes [69–71].

As will be shown in Section 3, for the interpretation of several light-scattering experiments in liquids far away from any convective instability we may neglect the effects of gravity by taking the limit $g \to 0$. In this limit (15) reduces to:

$$\tilde{S}_{\mathrm{NE}}(\boldsymbol{q}) = \left( \frac{\partial \mu}{\partial c} \right)_T \frac{(\nabla c_0)^2}{\nu\, D} \frac{q_{\|}^2}{q^6}, \tag{16}$$

which has been verified experimentally [51], as further discussed below. Note that (16), contrary to (15), does not include any divergence and is valid for any positive or negative value of $\nabla c_0$. This is expected since in the derivation of (16) gravity was neglected, so that the system is always in a stable quiescent state.

We remark that the derivation of the structure factor from (7) described above can just as easily be implemented without neglecting the time derivative $\partial/\partial t(\nabla^2 w)$ on the left-hand side of (7a). Instead of (16) we would then have obtained:

$$\tilde{S}_{\mathrm{NE}}(\boldsymbol{q}) = \left(\frac{\partial\mu}{\partial c}\right)_T \frac{(\nabla c_0)^2}{(\nu^2 - D^2)} \frac{\nu}{D} \frac{q_\parallel^2}{q^6}, \tag{17}$$

which for $\nu \gg D$ reduces to (16). We have preferred to perform the derivation in the way it was done to retain consistency with the hydrodynamic approximations as originally introduced by Velarde and Schechter [52].

As already pointed out, the various expressions derived here for the contribution of the nonequilibrium concentration fluctuations to the structure factor are valid for wave numbers $q$ such that $qL \ll 1$. When $qL$ becomes of order unity, finite-size effects need to be considered which depend to some extent on the nature of the boundary conditions [41,42]. Finite-size effects on the nonequilibrium concentration fluctuations in binary liquids will be considered in a future publication [47].

## 3 Experiments

During the past years several experimental results on nonequilibrium Soret-driven concentration fluctuations have appeared in the literature. The goal of this section is to make a comparison between experimental results and the theory developed in Sect. 2. Two different experimental techniques have been employed to study nonequilibrium concentration fluctuations in liquid mixtures: light scattering and, more recently, quantitative shadowgraph analysis. Experiments have been performed with binary liquid mixtures and with dilute polymer solutions. In the case of dilute polymer solutions, as long as the concentration is low enough to neglect entanglements, an hydrodynamic approach of the type described in Sect. 2 should be applicable.

### 3.1 Experimental techniques

A nonequilibrium enhancement of the concentration fluctuations in binary liquid systems has indeed been observed by small-angle light-scattering experiments [34,35,39,40,50,51]. A schematic representation of such a light-scattering experiment is shown in Fig. 2. The scattering medium is a thin horizontal fluid layer bounded by two parallel plates of high thermal conductivity. The temperatures of the plates $T_1$ and $T_2$ can be controlled independently so as to establish a temperature gradient across the fluid layer in the vertical direction, parallel or antiparallel to the direction of gravity. The horizontal plates are furnished with windows allowing a laser beam to propagate through the fluid in the direction

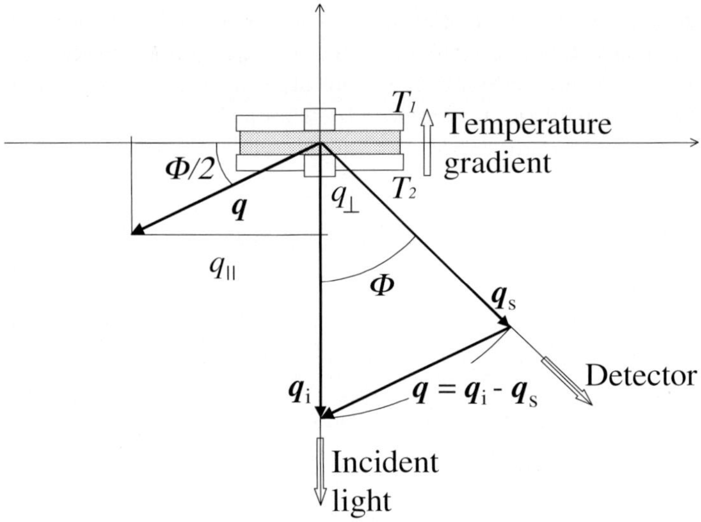


**Fig. 2.** Schematic representation of a low-angle nonequilibrium light-scattering experiment. $\boldsymbol{q}_\mathrm{i}$ is the wave vector of the incident light, $\boldsymbol{q}_\mathrm{s}$ is the wave vector of the scattered light. The magnitude $q = |\boldsymbol{q}_\mathrm{i} - \boldsymbol{q}_\mathrm{s}|$ of the scattering wave vector is related to the scattering angle $\Phi$ by $q = 2q_0 \sin(\Phi/2)$, where $q_0$ is the magnitude of the wave vector $\boldsymbol{q}_\mathrm{i}$ of the incident light inside the liuqid. For clarity, the scattering angle $\Phi$ has been exaggerated.

(anti)parallel to the temperature gradient. Light scattered over an angle $\Phi$ arises from fluctuations with a wave number $q$ such that [72]:

$$q = 2q_0 \sin(\Phi/2), \tag{18}$$

where $q_0$ is the wave number of the incident light inside the fluid medium. Due to the dependence of the nonequilibrium enhancement (16) of the structure factor on $q^{-4}$, to observe any nonequilibrium enhancement one needs to measure the scattered light at small wave numbers and, hence, at very small scattering angles $\Phi$. In practice this means that one measures the structure factor at $q_{\|} = q$ and $q_\perp = 0$, where $q_\perp$ is the component of the wave vector $\boldsymbol{q}$ along the vertical direction. The need for very small scattering angles complicates the experimental technique. To avoid uncertainties in the scattering angle due to laser beam bending inside the sample the two horizontal plates need to be parallel to a high degree and the laser beam has to be carefully aligned perpendicularly to the windows. In addition, at these small scattering angles stray scattering from window surfaces is dominant. Thus to obtain enough signal to background ratio great care is required to have always very clean windows.

Nonequilibrium concentration fluctuations can be observed by both static and dynamic light scattering. For static light scattering a photodiode array can be employed to measure the total intensity of light. For dynamic light scattering the light collected in the detector has to be analyzed with a correlator. In the experiments one first measures the equilibrium light scattering by maintaining the temperatures of the two horizontal plates at the same value, $T_1 = T_2$. Subsequently, a temperature gradient is established by independently changing the temperatures of the two plates bounding the fluid layer. Comparing the intensity of light collected in the detector (or the amplitude of the autocorrelation function of the scattered light in dynamic experiments) in thermal equilibrium and out of thermal equilibrium, experimental values for the nonequilibrium enhancement are obtained.

Recently nonequilibrium concentration fluctuations have also been investigated by a shadowgraph projection technique [70]. The experimental arrangement is similar to the one depicted in Fig. 2, but instead of a laser beam, an extended uniform monochromatic light source is employed to illuminate the sample. Then many shadowgraph images of a plane perpendicular to the temperature gradient are obtained with a charge coupled device (CCD) detector, which measures a spatial distribution of intensity $I(\boldsymbol{x})$, where here $\boldsymbol{x}$ is a two-dimensional position vector in the imaging plane. For each image a shadowgraph signal is defined by:

$$i(\boldsymbol{x}) = \frac{I(\boldsymbol{x}) - I_0(\boldsymbol{x})}{I_0(\boldsymbol{x})}, \tag{19}$$

where $I_0(\boldsymbol{x})$ is the blank intensity distribution, when there are no fluctuations in the index of refraction of the sample. In practice $I_0(\boldsymbol{x})$ is obtained by averaging over many shadowgraph images, so that fluctuations cancel out and the resulting $I_0(\boldsymbol{x})$ contains only contributions coming from nonuniform illumination of the sample. Using physical and geometrical optics one can demonstrate that the modulus squared of the two-dimensional Fourier transform of the shadowgraph signal, $|i(\boldsymbol{q})|^2$, taking an azimuthal average can be expressed as [70,73,74]:

$$\overline{|i(\boldsymbol{q})|^2} = \hat{T}(q)\ S(q_\parallel = q, q_\perp = 0). \tag{20}$$

The azimuthal average of $|i(\boldsymbol{q})|^2$, indicated by the overline, depends only on the modulus $q$ of the two-dimensional Fourier vector $\boldsymbol{q}$. The symbol $\hat{T}(q)$ represents an optical transfer function, which can be derived from the optical arrangement employed to produce the shadowgraph pictures; it includes contributions from the response of the CCD detector and from the dependence of the refractive index on temperature and concentration [70,74]. Therefore, from an analysis of the spectrum of the shadowgraph pictures one can determine the structure factor of the fluid in the plane $q_\perp = 0$.

We note that quantitative shadowgraph experiments so far performed by Brogioli et al. [70] refer to nonequilibrium concentration fluctuations not induced by the Soret effect, but induced by free diffusion. Nevertheless, as mentioned in Sect. 2.4, the final theoretical result, given by (15), depends only on the value of the stationary concentration gradient $\nabla c_0$. Hence, this equation remains valid independently of whether the concentration gradient is induced by a temperature gradient through the Soret effect or by any other process, as in a free-diffusion process. Equation (15) has indeed been employed successfully to analyze nonequilibrium concentration fluctuations in diffusion processes [69]. For these reasons, and also because this newly developed experimental technique is perfectly suited for the study of nonequilibrium concentration fluctuations induced by thermodiffusion we have briefly sketched the fundamentals of the shadowgraph method.

There is an equivalence between small-angle light-scattering and shadowgraph experiments, in the sense that both methods give us $S(q_{\|} = q, q_{\perp} = 0)$, where for light scattering $q$ is the scattering wave vector as given by (18), whereas for shadowgraph techniques $q$ is the modulus of the two-dimensional Fourier vector in the imaging plane. As mentioned by Bodenschatz et al., what it is measured in the experiments is a kind of vertical average of the fluctuations [61].

### 3.2 Nonequilibrium concentration fluctuations in binary liquid mixtures

Experimental measurements of the nonequilibrium enhancement of the concentration fluctuations induced by the Soret effect in binary liquid mixtures have been independently obtained at the University of Maryland [34,35,75,76] and at the University of Milan [39,40]. At the University of Maryland, Sengers and coworkers have performed nonequilibrium Rayleigh-scattering experiments in mixtures of liquid toluene and liquid n-hexane, far from the critical point of demixing. Because they used dynamic techniques, they were able to determine not only the enhancement of the scattering due to the nonequilibrium concentration fluctuations, but also the contribution from the nonequilibrium viscous and temperature fluctuations. In this mixture the Lewis number is about $Le = 3 \times 10^{-2}$ which is small enough for the viscous and temperature fluctuations to decay sufficiently fast so as to enable an independent determination of the nonequilibrium enhancement of the concentration fluctuations directly from the long-time behavior of the autocorrelation function of the scattered-light intensity. A compilation of the experimental values obtained for the magnitude of the nonequilibrium enhancement of the concentration fluctuations , designated by $A_c$ and to be identified with $\tilde{S}_{\mathrm{NE}}$ as given by (16) or (17), can be found in Tables IV-VI of [35] for various values of the temperature gradient $\nabla T_0$, the wave number $q$ and the average concentration $c$. In these experiments the wave number $q$ was large enough so that gravity effects could be neglected, but at the same time sufficiently small to identify $q$ with $q_{\|}$ and neglect $q_{\perp}$. It thus follows from (16) or (17) that $\tilde{S}_{\mathrm{NE}} = A_c$ should be proportional to $(\nabla T_0)^2/q^4$:

$$\tilde{S}_{\mathrm{NE}} \propto \frac{\nabla T_0^2}{q^4} . \tag{21}$$

As an illustration, we show in Fig. 3 observed values of $\tilde{S}_{\mathrm{NE}} = A_c$ as a function of $(\nabla T_0)^2/q^4$ obtained for two concentrations of the mixture toluene + n-hexane at 25°C [35]. The experiments confirm that the nonequilibrium enhancement of the concentration fluctuations is indeed directly proportional to $(\nabla T_0)^2$ and inversely proportional to $q^4$ [34,35,75,76]. Note that the $q^{-4}$ dependence for small $q$, verified by Fig. 3, means that the correlations in real space are long-ranged, encompassing the entire system not involving any intrinsic length

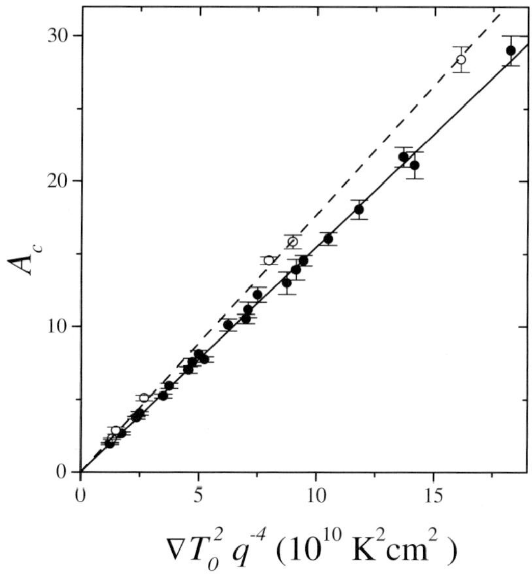


**Fig. 3.** Nonequilibrium enhancement of the concentration fluctuations $A_c$ as a function of $\nabla T_0^2 q^{-4}$ at two different concentrations of toluene in n-hexane. The open symbols are for a mole fraction of toluene $x = 0.5$ and the filled symbols are for a mole fraction of toluene $x = 0.75$. The solid lines represent linear fits to the data. From [35].

scale, only quenched by finite-size effects when $q$ becomes of the order $L^{-1}$, the finite height of the system acting as a natural cutoff for the wave vector of the fluctuations [41,42,47].

For a quantitative interpretation of the observed nonequilibrium enhancement of the concentration fluctuations one needs reliable values for the derivative $(\partial\mu/\partial c)_T$, the kinematic viscosity $\nu$, the mass diffusion coefficient $D$ and the Soret coefficient $S_T$. An accurate optical beam-bending method was developed to measure $D$ and the Soret coefficient $S_T$ of the liquid mixture [77], but data for $(\partial c/\partial\mu)_T$ and $\nu$ had to be deduced from the literature [35]. In spite of these efforts a quantitative confirmation of the accuracy of the prefactors in (16) and (17) has not yet been obtained for liquid mixtures [34,35,75,76], but it has been obtained for polymer solutions, as will be further discussed below [50,51].

Vailati and Giglio measured nonequilibrium concentration fluctuations induced by the Soret effect in a mixture of liquid aniline and cyclohexane [39,40]. Light-scattering experiments were performed in the homogeneous phase, but not very far from the critical consolute point. Due to critical slowing down, concentration fluctuations are dominant in this situation, assuring no contribution of temperature or viscous fluctuations to the total scattering intensity. This scenario corresponds to a negligibly small Lewis number and the light-scattering intensity should be given by (15). In addition, the experimental device used by Vailati and Giglio allowed them to measure the scattering intensity at angles sufficiently small to observe gravity effects. To simplify the analysis of the experiments, Vailati and Giglio introduced a "roll-off" scattering wave number defined as [39]:

$$q_{\mathrm{RO}}^4 = -\frac{\beta\, g\, \nabla c_0}{\nu\, D} = -\psi \frac{Ra}{Le} \frac{1}{L^4}. \tag{22}$$

Note again that the experiments are performed for $Ra < 0$ and $\psi > 0$, so that the right-hand side of (22) is positive and $q_{\mathrm{RO}}$ is a real quantity. With this definition, we obtain from (15):

$$\tilde{S}_{\mathrm{NE}} \propto \frac{\nabla T_0^2}{q^4 + q_{\mathrm{RO}}^4}. \tag{23}$$

Note that for large $q$ (23) reduces to (21). But, in the limit $q \to 0$, the $q^{-4}$ divergence in (21) becomes saturated and (23) predicts that a constant limiting value for $\tilde{S}_{\mathrm{NE}}$ will be reached [46].

The difference between the equilibrium and the nonequilibrium intensity of the scattered light, which is proportional to $\tilde{S}_{\mathrm{NE}}$ as measured by Vailati and Giglio [39] for a mixture of aniline and cyclohexane is shown in Fig. 4 as a function of the wave number $q$. The solid curve represents a fit to (23). Because of the small scattering angles covered in this experiment, the crossover from a $q^{-4}$ divergence to the gravitationally induced saturation at very small $q$ is clearly observed. From the fit of (23) to the experimental data, Vailati and Giglio deduced $q_{\mathrm{RO}} = 537\ \mathrm{cm}^{-1}$, which is close to the theoretical value $q_{\mathrm{RO}} = 410\ \mathrm{cm}^{-1}$

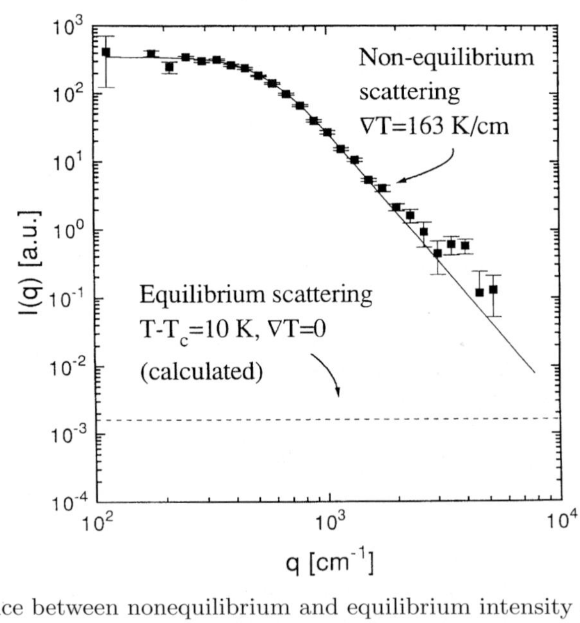


**Fig. 4.** Difference between nonequilibrium and equilibrium intensity of scattered light for a mixture of aniline and cyclohexane as a function of the scattering vector. The symbols represent experimental data obtained with $\nabla T_0 = 163\ \mathrm{K\ cm}^{-1}$. the solid curve represents a fit to (23). Reproduced with permission from [39].

they obtained from (22) using literature values for the thermophysical properties of this mixture. We should note that $\tilde{S}_{\mathrm{NE}}$ is plotted in Fig. 4 in arbitrary units. The authors did not have the resolution to determine the absolute magnitude of the nonequilibrium enhancement of the concentration fluctuations with sufficient accuracy to make a reliable check of the validity of the prefactor in (16) or (17).

### 3.3 Nonequilibrium concentration fluctuations in dilute polymer solutions

Experimental values for the nonequilibrium enhancement $\tilde{S}_{\mathrm{NE}}$ of concentration fluctuations have been obtained for dilute solutions of polystyrene ($M_W$ = 96 400) in toluene from low-angle dynamic light-scattering at the University of Maryland [51,50,78]. As described in Sect. 3.1, the nonequilibrium enhancements were determined by comparing the observed amplitude of the time correlation function measured in equilibrium and out of equilibrium. The experimental values thus obtained for $\tilde{S}_{\mathrm{NE}}$ at 25°C can be found in Table II of [50], where they are denoted again by $A_c$.

To make a quantitative comparison of the experimental data with the theoretical expression (16) for $\tilde{S}_{\mathrm{NE}}$, we have presented in Table 1 values for the various thermophysical properties of the polymer solutions at various polymer concentrations collected from the literature [50,79,80]. In this Table $S_T$ represents the Soret coefficient of the polymer in toluene as the solvent, as measured by Zhang et al. [80] by an optical beam-bending technique. The values for the solute expansion coefficient $\beta$ in Table 1 were calculated from an empirical correlation for the dependence of the density $\rho$ on the concentration, at 25°C, proposed by Scholte [79]. For the other properties, we refer to [50]. If we estimate the thermal expansion coefficient $\alpha$ of the polymer solution by the value corresponding to pure toluene at 25°C: $\alpha = 0.97 \times 10^{-3}$ K$^{-1}$, we note from (2) that the separation ratio $\psi$ is positive, ranging from $\psi = 0.121$ for $c = 0.25\%$ to $\psi = 1.101$ for $c = 4.00\%$. Hence, the polymer solution is indeed in a stable thermally conducting state when heated from above (i.e., $Ra < 0$). If we estimate the thermal diffusivity of the dilute solution by the value $D_{\mathrm{th}} = 0.89 \times 10^{-3}$ cm$^2$

**Table 1.** Thermophysical properties of a solution of polystyrene ($M_W$ = 96 400) in toluene at 25°C at several polymer concentrations $c$, as obtained from various literature sources [50].

| $c$ | $(\frac{\partial\mu}{\partial c})_T$ | $D$ | $\nu$ | $S_T$ | $\rho^{\mathrm{a}}$ | $\beta$ |
|---|---|---|---|---|---|---|
| (w/w) | cm$^2$ s$^{-2}$ | cm$^2$ s$^{-1}$ | cm$^2$ s$^{-1}$ | K$^{-1}$ | g cm$^{-3}$ | (w/w)$^{-1}$ |
| 0.005 | $7.48 \times 10^7$ | $5.16 \times 10^{-7}$ | $7.71 \times 10^{-3}$ | 0.218 | 0.8627 | −0.208 |
| 0.015 | $4.41 \times 10^7$ | $6.01 \times 10^{-7}$ | $10.8 \times 10^{-3}$ | 0.178 | 0.8645 | −0.209 |
| 0.020 | $4.16 \times 10^7$ | $6.51 \times 10^{-7}$ | $12.6 \times 10^{-3}$ | 0.167 | 0.8654 | −0.209 |
| 0.040 | $4.25 \times 10^7$ | $8.32 \times 10^{-7}$ | $21.4 \times 10^{-3}$ | 0.132 | 0.8690 | −0.209 |

[a] From Scholte [79].

s$^{-1}$ corresponding to pure toluene at 25°C, we conclude that $Le \simeq 7 \times 10^{-4}$. Hence the assumption of small Lewis number is well justified and (16) agrees indeed with the experimental data [50,51,78].

An elegant comparison between theory and experiment can be obtained in terms of a scaled universal representation of $\tilde{S}_{\mathrm{NE}}$ adopted by Brogioli et al. [70] for studying concentration fluctuations in free-diffusion processes. For this purpose we note that (15), with the definition (22) for $q_{\mathrm{RO}}$ and the small-angle approximation $q_{\|} \simeq q$, can be rewritten as:

$$\frac{\tilde{S}_{\mathrm{NE}}(s)}{\tilde{S}^{0}_{\mathrm{NE}}} = \frac{1}{1+s^4} \tag{24}$$

with

$$\tilde{S}^{0}_{\mathrm{NE}} = 1 + \left(\frac{\partial \mu}{\partial c}\right)_T \frac{(\nabla c_0)^2}{\nu D\, q^4_{\mathrm{RO}}}, \tag{25}$$

where $s = q/q_{\mathrm{RO}}$ is a scaled wave number. We note that $\tilde{S}^{0}_{\mathrm{NE}}$ given by (25) simply represents the nonequilibrium enhancement of the concentration fluctuations in the limit $q \to 0$.

The scale factors $\tilde{S}^{0}_{\mathrm{NE}}$ and $q_{\mathrm{RO}}$ in (24) depend on the thermophysical properties of the solution and on the value of the stationary concentration gradient $\nabla c_0$, which in turn depends on the applied stationary temperature gradient $\nabla T_0$ through the Soret coefficient $S_T$ in accordance with (1). For each pair of values $c$ and $\nabla T_0$, $q_{\mathrm{RO}}$ and $\tilde{S}^{0}_{\mathrm{NE}}$ can be calculated from the literature data presented in Table 1. If we then scale the experimental data for $\tilde{S}_{\mathrm{NE}}$ by $\tilde{S}^{0}_{\mathrm{NE}}$ and the wave number $q$ by $q_{\mathrm{RO}}$, a universal curve should be obtained according to (24). We have performed such an analysis for the experimental data obtained for the polystyrene solution in toluene [50] and the results are shown in Fig. 5. The insert in Fig. 5 shows in more detail the data in the actual range of experimental wave numbers. To make Fig. 5 more legible we have plotted only experimental data corresponding to three polymer concentrations, but there is no difference if the data obtained for all different polymer solutions are included. It should be emphasized that the curve in Fig. 5, unlike the ones in Fig. 3 or in Fig. 4, is not a fit of the experimental data but is just the universal function defined in (24). Although the experiment does not cover small enough wave numbers for the saturation effect due to gravity to be observed, Fig. 5 does show excellent agreement between theory and experiment. In fact, the light-scattering data obtained for the polystyrene solution in toluene represent the first accurate confirmation of the theory for the nonequilibrium concentration fluctuations, including the prefactor in (16).

### 3.4 Observation of nonequilibrium concentration fluctuations by shadowgraphy

Before closing we wish to comment shortly on the recent observation of nonequilibrium concentration fluctuations by Brogioli et al. from a quantitative shad-

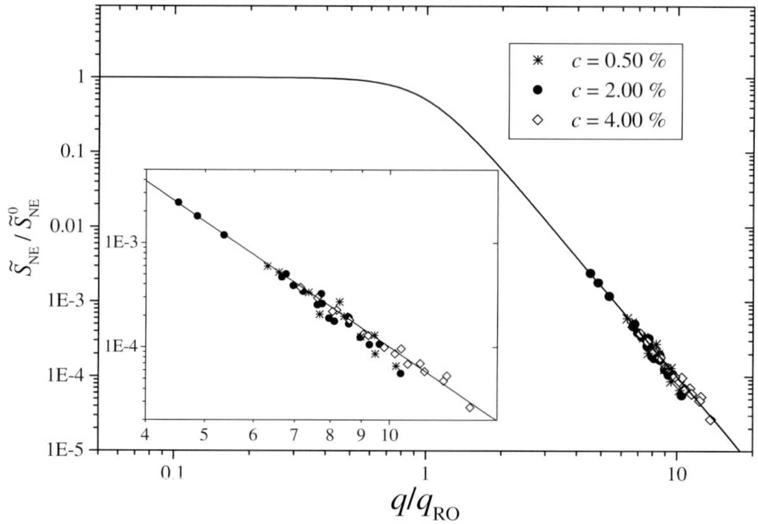


**Fig. 5.** Value of the scaled nonequilibrium enhancement of the concentration fluctuations as a function of the scaled scattering wave number for solutions of polystyrene in toluene at several concentrations. The solid curve represents the universal function (24). The insert shows the range were the actual experimental data have been obtained [50].

owgraph analysis [70]. As earlier mentioned, this work is concerned with concentration gradients induced not by the Soret effect, but by a free-diffusion process. The experimental setup is similar to the one depicted in Fig. 2 with the exception that both plates are maintained at the same temperature. Brogioli et al. studied free-diffusion in pure water of concentrated aqueous solutions of several solutes, including glycerol and the polymer PolyEthyleneGlycol. The initial state was prepared by filling the cell with the lighter liquid and carefully injecting the denser liquid from below, so that a sharp meniscus is formed. From this initial state free diffusion proceeds until an uniform concentration is reached inside the cell.

As commented earlier, nonequilibrium concentration fluctuations in this system should be described by (15), if one replaces the stationary concentration gradient $\nabla c_0$ by $\nabla c(z,t)$, the actual concentration gradient which will depend on the time $t$ and on the vertical coordinate $z$. This concentration gradient $\nabla c(z,t)$ can be calculated by solving the free-diffusion equation with a step function as initial condition. Averaging over $z$ one obtains a time-dependent vertically averaged structure factor, which is the quantity measured in the shadowgraph experiments. The averaged structure factor has an expression similar to (24), with a time-independent amplitude $\tilde{S}^0_{\mathrm{NE}}$, but with a time-dependent roll-off wave vector $q_{\mathrm{RO}}(t)$. The details of such a calculation can be found in the corresponding literature [49].

The important point is that Brogioli et al. [70] found excellent quantitative agreement between theory and experiments, both for polymer solutions and for binary liquid mixtures. We may thus conclude that the theoretical prediction (15) for the nonequilibrium concentration fluctuations has been confirmed experimentally. This recent quantitative confirmation of the working equation for nonequilibrium concentration fluctuations for free-diffusion processes in binary liquid mixtures makes additional quantitative measurements for liquid mixtures in the case when the concentration gradient is induced by thermal diffusion even more desirable.

## 4 Concluding remarks

The subject of Soret-driven nonequilibrium concentration fluctuations in binary liquid systems is very interesting indeed. In this chapter we have shown how the effects of nonequilibrium concentration fluctuations on the structure factor of a binary liquid can be derived from the linearized Boussinesq equations assuming that both the thermal diffusivity $D_{\mathrm{th}}$ and the kinematic viscosity $\nu$ are much larger than the mutual mass diffusion coefficient $D$. From the experiments performed so far as reviewed in this chapter it can be said that the theory is essentially verified. Two independent research groups have obtained equivalent results by using two different experimental methods. Nevertheless it would be desirable to obtain a better quantitative agreement for binary liquid mixtures subjected to a stationary temperature gradient. A very good knowledge of the thermophysical properties of the mixtures is required for this purpose. It would also be desirable to modify the theory for the nonequilibrium concentration fluctuations to take into account boundary conditions so as to extend its validity from the $Ra < 0$, $\psi > 0$ quadrant in the stability diagram of Fig. 1 to the full region where the conductive solution is stable.

## Acknowledgements

We have benefitted from a fruitful collaboration with R.W. Gammon in an earlier stage of this project. We are indebted to J.K. Platten for providing us with a copy of his publications. The research at the University of Maryland is supported by the Chemical Sciences, Geosciences and Biosciences Division of the Office of Basic Energy Sciences of the U.S. Department of Energy under Grant No. DE-FG-02-95ER-14509.

## References

1. H.J.V. Tyrrell: *Diffusion and Heat Flow in Liquids.* (Butterworths, London 1961)
2. S.R. de Groot, P. Mazur: *Non-Equilibrium Thermodynamics.* (North-Holland, Amsterdam 1962)
3. D.D. Fitts: *Nonequilibrium Thermodynamics.* (McGraw-Hill, New York 1962)
4. R. Haase: *Thermodynamics of Irreversible Processes.* (Addison-Wesley, Reading MA 1969)

5. J.-L. Lin, W.L. Taylor, W.M. Rutherford: 'Secondary Coefficients'. In: *Measurement of the Transport Properties of Fluids*, ed. by W.A. Wakeham, A. Nagashima, J.V. Sengers (Blackwell Scientific, Oxford 1991) pp. 323–387
6. R.S. Schechter, I. Prigogine, J.R. Hamm: Phys. Fluids **15**, 379 (1972)
7. R.S. Schechter, M.G. Velarde, J.K. Platten: 'The Two-Component Bénard Problem'. In: *Advances in Chemical Physics*, Vol. 26, ed. by I. Prigogine, S.A. Rice (Wiley, New York 1974) pp. 265–301
8. J.K. Platten, G. Chavepeyer: 'Finite Amplitude Instability in the Two-Component Bénard Problem'. In: *Advances in Chemical Physics*, Vol. 32, ed. by I. Prigogine, S. Rice (Wiley, New York 1975) pp. 281–322
9. J.K. Platten, J.C. Legros: *Convection in Liquids.* (Springer, Berlin 1984)
10. M.C. Cross, P.C. Hohenberg: Rev. Mod. Phys. **65**, 851 (1993)
11. St. Hollinger, M. Lücke: Phys. Rev. E **57**, 4238 (1998)
12. N.N. Bogoliubov: J. Phys. (USSR) **10**, 256, 265 (1946)
13. N.N. Bogoliubov: 'Problems of a Dynamical Theory in Statistical Physics'. In: *Studies in Statistical Mechanics I, Part A*, ed. by J. de Boer, G.E. Uhlenbeck (North-Holland, Amsterdam 1962) pp. 5–118
14. J.R. Dorfman, T.R. Kirkpatrick, J.V. Sengers: Annu. Rev. Phys. Chem. **45**, 215 (1994)
15. J.V. Sengers: 'Behavior of Viscosity and Thermal Conductivity of Fluids Near the Critical Point'. In: *Critical Phenomena*, ed. by M.S. Green, J.V. Sengers, National Bureau of Standards Miscellaneous Publication 273 (U.S. Governement Printing Office, Washington, DC 1966) pp. 165–178
16. M. Fixman: J. Chem. Phys. **47**, 2808 (1967)
17. K. Kawasaki: Ann. Phys. (NY) **61**, 1 (1970)
18. L.P. Kadanoff, J. Swift: Phys. Rev. **166**, 89 (1968)
19. P.C. Hohenberg, B.L. Halperin: Rev. Mod. Phys. **49**, 435 (1977)
20. Y. Pomeau, P. Résibois: Phys. Reports **19**, 63 (1975)
21. M.H. Ernst, E.H. Hauge, J.M.J. van Leeuwen: J. Stat. Phys. **15**, 7,23 (1976)
22. W.W. Wood: J. Stat. Phys. **57**, 675 (1989)
23. R.F. Fox: J. Phys. Chem. **86**, 2812 (1982)
24. T.R. Kirkpatrick, E.G.D. Cohen, J.R. Dorfman: Phys. Rev. A **26**, 995 (1982)
25. D. Ronis, I. Procaccia: Phys. Rev. A **26**, 1812 (1982)
26. R. Schmitz, E.G.D. Cohen: J. Stat. Phys. **40**, 431 (1985)
27. I. L'Heureux, I. Oppenheim: Physica **148A**, 503 (1988)
28. R. Schmitz: Phys. Reports **171**, 1 (1988)
29. B.M. Law, J.V. Sengers: J. Stat. Phys. **57**, 531 (1989)
30. J.V. Sengers, B.M. Law: 'Long-range Correlations in a Fluid out of Thermal Equilibrium'. In: *Lectures on Thermodynamics and Statistical Mechanics*, ed. by M. López de Haro, C. Varea (World Scientific, Singapore 1990) pp. 201–218
31. J.V. Sengers, B.M. Law: 'Thermal Fluctuations in a Quiescent Liquid far from Equilibrium'. In: *Nonlinear and Chaotic Phenomena in Plasmas, Solids and Fluids*, ed. by W. Rozmus, J.A. Tuszynski (World Scientific, Singapore 1991) pp. 574–581
32. B.M. Law, P.N. Segrè, R.W. Gammon, J.V. Sengers: Phys. Rev. A **41**, 816 (1990)
33. P.N. Segrè, R.W. Gammon, J.V. Sengers, B.M. Law: Phys. Rev. A **45**, 714 (1992)
34. P.N. Segrè, R.W. Gammon, J.V. Sengers: : Phys. Rev. E **47**, 1026 (1993)
35. W.B. Li, P.N. Segrè, R.W. Gammon, J.V. Sengers: Physica A **204**, 399 (1994)
36. G. Grinstein, D.-H. Lee, S. Sachdev: Phys. Rev. Lett. **64**, 1927 (1990)
37. G. Grinstein: J. Appl. Phys. **69**, 5441 (1991)

38. P.N. Segrè, R. Schmitz, J.V. Sengers: Physica A **195**, 31 (1993)
39. A. Vailati, M. Giglio: Phys. Rev. Lett. **77**, 1484 (1996)
40. A. Vailati, M. Giglio: Prog. Colloid. Polym. Sci. **104**, 76 (1997)
41. J.M. Ortiz de Zárate, R. Pérez Cordón, J.V. Sengers: Physica A **291**, 115 (2001)
42. J.M. Ortiz de Zárate, L. Muñoz Redondo: Euro. Phys. J., **21**, 135 (2001)
43. B.M. Law, J.C. Nieuwoudt: Phys. Rev. A **40**, 3880 (1989)
44. J.C. Nieuwoudt, B.M. Law: Phys. Rev. A **42**, 2003 (1990)
45. R.M. Velasco, L.S. García-Colín: J. Phys. A: Math. Gen. **24**, 1007 (1991)
46. P.N. Segrè, J.V. Sengers: Physica A **198**, 46 (1993)
47. J.V. Sengers, J.M. Ortiz de Zárate: Revista Mexicana de Física, to be published (2001)
48. R. Schmitz: Physica A **206**, 25 (1994)
49. A. Vailati, M. Giglio: Phys. Rev. E **58**, 4361 (1998)
50. W.B. Li, K.J. Zhang, J.V. Sengers, R.W. Gammon, J. M. Ortiz de Zárate: J. Chem. Phys. **112**, 9139 (2000)
51. W.B. Li, K.J. Zhang, J.V. Sengers, R.W. Gammon, J. M. Ortiz de Zárate: Phys. Rev. Lett. **81**, 5580 (1998)
52. M.G. Velarde, R.S. Schechter: Phys. Fluids **15**, 1707 (1972)
53. L.D. Landau, E.M. Lifshitz: *Fluid Mechanics.* (Addison-Wesley, Reading MA 1959)
54. D. Ronis, I. Procaccia, J. Machta: Phys. Rev. A **22**, 714 (1980)
55. S. Chandrasekhar: *Hydrodynamic and Hydromagnetic Stability.* (Clarendon, Oxford 1961)
56. R. Schmitz, E.G.D. Cohen: J. Stat. Phys. **39**, 285 (1985)
57. P.C. Hohenberg, J.B. Swift: Phys. Rev. A **46**, 4773 (1992)
58. C. Cohen, J.W.H. Sutherland, J.M. Deutch: Phys. Chem. Liquids **2**, 213 (1971)
59. W. Schöpf, W. Zimmermann: Phys. Rev. E **47**, 1739 (1993)
60. W. Hort, S.J. Linz, M. Lücke: Phys. Rev. A **45**, 3737 (1992)
61. E. Bodenschatz, W. Pesch, G. Ahlers: Annu. Rev. Fluid Mech. **32**, 709 (2000)
62. St. Hollinger, M. Lücke, H. W. Müller: Phys. Rev. E **57**, 4250 (1998)
63. M. Lücke, W. Barten, P. Büchel, F. Fütterer, St. Hollinger, Ch. Jung: 'Pattern Formation in Binary Fluid Convection and in Systems with Throughflow'. In: *Evolution of Spontaneous Structures in Dissipative Continuous Systems*, ed. by F.H. Busse, S.C. Müller, Lecture Notes in Physics, vol. m55 (Springer, Berlin 1998) pp. 128–196
64. A. La Porta, C.M. Surko: Phys. Rev. Lett. **80**, 3759 (1998)
65. M.C. Cross, K. Kim: Phys. Rev. A **37**, 3909 (1988)
66. J.C. Legros, J.K. Platten, P.G. Poty: Phys. Fluids **15**, 1383 (1972)
67. V. De Giorgio: Phys. Rev. Lett. **41**, 1293 (1978)
68. J.S. Turner: Annu. Rev. Fluid Mech. **17**, 11 (1985)
69. A. Vailati, M. Giglio: Nature **390**, 262 (1997)
70. D. Brogioli, A. Vailati, M. Giglio: Phys. Rev. E **61**, R1 (2000); J. Phys.: Condens. Matter **12**, A39 (2000)
71. P. Cicuta, A. Vailati, M. Giglio: Phys. Rev. E **62**, 4920 (2000)
72. B.J. Berne, R. Pecora: *Dynamic Light Scattering.* (Wiley, New York 1976)
73. M. Wu, G. Ahlers, D.S. Cannell: Phys. Rev. Lett. **75**, 1743 (1995)
74. J.R. de Bruyn, E. Bodenschatz, S.W. Morris, Y. Hu, D.S. Cannell, G. Ahlers: Rev. Sci. Instrum. **67**, 2043 (1996)
75. W.B. Li, P.N. Segrè, J.V. Sengers, R.W. Gammon: J. Phys.: Condens. Matter **6**, A119 (1994)

76. W.B. Li, J.V. Sengers, R.W. Gammon, P.N. Segrè: Int. J. Thermophys. **16**, 23 (1995)
77. K.J. Zhang, M.E. Briggs, R.W. Gammon, J.V. Sengers: J. Chem. Phys. **104**, 6881 (1996)
78. J.V. Sengers, R.W. Gammon, J. M. Ortiz de Zárate: 'Thermal-Diffusion Driven Concentration Fluctuations in a Polymer Solution'. In: *Computational Studies, Nanotechnology and Solution Thermodynamics*, ed. by M. Dadmun, D. Noid, B. Sumpter, Y. Melnichenko (Kluwer, New York 2000) pp. 37–44
79. Th.G. Scholte: J. Polym. Sci. Part A-2, Polym. Phys. **8**, 841 (1970)
80. K.J. Zhang, M.E. Briggs, R.W. Gammon, J.V. Sengers, J.F. Douglas: J. Chem. Phys. **111**, 2270 (1999)


## Supplementary glossary

| | |
|---|---|
| $\boldsymbol{r}(x, y, z)$ | position vector |
| $L$ | finite height of the fluid layer |
| $\theta(\boldsymbol{r}, t)$ | local fluctuation of the temperature |
| $\Gamma(\boldsymbol{r}, t)$ | local fluctuation of the concentration |
| $w(\boldsymbol{r}, t)$ | local fluctuation of the $z$-component of velocity |
| $\nabla T_0$ | stationary temperature gradient |
| $\nabla c_0$ | stationary concentration gradient induced by thermal diffusion |
| $g$ | gravitational acceleration constant |
| $\delta\mathsf{T}(\boldsymbol{r}, t)$ | random stress tensor |
| $\delta\boldsymbol{Q}(\boldsymbol{r}, t)$ | random heat flux |
| $\delta\boldsymbol{J}(\boldsymbol{r}, t)$ | random solute flux |
| $\Phi$ | scattering angle |
| $n$ | refractive index |
| $S$ | structure factor for Rayleigh scattering |
| $S_{\mathrm{E}}$ | structure factor in thermal equilibrium |
| $\tilde{S}_{\mathrm{NE}}$ | normalized nonequilibrium enhancement of the structure factor |
| $q_{\mathrm{RO}}$ | gravitationally induced "roll-off" wave number (22) |
| $s$ | scaled wave number $(q/q_{\mathrm{RO}})$ |

# Thermodiffusion in Ionic Solids - Model Experiments and Theory

Jürgen Janek[1], Carsten Korte[1], and Alan B. Lidiard[2]

[1] Physikalisch-Chemisches Institut der Justus-Liebig-Universität Gießen, Heinrich-Buff-Ring 58, 35392 Gießen, Germany
[2] The J.J. Thomson Physical Laboratory, University of Reading, Whiteknights, Reading RG6 6AF, UK

**Abstract.** The current state of theory and experimental work on thermotransport (diffusion in a temperature gradient) in solids and the heat of transport is reviewed and the results of the authors' own studies are summarized. Results from molecular dynamical simulations of vacancy migration in solid noble gases suggest that the heat of transport is temperature dependent and that it differs substantially from the migration enthalpy of the mobile point defect, in disagreement with some pre-existing models. Experimental results on mixed conducting solids support the theoretical results. The difficulties of thermal diffusion experiments are analysed, and possible experimental strategies are discussed.

## 1 Introduction

It has been known for more than a century that a temperature gradient not only induces heat transport but also induces the transport of matter (thermodiffusion, Ludwig-Soret effect) in systems containing mobile components (see Allnatt and Chadwick [1], Jost [2] and Allnatt and Lidiard [3] for some historical remarks on solid state thermodiffusion). Thus, originally homogeneous mixtures tend to separate in a temperature gradient. This effect is used quite effectively for the separation of gaseous isotopes in the so-called Clusius separation tube and is of current interest in the field of fluids and polymers, as the other contributions in this volume show.

Systematic studies of thermodiffusion and ionic thermopower in the solid state started with the experiments of Reinhold et al. on the thermopower and the Soret-effect of silver and copper halides and chalcogenides in a temperature gradient (Reinhold [4]). It was the time of Wagner and Schottky who established the field of defect thermodynamics, and a number of fundamental studies of defect-related crystal properties were performed [5]. However, due to ill-defined experimental conditions (mostly the applied temperature range included a miscibility gap), no reliable results on the Soret effect were obtained. Later, following new theoretical work by Howard and Lidiard [6], the measurement of thermopower became a standard tool in this study of point defects. Major contributions were reported by Christy [7], Haga [8] and Wagner [9].

Extensive work on thermomigration in metals was stimulated by experimental observations during electromigration studies in the 1960's. The need for high

W. Köhler and S. Wiegand (Eds.): LNP 584, pp. 146–183, 2002.

current densities in the electromigration experiments led to large temperature gradients and a superposition of both effects. A survey of the early studies in this field was given by Wever in 1973 [10]. More recent studies and surveys of the field are reported by Bocquet et al. [11] and Faupel [12].

Further interest in thermomigration in ionic compounds in the sixties and seventies arose from the growing importance of nuclear power generation. Caused by the strongly exothermic fission process the nuclear fuel pellets are subject to strong temperature gradients and degrade continuously by thermomigration of its components (e.g. see Bober and Schumacher [13], Millot and Gerdanian [14] and Janek and Timm [15]). With the increasing application of ceramic multi-component materials at high temperatures the interest in degradation processes driven by external potential gradients (either in chemical, electric, temperature or mechanical stress fields; cf. Schmalzried [16]) grew in general.

In parallel with the different experimental studies a number of theoretical approaches towards an understanding of thermodiffusion by defects in solids have been put forward, as summarized by Allnatt and Chadwick [1]. In the past few years, we started simultaneous theoretical and experimental work on different systems aiming for a better understanding. It is the purpose of the present review to summarize our work and our understanding of thermodiffusion in solids and to include references to related work. In doing so we will not review all the necessary formal relations but instead will concentrate on the underlying physical ideas. Before going into detail, some specific aspects of diffusion in the solid state will be pointed out, in order to help those readers who are less familiar with the solid state and with defect theory.

In the formal description of matter transport by diffusion, a temperature gradient creates a driving force which adds to the driving forces like concentration gradients or an electric field. Thus, if one considers a system with mobile components $k$, the well known empirical Fick's law is expanded by that temperature gradient term:

$$\boldsymbol{J}_k = n_k \boldsymbol{v}_k = -n_k \left( \tilde{D}_k \frac{\nabla n_k}{n_k} + D_k^{\mathrm{T}} \frac{\nabla T}{T} \right) \tag{1}$$

In eq. (1) the chemical diffusion coefficient $\tilde{D}_k$ relates the molar flux $\boldsymbol{J}_k$ of component $k$ to the molar concentration gradient $\nabla n_k$, and the thermodiffusion coefficient $D_k^{\mathrm{T}}$ relates the flux of $k$ to the temperature gradient $\nabla T$. The sum of the gradient terms in the bracket can be interpreted as the negative of the velocity $\boldsymbol{v}_k$ of the mobile component $k$. Rewriting that term according to the relation

$$\boldsymbol{v}_k = b_k \boldsymbol{F} \tag{2}$$

which defines the mobility $b_k$ of the mobile component subject to a force $\boldsymbol{F}$, we obtain:

$$\boldsymbol{v}_k = -\left( \tilde{b}_k RT \nabla \ln n_k + b_k^{\mathrm{T}} \nabla (RT) \right) \tag{3}$$

with $\tilde{b}_k \equiv \tilde{D}_k / RT$ denoting the "chemical" mobility and with $b_k^{\mathrm{T}} \equiv D_k^{\mathrm{T}} / RT$ denoting the "thermodiffusive" mobility. The empirical diffusion coefficients $\tilde{D}_k$ and $D_k^{\mathrm{T}}$ are related to the same single (self) diffusion coefficent $D_k$. This self diffusion coefficient in turn is directly related to the *microscopic jump processes* of the mobile component [3]. In the case of chemical diffusion, the self diffusion coefficient is multiplied (or better: amplified) by the so-called thermodynamic factor $((\partial \ln a_k/\partial \ln n_k)_T)$. In the case of thermomigration the self diffusion coefficient is multiplied by the so-called heat of transport (normalized by $RT$), $Q^{*\prime}_k/RT$, as can be seen from the flux equation

$$\boldsymbol{J}_k = -n_k \left[ D_k \left( \frac{\partial \ln a_k}{\partial \ln n_k} \right)_{\mathrm{T}} \frac{\nabla n_k}{n_k} + D_k \frac{Q^{*\prime}_k}{RT} \frac{\nabla T}{T} \right] \tag{4}$$

which corresponds to eq. (1) but is derived in the framework of linear irreversible thermodynamics. Thus, $D_k$ is modified by an "amplifying" factor, depending on the driving force. The thermodynamic factor is well understood within thermodynamics of point defects. In contrast, the heat of transport is not yet fully understood, but see section 3.

As we shall show in section 2, the theory of atomic transport in crystalline solids relies heavily on the ideas of lattice defects, viz. vacant sites ("vacancies") and atoms occupying interstitial sites in the atomic crystal lattice ('interstitials'). These defects move from site to site as a result of local fluctuations in energy and momentum which cause the displacement in question and then subside. Such thermally activated movements are generally referred to as "jumps" or "hops". They are characterized by a single migration enthalpy (the "activation barrier") for each individual jump configuration. They are normally rare events, i.e. the average interval between successive jumps of any particular defect is long compared to the period of the lattice vibrations.

In the case of vacancies the movement is really the motion of a neighbouring atom into the vacant site. For interstitial atoms different kinds of movement are possible, depending upon the system. Thus an interstitial solute atom, like C or N in $\alpha$-Fe (which has a body-centred cubic lattice) jumps directly from one interstitial site to another. An interstitial of the same species as the host lattice (called a "self-interstitial") may also move directly between interstitial sites, but in most cases does so in a way which directly involves another atom of the host lattice. For example in silver halides the cation interstitials move by what is called an "interstitialcy mechanism" in which the interstitial cation moves to a neighbouring normal lattice site at the same time pushing the ion initially at that site on towards the next interstitial position. A similar mechanism operates in the anion-sublattice of compounds with the fluorite structure.

In summary, the correlation between microscopic and macroscopic transport processes is well understood (see Allnatt and Lidiard [3] for a detailed description), and in simple cases a single microscopic process is sufficient to interpret macroscopic mass fluxes.

However, there are also specific features which may complicate diffusion studies in the solid state. In addition to equilibrium defects like vacancies and interstitials, each crystal contains non-equilibrium defects of different dimensionality (dislocations, grain boundaries, pores, inclusions). These extended defects provide fast-diffusion paths and may disturb the interpretation of diffusion studies on the basis of point defect motion. Less severe is the presence of elastic stress due to the rigidity of the crystal lattice. Any local change by concentration or temperature gradients causes elastic strain which will influence the dynamics of the lattice. In our experimental studies we neglect both aspects.

The paper is organized as follows: In section 2 we summarize briefly the results for the relation between different fluxes and forces in the framework of linear irreversible thermodynamics and discuss the definition of the heat of transport and its relation to measurable quantities. In section 3 we present a summary of the current theory of the heat of transport and discuss existing models and their results. Then in section 4 we give an account of experimental studies of thermodiffusion in crystalline solids, focussing on the determination of heats of transport in nonmetalic systems and their analysis. Finally, we give some brief comments on the use of thermopower measurements for the determination of defect concentrations, again emphasizing the influence of the heat of transport.

## 2 Phenomenology – flux equations and the heat of transport

In order to find out what are the principal kinds of experiments that can be performed in the study of non-isothermal transport processes, it is necessary to analyse the phenomena in the framework of irreversible thermodynamics. By formulating all independent (linear) flux equations which are required for a complete description of the entropy production one finds the number of independent transport coefficients, and thus, also the number of independent experimental effects. For a full description of this topic see [17].

In this section we will give a brief analysis of thermodiffusion in ionic solids in the framework of linear irreversible thermodynamics. There are three parts to this analysis: (i) a brief summary of the formal macroscopic theory, (ii) an account of the relevant defect thermodynamics and (iii) bringing the first two parts together to yield a theory for the analysis of experimental measurements. This treatment leads to the conclusion that various experiments can be performed in order to investigate thermodiffusion. As in liquids the diffusion of a mobile component in a temperature gradient can be studied either by direct measurement of its local concentration or its local chemical potential (and properties related to these). In the case of charged particles the electric potential difference across a non-isothermal system can also be measured (thermopower [9], see section 4.3). Additionally, measurements of the Peltier heat can be performed. However, as we will see, only two non-isothermal transport effects remain independent for a system with two charge carriers representing two independent off-diagonal transport coefficients in the transport matrix.

### 2.1 Phenomenology – linear irreversible thermodynamics

The formal analysis of transport phenomena is based upon the correct choice of forces and fluxes which have to describe the entropy production correctly (using vector/matrix notation):

$$\dot{\sigma}T = \boldsymbol{J}\boldsymbol{X} \tag{5}$$

Here $\dot{\sigma}$ is the rate of production of entropy per unit volume, $T$ the absolute temperature, $\boldsymbol{J}$ is a row matrix of fluxes and $\boldsymbol{X}$ a column matrix of the corresponding forces. Linear flux equations relate the thermodynamic forces to the corresponding independent fluxes in the linear approximation:

$$\boldsymbol{J} = \boldsymbol{L}\boldsymbol{X} \tag{6}$$

The Onsager theorem tells us that the matrix $\boldsymbol{L}$ is symmetric (subject to conditions which are satisfied in all the cases considered (see [17])). Two principal sets of forces and fluxes are used in the description of non-isothermal transport processes [3] [1]: a) Starting with the total heat flux, $\boldsymbol{J}_Q$, the forces result as

$$\boldsymbol{X}_Q = -\frac{\nabla T}{T} \quad \boldsymbol{X}_k = -T\nabla\left(\frac{\tilde{\mu}_k}{T}\right) \tag{7}$$

with $\tilde{\mu}_k = \mu_k + z_k F\phi$ denoting the electrochemical potential of a charged component $k$. b) Introducing the "reduced" heat flux, $\boldsymbol{J}'_Q$, by the transformation $\boldsymbol{J}'_Q = \boldsymbol{J}_Q - \sum_k^n h_k \boldsymbol{J}_k$ one obtains the following alternative forces:

$$\boldsymbol{X}'_Q = -\frac{\nabla T}{T} \quad \boldsymbol{X}'_k = -(\nabla\tilde{\mu}_k)_T = -\nabla\tilde{\mu}_k - s_k\nabla T \tag{8}$$

Here $\tilde{\mu}_k$, the electrochemical potential of component $k$ is regarded as a function of composition, pressure and temperature and $(\nabla\tilde{\mu}_k)_T$ indicates the spatial gradient of $\tilde{\mu}_k$ at constant temperature. The material fluxes $\boldsymbol{J}_i$ remain unchanged, i.e. $\boldsymbol{J}'_i = \boldsymbol{J}_i$. It is easily verified that the rate of entropy production by eq. (5) is the same for these two sets of fluxes and forces.

Since it is the reduced heat of transport which controls the Soret effect we will make use of the second set in eq. (8) for the discussion of experiments. Eq. (7) and the heat flux $\boldsymbol{J}_Q$ is of more direct use in computations, since it is based on the internal energy rather than on heat. The heats of transport of the mobile components are introduced as combinations of different transport coefficients via the mathematical transformation

$$L'_{kQ} = \sum_{i=1}^{n} L_{ki}\, Q_i^{*\prime} \tag{9}$$

and enter into the flux equations as

[1] A problem which has frequently led to confusion in literature concerns the definition of the heat flux, and thus, the heat of transport. Although alternative representations of these heat flow phenomena are possible (see. e.g. de Groot [17], especially §52), only the heat flux $\boldsymbol{J}_Q$ and the so-called reduced heat flux $\boldsymbol{J}'_Q$ are of relevance in the present context.

$$\boldsymbol{J}_k = \sum_{i=1}^{n} L_{ki}(\boldsymbol{X}'_i + Q_i^{*\prime}\, \boldsymbol{X}'_Q) \tag{10}$$

$$\boldsymbol{J}'_Q = \sum_{i=1}^{n} Q_i^{*\prime}\, \boldsymbol{J}'_i + \left( L'_{QQ} - \sum_{i=1}^{n} L'_{iQ}\, Q_i^{*\prime} \right) \boldsymbol{X}'_Q \tag{11}$$

For further details of this transformation see [17] and [18]. Here it is sufficient to note, that $\boldsymbol{J}'_k \equiv \boldsymbol{J}_k$ and $L'_{ik} \equiv L_{ik}$, but that $L'_{iQ}$, $L'_{QQ}$ and $Q_i^{*\prime}$ differ from $L_{iQ}$, $L_{QQ}$ and $Q_i^*$.

The use of heats of transport provides an intuitive approach to the interpretation of thermodiffusion. In an isothermal system ($\boldsymbol{X}'_Q = 0$) eq. (11) can be simplified to an expression which gives a direct relation between the reduced heat flux $\boldsymbol{J}'_Q$ and the matter fluxes $\boldsymbol{J}_i$:

$$\boldsymbol{J}'_Q = \sum_{i=1}^{n} Q_i^{*\prime}\, \boldsymbol{J}_i \tag{12}$$

The reduced heat of transport $Q_i^{*\prime}$ of a mobile component $i$ is the amount of heat in J which is transfered by a unit flux of the corresponding component in an isothermal system.

It is the reduced heat of transport which is determined experimentally. Rewriting eq. (10) by the use of the differential

$$\nabla \mu_i = \left( \frac{\partial \mu_i}{\partial n_i} \right)_T \nabla n_i + \left( \frac{\partial \mu_i}{\partial T} \right)_{n_i} \nabla T \tag{13}$$

and eq. (8) we obtain

$$\boldsymbol{J}_k = - \sum_{i=1}^{n} \left[ L_{ki} \left( \frac{\partial \mu_i}{\partial n_i} \right)_T \nabla n_i + L_{ki} z_i F \nabla \phi + L_{ki} Q_i^{*\prime} \frac{\nabla T}{T} \right] \tag{14}$$

By introducing the thermal diffusion coefficient $D_{ki}^T$, the chemical diffusion coefficient $\tilde{D}_{ki}$ and the electric conductivity $\sigma_{ki}$ of the component $k$

$$D_{ki}^T = \frac{L_{ki} Q_i^{*\prime}}{n_i} \qquad \tilde{D}_{ki} = L_{ki} \left( \frac{\partial \mu_i}{\partial n_i} \right)_T \qquad \sigma_{ki} = L_{ki} (z_i F)^2 \tag{15}$$

eq. (14) can be written as

$$\boldsymbol{J}_k = - \sum_{i=1}^{n} \left[ \tilde{D}_{ki} \nabla n_i + \frac{\sigma_{ki}}{z_i F} \nabla \phi + D_{ki}^T n_i \frac{\nabla T}{T} \right] \tag{16}$$

If no concentration gradient $\nabla n_i$ and no electric field $\nabla \phi$ are present, the thermal diffusion coefficient $D_{ki}^T$, equivalently the reduced heats of transport $Q_i^{*\prime}$, determine the value and the sign of the flux of the component $i$ in a non-isothermal system. It has to be noted that the off-diagonal coefficents $L_{ki}$ with $k \neq i$ may sometimes be neglected, which simplifies the flux equations considerably (see Janek et al. [19] [20]).

Rewriting eq. (10) with insertion of eq. (8) one obtains

$$\boldsymbol{J}_k = -\sum_{i=1}^{n} L_{ki} \left[ \nabla\tilde{\mu}_i + \left( s_i + \frac{Q_i^{*\prime}}{T} \right) \nabla T \right] \tag{17}$$

Both eqs. (16) and (17) can be directly used for the discussion of the possible experimental arrangements [9].

## 2.2 Mixed conductors as model systems – defect thermodynamics

In this section we describe the required thermodynamic functions. Those are provided by statistical thermodynamics of defect solids [3]. Here we restrict our analysis to a (nonstoichiometric) mixed conducting system $A_{\nu+\delta}X$ with Frenkel disorder in one sublattice, and neglect any defect associates or defects with variable charge. In principle, the presence of associates or defects with variable charge can be included in the formalism in a straightforward manner [3]. A corresponding analysis for the case of electromigration in semiconducting oxides has been reported by Yoo et al. [19] and Janek [21], see also Wagner [22].

In such a system we have to consider four mobile defects: interstitials ("i" $\equiv$ $A_i^{z\cdot} - V_i^{\times}$), vacancies ("v" $\equiv$ $V_A^{z\prime} - A_A^{\times}$), electrons in the conduction band ("n" $\equiv$ $e_c^{\prime} - h_c^{\times}$) and electron holes in the valence band ("p" $\equiv$ $h_v^{\cdot} - e_v^{\times}$) with $z$ defining the charge number of $A^{z+}$ cations. Their electrochemical potentials are coupled by internal defect equilibria :

$$\tilde{\mu}_{\mathrm{i}} + \tilde{\mu}_{\mathrm{v}} = 0 \tag{18}$$

$$\tilde{\mu}_{\mathrm{n}} + \tilde{\mu}_{\mathrm{p}} = 0 \tag{19}$$

and the individual molar fractions $x_{\mathrm{i}}$, $x_{\mathrm{v}}$, $x_{\mathrm{n}}$ and $x_{\mathrm{p}}$ depend on the deviation $\delta$ from stoichiometry ($x_k = n_k V_m^{A_\nu X}$; $k = \mathrm{i, v, n, p}$; $n_k$ = molar concentration, $V_m^{A_\nu X}$ = molar volume of the compound $A_\nu X$):

$$\delta = x_{\mathrm{i}} - x_{\mathrm{v}} = \frac{1}{z}(x_{\mathrm{n}} - x_{\mathrm{p}}) \tag{20}$$

where $z$ is the charge number of the cations. If we restrict ourselves to small defect concentrations, i.e. small deviations from stoichiometry, we can assume an ideal solution of defects in the crystal lattice when (18) and (19) become equivalent to simple mass action laws:

$$x_{\mathrm{i}}\, x_{\mathrm{v}} = K_{\mathrm{F}} \qquad K_{\mathrm{F}} = e^{-\Delta G_{\mathrm{F}}^0/RT} \tag{21}$$

$$x_{\mathrm{n}}\, x_{\mathrm{p}} = K_{\mathrm{e}} \qquad K_{\mathrm{e}} = e^{-\Delta G_{\mathrm{e}}^0/RT} \tag{22}$$

We then introduce two dimensionless concentration variables $\alpha_{\mathrm{i}}$ and $\alpha_{\mathrm{n}}$:

$$\alpha_{\mathrm{i}} = \frac{x_{\mathrm{i}}}{x_{\mathrm{i}}^{\#}} = \left( \frac{x_{\mathrm{v}}}{x_{\mathrm{v}}^{\#}} \right)^{-1} \qquad \alpha_{\mathrm{n}} = \frac{x_{\mathrm{n}}}{x_{\mathrm{n}}^{\#}} = \left( \frac{x_{\mathrm{p}}}{x_{\mathrm{p}}^{\#}} \right)^{-1} \tag{23}$$

with $x_{\rm i}^{\#}$, $x_{\rm v}^{\#}$, $x_{\rm n}^{\#}$ and $x_{\rm p}^{\#}$ as the molar (site) fractions at the stoichiometric point ($\delta = 0$, intrinsic concentrations). The relation between the variables $\alpha_{\rm i}$ and $\alpha_{\rm n}$ and the deviation $\delta$ from stoichiometry follows from eq. (20), (21) and (22):

$$\ln \alpha_{\rm i} = \operatorname{arsinh} \frac{\delta}{2K_{\rm F}^{1/2}} \qquad \ln \alpha_{\rm n} = \operatorname{arsinh} \frac{z\delta}{2K_{\rm e}^{1/2}} \tag{24}$$

In an ideal solution the partial molar enthalpies $h_k$ of the lattice defects are constant:

$$h_k = h_k^0 \qquad k = {\rm i, v, n, p} \tag{25}$$

The chemical potentials $\mu_k$ and the partial molar entropies $s_k$ can be expressed as functions of $\alpha_{\rm i}$ and $\alpha_{\rm n}$:

$$s_{\rm i} = \frac{s_{\rm i}^0 - s_{\rm v}^0}{2} + \frac{\Delta H_{\rm F}^0}{2T} - R \ln \alpha_{\rm i} \qquad \mu_{\rm i} = \frac{\mu_{\rm i}^0 - \mu_{\rm v}^0}{2} + RT \ln \alpha_{\rm i} \tag{26}$$

$$s_{\rm v} = \frac{s_{\rm i}^0 - s_{\rm v}^0}{2} + \frac{\Delta H_{\rm F}^0}{2T} + R \ln \alpha_{\rm i} \qquad \mu_{\rm v} = \frac{\mu_{\rm i}^0 - \mu_{\rm v}^0}{2} - RT \ln \alpha_{\rm i} \tag{27}$$

$$s_{\rm n} = \frac{s_{\rm n}^0 - s_{\rm p}^0}{2} + \frac{\Delta H_{\rm e}^0}{2T} - R \ln \alpha_{\rm n} \qquad \mu_{\rm n} = \frac{\mu_{\rm n}^0 - \mu_{\rm p}^0}{2} + RT \ln \alpha_{\rm n} \tag{28}$$

$$s_{\rm p} = \frac{s_{\rm n}^0 - s_{\rm p}^0}{2} + \frac{\Delta H_{\rm e}^0}{2T} + R \ln \alpha_{\rm n} \qquad \mu_{\rm p} = \frac{\mu_{\rm n}^0 - \mu_{\rm p}^0}{2} - RT \ln \alpha_{\rm n} \tag{29}$$

For a discussion of ways to derive the necessary thermodynamic functions for other types of solids see Allnatt and Lidiard ([3], especially ch. 3).

## 2.3 Mixed conductors as a model system – non-isothermal transport

Four flux equations are sufficient to describe mass and charge transport in a system as introduced in the preceeding section 2.2. For simplicity we shall here carry the analysis through neglecting the off-diagonal transport coefficients, $L_{ik}$ ($i \neq k$). However, this is not essential and much the same general conclusions follow if we retain them:

$$\boldsymbol{J}_k = -L_k \left[ \nabla \tilde{\mu}_k + \left( s_k + \frac{Q_k^{*\prime}}{T} \right) \nabla T \right] \qquad k = {\rm i, v, n, p} \tag{30}$$

In experiments we cannot distinguish between the individual contributions of interstitials (i) and vacancies (v) to the transport of ions or of electrons in the valence band (n) and electron holes (p) to the transport of electrons. Rather, they are coupled by the internal defect equilibria (18) and (19) and we have to reduce the number of fluxes to the independent fluxes of ionic and electronic charge carriers $A^{z+}$ and $e^-$. These fluxes of ions and electrons are linear combinations of the defect fluxes:

$$\boldsymbol{J}_{{\rm A}^{z+}} = \boldsymbol{J}_{\rm i} - \boldsymbol{J}_{\rm v} \tag{31}$$

$$\boldsymbol{J}_{{\rm e}^-} = \boldsymbol{J}_{\rm n} - \boldsymbol{J}_{\rm p} \tag{32}$$

As exemplified by Korte [18] these operations can be done quite effectively by matrix operations. We then obtain two flux equations which formally appear like eq. (30):

$$\boldsymbol{J}_k = -L_k \left[ \nabla\tilde{\mu}_k + \left( \frac{Q_k^{*\prime}}{T} + s_k \right) \nabla T \right] \quad k = \mathrm{A}^{z+}, \mathrm{e}^- \tag{33}$$

The transport coefficients of the charge carriers $\mathrm{A}^{z+}$ and $\mathrm{e}^-$ are related to those of the components $k$ (building units) by:

$$L_{\mathrm{A}^{z+}} = L_\mathrm{i} + L_\mathrm{v} \tag{34}$$

$$L_{\mathrm{e}^-} = L_\mathrm{n} + L_\mathrm{p} \tag{35}$$

with their electrochemical potential gradients as:

$$\nabla\tilde{\mu}_{\mathrm{A}^{z+}} = \nabla\tilde{\mu}_\mathrm{i} = -\nabla\tilde{\mu}_\mathrm{v} \qquad \nabla\tilde{\mu}_{\mathrm{e}^-} = \nabla\tilde{\mu}_\mathrm{n} = -\nabla\tilde{\mu}_\mathrm{p} \tag{36}$$

and with the heats of transport for use in eq. (33) as:

$$\begin{aligned} Q^{*\prime}_{\mathrm{A}^{z+}} &= \frac{1}{2}(Q^{*\prime}_\mathrm{i} + Q^{*\prime}_\mathrm{v} + \Delta H^0_\mathrm{F}) \tanh \ln(\psi_\mathrm{i}^{1/2}\alpha_\mathrm{i}) \\ &\quad - \frac{1}{2}\Delta H^0_\mathrm{F} \tanh \ln \alpha_\mathrm{i} + \frac{1}{2}(Q^{*\prime}_\mathrm{i} + Q^{*\prime}_\mathrm{v}) \end{aligned} \tag{37}$$

$$\begin{aligned} Q^{*\prime}_{\mathrm{e}^-} &= \frac{1}{2}(Q^{*\prime}_\mathrm{n} + Q^{*\prime}_\mathrm{p} + \Delta H^0_\mathrm{e}) \tanh \ln(\psi_\mathrm{n}^{1/2}\alpha_\mathrm{n}) \\ &\quad - \frac{1}{2}\Delta H^0_\mathrm{e} \tanh \ln \alpha_\mathrm{n} + \frac{1}{2}(Q^{*\prime}_\mathrm{n} + Q^{*\prime}_\mathrm{p}) \end{aligned} \tag{38}$$

Thus, the reduced heats of transport of the charge carriers $\mathrm{A}^{z+}$ and $\mathrm{e}^-$ are linear functions of the reduced heats of transport of the underlying mobile defects, weighted by their molar fractions and the mobility ratios $\psi_\mathrm{i} = u_\mathrm{i}/u_\mathrm{v}$ and $\psi_\mathrm{n} = u_\mathrm{n}/u_\mathrm{p}$. They also depend on the reaction enthalpies $\Delta H^0_\mathrm{F}$ and $\Delta H^0_\mathrm{e}$ of the relevant defect forming equilibria. In a thermally or chemically inhomogeneous system, matter transport by mobile lattice defects always requires local recombination or formation reactions of defects in order to maintain local equilibrium. Correspondingly, the heat balances of these defect reactions appear in the heat of transport of ionic or electronic charge carriers [23]. This is shown for a hypothetical material in fig. (1).

Based on the defect model, the chemical potentials and partial molar entropies of the charge carriers $\mathrm{A}^{z+}$ and $\mathrm{e}^-$ can also be expressed in terms of defect properties:

$$\mu_{\mathrm{A}^{z+}} = \frac{\mu^0_\mathrm{i} - \mu^0_\mathrm{v}}{2} + RT \ln \alpha_\mathrm{i} \tag{39}$$

$$s_{\mathrm{A}^{z+}} = \frac{s^0_\mathrm{i} - s^0_\mathrm{v}}{2} + \frac{\Delta H^0_\mathrm{F}}{2T} \tanh \ln \alpha_\mathrm{i} - R \ln \alpha_\mathrm{i} \tag{40}$$

$$\mu_{\mathrm{e}^-} = \frac{\mu^0_\mathrm{n} - \mu^0_\mathrm{p}}{2} + RT \ln \alpha_\mathrm{n} \tag{41}$$

$$s_{\mathrm{e}^-} = \frac{s^0_\mathrm{n} - s^0_\mathrm{p}}{2} + \frac{\Delta H^0_\mathrm{e}}{2T} \tanh \ln \alpha_\mathrm{n} - R \ln \alpha_\mathrm{n} \tag{42}$$

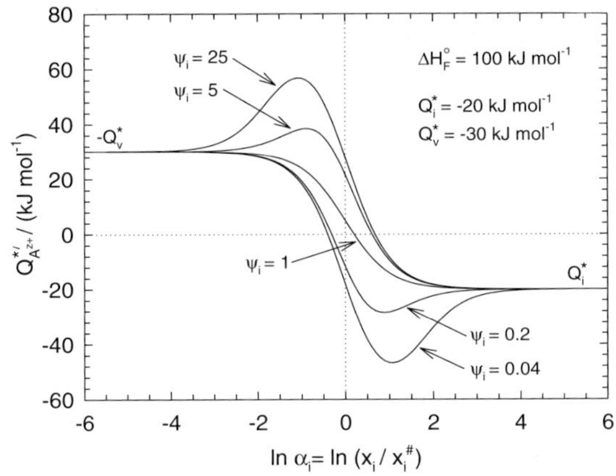


**Fig. 1.** Calculated data for heat of transport of ions in an ionic crystal with Frenkel disorder. An extremum is always observed if the mobility ratio of interstitials and vacancies is different from zero. It represents the reaction heat of internal defect formation/recombination.

Eqs. (37)–(42) provide the equations needed to apply the two flux equations (33).

At this point we may remark that, although eqns. (33) et seq. have been obtained by neglecting the off-diagonal transport coefficients, their generalisation to include these is straightforwardly obtained and understood. Thus, in place of eq. (33), we shall obtain a pair of equations containing additional off-diagonal terms in the manner of eq. (17). Onsager's reciprocal relations hold amoung the new transport coefficients. Hence we have arrived at the point where we have three independent fluxes, the heat flux and the two material fluxes $\boldsymbol{J}_k$ ($k = \mathrm{A}^{z+}$, $\mathrm{e}^-$). This tells us that the corresponding transport coefficient matrix comprises 9 transport coefficients of which only 6 are independent (due to Onsager's reciprocity relations). In practice a number of distinct transport phenomena exists of which not more than six can therefore be independent (see section 4). For some of these we may have to insert also the Nernst-Planck condition of zero electrical current (open circuit conditions). For the present model this requires:

$$\boldsymbol{J}_\mathrm{A} = \boldsymbol{J}_{\mathrm{A}^{z+}} = \frac{1}{z}\boldsymbol{J}_{\mathrm{e}^-} \tag{43}$$

This leads to a further reduction of the two flux equations in eq. (33) to a single flux equation:

$$\boldsymbol{J}_\mathrm{A} = -\mathsf{L}_\mathrm{A}\left[\nabla\mu_\mathrm{A} + \left(s_\mathrm{A} + \frac{Q^{*\prime}}{T}\right)\nabla T\right] \tag{44}$$

with the transport coefficient $\mathsf{L}_\mathrm{A}$:

$$\mathsf{L}_\mathrm{A} = \frac{L_{\mathrm{A}^{z+}} L_{\mathrm{e}^-}}{z^2 L_{\mathrm{A}^{z+}} + L_{\mathrm{e}^-}} \tag{45}$$

the reduced heat of transport $Q^{*\prime}_\mathrm{A}$:

$$Q^{*\prime}_\mathrm{A} = Q^{*\prime}_{\mathrm{A}^{z+}} + zQ^{*\prime}_{\mathrm{e}^-} \tag{46}$$

and the chemical potential gradient $\nabla\mu_{\mathrm{A}}$ and partial molar entropy $s_{\mathrm{A}}$:

$$\nabla\mu_{\mathrm{A}} = \nabla\tilde{\mu}_{\mathrm{A}^{z+}} + z\nabla\tilde{\mu}_{\mathrm{e}^-} \qquad s_{\mathrm{A}} = s_{\mathrm{A}^{z+}} + z s_{\mathrm{e}^-} \tag{47}$$

As remarked previously, the above equations, for simplicity, have been obtained by neglecting the off-diagonal coefficients. However, the same form as in eq. (44) is obtained when they are included, although eq. (45) for $\mathsf{L}_{\mathrm{A}}$ is replaced by a slightly more complicated expression [24]. Summarizing the complete section we note that no new independent transport coefficients can be obtained. Rather eqs. (45) and (46) contain the six independent transport coefficients which have been defined above. Eqs. (44) and (33) are the relevant relations for the experimental study of thermal diffusion in mixed conducting compounds. Clearly, other systems with different defect structures can be described by an analogous formalism. The final equations that contain the experimental observables, always appear like eq. (33) for neutral components or like eq. (44) for charged particles (see also [9]).

## 3 Theory of the heat of transport

We have already seen that the processes of atomic thermotransport have long been of interest in solid state physical science [1] [9] [25] [26]. To understand these processes at the atomic level requires theories of the transport coefficients $L_{ik}$ and of the heats of transport. The theory of the former is well developed both in their relation to defect models and, at a deeper level, to atomic forces [3]. We shall not give much furter attention to it here and will take its results for granted. On the other hand, the progress in understanding the heat of transport in solids has been very slow and there is still no extensively tested way to calculate this parameter for specific solids. In this respect the situation is entirely different from that for other defect properties [27]. However, significant progress has been made recently (Jones et al. [28] [29] [30] [31]) and in this chapter we shall briefly describe the present position as well as previous work.

### 3.1 Atomic movements under isothermal conditions (uniform T, P)

The theory used to describe these various defect movements under isothermal conditions is a version of the transition state theory of molecular reactions modified to describe atomic re-arrangements in crystal lattices. The principal result is that the average frequency of movement of a defect is of the form

$$\omega = \nu_{\mathrm{D}} \cdot \exp\left(-\frac{g_{\mathrm{m}}}{kT}\right) \tag{48}$$

in which the pre-exponential factor is a certain lattice frequency - often taken to be approximately the Debye frequency - while the quantity $g_{\mathrm{m}}$ in the exponent is a Gibbs free energy of activation. Both quantities are characteristic of the particular type of jump in question and may be calculated when the potential energy function of the solid is known as a function of atomic positions (Harding [27]). We emphasise that eqn. (48) applies under equilibrium conditions. It is derived by using statistical mechanics to calculate the proportion of defects passing through the transition state, or configuration, in unit time under conditions of uniform temperature and pressure. It has been widely used in the theory of atomic transport under these same conditions, which, of course, are the common conditions of experiments in this field.

### 3.2 Atomic movements in a temperature gradient

As already emphasized, atomic transport in a temperature gradient is characterised by the heat of transport $Q^*$. In the simplest case the flux of a species $k$ in a solution is given by eq. (49)

$$\boldsymbol{J}_k = -L_k \left[ T\nabla\left(\frac{\mu_k}{T}\right) + Q_k^* \frac{\nabla T}{T} \right] \tag{49}$$

in which $\mu_k$ is the chemical potential of the species $k$ and $Q_k^*$ is the corresponding heat of transport. Under some circumstances a related quantity – the reduced heat of transport – is used, cf. section 2 and [3]. This is obtained from $Q_k^*$ as

$$Q_k^{*\prime} = Q_k^* - h_k \tag{50}$$

in which $h_k$ is the partial enthalpy of $k$. It has to be noted that, of course, eq. (49) is equivalent to eq. (1), (4) and (33) but makes use of the phenomenological transport coefficient $L_k = D_k c_k / RT$, see 3.1. In a pure substance exhibiting vacancy disorder one may describe the thermal diffusion process as a flow of atoms or equivalently as an opposite flow of vacancies. It follows that the heat of transport of the vacancies is the negative of that of the atoms; this is true for both the heat of transport itself and the reduced heat of transport in eq. (50).

It is physically obvious that in a temperature gradient, $\nabla T$, we would expect the rate of jumps in the same direction as the gradient to differ from that in the opposite direction by an amount which, in lowest order, will be proportional to the magnitude of $\nabla T$. Following Howard and Manning [32], we may then write in place of (48)

$$\omega(\boldsymbol{s}) = \omega_0 \left[ 1 - \frac{q^{*\prime}}{2kT^2} (\boldsymbol{s}\ \nabla T) \right] \tag{51}$$

in which $\omega_0$ is the jump frequency of a uniform system at temperature $T$ and $\boldsymbol{s}$ is the atomic jump vector. The quantity $q^{*\prime}$ which we have introduced here has the units of energy ($Q^{*\prime} = N_A\, q^{*\prime}$) and is such that when positive the jump is more likely to be from the hotter to the cooler region, i.e. down the gradient. In the case of self-diffusion via vacancies it turns out to be just the reduced heat of transport of the atoms (= – the reduced heat of transport of the vacancies). In other more complicated cases, e.g. solute diffusion via vacancies, kinetic theory shows that the measurable reduced heat of transport is a combination of several quantities like $q^{*\prime}$, one for each of the different types of atomic jump affected by the presence of the solute (Howard and Manning [32], Allnatt and Okamura [33]). This combination comes about from the occurrence of correlations between successive jumps of the individual solute atoms – a general feature of defect induced atomic movements. This should be remembered when analysing Soret effect measurements in solid solutions and alloys.

### 3.3 Atomic theories of $q^{*\prime}$

As already indicated, kinetic theory allows us to bridge the gap between the Ansatz (51) from defect theory and the phenomenological theory of non-equilibrium thermodynamics. As it stands eq. (51) is also phenomenological in character, since, although the form is undoubtedly correct, in itself it provides no indication of how the parameter $q^{*\prime}$ may be obtained from a knowledge of atomic forces. This challenge has existed for many years, but relatively little progress has been made until quite recently. In this it stands in sharp contrast to the problem of calculating defect properties (e.g. configurations, activation energies and other characteristic thermodynamic quantities) from a knowledge of atomic forces and/or electronic theory [27]. In this section we shall review the attemps which have been made to provide theories of $q^{*\prime}$.

**From transition state theory.** It is perhaps natural to look first at the theory of thermally activated transitions to see if the presence of a temperature gradient can be incorporated in some way with a view to deriving the form (51) and so arriving at a prescription for calculating $q^{*\prime}$. However, this has not so far been done in any rigorous way, although many intuitive arguments of this kind have been advanced over the past 50 years. The first of these and one which is still widely quoted is due to Wirtz [34] from 1943. He argued that the free energy of activation, $g_{\mathrm{m}}$, is essentially composed of three parts; (a) one, $g_{\mathrm{m},1}$, which must be supplied at the initial position of the atom which jumps, (b) a second, $g_{\mathrm{m},2}$, which must be supplied to open the 'gate' of atoms which provide the principal barrier to the motion of the atom from its old to its new position and (c) a third, $g_{\mathrm{m},3}$, which must be supplied at the new position to move the neighbouring atoms outwards from the new site and so allow the jumping atom to settle there. By assigning different mean temperatures to these three positions in a way compatible with the assumed temperature gradient he was able to derive the relation

$$q^{*\prime} = h_{\mathrm{m},1} - h_{\mathrm{m},3} \tag{52}$$

in which $h_{\mathrm{m},i}$ is the enthalpy corresponding to $g_{\mathrm{m},i}$, so that the total enthalpy of activation is

$$h_{\mathrm{m}} = h_{\mathrm{m},1} + h_{\mathrm{m},2} + h_{\mathrm{m},3} \tag{53}$$

This leads to the conclusion

$$q^{*\prime} \leq |h_{\mathrm{m}}| \tag{54}$$

a result which is sometimes satisfied by values determined experimentally and sometimes not. Other versions of the Wirtz argument were given by Brinkman [35] and LeClaire [36].

An obvious extension of the Wirtz argument would be to suppose that the positions at which $g_{\mathrm{m},1}$ and $g_{\mathrm{m},3}$ are supplied lie farther away from the saddle point position of the jumping atom than the original and final sites. If the ratio of distances is $\beta$ then in place of (52) we shall have

$$q^{*\prime} = \beta(h_{\mathrm{m},1} - h_{\mathrm{m},3}) \tag{55}$$

and we escape from the limitation (54). This idea was first put forward by Girifalco [37]. It is interesting to note that comparable theoretical models have been proposed for thermodiffusion in liquids [38] [39]. Here the activation energy of viscous flow has been related to the heat of transport.

It is probably fair to say that none of the subsequent arguments of the same kind have been significantly more successful than the original Wirtz argument. The limitations of all of them are (a) that they are intuitive extensions of isothermal transition state theory and (b) that they provide no link with lattice structure or atomic forces. Later work (see sections below, "From lattice dynamics" and "Approach via molecular dynamics computer simulation") indicates that $\beta$ may in fact be about 2, at least in self-diffusion.

**From lattice dynamics.** At the end of the 1950's attempts were made, notably by Rice and co-workers, to particularise the transition-state theory of atomic jumps by starting from the theory of lattice vibrations and evaluating the frequency of occurrence of certain critical configurations identified as the saddle points for the transition. A vacancy jump was assumed to occur when the displacement of an atom neighbouring the vacancy exceeded some critical value in the direction of the vacant site. As applied by Allnatt and Rice [40] this again led to a result like eq. (54). Later Schottky [41] again applied this approach to the problem of vacancy jumps in a temperature gradient. He evaluated the rates of vacancy jumps up and down the gradient and from the difference obtained the heat of transport. Although he only studied a one-dimensional model, his results emphasised the importance of the relaxation of the lattice following the jump of the atom to its new position. In particular, for vacancy self-diffusion he derived an expression for the ratio of the reduced heat of transport for vacancies to the corresponding activation enthalpy of the form

$$\frac{q_{\mathrm{v}}^{*\prime}}{h_{\mathrm{m}}} = -\frac{2\bar{v}\tau}{s} \tag{56}$$

in which $\bar{v}$ is a mean velocity defined by reference to the vibrational modes active in promoting the jump, $\tau$ is the relaxation time for the disturbance to subside after the jump and $s$ is the jump distance (in this case the nearest neighbour distance). Unfortunately, the theory did not contain within it the means to estimate $\bar{\nu}$ and $\tau$. Nevertheless, Schottky was able to provide plausible estimates for these which led to the results in Table 1 for the noble metals (at their melting points). Furthermore, since phonon relaxation times decrease with increasing temperature, (56) leads us to expect a similar decrease in the

**Table 1.** Ratios of reduced heats of transport and migration enthalpies of vacancies at the melting point ($q_{\mathrm{v}}^{*\prime}/h_{\mathrm{m,v}}$, theoretical estimations according to Schottky [41]).

| $\tau$ used | Cu | Ag | Au |
|---|---|---|---|
| $\tau = const.$ | –2.84 | –3.90 | –3.17 |
| $\tau = A\omega^{-2}$ | –2.34 | –3.69 | –3.22 |

magnitude of the ratio $q_{\mathrm{v}}^{*\prime}/h_{\mathrm{m}}$ and, thus, since $h_{\mathrm{m}}$ generally varies rather little with temperature, a decrease in magnitude of $q_{\mathrm{v}}^{*\prime}$. (As we shall see below recent molecular dynamics simulations confirm this prediction.) This work of Schottky's was significant in that it freed thinking from the limitations of the transition state approach, especially by relaxing the conclusion (54). However, by making $q_{\mathrm{v}}^{*\prime}$ directly proportional to $\tau$ it also appeared to leave something out, as was argued by Huntington [42] and by Boswarva and Tan [43].

A more elaborate analysis for a three-dimensional model by Gillan [44], in fact, obtained formal expressions for the additional terms. Although Gillan's papers showed considerable insight into the problem, the final results still contained unknown parameters (relaxation times, etc.). The formal expressions were subsequently evaluated (Gillan [45]) for a model simple cubic lattice, with the result that the ratio $q_{\mathrm{v}}^{*\prime}/h_{\mathrm{m}}$ was given as the sum of two terms one proportional to a decay time $\tau$ and the other not. This result thus offered the prospect of combining the ideas of Wirtz and of Schottky.

However, the most significant aspect of this work was the basic idea of studying an isothermal system and using the relation between heat flow and matter flow under isothermal conditions to obtain the heat of transport, viz.

$$\boldsymbol{J}_q = Q_k^* \boldsymbol{J}_k \tag{57}$$

That the heat of transport so defined is identical to that in (49) is a consequence of the Onsager reciprocal relations [3]. From eq. (57) Gillan [44] arrived at the following relation for the heat of transport of atoms jumping into vacancies

$$Q^* = \frac{2}{s} \int_0^\infty \bar{q}(t)dt \tag{58}$$

in which $\bar{q}(t) \equiv \langle \boldsymbol{s} \cdot \boldsymbol{J}_q \rangle$ is the average over very many atom jumps of the heat current at time $t$ after the jumping atom crosses the saddle point. As before $\boldsymbol{s}$ is the jump vector and $s$ its magnitude. Since, in arriving at this expression, $\boldsymbol{J}_k$ is expressed in terms of atomic or defect jumps in the lattice, it follows that $\bar{q}(t)$ must also be referred to the lattice frame of reference. This is an important point (Jones et al. [29] [31]). Although eq. (58) was used as the basis of the phonon calculations by Gillan [44] [45], its greatest value to date has been as the foundation of the molecular dynamics computer simulations which we now consider.

**Approach via molecular dynamics computer simulation.** Modern statistical mechanics in the form of Green-Kubo linear response theory in principle provides an exact prescription for the calculation of the heat of transport from the integral over all time $t$ of the time-correlation function between the flow of the material species in question at time 0 and the heat flow at time $t$. Although it is practicable to use this method in conjunction with molecular dynamics to compute the transport coefficients of liquids (see e.g. Allen and Tildesley [46] or Hansen and McDonald [47]), the rarity of thermally activated atomic jumps renders it impracticable for most solids. Herein lies the great value of eq. (58), which, although not mathematically exact, is believed to be quite accurate [48]. Its feasibility is made possible by a device for the 'preparation' of the system in a representative set of transition states which then serve as the starting points for the many jumps over which the average heat current is obtained. Some initial investigations along these lines were made by Gillan and Finnis [49].

Following previous computer studies of atom jumps into vacancies in a Lennard-Jones model of solid Ar (f.c.c.) by Hoheisel and co-workers [50] [51] [52] [53], Jones et al. [28] [29] [31] have recently made many similar computations to evaluate the average heat current $\bar{q}_{\mathrm{cm}}(t)$ in the direction of the atomic jump and its time integral

$$\chi_{\mathrm{cm}}(t) = \int_0^t \bar{q}_{\mathrm{cm}}(t')dt' \qquad (59)$$

Similar studies of a pair potential model of Au are also in progress [54]. In both cases the studies have been made at several temperatures up to some 70 % of the melting temperature. Since in a molecular dynamics simulation momentum is conserved (maintained at zero), the average heat current is necessarily obtained relative to the centre-of-mass and we have added subscripts to the above quantities to emphasise that fact. The quantity $\chi(t)$ is actually the difference between the average spatial energy moment at time $t$ and at time 0. This follows from the form of the dynamical expression for $\boldsymbol{J}_{\mathrm{q}}$ [46]. The relation of $\chi(t)$ to $\bar{q}(t)$ is analogous to that between polarisation and electrical current in electromagnetic theory.

The nature of the results is best understood by reference to Figure 2 which is a schematic diagram of $\bar{q}_{\mathrm{cm}}(t)$ and $\chi_{\mathrm{cm}}(t)$. Time subsequent to the passage of the jumping atom through the transition state is divided into three periods, I, II and III. Period I is roughly the time taken for the jumping atom to reach its new site ($\sim$ 1–2 ps, cf. Figure 2). During period I the average current $\bar{q}_{\mathrm{cm}}(t)$ is initially negative and then rapidly becomes positive. Period II is the subsequent period during which the lattice oscillations stimulated by the jump decay and period III is all times thereafter.

Depending upon the temperature and the size of the sample of trajectories, the computed $\bar{q}_{\mathrm{cm}}(t)$ and $\chi_{\mathrm{cm}}(t)$ will show fluctuations about the smooth curves drawn in Figure 2. As might be expected, these fluctuations are larger the higher the temperature and the smaller the sample of trajectories. In the most favourable cases where these fluctuations are small, we find that the approach

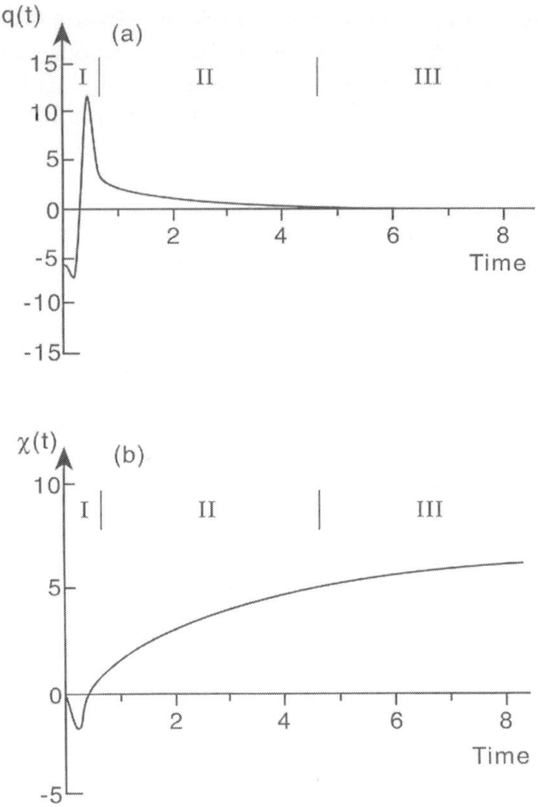


**Fig. 2.** Schematic diagram for a jump of a host atom into a vacant site showing (a) the average heat flow $\bar{q}_{\rm cm}(t)$ in the direction of the atomic jump as a function of time from the instant the atom passes through the transition state and (b) $\chi_{\rm cm}(t)$, the integral of $\bar{q}_(t)$ from time zero to time $t$ as a function of $t$ (In both cases the units are the usual dimensionless units appropriate to the Lennard-Jones potential [46]). The statistical fluctuations which are visible in actual simulations are not shown in this diagram. The qualitative features shown here are believed to be common to self-diffusion jumps into vacancies generally, but they change if the mass of the jumping atom is significantly changed

of $\chi_{\rm cm}(t)$ to its asymptotic value is closely exponential and is characterised by a single relaxation time, $\tau$. This relaxation time decreases as the temperature is raised. The contribution of this exponentially decaying contribution (period II) to the heat of transport is considerable, in agreement with the earlier lattice phonon calculations. Unlike those, however, these computations indicate what the relaxation times actually are.

To obtain the heat of transport from the asymptotic value of $\chi_{\rm cm}(t)$ we need to correct for the relative motion of the centre-of-mass and the atomic lattice. This correction is easily obtained in terms of (a) the mass of the jumping atom relative to that of the host atoms and (b) the self-energy of the host atoms, $u$.

**Table 2.** Vacancy heats of transport for self diffusion in solid Ar obtained from computer simulation (MD) [31].

| $T$/K | $u$/kJ mol$^{-1}$ | $Q_{\rm v}^{*({\rm cm})}$/kJ mol$^{-1}$ | $Q_{\rm v}^{*(1)}$/kJ mol$^{-1}$ | $Q_{\rm v}^{*(1)\prime}$/kJ mol$^{-1}$ |
|---|---|---|---|---|
| 30 | –7.44 | –12 | –4.6 | –13 |
| 40 | –7.19 | –7.3 | –0.1 | –8.6 |
| 60 | –6.8 | –6.8 | –0.2 | –8.6 |

N.B. Values relative to the Lennard-Jones well-depth are very closely the same as the numbers entered here.

Some results are given in Table 2. These have been obtained for a Lennard-Jones model of solid Ar. The values given are for the vacancies since in self-diffusion it is common to specify the vacancy heats of transport. Those for the atoms are just the negative of the quantities given. We comment as follows.

(i) Since there are no experimental values for Ar, it is potentially useful to express the calculated quantities as multiples of the Lennard-Jones well depth, $\varepsilon$, in the expectation that such dimensionless values may also be applicable to other substances of the same structure. The well depth of the potential used for these Ar calculations turns out to be very close to 1 kJ/mol so that the values for $Q_{\rm v}^{*\prime}/\varepsilon$ are the same as those given to within less than 1 %.

(ii) The values for the vacancies are negative and thus those for the atoms are positive. In magnitude they are roughly twice the activation energy for vacancy jumps. This implies that the corresponding flux of atoms will be down the temperature gradient, i.e. from the hotter to the cooler region. For vacancies the opposite will be true. Although, we know of no experimental results on solid Ar with which to directly compare our prediction, the sign and relative magnitude are broadly in line with such results for other monatomic substances.

(iii) Since the activation energy $h_{\rm m,v}$ is the amount of energy which must be supplied to cause the jump to occur, one might ask how the heat of transport can be as much as twice this. The answer lies with eqn. (58) and the interpretation of $\chi_{\rm cm}(t)$ as an energy moment. This shows that this part of the heat of transport depends not only on $h_{\rm m,v}$ but also on the mean distance over which this activation energy moves relative to the jump distance $s$. Hence this contribution can be considered as $\beta h_{\rm m,v}$, whence $\beta$ can be identified with the factor introduced in connection with the Wirtz theory in eq. (55). The results for self-diffusion thus give $\beta \approx 2$.

(iv) In many experiments on metals in which a flux of vacancies occurs in a temperature gradient, the directly measured quantity is not $Q_{\rm v}^{*(1)\prime}$ but $Q_{\rm v}^{*(1)} = Q_{\rm v}^{*(1)\prime} + h_{\rm f,v}$, the magnitude of which is much smaller [25]. This may account for many of the difficulties encountered in early attempts to observe these vacancy fluxes.

(v) The magnitude of the values of $Q_{\rm v}^{*(1)\prime}$ given in Table 2 decreases significantly with increasing temperature. This is largely due to the decrease in the relaxation time for the current in period II, which falls like $T^{-1}$. The effect on $Q_{\rm v}^{*(1)}$ is relatively more pronounced, although the origin is essentially the same.

Finally we note that there is a positive isotope effect on the reduced heat of transport for the atom, i.e. increasing (decreasing) the mass of the jumping atom increases (decreases) its heat of transport. The ratio $\delta(Q^{*\prime}/h_{\rm m,v})/(\delta m/m)$ is roughly 1. We emphasise that this applies to a vacancy mechanism and that it is subject to a qualification relating to correlation effects of the same kind as that mentioned in section 3.3 "From transition state theory" in connection with solute diffusion.

### 3.4 Metals: electron wind effect

We have so far discussed the heat of transport in purely atomic terms. This should be adequate for the noble gas solids and, by simple extension, to ionic solids and many oxides. However, in metals the existence of a temperature gradient can lead to a current of the valence electrons which in turn can bias the atomic jumps through collisions with the atoms in their saddle point configurations. The theory of the effect has been considered by Gerl [55] and other authors subsequently. For the jumps of solute atoms of type $k$ into vacancies its contribution to the heat of transport can be expressed in terms of the specific absolute thermoelectric power, $\epsilon_k$ and the effective charge of the solute, $Z_k^*$ as it would be obtained in an electromigration experiment

$$Q_{k,\rm el}^* = -\epsilon_k T Z_k^* e \tag{60}$$

The estimated magnitude of this contribution for many solutes in the noble metals (although not all) is quite small (10 kJ/mol - 20 kJ/mol). Since, by the results described in section 3.3, "From lattice dynamics", we expect the atomic reduced heat of transport to be about ten times this amount it is clear that the electron wind term is not generally of major importance, although there may be cases where it is. In particular, Hehenkamp and co-workers have been able to represent the variation of the solute heat of transport with concentration in dilute noble metal alloys as arising from this term, which thereby accounts for about half the total in the limit of infinite dilution (Faupel [12]).

## 4 Non-isothermal transport experiments – experimental determination of the heat of transport

As outlined above in section 2.3, six independent transport phenomena exist for a binary mixed conductor of which three can be denoted as *primary* phenomena (ionic conduction, electronic conduction, heat conduction), since they relate principally to the diagonal transport coefficients ($L_{ik}$ with $i = k$). The remaining three effects may be denoted as *secondary* or *cross* effects, since they relate principally to the off-diagonal transport coefficients ($L_{ik}$ with $i \neq k$). One of these describes the isothermal kinetic interaction between ionic and electronic fluxes and is neglected in our present discussion. The interested reader is referred to Yoo et al. [19] and Janek [20] [21].

The two remaining phenomena contain the two independent heats of transport of ions and electrons and are related to a number of different transport experiments [56]. In essence, one can perform measurements on either electronic or ionic thermopower [9], on the Soret effect of the neutral mobile component (e.g. [57]), on the electronic and ionic Peltier effect and on thermodiffusion in an open system [24] [58]. We will concentrate in the following only on those experiments which offer direct information on the heat of transport.

The main problem in the analysis of thermodiffusion in solid materials is the relation of measurable macroscopic observables to the heats of transport of mobile point defects which are accessible by theoretical calculations. This relation was discussed in section 2, and the theory of defect heats of transport was summarized in section 3. It is in principle not possible to measure the heat of transport of a charged point defect alone, any more than it is for other thermodynamic quantities of charged particles (cf. [59]). In ionic systems one necessarily determines either the heat of transport of neutral combinations of point defects (chemical components) or the thermopower (entropy of transport) as a sum of the partial entropy $s_k$ and $Q_k^{*\prime}/T$. Of course, the same problem occurs in the liquid state: In liquid electrolytes, also only the heat of transport of neutral combinations of ions (mostly identified as the electrolyte itself) can be determined experimentally (Haase, [60]). Solutions and mixtures of electrically uncharged species do not suffer from this problem (cf. other contributions to this volume).

In ionic crystals these neutral combinations are often the equivalent of neutral atoms, being composed of an ion and the dominating electronic charge carrier. Thus, in order to evaluate the heat of transport of an ion one has to introduce suitable estimates of the electronic heat of transport. Whereas reasonable estimates can be made for delocalized charge carriers with high mobility [61], corresponding estimates for polarons are more difficult.

A further question concerns the ionic heat of transport itself. In ionic crystals, ionic species may move via different mechanisms in regular and interstitial sublattices. Thus, the ionic heat of transport is necessarily a function of the heats of transport of these individual mobile defects and their mobilities. The route from a suitable experiment to the evaluation of individual defect heats of transport is straightforward but relies on the knowledge of defect structure and on reasonable estimates of electronic properties, cf. remarks in section 3.

In fig. (3) these relations are depicted schematically. Vertically one finds the hierarchy of descriptions of the crystal: The top-level contains the properties of neutral *components*, i.e. chemical elements in most cases (e.g. neutral oxygen atoms in the case of oxygen ion conductors). These can be determined experimentally. The next level contains the properties of charge carriers (*building units*

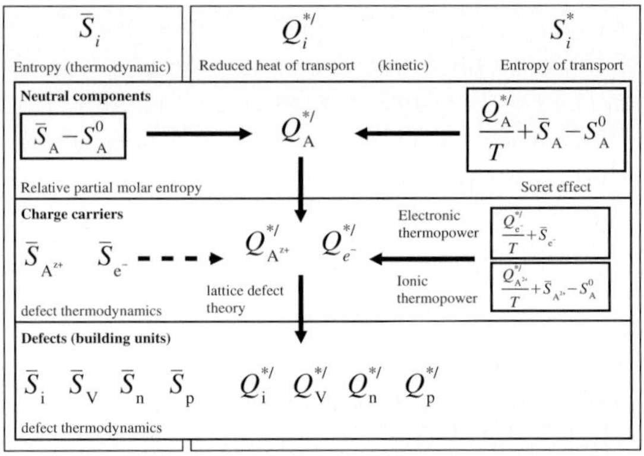


**Fig. 3.** Survey of experimentally accessible non-isothermal transport properties (framed in boxes). In vertical order the different levels of description of crystal constituents are depicted: neutral components → ions, electrons → defects

in terms of defect thermodynamics [62], i.e. ions and electrons. These are themselves functions of the properties of individual point defects (*structure elements* in terms of defect thermodynamics). Horizontally we distinguish between the partial entropy $s_k$ (of components, charge carriers and defects) as a purely thermodynamic property, between the entropy of transport $S_i^* = s_i + Q_i^{*\prime}/T$ and the (reduced) heat of transport $Q_i^{*\prime}$. Only quantities in framed boxes are directly accessible by experiment.

In the following we analyse some different possible experiments and give reference to corresponding experimental data. We do so under the five headings: "Thermodiffusion in closed systems (Ludwig-Soret effect)", "Thermodiffusion in open systems", "Thermopower", "Short-Circuited Thermocells" and "Peltier effect".

### 4.1 Thermodiffusion in closed systems (Ludwig-Soret effect)

The working equation for the study of the Soret effect is easily derived from eq. (44), as follows. As a consequence of thermal diffusion, a temperature gradient leads to the establishment of a definite concentration gradient in the stationary state - provided that the system is finite and closed for matter exchange at its ends. The stationary concentration gradient $\nabla n_{\mathrm{A}}$ of a component A for this steady state ($\boldsymbol{J}_{\mathrm{A}} = 0$) follows from eq. (44) by using the total differential of $\mu_{\mathrm{A}}$ (see. eq. 13) as:

$$\nabla n_{\mathrm{A}}\big|_{\boldsymbol{J}_{\mathrm{A}}=0} = -\frac{Q_{\mathrm{A}}^{*\prime}}{\left(\frac{\partial \mu_{\mathrm{A}}}{\partial n_{\mathrm{A}}}\right)_T} \frac{\nabla T}{T} \tag{61}$$

In materials with small concentration variations (i.e. deviations from stoichiometry), the stationary chemical potential gradient (from eq. (44)) rather than the concentration gradient is directly accessible by experiment [63]:

$$\nabla \mu_A|_{J_A=0} = -\left(\frac{Q_A^{*\prime}}{T} + s_A\right) \nabla T \tag{62}$$

An ever present difficulty in measurements of the Soret effect is the strict realization of closed surfaces ($\boldsymbol{J}_i = 0$), and certainly, most reported experiments on oxides suffer more or less seriously from this drawback (cf. Millot [64] and section 4.4). With respect to the evaluation of the heat of transport, the application of eq. (61) requires precise data for the thermodynamic factor whereas the use of eq. (62) requires knowledge of the partial entropy $s_A$. Thus, in any case the thermodynamics of the system under investigation has to be known for the evaluation of $Q_A^{*\prime}$.

For historical reasons and technological interest the Soret effect has primarily been studied in metal systems, often being related to simultaneous studies of electromigration. Only a restricted number of studies have been performed in ionic crystals, and most of these are related to materials which are of importance in nuclear technology. To date, no systematic review of all these experimental studies is available. Therefore, some essential results are summarized.

**Metals.** Thermomigration in metal systems has been studied extensively in the fifties and sixties. A major review which covers all studies up to the seventies has been provided by Wever [10]. A more recent brief account with tables of data has been given by Bocquet et al. [11]. A more recent review on atomic transport in metals in general by Faupel [12] includes a brief section on thermodiffusion. As stated in section 3.4, Faupel concludes that (a) the electron wind gives a major contribution to the total heat of transport and (b) essentially determines its concentration dependence at low concentrations. After subtraction of this contribution a concentration independent heat of transport results. However, the results on this phonon contribution so far disclose no general relation with the migration enthalpy.

**Ionic crystals.** The study of thermal diffusion in binary mixed conductors with variable composition shows some major advantages compared to metal systems. Firstly, the concentration of electronic charge carriers is considerably smaller, and thus the so-called "electron wind" plays no role in the interpretation of experimental results, cf. section 3.4. Secondly, the concentration of mobile point defects can be controlled very effectively either by "self-doping" (deviation from stoichiometry) with one of the crystal components or by doping with suitable impurities [63]. As is emphasized in the following, the measurement of $Q^{*\prime}$ as a function of defect concentration is of major importance for the quantitative interpretation of thermal diffusion experiments.

Studies of the Soret effect in mixed conductors by means of chemical potential probes (galvanic cells) have been reported only for a few compounds (e. g. $CeO_2$ [64] [65], $\beta$-$Ag_2S$ [66], $\beta$-$Ag_2Se$ [67] [68], $FeAl_2O_4$ [69], $Ag_{1-x}TiS_2$ [70][71], $\alpha$-AgI

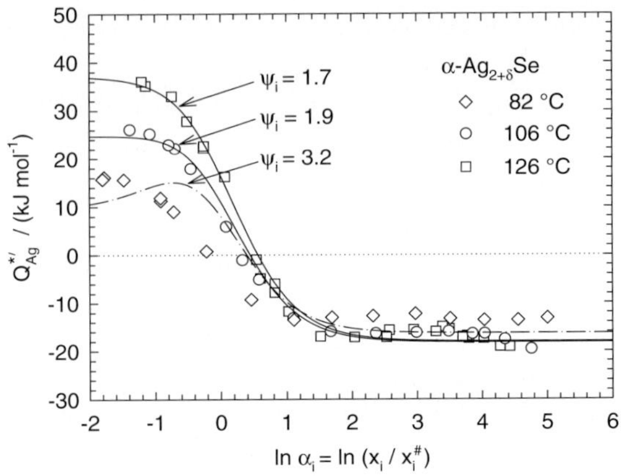


**Fig. 4.** Experimental results for the reduced heat of transport of silver metal in $\alpha$-$Ag_{2+\delta}Se$ as a function of the defect concentration (after Korte [76]). The maximum in the heat of transport reflects the Frenkel defect equilibrium which couples together the concentrations of interstitials and vacancies; cf. eq. (38) in section 2.3

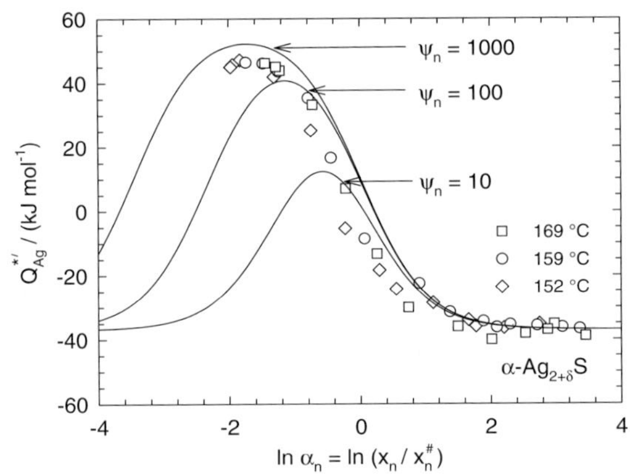


**Fig. 5.** Experimental results for the reduced heat of transport of silver metal in $\alpha$-$Ag_{2+\delta}S$ as a function of the defect concentration (after Korte [61]). The maximum of the heat of transport reflects the electron/hole defect equilibrium which couples together the concentrations of electrons and electron holes.

[72], $ZrO_2$ (+ $Y_2O_3$) [73]). Direct measurements of the composition gradient have been reported for $(U,Pu)O_2$ which exposes a sufficiently large range of homogeneity [74]. In other cases radiotracer techniques can also be used for the same purpose [75].

In order to improve the experimental knowledge on thermal diffusion and the heat of transport in crystalline compounds, we started systematic investigations of the non-isothermal transport properties of mixed-conducting model systems ($\alpha$-$Ag_2S$ [61], $\alpha$-$Ag_2Se$ [76], $\beta$-$Ag_2S$ [24] [57] [58]). Present work is focussed on thermal diffusion in oxides [77] [78] [79]. It is worth noting that in mixed conduc-

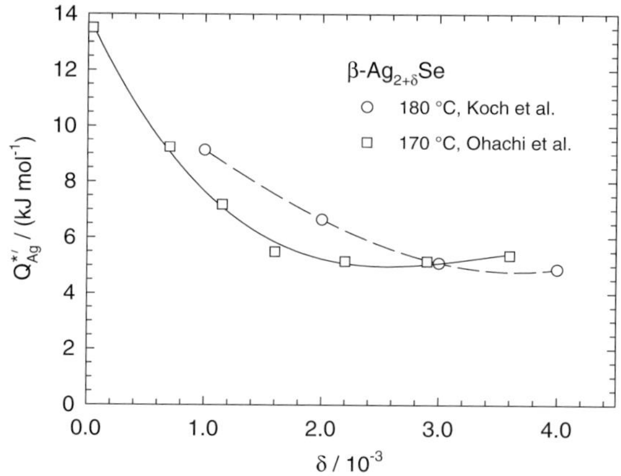


**Fig. 6.** Experimental results for the reduced heat of transport of silver metal in $\beta$-$Ag_{2+\delta}Se$ as a function of the defect concentration (deviation from stoichiometry). The data is compiled from Koch [67] and Ohachi [68] and suggest that a maximum of the heat of transport (according to eq. (38)) exists close to the stoichiometric point.

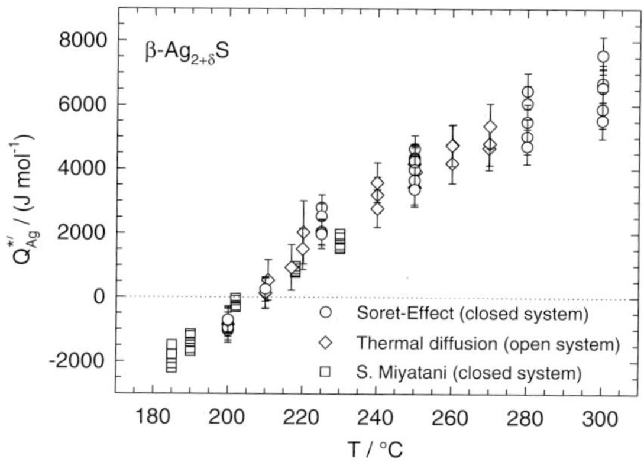


**Fig. 7.** Experimental results for the reduced heat of transport of silver metal in $\beta$-$Ag_{2+\delta}S$ as a function of temperature (after Janek / Korte [57] [58] and Miyatani [66]).

tors the Soret effect can be determined by suitable time-resolved measurements of one partial (ionic or electronic) thermopower (see [80]). A further remark on this is included in section 4.3 which deals with thermopower measurements.

Of particular importance for the support of the theoretical results in section 3 are our recent experiments on the Soret effect in cuprous oxide, $Cu_{2-\delta}O$, by solid state galvanic cells [77]. The heat of transport of the mobile copper component is determined experimentally as $Q^{*\prime}_{Cu} \approx 180$ kJ/mol. On the basis of suitable assumptions for the electronic heat of transport and for the defect structure this value can be interpreted as the vacancy heat of transport, $Q^{*\prime}_{v} \approx -180$ kJ/mol. The migration enthalpy of copper vacancies equals $h_{m,v} \approx 90$ kJ/mol. Thus, the result for this binary oxide with a simple defect structure supports our theoretical rule of thumb, $Q_v \approx -2h_{m,v}$.

### 4.2 Thermodiffusion in open systems

The most direct experimental determination of $Q_{\mathrm{A}}^{*\prime}$ is based on the measurement of the stationary matter flux in a temperature gradient under the condition that no concentration gradient is present. In this case we can directly apply eqs. (4) or (44), which can be rewritten as a function of the concentration gradient $\nabla n_{\mathrm{A}}$ of excess metal A and $\nabla T$, by using the total differential of the chemical potential of A:

$$\boldsymbol{J}_{\mathrm{A}} = -\mathsf{L}_{\mathrm{A}} \left( \frac{\partial \mu_{\mathrm{A}}}{\partial n_{\mathrm{A}}} \right)_T \nabla n_{\mathrm{A}} - \mathsf{L}_{\mathrm{A}} Q_{\mathrm{A}}^{*\prime} \frac{\nabla T}{T} \tag{63}$$

In a specimen which is homogeneous with respect to the concentration of metal A ($\nabla n_{\mathrm{A}} = 0$), a flux $\boldsymbol{J}_{\mathrm{A}}$ of the metal component A is only observed in the case of a non-vanishing heat of transport:

$$\boldsymbol{J}_{\mathrm{A}}\big|_{\nabla n_{\mathrm{A}}=0} = -\mathsf{L}_{\mathrm{A}} Q_{\mathrm{A}}^{*\prime} \frac{\nabla T}{T} \tag{64}$$

The realisation of such experiment is difficult since real systems are finite, and thus, are subject to the establishment of a concentration gradient. Only by the use of electrochemical methods is it possible to make use of eq. (64). A detailed discussion of stationary thermal diffusion experiments in virtually homogeneous specimens can be found in [24], while experiments are reported in [58].

If no appropriate realisation of a Soret experiment is possible, e.g. due to permeable walls of the system, the ends of a non-isothermal ionic crystal can be equilibrated with the gaseous non-metal component. The formal boundary conditions are changed, and the resulting flux

$$\boldsymbol{J}_{\mathrm{A}} = -\mathsf{L}_{\mathrm{A}} \left[ \frac{Q_{\mathrm{A}}^{*\prime}}{T} + \frac{1}{\nu} \left( \frac{1}{2} s_{\mathrm{X}_2} - s_{\mathrm{X}} + s_{\mathrm{A}_\nu \mathrm{X}} \right) \right] \nabla T \tag{65}$$

of the mobile component includes now two contributions: (a) the thermal diffusion flux determined by the heat of transport, and (b) a chemical diffusion flux which is determined by a combination of entropies. Since the entropies are not known a priori, eq. (65) does not offer a simple route to the evaluation of $Q_{\mathrm{A}}^{*\prime}$. Rather we emphasize that a non-isothermal specimen, which at its surfaces is in local equilibrium with the surrounding gas atmosphere, exhibits both thermal diffusion ($\boldsymbol{J} \sim \nabla T$) and chemical diffusion ($\boldsymbol{J} \sim \nabla n$). The observation of thermal diffusion proper requires the complete isolation of the specimen from the surroundings which normally act as a chemical potential buffer.

To our knowledge the results for the heat of transport of silver metal in $\beta$-$\mathrm{Ag}_{2-\delta}\mathrm{S}$, which are included in Fig. (7), are the only available data from a thermodiffusion experiment in an open system. The data agree perfectly with results from a Soret experiment, and thus, validate the experimental approach.

### 4.3 Thermopower

Thermoelectricity denotes the existence of a thermal emf (electromotive force) along a non-isothermal electric conductor. Thermopower denotes the differential of this emf with respect to the applied temperature difference. In mixed conductors one has to distinguish between the partial ionic and electronic thermopowers $\varepsilon_{\mathrm{A}^{z+}}$ and $\varepsilon_{\mathrm{e}^-}$. The pure ionic thermopower of $\mathrm{A}_{\nu+\delta}\mathrm{X}$ can be measured with probes which are reversible only for ionic charge carriers ($\mathrm{A}^{z+}$-conducting solid electrolyte with metal A as reference). The pure electronic thermopower has to be measured with probes which are reversible only for electronic charge carriers (inert metal, e.g. platinum) [9] [80]:

$$\varepsilon_{\mathrm{A}^{z+}} = \lim_{\Delta T \to 0} \frac{\Delta\phi_{(\mathrm{I})}}{\Delta T} = \frac{1}{zF}\left(\frac{d\tilde{\mu}_{\mathrm{A}^{z+}}}{dT} + s_{\mathrm{A}}^{\circ}\right) \tag{66}$$

$$\varepsilon_{\mathrm{e}^-} = \lim_{\Delta T \to 0} \frac{\Delta\phi_{(\mathrm{II})}}{\Delta T} = -\frac{1}{F}\frac{d\tilde{\mu}_{\mathrm{e}^-}}{dT} \tag{67}$$

$d\tilde{\mu}_{\mathrm{A}^{z+}}/dT$ ($d\tilde{\mu}_{\mathrm{e}^-}/dT$) represents the change of the electrochemical potential of ionic (electronic) charge carriers with temperature along the nonisothermal $\mathrm{A}_{\nu+\delta}\mathrm{X}$ sample. $s_{\mathrm{A}}^{\circ}$ is the standard entropy of pure metal A (as chemical potential reference in the ionic probes) and $F$ is the Faraday constant.

It is important to note that the thermopower of compounds will change with time as thermodiffusion of the mobile component proceeds. From an analysis of the change of thermopower with time, $Q_{\mathrm{A}}^{*\prime}$ can directly be obtained. In the very beginning after establishment of a temperature gradient, the $\mathrm{A}_{\nu+\delta}\mathrm{X}$ specimen is still homogenous in composition ($\nabla n_{\mathrm{A}} = 0$). The resulting ionic and electronic thermopower $\varepsilon_{\mathrm{A}^{z+}}|_{\nabla n_{\mathrm{A}}=0}$ and $\varepsilon_{\mathrm{e}^-}|_{\nabla n_{\mathrm{A}}=0}$ can be calculated from eq. (33), by using the condition of coupled ionic and electronic fluxes ($\boldsymbol{J}_{\mathrm{A}^{z+}} = 1/z\,\boldsymbol{J}_{\mathrm{e}^-}$):

$$\varepsilon_{\mathrm{A}^{z+}}|_{\nabla n_{\mathrm{A}}=0} = -\frac{1}{zF}\left(s_{\mathrm{A}^{z+}} - s_{\mathrm{A}}^{\circ} + t_{\mathrm{A}^{z+}}\frac{Q_{\mathrm{A}^{z+}}^{*\prime}}{T} - z\cdot t_{\mathrm{e}^-}\frac{Q_{\mathrm{e}^-}^{*\prime}}{T}\right) \tag{68}$$

$$\varepsilon_{\mathrm{e}^-}|_{\nabla n_{\mathrm{A}}=0} = \frac{1}{F}\left(s_{\mathrm{e}^-} - \frac{t_{\mathrm{A}^{z+}}}{z}\frac{Q_{\mathrm{A}^{z+}}^{*\prime}}{T} + t_{\mathrm{e}^-}\frac{Q_{\mathrm{e}^-}^{*\prime}}{T}\right) \tag{69}$$

To simplify the expressions, the electric transference numbers $t_{\mathrm{A}^{z+}} \equiv z^2L_{\mathrm{A}^{z+}}/(z^2L_{\mathrm{A}^{z+}} + L_{\mathrm{e}^-})$ and $t_{\mathrm{e}^-} \equiv L_{\mathrm{e}^-}/(z^2L_{\mathrm{A}^{z+}} + L_{\mathrm{e}^-})$ have been introduced. If the steady state is reached, the ionic and electronic thermopower $\varepsilon_{\mathrm{A}^{z+}}|_{\boldsymbol{J}_{\mathrm{A}}=0}$ and $\varepsilon_{\mathrm{e}^-}|_{\boldsymbol{J}_{\mathrm{A}}=0}$ can be calculated from eq. (33) under the condition of vanishing fluxes ($\boldsymbol{J}_{\mathrm{A}^{z+}} = 0$ and $\boldsymbol{J}_{\mathrm{e}^-} = 0$):

$$\varepsilon_{\mathrm{A}^{z+}} \equiv \varepsilon_{\mathrm{A}^{z+}}|_{\boldsymbol{J}_{\mathrm{A}}=0} = -\frac{1}{zF}\left(\frac{Q_{\mathrm{A}^{z+}}^{*\prime}}{T} + s_{\mathrm{A}^{z+}} - s_{\mathrm{A}}^{\circ}\right) \tag{70}$$

$$\varepsilon_{\mathrm{e}^-} \equiv \varepsilon_{\mathrm{e}^-}|_{\boldsymbol{J}_{\mathrm{A}}=0} = \frac{1}{F}\left(\frac{Q_{\mathrm{e}^-}^{*\prime}}{T} + s_{\mathrm{e}^-}\right) \tag{71}$$

In the following, we always denote the stationary partial thermopowers ($\boldsymbol{J}_{\mathrm{A}} = 0$) as $\varepsilon_{\mathrm{A}^{z+}}$ and $\varepsilon_{\mathrm{e}^-}$, in order to simplify the equations. The difference between the thermopower in the initial state (eq. (68) and (69)) and the thermopower in the steady state (eq. (70) and (71)) is determined by the reduced heat of transport $Q_A^{*\prime}$ of the mobile component A [9] [80]:

$$\varepsilon_{A^{z+}}|_{\nabla n_A=0} - \varepsilon_{A^{z+}} = \frac{t_{e^-}}{zF} \cdot \frac{Q_A^{*\prime}}{T} \tag{72}$$

$$\varepsilon_{e^-}|_{\nabla n_A=0} - \varepsilon_{e^-} = -\frac{t_{A^{z+}}}{zF} \cdot \frac{Q_A^{*\prime}}{T} \tag{73}$$

Thus, for materials with either small electronic ($t_{A^{z+}} \simeq 1$) or small ionic transference number ($t_{e^-} \simeq 1$), $Q_A^{*\prime}$ can be evaluated directly from suitable measurements of the corresponding thermopower without the need for additional data (see [80] for a more extensive formal treatment and for experimental examples, see Yoo [73] for a corresponding experiment on YSZ). Equivalent experiments on the initial and stationary thermopower of liquid electrolytes have already been reported by Agar [81] and Tyrrell [82].

At this point, we have in principle derived all necessary formal relations for the evaluation of the heat of transport of the neutral mobile component. But it is also instructive to ask for the contribution of the defect heats of transport to the thermopower of ions and electrons. As already shown above for the ionic and electronic heats of transport $Q_{A^{z+}}^{*\prime}$ and $Q_{e^-}^{*\prime}$, the partial thermopower $\varepsilon_{A^{z+}}$ and $\varepsilon_{e^-}$ (eqs. (70) and (71)) can also be expressed in terms of the concentration variables $\alpha$ and mobility ratios $\psi$ (see fig. 8 for $\varepsilon_{A^{z+}} = f(\ln \alpha_i)$):

$$\begin{aligned} \varepsilon_{A^{z+}} = -\frac{R}{zF} \Bigg[ & \frac{Q_i^{*\prime} + Q_v^{*\prime} + \Delta H_F^\circ}{2RT} \tanh \ln(\alpha_i \psi_i^{1/2}) \\ & - \ln \alpha_i + \frac{Q_i^{*\prime} - Q_v^{*\prime}}{2RT} + \frac{s_i^\circ - s_v^\circ}{2R} - \frac{s_A^\circ}{R} \Bigg] \end{aligned} \tag{74}$$

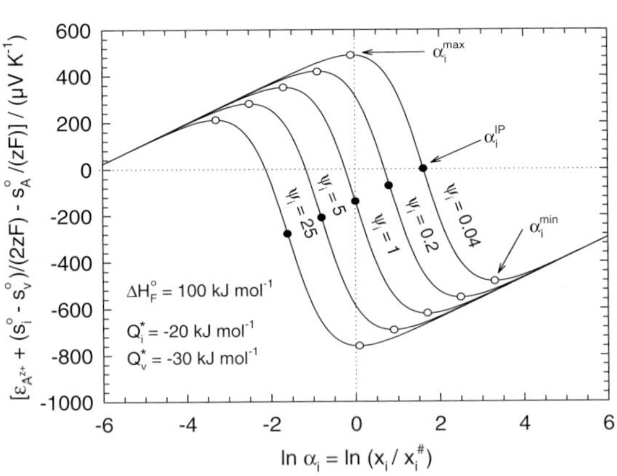


**Fig. 8.** Theoretical ionic thermopower in a mixed ionic/electronic model system. The thermopower curve deforms around the stoichiometric point as a function of the mobility ratio of interstitials and vacancies and shifts vertically as function of the defect heats of transport and the standard defect entropies.

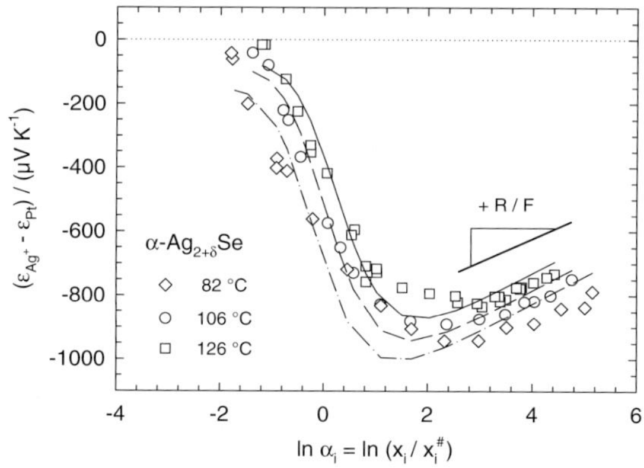


**Fig. 9.** Experimental results for the ionic thermopower of $\alpha$-$Ag_{2+\delta}Se$ as a function of the defect concentration (deviation from stoichiometry). The data agree completely with the theoretical curve in Fig. 8 (after [76]).

$$\varepsilon_{e^-} = \frac{R}{F}\left[\frac{Q_n^{*\prime} + Q_p^{*\prime} + \Delta H_e^\circ}{2RT}\tanh\ln(\alpha_n\psi_n^{1/2}) - \ln\alpha_n + \frac{Q_n^{*\prime} - Q_p^{*\prime}}{2RT} + \frac{s_n^\circ - s_p^\circ}{2R}\right] \tag{75}$$

The resulting curves essentially reflect the superposition of a $\tanh(\alpha_i\psi_i^{1/2})$-function and a $\ln(\alpha_i)$-function, as depicted in fig. (8) for a hypothetical material. The inflection points of the S-shaped thermopower curves in fig. 8 are situated near the stoichiometric composition at $\alpha_i^{IP} = \psi_i^{-1/2}$ and $\alpha_n^{IP} = \psi_n^{-1/2}$. The positions of the extrema of the ionic thermopower $\varepsilon_{A^{z+}}$ also depend on $\psi_i$ as well as on the heats of transport $Q_i^{*\prime}$ and $Q_v^{*\prime}$ and the defect formation enthalpy $\Delta H_F^\circ$:

$$\ln(\alpha_i^{Max}\psi_i^{1/2}) = -\text{arcosh}\left[(Q_i^{*\prime} + Q_v^{*\prime} + \Delta H_F^\circ)/2RT\right]^{1/2} \tag{76}$$

$$\ln(\alpha_i^{Min}\psi_i^{1/2}) = +\text{arcosh}\left[(Q_i^{*\prime} + Q_v^{*\prime} + \Delta H_F^\circ)/2RT\right]^{1/2} \tag{77}$$

For the electronic thermopower one obtains for the position of the extrema:

$$\ln(\alpha_n^{Max}\psi_n^{1/2}) = +\text{arcosh}\left[\left(Q_n^{*\prime} + Q_p^{*\prime} + \Delta H_e^\circ\right)/2RT\right]^{1/2} \tag{78}$$

$$\ln(\alpha_n^{Min}\psi_n^{1/2}) = -\text{arcosh}\left[\left(Q_n^{*\prime} + Q_p^{*\prime} + \Delta H_e^\circ\right)/2RT\right]^{1/2} \tag{79}$$

Thus, if a complete thermopower curve ($\varepsilon_{A^{z+}} = f(\ln\alpha_i)$) as a function of the charge carrier concentration is available, together with mobility data, the sum $Q_i^{*\prime} + Q_v^{*\prime} + \Delta H_F^\circ$ can be determined easily. The is a generalisation of the result first obtained by Howard and Lidiard [6].

Utilizing eq. (74), an often applied technique for the determination of heats of transport of ionic defects is the comparison of the ionic thermopower of pure and doped crystals [6] [8] [83] [84] [85]. By doping a binary compound $A_{\nu+\delta}X$

(which exhibits Frenkel disorder) with cations $B^{z+1}$ of the charge $z+1$, additional cation vacancies are introduced:

$$B_{\nu z}X_{z+1} = \nu\, z\, B_A^{\bullet} + \nu V_A^{z\prime} + (z+1)\, X_X^{\times} \tag{80}$$

The difference between the ionic thermopower (steady state) of a doped crystal and a pure crystal results from eq. (74) in the following expression[2]:

$$\varepsilon_{A^{z+}}(\ln \alpha_i) - \varepsilon_{A^{z+}}(\ln \alpha_i = 0) = \\ -\frac{R}{zF}\left[\frac{Q_i^{*\prime} + Q_v^{*\prime} + \Delta H_F^{\circ}}{RT}\,\frac{\psi_i}{\psi_i + 1}\,\frac{1-\alpha_i^{-2}}{\psi_i + \alpha_i^{-2}} - \ln \alpha_i\right] \tag{82}$$

For a heavily doped crystal one can safely assume that the concentration of cation interstitials is negligibly small compared to the concentration of cation vacancies ($\alpha_i \ll 1$), and eq. (82) can be further simplified:

$$\varepsilon_{A^{z+}}(\ln \alpha_i) - \varepsilon_{A^{z+}}(\ln \alpha_i = 0) \simeq \\ -\frac{R}{zF}\left[\frac{Q_i^{*\prime} + Q_v^{*\prime} + \Delta H_F^{\circ}}{RT}\,\frac{\psi_i}{\psi_i + 1} - \ln \alpha_i\right] \tag{83}$$

If the mobility ratio $\psi_i$ and the relative defect concentration $\alpha_i = 1/\alpha_v$ are known, the sum of the heats of transport of the individual defects and their formation enthalpy ($Q_i^{*\prime} + Q_v^{*\prime} + \Delta H_F^{\circ}$) can be evaluated [84] [85]. For high dopant concentrations, the association between charged impurity ions and vacancies has to be taken into account, and the formal relations have to be modified accordingly [85]. Examples of experimental studies using the above equations have been reported for silver halides by Christy et. al. [7], by Haga [8] and by Corish and Jacobs [86]. The conclusions of Christy [7] are shown in fig. 10. A strong temperature dependence of the sum $Q_i^{*\prime} + Q_v^{*\prime}$ may be noted. Insofar as $Q_v^{*\prime}$ is expected to be negative in sign and larger in magnitude than $Q_i^{*\prime}$ these results would imply that the magnitude of $Q_v^{*\prime}$ increases with increasing temperature, contrary to the results described in section 3. Corish and Jacobs have argued that these results are made unreliable by use of earlier ionic conductivity measurements and they have proposed a new analysis on the basis that the individual defect heats of transport and associated defect entropies are independent of temperature [86]. This approach has been used for the analysis of data on other ionic conductors too (Allnatt and Lidiard [3], especially §11.8.1). However plausible as this approach is, we emphasize that there is no rigorous way to separate the measured sum $Q_i^{*\prime} + Q_v^{*\prime}$ into its component parts. More recently, a related study of the Soret effect (section 4.1) in AgCl [80] allowed the inference of the heat

[2] Eq. (82) can also formulated as:

$$\varepsilon_{A^{z+}}(\ln \alpha_i) - \varepsilon_{A^{z+}}(\ln \alpha_i = 0) = \\ -\frac{R}{zF}\left[\frac{Q_i^{*\prime} + Q_v^{*\prime} + \Delta H_F^{\circ}}{2RT}\left[\tanh \ln(\alpha_i \psi_i^{1/2}) - \tanh \ln \psi_i^{1/2}\right] - \ln \alpha_i\right] \tag{81}$$

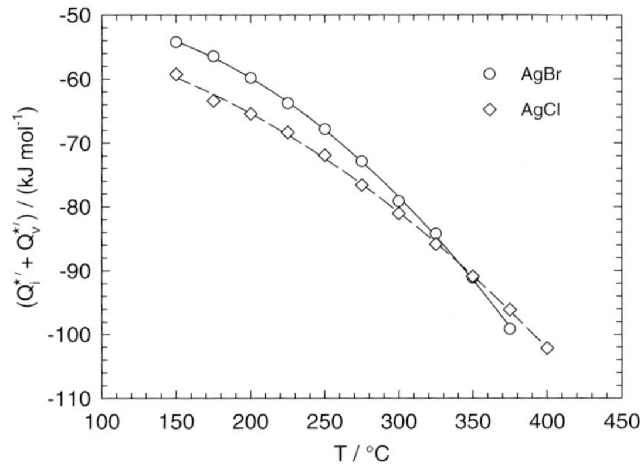


**Fig. 10.** Results for the sum of the reduced heats of transport of interstitials and vacancies in AgCl and AgBr as a function of temperature (after Christy [7]).

of transport of the Ag component. Insofar as the contribution of the electrons to this $Q^{*\prime}_{\mathrm{Ag}}$ is small (as expected for delocalized electrons), these measurements in combination with results such as those in fig. 10 allow the estimation of $Q^{*\prime}_{\mathrm{i}}$ and $Q^{*\prime}_{\mathrm{v}}$. At the temperature of the Soret measurements (300 °C) $Q^{*\prime}_{\mathrm{v}}$ is –77 kJ mol$^{-1}$ which is close to minus twice the heat of activation for vacancies. Clearly, further studies of this and other systems by this combination of techniques are desirable.

A comment should also be made on a different formal approach to the thermopower. Whereas the use of the electrochemical potential, and consequently the definition of the absolute thermopower, is favored in modern physical chemistry (cf. Wagner [9]) in earlier work the thermopower was seperated into a "homogenous" part (electrical potential gradient within the specimen) and a "heterogenous" part (electric potential difference at the phase boundary with the electrodes). Attempts were made to estimate the second of these independently until Howard and Lidiard [6] proposed that both parts should be analysed consistently on the basis of defect theory. At the same time in the physics of solids thermopower is generally analysed in terms of Peltier and Thomson heat coefficients. Clearly both methods correctly applied merely provide alternative descriptions of the same physical phenomenon.

### 4.4 Short-circuited thermocells

The realisation of Soret experiments on solids at high temperatures leads to considerable practical difficulties. Most critical is the encapsulation of the specimen with an inert and impermeable cover. Thermal diffusion experiments on e. g. $UO_2$ have mostly been performed with oxide pellets in a zircalloy cover. The metal cover provides an electronic short-circuit, and the simple derivation of eqs. (61) and (62) by using the zero-flux condition fails [14] [64] [65]. If an electronic short-circuit for a mixed cation and electron conducting compound $A_{\nu+\delta}X$ by an inert metal cover is taken into account, we obtain a modified expression:

$$\nabla\mu_{\mathrm{A}}\Big|_{\boldsymbol{J}_{\mathrm{A}}=0}^{\mathrm{sc}} = -\left(\frac{Q_{\mathrm{A}}^{*\prime}}{T} + s_{\mathrm{A}}\right)\nabla T + zF\frac{A_{\mathrm{Me}}\sigma_{\mathrm{e}^-}^{\mathrm{Me}}}{A_{\mathrm{A}_\nu\mathrm{X}}\sigma_{\mathrm{e}^-}^{\mathrm{A}_\nu\mathrm{X}} + A_{\mathrm{Me}}\sigma_{\mathrm{e}^-}^{\mathrm{Me}}}\left(\varepsilon_{\mathrm{e}^-}^{\mathrm{A}_\nu\mathrm{X}} - \varepsilon_{\mathrm{e}^-}^{\mathrm{Me}}\right)\nabla T \tag{84}$$

for the chemical potential gradient in the steady state. $A_{\mathrm{A}_\nu\mathrm{X}}$ and $A_{\mathrm{Me}}$ are the cross sectional areas of the $\mathrm{A}_{\nu+\delta}\mathrm{X}$-specimen and the short-circuiting metal cover. $\sigma_{\mathrm{e}^-}^{\mathrm{A}_\nu\mathrm{X}}$, $\sigma_{\mathrm{e}^-}^{\mathrm{Me}}$, $\varepsilon_{\mathrm{e}^-}^{\mathrm{A}_\nu\mathrm{X}}$ and $\varepsilon_{\mathrm{e}^-}^{\mathrm{Me}}$ are the electronic conductivities and thermopowers of the compound $\mathrm{A}_{\nu+\delta}\mathrm{X}$ and the cover metal Me, respectively. The stationary composition gradient within a short-circuited specimen is then:

$$\nabla n_{\mathrm{A}}\Big|_{\boldsymbol{J}_{\mathrm{A}}=0}^{\mathrm{sc}} = -\frac{1}{\left(\frac{\partial\mu_{\mathrm{A}}}{\partial n_{\mathrm{A}}}\right)_T}\left[\frac{Q_{\mathrm{A}}^{*\prime}}{T} - zF\frac{A_{\mathrm{Me}}\,\sigma_{\mathrm{e}^-}^{\mathrm{Me}}}{A_{\mathrm{A}_\nu\mathrm{X}}\,\sigma_{\mathrm{e}^-}^{\mathrm{A}_\nu\mathrm{X}} + A_{\mathrm{Me}}\,\sigma_{\mathrm{e}^-}^{\mathrm{Me}}}(\varepsilon_{\mathrm{e}^-}^{\mathrm{A}_\nu\mathrm{X}} - \varepsilon_{\mathrm{e}^-}^{\mathrm{Me}})\right]\nabla T \tag{85}$$

Comparing this expression with the Soret effect under zero-flux conditions (eq. (61)), it is reasonable to define a "practical" heat of transport (cf. Bober and Schumacher [13]),

$$Q_{\mathrm{A}}^{\mathrm{exp}} = Q_{\mathrm{A}}^{*\prime} - zF\frac{A_{\mathrm{Me}}\,\sigma_{\mathrm{e}^-}^{\mathrm{Me}}}{A_{\mathrm{A}_\nu\mathrm{X}}\,\sigma_{\mathrm{e}^-}^{\mathrm{A}_\nu\mathrm{X}} + A_{\mathrm{Me}}\,\sigma_{\mathrm{e}^-}^{\mathrm{Me}}}\left(\varepsilon_{\mathrm{e}^-}^{\mathrm{A}_\nu\mathrm{X}} - \varepsilon_{\mathrm{e}^-}^{\mathrm{Me}}\right)T \tag{86}$$

which takes the contribution from the electronic short-circuit into account. If we assume that the thermopower $\varepsilon_{\mathrm{e}^-}^{\mathrm{Me}}$ of the metal Me is negligibly small compared to the electronic thermopower of the semiconducting oxide[3] and the conductance ($\sim A_{\mathrm{Me}}\,\sigma_{\mathrm{e}^-}^{\mathrm{Me}}$) of the metal cover is high compared to that of the $\mathrm{A}_{\nu+\delta}\mathrm{X}$ specimen, we obtain:

$$Q_{\mathrm{A}}^{\mathrm{exp}} \simeq Q_{\mathrm{A}}^{*\prime} - zF\,\varepsilon_{\mathrm{e}^-}^{\mathrm{A}_\nu\mathrm{X}}\,T \tag{87}$$

In the case of the oxygen (anion) conducting uranium oxide ($z = -2$ and $\nu = 2$), the "practical" heat of transport $Q_{\mathrm{O}}^{\mathrm{exp}}$ results as:

$$Q_{\mathrm{O}}^{\mathrm{exp}} \simeq Q_{\mathrm{O}}^{*\prime} + 2F\,\varepsilon_{\mathrm{e}^-}^{\mathrm{UO_2}}\,T \tag{88}$$

Thus, the practical heat of transport $Q_{\mathrm{O}}^{\mathrm{exp}}$ which is determined experimentally from nuclear fuel pellets (short-circuited thermocells) differs from the oxygen heat of transport $Q_{\mathrm{O}}^{*\prime}$ by a term which includes the electronic thermopower of the oxide. As long as the electronic short-circuit leads to a negligible flux of electrons through the oxide at all ($A_{\mathrm{Me}}\,\sigma_{\mathrm{e}^-}^{\mathrm{Me}} \gg A_{\mathrm{A}_\nu\mathrm{X}}\,\sigma_{\mathrm{e}^-}^{\mathrm{A}_\nu\mathrm{X}}$) and the metal thermopower is small, the stationary composition gradient $\nabla\delta$ is determined by the transport properties of the oxide itself. A corresponding analysis of thermal diffusion in nuclear fuel pellets can be found in [15].

[3] which is justified for conductors with low electronic charge carrier concentrations.

## 4.5 Peltier effect

The Peltier effect leads to the heating or cooling of the interface between two different conductors when an electric current crosses the interface. In mixed conductors, one can distinguish an ionic and an electronic Peltier effect. Generally any ionic Peltier effect will only lead to minor temperature effects due to the small ionic current densities. Regarding the separation of the transfered heat into a kinetic and a thermodynamic contribution, it is, however, of interest to ask for the magnitude of the different contributions.

If one assumes local equilibrium at the interface, the heat effect can be expressed as:

$$\frac{1}{A}\frac{dQ}{dt} = T\ [S_k^*(\beta) - S_k^*(\alpha)]\ J_k \tag{89}$$

with $J_k$ denoting the molar flux of charge carriers $k$ across the interface, and with $S_k^*$ representing the entropy of transport of $k$. Introducing the definition of the absolute thermopower, $\varepsilon_k \equiv -1/(z_k F) \cdot S_k^*$, and the partial current density $i_k = (z_k F) J_k$, one arrives at the well known relation

$$\frac{1}{A}\frac{dQ}{dt} = -T\ [\varepsilon_k(\beta) - \varepsilon_k(\alpha)]\ i_k = -[\pi_k(\beta) - \pi_k(\alpha)]\ i_k = \Pi_{k,\alpha|\beta}\, i_k \tag{90}$$

where we have introduced the Peltier coefficient $\pi_k(\alpha) \equiv T\,\varepsilon_k(\alpha)$ of the individual phase $\alpha$ and the Peltier coefficient $\Pi_{k,\alpha|\beta} \equiv \pi_k(\beta) - \pi_k(\alpha)$ of the interface $\alpha|\beta$ (transfer of charge carriers $k$).

It is worth noting two points: For the estimation of the Peltier effect, it is sufficient to know the absolute thermopowers of the two bulk phases. Correspondingly, the experimental study of the Peltier effect does not by itself allow the determination of the heats of transport. Rather, the heats of transport of the transfering charge carriers represent only a part of the Peltier effect:

$$\frac{1}{A}\frac{dQ}{dt} = [T\,(s_k(\beta) - s_k(\alpha)) + (Q_k^{*\prime}(\beta) - Q_k^{*\prime}(\alpha))]\ J_k \tag{91}$$

Since it is impossible to determine either the partial entropies or the individual heats of transport, one has to introduce additional assumptions in order to obtain further insight. From the analysis of the nonisothermal transport properties of $\alpha$-$Ag_{2+\delta}S$ [61], we derive the following data for the different contributions to the ionic (absolute) thermopower at $T = 433$ K: The entropy of transfer results from thermopower measurements as $s_{Ag^+} + Q_{Ag^+}^{*\prime}/T \simeq 81.5$ J mol$^{-1}$ K$^{-1}$, and the partial entropy of silver ions can be estimated from measurements of the metal entropy as $s_{Ag^+} \simeq 167$ J mol$^{-1}$ K$^{-1}$. The "kinetic" entropy of the ions results therefore as $Q_{Ag^+}^{*\prime}/T \simeq -85.5$ J mol$^{-1}$ K$^{-1}$. In accord with eq. (91), one has to conclude that the heat transfer due to the heat of transport ($Q_{Ag^+}^{*\prime}$) represents a non-negligible contribution to the Peltier effect.

## 4.6 Ionic thermopower: a measure of defect concentrations

As in semiconductor physics, thermopower measurements can be used for the study of charge carrier concentrations. In mixed conductors, both electronic and ionic thermopower (eqs. (74) and (75)) can be measured independently and therefore have to be distinguished. Furthermore, mixed conducting compounds are subject to thermodiffusion and the thermopower reflects the diffusion process in its time-dependence.

In general, thermopower includes a thermodynamic part (partial entropy of charge carriers) and a kinetic part (heat of transport). The entropy is always concentration dependent which is the basis for the use of thermopower in semiconductor physics. The heat of transport is principally concentration dependent, if local defect formation equilibria are involved in the transport process, i.e. far in the extrinsic regime. The absolute thermopower gives no direct information on absolute charge carrier concentrations. It includes standard entropies and heats of transport of defects which are not known a priori. Either these quantities are estimated from a reasonable model or one restricts the analysis to changes of the charge carrier concentration. A reliable statement on the n- or p-type character (v- or i-type) of a semiconductor (of an ionic conductor) can only be given for extrinsic materials with sufficiently high charge carrier concentrations (compared to the intrinsic case). This is clearly shown in fig. (8). Clear examples of the relation between defect concentration and thermopower in ionic crystals can be found in the studies by Christy and others [7] [87] [88], who, following the theoretical analysis of Howard and Lidiard [6], were the first to combine lattice point defect theory with thermopower measurements.

Another distinct example concerns the application of thermopower measurements to the study of heterogeneous conductors [89]. The contribution of inhomogeneous space charge regions to the measurable thermopower of a macroscopic sample is analysed and calculated for an ionic crystal with Frenkel disorder. It is shown that thermopower measurements can indeed supply supplementary information on defect concentrations in space charge regions. Since the approach is phenomenological, inhomogeneous semiconductors can be treated in the same way.

## 5 Conclusions

In this chapter we have summarized the results of our combined theoretical and experimental efforts to obtain a better understanding of the heat of transport as an intriguing problem in solid state kinetics. In both approaches we have necessarily to concentrate on the simplest solids, in order to obtain results which can be interpreted in terms of individual point defects. There are two main parts to the theory: The first is the description and classification of thermotransport phenomena by the application of the principles of non-equilibrium thermodynamics to systems where atomic and ionic mobilty result from the presence of point defects (vacancies and interstitials). This synthesis was described in section 2. It allows the prediction of experimentally measurable quantities as function of stoichiometry and the concentration of foreign ions, and thus the inference of heats of transport from such measurements. In general, this part of the theory is fruitful and successful, as shown by the experimental results described and analysed in section 4.

The second part of the theory is the representation of the heats of transport of specific types of defects in terms of interatomic forces and motions. As shown by our account in section 3 this part is at a much earlier stage of development. Thus, our computer experiments focus on noble gas crystals and vacancy jumps therein. As yet there are no experiments on thermodiffusion in these systems but we believe that the physics behind our results and their analysis is independent of the type of chemical bonding in a crystal. Especially the computer experiments so far offer clear and fascinating insight into the heat fluxes associated with defect jumps.

Our physical experiments focus on mixed-conducting materials with small deviation from stoichiometry. Here we can apply electrochemical techniques which offer the necessary precision for the determination of small concentration variations. Apart from many details given above, the main result of these studies is an agreement on the fact that the heat of transport of vacancies equals roughly minus twice the migration enthalpy. This result of the MD simulations is supported so far by experimental results for vacancy motion in $Cu_{2-\delta}O$ and AgCl. Thus, we have to conclude that all previous theoretical models which result in reduced heats of transport being smaller than or equal to the migration enthalpy are most probably incorrect.

The microscopic picture for the understanding of the heat of transport has been discussed in detail in section 3. In broad terms, our results show that when a particle (moving by a vacancy mechanism) makes an elementary jump over one lattice distance, the necessary activation energy is transported approximately along twice the jump distance. In simple words, the lattice supplies the jumping particle with activation energy from its back and the landing particle gives this energy off in front of its landing lattice site. Of course, this picture may be completely different for interstitial jumps.

It would be of great value to extend the computer experiments to other systems, in particular to interstitial defects. The same is true for "real" physical experiments. The study of thermodiffusion and the heat of transport is an excellent example of the need for and the value of combined theoretical and experimental efforts. Nevertheless we believe that our current state of knowledge already represents significant progress compared to the early years of thermodiffusion studies in the solid state.

## Acknowledgements

We wish to record our appreciation of many fruitful discussions with A.R. Allnatt (London, Ontario), P.J. Grout (Oxford), P.W.M. Jacobs (London, Ontario) and H. Schmalzried (Hannover). Financial support by the DFG (Ja 648/3-1), the British Council and the DAAD is gratefully acknowledged.

## References


1. A.R. Allnatt, A.V. Chadwick: Chem. Rev. **67**, 681 (1967)
2. W. Jost: *Diffusion in Solids, Liquids and Gases* (Academic Press, New York 1960)
3. A.R. Allnatt, A.B. Lidiard: *Atomic Transport in Solids* (Cambridge University Press, Cambridge 1993)
4. H. Reinhold: Z. Phys. Chem. A **141**, 137 (1929)
5. C. Wagner, W. Schottky: Z. Phys. Chem. B **11**, 163 (1930)
6. R.E. Howard, A.B. Lidiard: Phil. Mag. **2**, 1462 (1957)
7. R.W. Christy: J. Chem. Phys. **34**, 1148 (1961)
8. E. Haga: J. Phys. Soc. Japan **13**, 1090 (1958); **14**, 992 (1959); **15**, 1949 (1960)
9. C. Wagner: Prog. Solid State Chem. **7**, 1 (1972)
10. H. Wever: *Elektro- und Thermotransport in Metallen* (Julius Ambrosius Barth, Leipzig 1973)
11. J.-L. Bocquet, G. Bríbec, Y. Limoge in: *Physical Metallurgy* 3rd ed., vol. 1, ed. by R.W. Cahn, P. Haasen (North-Holland, Amsterdam 1983) p. 385
12. F. Faupel: Phil. Mag. A **65**, 1287 (1992)
13. M. Bober, G. Schumacher: Adv. Nucl. Sci. Technol. **7**, 121 (1973)
14. F. Millot, P. Gerdanian: J. Nucl. Mater. **92**, 257 (1980)
15. J. Janek, H. Timm: J. Nucl. Mater. **255**, 116 (1998)
16. H. Schmalzried: *Chemical Kinetics of Solids* (VCH, Weinheim 1995)
17. S. R. de Groot: *Thermodynamik irreversibler Prozesse*, vol. 18/18a (Bibliographisches Institut, Mannheim 1960)
18. C. Korte: Nichtisotherme Transportprozesse in gemischtleitenden Ionenkristallen. Ph. D. Thesis, University of Hannover (1997)
19. H.I. Yoo, H. Schmalzried, M. Martin, J. Janek: Z. Phys. Chem. N. F. **168**, 129 (1990)
20. J. Janek: Ber. Bunsenges. Phys. Chem. **98**, 1213 (1994)
21. J. Janek: Dynamische Wechselwirkungen bei Transportvorgängen in Übergangsmetalloxiden. Ph. D. Thesis, University of Hannover (1992)
22. C. Wagner: Progress in Solid State Chemistry (Part 1) **10**, 3 (1975)
23. J. Janek, C. Korte: Z. Phys. Chem. **196**, 187 (1996)
24. J. Janek: Ber. Bunsenges. Phys. Chem. **99**, 920 (1995)
25. R.E. Howard, A.B. Lidiard: Rep. Prog. Phys. **27**, 1610 (1964)
26. M.J. Gillan: "Diffusion in a temperature gradient". In: *Mass Transport in Solids* eds. F. Bénière and C.R.A. Catlow (Plenum, New York and London 1983) pp. 227-250
27. J.H. Harding: Rep. Progr. Phys. **53**, 1403 (1990)
28. C. Jones, P.J. Grout, A.B. Lidiard: Phil. Mag. Letts. **74**, 217 (1996)
29. C. Jones, P.J. Grout, A.B. Lidiard: Ber. Bunsenges. Phys. Chem. **101**, 1232 (1997)
30. C. Jones, D. Phil. Thesis, University of Oxford (1997)
31. C. Jones, P.J. Grout, A.B. Lidiard: Phil. Mag, A **79**, 2051 (1999)
32. R.E. Howard, J.R. Manning: J. Chem. Phys. **36**, 910 (1962)
33. A.R. Allnatt, Y. Okamura: J. Phys. **C17**, 5137 (1984)
34. K. Wirtz: Phys. Zeits. **44**, 221 (1943)
35. J.A. Brinkman: Phys. Rev. **93**, 345 (1954)
36. A.D. LeClaire: Phys. Rev. **93**, 344 (1954)
37. L.A. Girifalco: Phys. Rev. **128**, 2630 (1962)
38. S. Prager, H. Eyring: J. Chem. Phys. **21**, 1347 (1953)
39. E. L. Dougherty, H. G. Drickamer: J. Phys. Chem. **59**, 443 (1995)

40. A.R. Allnatt, S.A. Rice: J. Chem. Phys. **33**, 573 (1960)
41. G. Schottky: Phys. Stat. Solidi **8**, 357 (1965)
42. H.B. Huntington: J. Phys. Chem. Solids **29**, 1641 (1968)
43. I.M. Boswarva, P.L. Tan: Phys. Stat. Solidi **50**, 141 (1972)
44. M.J. Gillan: J. Phys. C **10**, 1641 (1977)
45. M.J. Gillan: J. Phys. C **10**, 3051 (1977)
46. M.P. Allen, D.J. Tildesley: *Computer Simulation of Liquids* (Oxford University Press, Oxford 1987)
47. J.-P. Hansen, I.R. McDonald: *Theory of Simple Liquids*, 2nd edn. (Academic Press, London 1986)
48. A. R. Allnatt: J. Phys. A **34**, 7441 (2001)
49. M.J. Gillan, M.W. Finnis: J. Phys. C **11**, 4469 (1978)
50. K.-D. Becker, C. Hoheisel, J. Chem. Phys. **77**, 5108 (1982)
51. R. Vogelsang, C. Hoheisel, Z. Naturf. a **41**, 1051 (1986)
52. R. Vogelsang, C. Hoheisel, J. Phys. Chem. **20**, 5923 (1987)
53. R. Vogelsang, C. Hoheisel, J. Phys. Chem. **20**, 5933 (1987)
54. P. Grout, private communication
55. M. Gerl: J. Phys. Chem. Solids **28**, 725 (1967)
56. J. Janek, C. Korte: "Thermal diffusion in mixed conductors". In: *Proc. Third Inter. Symp. Ionic and Mixed Conducting Ceramics*, Hrsg.: T. A. Ramanarayanan, The Electrochemical Society, Proc. Vol. **97-24**, 304-328 (1998)
57. C. Korte, J. Janek: Ber. Bunsenges. Phys. Chem. **100**, 425 (1996)
58. J. Janek, C. Korte: Ber. Bunsenges. Phys. Chem. **99**, 932 (1995)
59. E.A. Guggenheim: *Thermodynamics* (North-Holland, Amsterdam 1967)
60. R. Haase: "Thermodynamik der Irreversiblen Prozesse". In: *Fortschritte der Physikalischen Chemie, vol. 8*, editor: W. Jost / Göttingen (Dr. Dietrich Steinkopff Verlag, Darmstadt 1963)
61. C. Korte, J. Janek: J. Phys. Chem. Solids **58**, 623 (1997)
62. H. Schmalzried, A. Navrotsky: *Festkörperthermodynamik* (Verlag Chemie, Weinheim 1975)
63. C. Korte, J. Janek, H. Timm: Solid State Ionics **101-103**, 465 (1997)
64. F. Millot, P. Gerdanian: J. Phys. Chem. Sol. **43(6)**, 501 (1982)
65. F. Millot, P. Gerdanian: J. Nucl. Mater. **116**, 55 (1983)
66. S. Miyatani: J. Phys. Soc. Japan **24**, 328 (1968)
67. W. Koch, H. Rickert, G. Schlechtriemen: Solid State Ionics **9/10**, 1197 (1983)
68. T. Ohachi, T. Taniguchi: Solid State Ionics **3/4**, 89 (1981)
69. W.T. Petuskey, H.K. Bowen: J. Americ. Cer. Soc. **64**, 611 (1981)
70. A. Honders, E.W.H. Young, J.H.W. de Witt, G.H.J. Broers: Solid State Ionics **8**, 115 (1983)
71. A. Honders, A.J.H. Hintzen, J.M. der Kinderen, J.H.W. de Witt, G.H.J. Broers: Solid State Ionics **9/10**, 1205 (1983)
72. D.-L. Kim, H.-I. Yoo: Solid State Ionics **81**, 135 (1995)
73. H.-I. Yoo, J.H. Hwang: J. Phys. Chem. Solids **53**, 973 (1992)
74. C. Sari, G. Schumacher: J. Nucl. Mater. **61**, 192 (1976)
75. J.-A. Crolet: Phil. D. Thesis, Université de Paris, Orsay
76. C. Korte, J. Janek: Z. Phys. Chem. **206**, 129 (1998)
77. H. Timm, J. Janek: "Ionic thermopower and Soret effect in $Cu_{2+\delta}O$: An experimental study", to be published (2001)
78. H. Timm, J. Janek: "Demixing and Pore Migration in Oxide Solid Solutions in Temperature Gradients", to be published (2001)

79. H. Timm: Thermodiffusion in binären und ternären Oxiden. Ph. D. Thesis, Universty of Hannover (1999)
80. J. Janek, C. Korte: Solid State Ionics **92**, 193 (1996)
81. J.N. Agar, W.G. Breck: Trans. Farady Soc. **53**, 167 (1957)
82. H.J.V. Tyrrell: *Diffusion and Heat Flow in Liquids* (Butterworth, London 1961) pp. 183, 262
83. L. Patrick, A.W. Lawson: J. Chem. Phys. **22**(6), 1492 (1954)
84. R.E. Howard: Some problems concerning defects in solids. D. Phil. Thesis, Oxford (1957)
85. R.E. Howard, A.B. Lidiard: Disc. Faraday Soc. **23**, 113 (1957)
86. J. Corish, P.W.M. Jacobs: J. Phys. C **6**, 57 (1973)
87. R.W. Christy, E. Fukushima, H.T. Li: J. Chem. Phys. **30**, 136 (1959)
88. Y.W. Hsueh, R.W. Christy, R.C. Mueller: Technical Report No. 3 on Contract Nonr-2799(00) NR 032-460 (Office of Naval Research, Washington)
89. M. Vennekamp, J. Janek: Solid State Ionics **118**, 43 (1999)

## Supplementary glossary

| | |
|---|---|
| $n_k$ | molar concentration of the component (species) $k$ in mol l$^{-1}$ |
| $a_k$ | thermodynamic activity of the component (species) $k$ |
| $V_{\mathrm{m}}^{\mathrm{A}_\nu\mathrm{X}}$ | molar volume of the compound $A_\nu X$ in m$^3$ mol$^{-1}$ |
| $\mu_k$ | chemical potential of the component (species) $k$ in J mol$^{-1}$ |
| $\tilde{\mu}_k$ | electrochemical potential of the component (species) $k$ in J mol$^{-1}$ |
| $z_k$ | formal charge number of the component (species) $k$ |
| $\phi$ | electric (Galvani) potential in V |
| $h_k$ | partial molar enthalpy of the component (species) $k$ in J mol$^{-1}$ |
| $s_k$ | partial molar entropy of the component (species) $k$ in J mol$^{-1}$ |
| $v_k$ | velocity of the mobile component (species) $k$ in m s$^{-1}$ |
| $\boldsymbol{F}$ | mechanical force in N |
| $u_k$ | electrochemical mobility of the component (species) $k$ in m (s V)$^{-1}$ |
| $b_k$ | mobility of the component (species) $k$ in m (s N)$^{-1}$ |
| $b_k^{\mathrm{T}}$ | thermodiffusive mobility of the component (species) $k$ in m (s N)$^{-1}$ |
| $Q_k^{*\prime}$ | reduced heat of transport of the component (species) $k$ in J mol$^{-1}$ |
| $S_k^*$ | entropy of transport of the component (species) $k$ in J mol$^{-1}$ K$^{-1}$ |
| $\boldsymbol{J}_k$ | molar flux $\boldsymbol{J}_k$ of the component (species) $k$ in mol m$^{-2}$ s$^{-1}$ |
| $\boldsymbol{X}_k$ | thermodynamic force on the component (species) flux $\boldsymbol{J}_k$ in N mol$^{-1}$ |
| $\boldsymbol{J}_Q$ | heat flux $\boldsymbol{J}_Q$ in J m$^{-2}$ s$^{-1}$ |
| $\boldsymbol{X}_Q$ | thermodynamic force on the heat flux $\boldsymbol{J}_Q$ in m$^{-1}$ |
| $L_{ki}$ | phenomenological transport coefficient |
| $F$ | Faraday's constant, 96484.8 C mol$^{-1}$ |
| $R$ | gas constant, 8.31441 J (K mol)$^{-1}$ |
| $\sigma_{ki}$ | conductivity in S m$^{-1}$ |
| $\tilde{D}_{ki}$ | chemical diffusion coefficient in m$^2$ s$^{-1}$ |
| $D_k$ | self diffusion coefficient of the component (species) $k$ in m$^2$ s$^{-1}$ |
| $D_k^{\mathrm{T}}$ | thermodiffusion coefficient of the component (species) $k$ in m s$^{-1}$ |
| $\delta$ | deviation from stoichiometry |

$\Delta H_{\mathrm{F}}^{0}$ reaction enthalpy of Frenkel defect formation in J mol$^{-1}$
$\Delta H_{\mathrm{e}}^{0}$ reaction enthalpy of electron/electron hole formation in J mol$^{-1}$
$x_k$ molar (site) fraction of the mobile defect $k = \mathrm{i, v, n, p}$
$\psi_i$ mobility ratio $u_{\mathrm{i}}/u_{\mathrm{v}}$ of interstitials and vacancies
$\psi_n$ mobility ratio $u_{\mathrm{n}}/u_{\mathrm{p}}$ of electrons and electron holes
$\alpha_{\mathrm{i}}$ molar (site) fraction of interstitials $x_{\mathrm{i}}$ normalized to the intrinsic concentration $x_{\mathrm{i}}^{\#}$, i.e. $x_{\mathrm{i}}/x_{\mathrm{i}}^{\#}$
$\alpha_{\mathrm{n}}$ molar (site) fraction of electrons $x_{\mathrm{n}}$ in the conduction band normalized to the intrinsic concentration $x_{\mathrm{n}}^{\#}$, i.e. $x_{\mathrm{n}}/x_{\mathrm{n}}^{\#}$
$\delta$ deviation of stoichiometry
$\nu$ stoichiometric coefficient
$t_k$ electric transference number of the mobile charge carrier $k = \mathrm{A}^{z+}, \mathrm{e}^-$
$\epsilon_k$ absolute thermoelectric power of the mobile charge carrier $k = \mathrm{A}^{z+}, \mathrm{e}^-$
$\omega$ average frequency of defect jumps in s$^{-1}$
$\nu_{\mathrm{D}}$ Debye frequency or characteristic lattice frequency in s$^{-1}$
$g_{\mathrm{m,k}}$ Gibbs free energy of activation for the jump of defect $k$
$h_{\mathrm{m,k}}$ migration enthalpy of mobile defect $k$
$q^{*\prime}$ reduced heat of transport of one particle in units of energy (J), $Q^{*\prime} = N_{\mathrm{A}}\, q^{*\prime}$
$\tau$ relaxation time of a local lattice disturbance in s
$\bar{v}$ mean velocity of phonons in m s$^{-1}$
$s$ jump distance in m
$u$ energy per atom of the molecular dynamics cell

# Hip, Hip, Soret!

Florian Müller-Plathe

Max-Planck-Institut für Polymerforschung,
Ackermannweg 10, D-55128 Mainz, Germany

Many years ago, Ludwig did detect
the behaviour now called the Soret effect.
Sodium sulphate in eighteen-fifty-six
did not obediently follow Fick's
first law. But if he cooled down one side
the salt went left, the water to the right.
He was surprised in every way.
Hip, hip, Soret!

Soret gave the effect his name,
even though it was still the same
phenomenon: In an electrolyte
some ions go left, the others to the right.
Predict which way which ions go?
Forget it. Charles Soret don't know.
But let's shout anyway:
Hip, hip, Soret!

Salt solutions and organic
mixtures, molten salts – don't panic! –,
polymers, alloys, and gases too
show the effect. It is not taboo
in something as ridiculous
as Ising spins. It's ubiquitous.
Thermal diffusion rules o.k.!
Hip, hip, Soret!

The effect, though, lacks explanations.
This goes on since generations.
Onsager's linear theory
links a flux of matter to gradient $T$.
But while its success was spectacular,
it is, alas!, not molecular.
He left the world in disarray.
Hip, hip, Soret!

There still is need to resolve the topic.
The reason must be microscopic.
A small difference of the atomic potential
in thermal diffusion must be influential.
But understand we do not quite
why A goes left and B to the right.
That continues to this day.
Hip, hip, Soret!

After two centuries of experimentation
the world now turns to simulation.
Latest is the quest of young Dirk Reith
for what drives A to the left and B to the right.
His simple model avoids confusion
and solves the mysteries of thermal diffusion.
Believe me when I say:
Hip, hip, Soret!

In four lines now I summarise
his results for you to memorise:
On the arctic end we find compacting
the heavy, the small and the most interacting.
The tropical beaches populate meanwhile
the light, the big and the volatile.
And this is where we stay.
Hip, hip, Soret!



W. Köhler and S. Wiegand (Eds.): LNP 584, pp. 184–185, 2002.

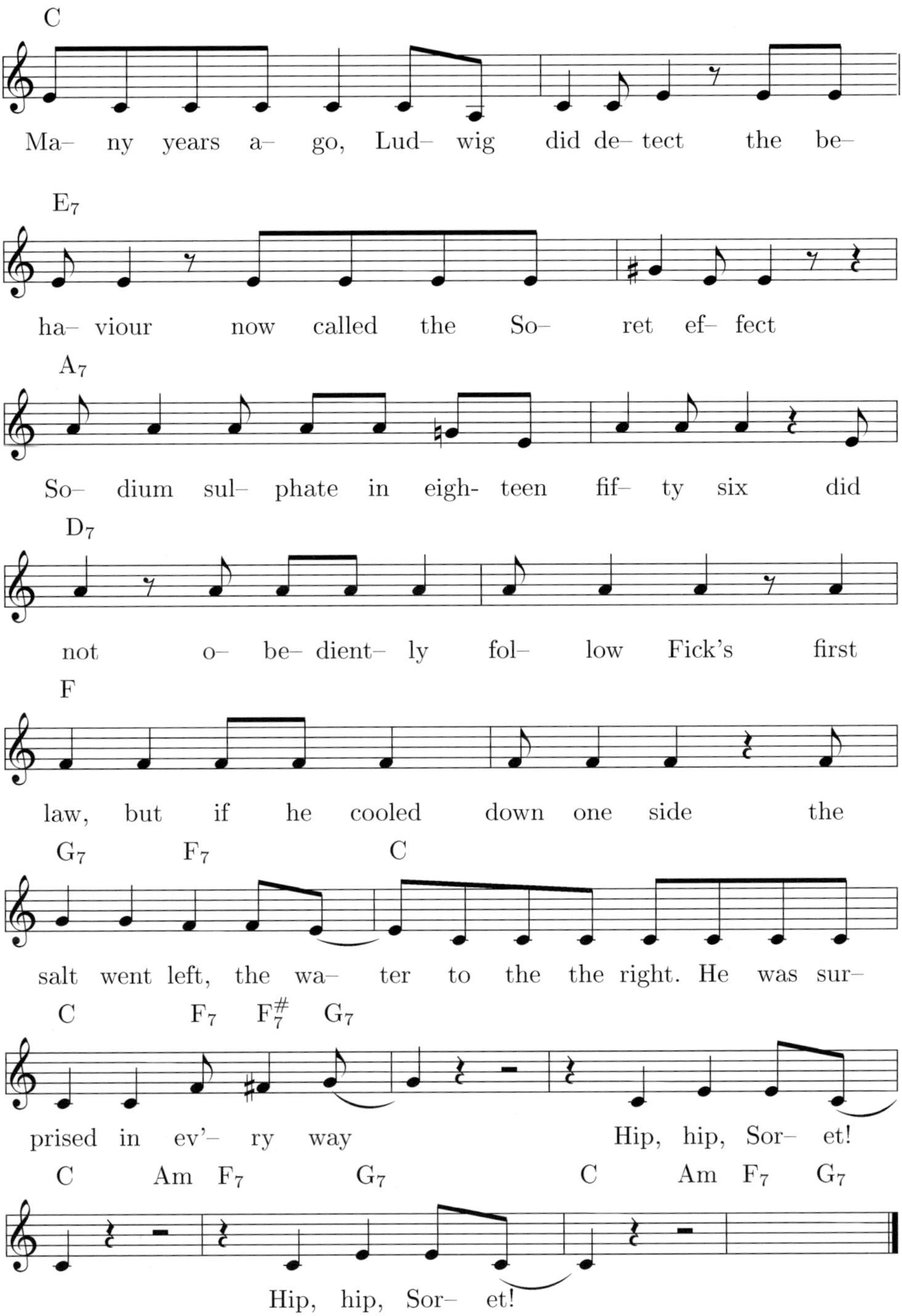
C
Ma– ny years a– go, Lud– wig did de– tect the be–
E7
ha– viour now called the So– ret ef– fect
A7
So– dium sul– phate in eigh- teen fif– ty six did
D7
not o– be– dient– ly fol– low Fick's first
F
law, but if he cooled down one side the
G7 F7 C
salt went left, the wa– ter to the the right. He was sur–
C F7 F#7 G7
prised in ev'– ry way Hip, hip, Sor– et!
C Am F7 G7 C Am F7 G7
Hip, hip, Sor– et!

# Part II

# Experimental Techniques and Special Systems

# Measurement of Transport Coefficients by an Optical Grating Technique

Simone Wiegand[1] and Werner Köhler[2]

[1] Max-Planck-Institut für Polymerforschung, D-55021 Mainz, Germany
[2] Physikalisches Institut, Universität Bayreuth, D-95440 Bayreuth, Germany

**Abstract.** Principle and applications of the holographic grating technique *thermal diffusion forced Rayleigh scattering* (TDFRS) are summarized and discussed in the context of other optical methods employed for the investigation of the Soret effect. The set-up and homodyne and heterodyne detection schemes are explained and the phenomenological model and the working equations are summarized. After briefly reviewing the use of TDFRS in the field of polymer characterization, this contribution concentrates on two aspects of the TDFRS technique. First, the influence of convection and flow is discussed and it is shown that even in the presence of convection, it is still possible to determine the correct thermal diffusion coefficient $D_T$, although a correct determination of the mutual diffusion coefficient $D$ and the Soret coefficient $S_T$ is no longer possible. Second, the asymptotic behavior of the Soret coefficient $S_T$ in a critical binary mixture of aniline/cyclohexane is determined. Modification of the set-up in order to study convection behavior in one case or strongly scattering samples in the other case are discussed.

## 1 Introduction

Quite a number of different experimental methods have been employed for the investigation of the Ludwig-Soret effect [1,2], and many of the more recent experiments are based on optical techniques.

Conceptually the most simple experiment is a diffusion cell which is heated from above and cooled from below. The concentration gradient induced by the temperature gradient is detected by deflection of a laser beam which passes through the cell in a direction parallel to the top and bottom plates. In most cases, the time dependence of the deflection angle contains a fast contribution stemming from the temperature and a slow contribution from the concentration gradient. Beam deflection is caused by changes of the refractive index of the liquid associated with the nonuniform temperature and concentration.

Diffusion cells with optical detection have already been utilized by Meyerhoff and Nachtigall, who employed a Schlieren technique [3,4]. More recently, laser beam deflection has been used by Giglio and Vendramini [5,6], Kolodner et al. [7], and Zhang et al. [8]. Since the diffusion length, the distance between the two plates, is of the order of a few millimeters to one centimeter, establishment of equilibrium is rather slow, especially for systems with small diffusion coefficients such as polymer solutions and systems close to the critical point.

Diffusion cells with beam deflection utilize optical effects only for detection. The nonuniform temperature distribution is established by heating at the bound-

W. Köhler and S. Wiegand (Eds.): LNP 584, pp. 189–210, 2002.

aries. Into the same category falls laser Doppler velocimetry within a thermogravitational column as reported by Platten et al. [9]. Contrary to the diffusion cell, where convection must be minimized, the heated and cooled plates are oriented vertically in order to enforce convection.

Conceptually different are experiments where light is not only used for detection but also for heating of the fluid. Detection may be accomplished by a separate readout beam or by monitoring the time-dependent changes of a suitable property of the transmitted heating beam directly.

In thermal lensing experiments, the transmitted laser beam heats the sample along its path and thus creates a lense which diffracts the beam. Since, as a rule, $(\partial n/\partial T)_{p,c} < 0$ in liquids, a thermal lense with a negative focal length is formed, causing beam divergence. Much work has been devoted to this purely thermal effect [10], but only few authors have investigated the influence of the Soret effect, which adds a contribution stemming from the concentration gradient to the lensing effect [11,12].

While traditional thermal lensing experiments impose only weak perturbations, stronger intensities may lead to a significant redistribution of the components within the path of the laser beam and to nonlinear feedback, if the absorbers themselves are subject to thermodiffusion [13–16]. Strong effects have been observed in case of magnetic colloids [16]. Due to the intense laser heating, such experiments are difficult to perform without convective perturbations [17,18].

In 1978, Thyagarajan and Lallemand [19] observed a slow contribution stemming from the Ludwig-Soret effect in forced Rayleigh scattering experiments on the binary liquid mixture $CS_2$/Ethanol, which had previously been predicted by Allain and Lallemand [20]. Later, Pohl studied a critical mixture of 2,6-lutidine/$H_2O$ with the same technique [21]. During the last years, we have improved this transient holographic grating technique [22,23] and employed it for the investigation of different systems such as liquid mixtures in equilibrium [24] and near the critical point [25], mono- and polydisperse polymer solutions [26], and colloidal suspensions [18,27]. Recently, studies on magnetic colloids have been reported utilizing forced Rayleigh scattering [28,29] and also self-diffraction [30]. Since thermodiffusion, also called thermal diffusion, contributes to the forced Rayleigh scattering signal, we call this technique *thermal diffusion forced Rayleigh scattering* (TDFRS).

In principle, TDFRS is simple. Two laser beams intersect within the sample and create a holographic interference grating. Due to the finite absorption coefficient of the liquid, which can be enhanced by adding a spurious amount of an inert dye such as quinizarin, energy is absorbed from the light field and the sample is heated with the spatial periodicity of the grating, which is typically a few micrometers. The temperature gradients within this thermal grating give – via the Ludwig-Soret effect – rise to the buildup of a concentration modulation with the same spatial periodicity. Both the temperature and the concentration grating are accompanied by a refractive index grating via $(\partial n/\partial T)_{p,c}$ and $(\partial n/\partial c)_{p,T}$, respectively. The time dependent diffraction efficiency of the two superimposed refractive index gratings is read out by Bragg diffraction of a third laser beam.

The mesoscopic diffusion length within the grating reduces the diffusion time to the millisecond range, which is easily accessible and about six orders of magnitude faster than in case of macroscopic diffusion cells. Due to the high sensitivity of holographic grating techniques, the perturbations of the samples can be kept very low. Typically, the amplitude of the temperature grating is of the order of 10 to 100 μK (see Eq. 10) and the relative change of concentration is below $10^{-5}$. The short diffusion time constants together with the very subtle perturbations allow for measurements very close to thermal equilibrium without the need for taking temperature or concentration dependences of system parameters, such as viscosity or diffusion coefficient, into account. Due to the periodicity of the grating, which contains only a single Fourier component, the working equations take a particularly simple form with a single-exponential time dependence for a system with a single mass diffusion coefficient. The thermal diffusivity $D_{\mathrm{th}}$ and the mass diffusion coefficient $D$ are obtained without any additional information. For the thermal diffusion coefficient $D_{\mathrm{T}}$ and the Soret coefficient $S_{\mathrm{T}} = D_{\mathrm{T}}/D$, the two contrast factors $(\partial n/\partial T)_{p,c}$ and $(\partial n/\partial c)_{p,T}$ must be measured separately, which can usually be done with a high accuracy (Sec. 2).

The heterodyne diffraction efficiency in TDFRS shows some similarities to the electric field autocorrelation function in photon correlation spectroscopy (PCS). One important difference is that the TDFRS signal stems from time dependent excitation patterns imposed by the experiment. This can be of advantage, for example, for solutions of polydisperse polymers, where excitation energy can be shifted into certain frequency ranges of interest by employing tailored pseudorandom binary sequences in combination with correlation techniques and digital filtering [31].

The application of TDFRS to dilute polymer solutions and the proper treatment of polydispersity has been described in detail in [22,23,26,32]. When compared to PCS, even small amounts of low molar mass material remain visible, which are masked in PCS by the strong scattering power of high molar mass material. Molar mass distribution functions are obtained from an inverse Laplace transformation of the heterodyne diffraction efficiency together with the scaling relation for the molar mass dependence of the diffusion coefficient, $D \propto M^{-b}$. Alternatively, the polydispersity of the dissolved polymer can be calculated from the second cumulant of the time-dependent diffraction efficiency and the scaling exponent $b$.

The usefulness for analytical purposes is demonstrated in Fig. 1, where a dilute solution of a bimodal polystyrene sample with molar masses of 825 g/mol and 250 kg/mol in toluene at concentrations of 0.0166 and 0.00228 weight percent, respectively, has been measured by PCS and TDFRS employing both a step-excitation and a pseudostochastic random excitation sequence. The latter gives direct access to the linear response function $g(t)$ of the concentration mode [31]. Despite of its higher concentration, the low molar mass component is not resolved in the PCS but clearly visible in the two TDFRS measurements.

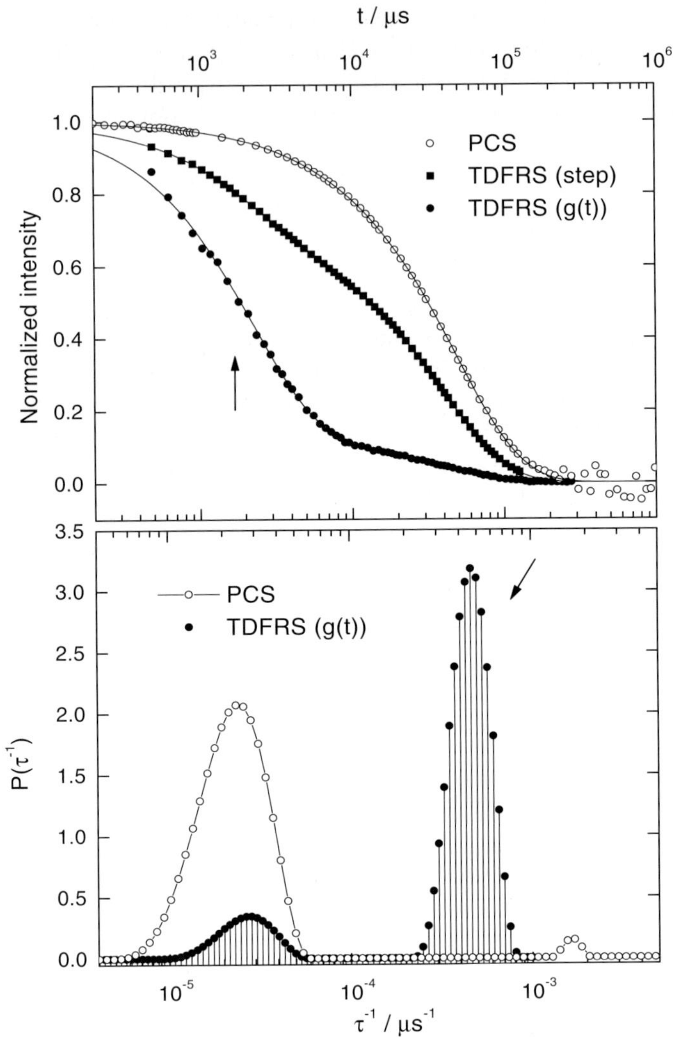


**Fig. 1.** Measurement of bimodal polystyrene in toluene with PCS and TDFRS with step excitation and optimized pseudorandom binary sequence ($g(t)$). The two modes are clearly resolved in the TDFRS decay functions and in the rate distribution, while PCS is only sensitive to the strongly scattering high molar mass component. The fast mode stemming from the light component is marked by arrows. From Ref. [31].

The available experimental data base on Soret coefficients is remarkably small and inconsistent, and there is only a very limited number of systems that have been investigated with different techniques and by different groups. One example are toluene/n-hexane mixtures. These mixtures have been extensively studied using small angle light scattering (SALS) [33], thermogravitational columns

[34,35], diffusion cell with optical beam deflection [8], and TDFRS [24]. Finally, an agreement within a few percent has been achieved for the Soret coefficient [8,35] as measured by the last three mentioned experiments. Toluene/n-hexane serves, however, also as example for the difficult experimental situation. First of all, the good agreement has only been reached after this particular system had drawn attention and after experiments have been repeated. There was no agreement between the initially available data. Second, the deviation of the Soret coefficients obtained by small angle light scattering from nonequilibrium fluctuations is still not understood. Fig. 2 summarizes the initial data and the finally achieved agreement.

Besides toluene/n-hexane mixtures, ethanol/water mixtures [7,8,24] and dilute solutions of polystyrene in toluene [3,4,36–38] are the only other systems known to the authors where substantially different experimental techniques, including TDFRS, have been employed and where consistent results have been obtained by different authors.

All properties summarized in the preceeding paragraphs make TDFRS a tool that is well suited for the measurement of transport coefficients and the investigation of the Soret effect in liquid mixtures. There are, however, also some drawbacks and problems of the method. First of all, there is almost always the need for an enhancement of absorption by addition of an inert dye. While the dye is considered only a minor perturbation in most cases, it has been surprisingly difficult to find an inert dye, which does not contribute to the diffraction efficiency due to some kind of photoreaction, in case of aqueous solutions [39].

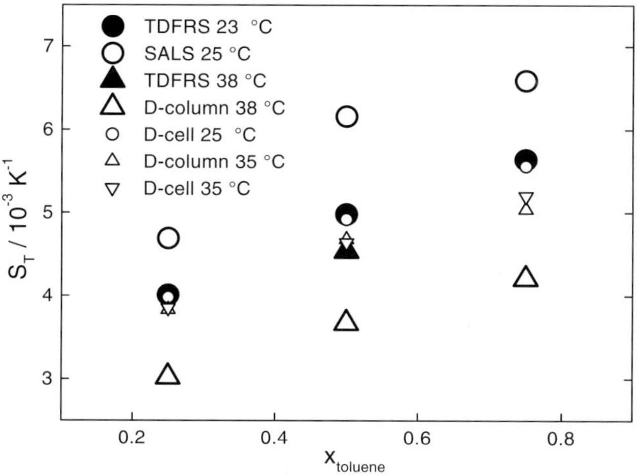


**Fig. 2.** Soret coefficient of toluene in n-hexane as obtained by small angle light scattering (SALS) [33], thermal diffusion cell (D-cell) [8], thermogravitational column (D-column) [34,35], and TDFRS [24] experiments at various temperatures plotted against the toluene mole fraction $x_{\text{toluene}}$. The large symbols indicate the initial results. From Ref. [23].

Furthermore, the change of composition is not truly analyzed in TDFRS and other optical experiments. Only the change of the refractive index is measured. Due to the minute temperature and concentration changes, nonlinearities of the contrast factors can safely be neglected. Problems may, however, arise in some special cases, where $n$ has a maximum or minimum at a certain temperature or composition. In the first case, the built-in calibration gets lost, and in the second case, the concentration signal is no longer visible. Both situations make the determination of $D_T$ and $S_T$ impossible. For most binary liquids, there is a unique correlation with the composition, but for systems containing three or more components, a unique deconvolution may be difficult or even impossible, but this has not yet been investigated systematically.

After this introduction we will first review the experimental principle of TDFRS, including heterodyne detection, and the simple model employed for a phenomenological description of the experiments. Then, we will specifically address the problem of convection which has not yet been described in detail elsewhere. The last chapter is devoted to measurements in the vicinity of a critical point.

## 2 Experimental aspects

### 2.1 Setup

The main part of a typical TDFRS setup as employed for the more recent work in our groups is sketched in Fig. 3.

An argon-ion laser ($\lambda_w = 488$ nm), which is polarized perpendicular to the optical table, is used for writing. Its beam is expanded to a diameter of approximately 8 mm and split into two beams of equal intensity. After passing through a set of optical components, the two beams are brought to intersection within the sample under an angle $\theta$.

After the beam splitter, one of the two beams then passes through a half-wave plate, which rotates its polarization by 90°. A Pockels cell is used as a switchable

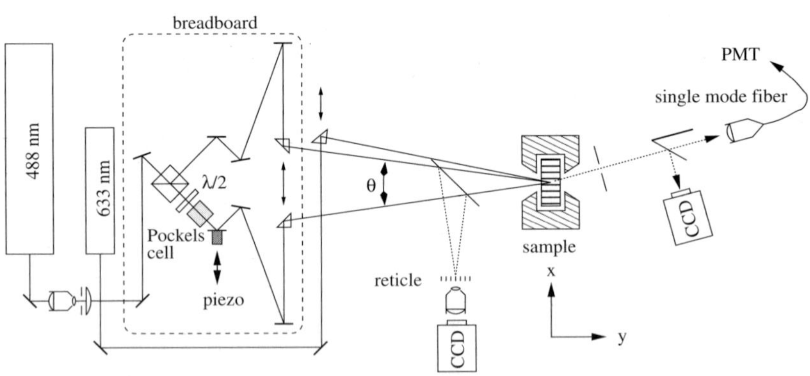


**Fig. 3.** Sketch of a TDFRS setup. From Ref. [22].

half-wave plate and rotates the polarization back to vertical in the on-state. In the off-state, when the polarization directions of the two beams are orthogonal, there is a uniform illumination of the sample without interference grating. The mirror mounted on a piezo actuator allows for phase adjustment of the grating and, hence, the diffracted beam. This is necessary for heterodyne detection, where the phase angles of the diffracted signal and the coherent reference wave must be matched (see below).

In principle, the fringe spacing of the grating, $d = 2\pi q^{-1}$, can be calculated from the angle $\theta$ via

$$q = \frac{4\pi}{\lambda_{\rm w}} \sin\frac{\theta}{2} \ . \tag{1}$$

For small angles of a few degrees, corresponding to fringe spacings of a few micrometers, it is difficult to measure $\theta$ with sufficient accuracy of better than 1 percent. Accurate results are obtained by projecting the grating together with a calibrated reticle onto a CCD camera.

A HeNe laser ($\lambda_{\rm r} = 632.8$ nm) is employed for the measurement of the diffraction efficiency of the grating. The diffracted beam and the coherent background are coupled into a single-mode optical fiber, which is connected to a photo multiplier tube operating in single photon counting mode. The single-mode fiber is ideally suited for heterodyne detection schemes, since it suppresses the incoherent background almost completely with low loss for the signal beam. In practice, coupling efficiencies of the order of 80 percent are achieved.

## 2.2 Heterodyne detection schemes

Because of the unavoidable background at low scattering angles, the higher detection efficiency for weak signals, and the robustness against perturbations, we exclusively employ heterodyne detection schemes [40] as sketched in Fig. 4. The concept in Fig. 4a has already been described by e.g. Thyagarajan and Lallemand [19] and Pohl [21]. The coherent reference wave is split off the readout beam and brought to interference with the signal on the detector. In this configuration, phase matching of the two beams can be achieved as shown in Fig. 4a by adjusting the phase of the reference. Such a configuration is more difficult

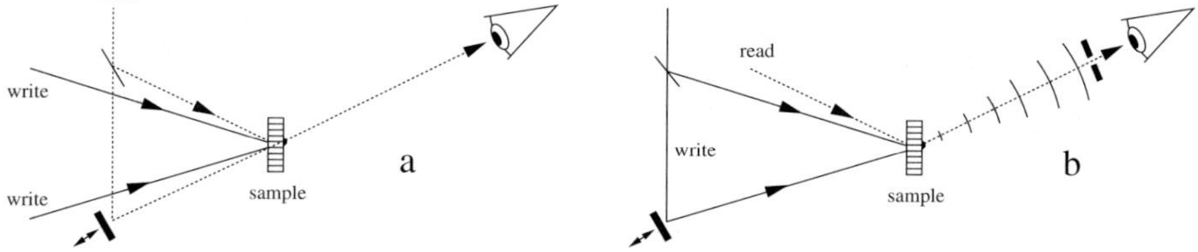


**Fig. 4.** Heterodyne detection schemes: (a) phase control of reference beam and (b) phase control of the diffracted beam with reference provided by local oscillator. From Ref. [40].

to align than the one in Fig. 4b, but it is the preferred geometry if a strong reference wave is required as in case of critical mixtures close to the critical point [25]. In case of strongly scattering samples, it turned out to be advantageous to adjust the intensity of the reference wave to the scattering of the sample.

The setup in Fig. 4b is easier to align [40]. Scratches on the cuvette provide a local oscillator for the reference wave as proposed by Miles et al. [41]. Now, phase matching is achieved by adjusting the phase angle of the grating and, hence, the phase angle of the diffracted beam. Other advanced heterodyne detection schemes for holographic gratings with phase modulation have been described by Pinsl et al. [42] and by Odear et al. [43].

Generally, the detected signal $S_\mathrm{m}$ contains both homodyne and heterodyne contributions:

$$S_\mathrm{m}(\varphi) \propto \left|E_\mathrm{c} + E_\mathrm{s}\mathrm{e}^{i\varphi}\right|^2 + E_\mathrm{inc}^2 = E_\mathrm{s}^2 + 2E_\mathrm{s}E_\mathrm{c}\cos\varphi + E_\mathrm{c}^2 + E_\mathrm{inc}^2 \qquad (2)$$

$E_\mathrm{c}$ and $E_\mathrm{s}$ are the electric field amplitudes of the coherent background and the diffracted beam, respectively, and $E_\mathrm{inc}$ is the electric field amplitude of the incoherent background. $\varphi$ is the phase shift between signal and reference. Homodyne ($S_\mathrm{hom}$) and heterodyne ($S_\mathrm{het}$) signals are separated by measuring at two different phase shifts:

$$S_\mathrm{hom} = \frac{1}{2}\left[S_\mathrm{m}(\varphi) + S_\mathrm{m}(\varphi+\pi)\right] - S_\mathrm{b} \propto E_\mathrm{s}^2 \qquad (3)$$

$$S_\mathrm{het} = \frac{1}{2}\left[S_\mathrm{m}(\varphi) - S_\mathrm{m}(\varphi+\pi)\right] \propto 2E_\mathrm{s}E_\mathrm{c}\cos\varphi \qquad (4)$$

$S_\mathrm{b} \propto E_\mathrm{c}^2 + E_\mathrm{inc}^2$ is the background intensity measured during the absence of a diffracted signal. $S_\mathrm{het}$ is maximized for $\varphi = 0$. Details about the adjustment and stabilization of the phase angle $\varphi$ can be found in Ref. [40] together with a discussion of optimized experimental parameters.

## 2.3 Sample preparation and contrast factors

For sample preparation, similar considerations are valid as in case of light scattering experiments, except that TDFRS experiments are much less susceptible to dust-like impurities. Optical path lengths are typically 100 to 200 μm, and the actual sample volume needed is only a few microliters. The optical density is adjusted to a value of approximately 2 for 1 cm optical path length by adding spurious amounts of an inert dye like quinizarin.

For $(\partial n/\partial T)_{p,c}$, interferometers as described in [44,45] and for $(\partial n/\partial c)_{p,T}$, either a scanning Michelson interferometer [45] or an Abbé-refractometer (ATG, resolution $\Delta n = 0.0002$) are employed. For most mixtures of liquids with significantly different $n$ ($\Delta n = 0.01...0.1$), $(\partial n/\partial c)_{p,T}$ can usually be measured with an accuracy of the order of 1 percent, and the accuracy of $(\partial n/\partial T)_{p,c}$ is even better.

## 3 Phenomenological model

Detailed derivations of the diffracted signal can be found elsewhere [19,22,46]. Here, we only want to summarize the important steps in order to establish the notation and the working equations for the evaluation of the experiments.

The mathematical description is particularly simple, since the experiment is set up in such a way that, due to first order Bragg diffraction of the readout beam, only a single Fourier component of the concentration modulation created within the sample contributes to the diffracted intensity. This corresponds to the fundamental spatial frequency of the optical grating, which contains most of the excitation energy. Under ideal experimental conditions, no other Fourier components are excited at all, and even if they were, they would not contribute to the signal.

Naturally, we start with the optical interference grating which is formed inside the sample volume by the intersecting laser beams. Due to the low optical absorption, we do not have to consider attenuation of the beams on their way through the fluid and a one-dimensional description is sufficient. The $x$-axis points perpendicular to the optical symmetry axis and lies within the plane spanned by the two writing beams. With these assumptions, the periodic intensity distribution within the grating is

$$I(x,t) = I_0 + I_{\mathrm{q}}(t)\,\mathrm{e}^{iqx} \ . \tag{5}$$

The space and time dependence of the temperature within the sample is obtained from the one-dimensional heat equation

$$\frac{\partial T(x,t)}{\partial t} = D_{\mathrm{th}} \frac{\partial^2 T(x,t)}{\partial x^2} + \frac{\alpha_\lambda}{\rho c_{\mathrm{p}}} I(x,t) \ . \tag{6}$$

$\alpha_\lambda$ is the optical absorption coefficient, $\rho$ the density, and $c_{\mathrm{p}}$ the specific heat at constant pressure. For simplicity, loss of energy due to luminescence has not been taken into account. Eq. 6 is solved by:

$$T(x,t) = T_0 + T_{\mathrm{m}}(t) + T_{\mathrm{q}}(t)\,\mathrm{e}^{iqx} \tag{7}$$

$$T_{\mathrm{m}}(t) = \frac{\alpha_\lambda I_0}{\rho c_{\mathrm{p}}}\, t \tag{8}$$

$$T_{\mathrm{q}}(t) = \frac{\alpha_\lambda}{\rho c_{\mathrm{p}}} \int_{-\infty}^{t} dt' I_{\mathrm{q}}(t')\,\mathrm{e}^{-(t-t')/\tau_{\mathrm{th}}} \tag{9}$$

$T_0$ is the initial temperature and $T_{\mathrm{m}}(t)$ is the initial rise of the mean sample temperature, which eventually comes to rest due to the heat loss from the cuvette to the laboratory environment. This heat loss is not included in Eq. 6. $T_{\mathrm{m}}(t)$ may be the cause of unwanted convection, if the experimental parameters are not chosen properly. The possible influence of convection will be discussed in Section 4. The temperature diffusion time constant is $\tau_{\mathrm{th}} = (D_{\mathrm{th}} q^2)^{-1}$.

The stationary modulation depth of the temperature grating for full modulation of the optical interference grating is

$$\delta T = \alpha_\lambda I_0 \tau_{\mathrm{th}} (\rho c_{\mathrm{p}})^{-1} . \tag{10}$$

For realistic parameters ( $\alpha_\lambda = 4\ \mathrm{cm}^{-1}$, $I_0 = 0.2\ \mathrm{W\,cm}^{-2}$, $\tau_{\mathrm{th}} = 20\,\mu\mathrm{s}$, $\rho = 0.87\ \mathrm{g\,cm}^{-3}$, $c_{\mathrm{p}} = 1.7\ \mathrm{J(gK)}^{-1}$) a modulation depth of only 11 $\mu$K is obtained.

Coupling between concentration and temperature gradients via the Ludwig-Soret effect is accounted for by an extension of Fick's second law of diffusion:

$$\frac{\partial c(x,t)}{\partial t} = D \frac{\partial^2 c(x,t)}{\partial x^2} + D_{\mathrm{T}}\, c(x,t)[1 - c(x,t)] \frac{\partial^2 T(x,t)}{\partial x^2} \tag{11}$$

For weak perturbations, $c(x,t)[1 - c(x,t)] \approx c_0[1 - c_0]$, and Eq. 11 is solved by:

$$c(x,t) = c_0 + c_{\mathrm{q}}(t)\, \mathrm{e}^{iqx} \tag{12}$$

$$c_{\mathrm{q}}(t) = -q^2 D_{\mathrm{T}} c_0 (1 - c_0) \int_{-\infty}^{t} dt' T_{\mathrm{q}}(t')\, \mathrm{e}^{-(t-t')/\tau} \tag{13}$$

For now, we are only interested in the spatially periodic part, and all non-periodic contributions will be neglected. Both the temperature and the concentration grating modulate the refractive index:

$$n(x,t) = n_0 + n_{\mathrm{q}}(t)\, \mathrm{e}^{iqx} \tag{14}$$

$$n_{\mathrm{q}}(t) = \left(\frac{\partial n}{\partial T}\right)_{p,c} T_{\mathrm{q}}(t) + \left(\frac{\partial n}{\partial c}\right)_{p,T} c_{\mathrm{q}}(t) \tag{15}$$

The heterodyne diffraction efficiency $\zeta_{\mathrm{het}}(t)$ is simply proportional to the modulation depth of the phase grating, $\zeta_{\mathrm{het}}(t) \propto n_{\mathrm{q}}(t)$, and the homodyne $\zeta_{\mathrm{hom}}(t) \propto |n_{\mathrm{q}}(t)|^2$.

The advantage of the formulation of the system response to the time dependent amplitude $I_{\mathrm{q}}(t)$ of the holographic grating according to Eqs. 5 - 15 is that $I_{\mathrm{q}}(t)$ has not yet been specified. Hence, the formalism can be applied to holographic gratings with arbitrary time dependent amplitudes, like pseudo-stochastic binary sequences [22,31]. The heterodyne diffraction efficiency for a simple step-excitation at $t = 0$ after normalization to the thermal signal is

$$\zeta_{\mathrm{het}}(t) = 1 - \mathrm{e}^{-t/\tau_{\mathrm{th}}} - \frac{A}{\tau - \tau_{\mathrm{th}}} \left[ \tau(1 - \mathrm{e}^{-t/\tau}) - \tau_{\mathrm{th}}(1 - \mathrm{e}^{-t/\tau_{\mathrm{th}}}) \right] . \tag{16}$$

The amplitude factor $A$ is given by

$$A = \left(\frac{\partial n}{\partial c}\right)_{p,T} \left(\frac{\partial n}{\partial T}\right)_{p,c}^{-1} S_{\mathrm{T}} c_0 (1 - c_0) . \tag{17}$$

Since usually $(\partial n/\partial T)_{p,c} < 0$, the amplitude $A$ is determined by the product $(\partial n/\partial c)_{p,T}\, S_{\mathrm{T}}$ and can be either positive or negative. There is no limitation of the method with respect to the sign of $S_{\mathrm{T}}$.

# 4 Influence of gravity and convection

There is an ongoing debate about the influence of gravity induced convection on the measurement of Soret and thermal diffusion coefficients. The notion that gravity makes such experiments difficult or even impossible has been fueled by the large discrepancy between results obtained by different authors even for presumably simple liquid mixtures. The influence of natural convection under microgravity on the measurement of mass transport coefficients in liquids has been discussed by Garandet et al. [47]. Experiments under microgravity conditions have recently been performed [48], and a measurement campaign has been started by the European Group on Thermodiffusion Research (E.G.T.R.) at the Fontainebleau meeting in summer 1999 with the aim to establish a reliable base of Soret coefficients for the three binary mixtures of dodecane, isobutylbenzene, and tetraline. As of today, however, no final results are available. The whole problem of gravity induced convection is discussed in more detail in the contribution of Platten et al. in this volume [9]. In the following paragraphs we will argue that

- a well-designed TDFRS experiment does not suffer from convection-induced errors and
- even in the presence of convection or flow, the correct thermal diffusion coefficients but not the correct diffusion and Soret coefficients are obtained.

The influence of convection on TDFRS experiments has been investigated in Ref. [40]. Convection is induced by the buoyancy of the liquid volume heated by the laser beam. Within a simple model (Fig. 5a), which assumes heating throughout the sample volume illuminated by the laser, heat transport through the sample windows to the ambient air, and negligible heat conduction within the plane of the windows, it has been shown that convection does not cause problems as long as

$$\frac{\alpha_\lambda I s^3 \tau}{\eta d} \leq L_c \ . \tag{18}$$

$I$ is the intensity of the laser beam used for writing, $\tau$ the diffusion time constant, $\eta$ the viscosity of the fluid, and $s$ the thickness of the cuvette. The constant $L_c$ has been estimated to $L_c \approx 200\ \mathrm{m^2s^{-1}}$. Eq. 18 is derived by assuming that the product of the velocity of the convective current times the diffusion time constant, $u_c\tau$, must be compared to the fringe spacing of the grating.

Since the main cause for the temperature rise within the sample is the low thermal conductivity of the optical windows of the cuvette, it can significantly be reduced by employing sapphire windows with a more than 20-fold higher thermal conductivity than quartz glass. Additionally, the influence of convection can be reduced by a careful symmetric alignment as sketched in Fig. 5b. Since the writing beam is expanded, but not the readout beam, only the center of the illuminated spot is sensed, where the velocity component perpendicular to the fringes of the grating, $u_{c\perp}$, is only small. $u_{c\|}$ does not lead to signal distortion for moderate convection. The presence of signal distortion due to convection is easily checked by variation of the laser power.

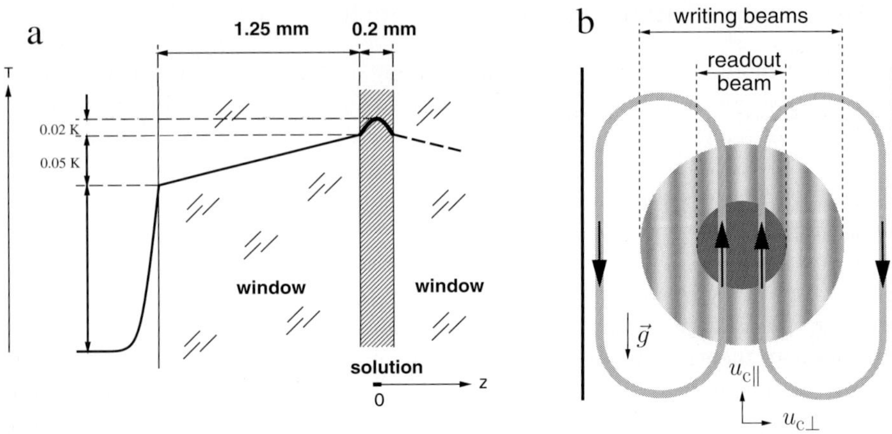


**Fig. 5.** a) Temperature distribution within sample cell (cross section). The z-direction points along the optical axis. b) Convection roles (front view). The grating exists within the spot illuminated by the writing beams (≈ 8 mm diameter). The central spot has a diameter of approximately 1 mm and is selected by the readout laser beam.

Normally, TDFRS experiments are designed such that Eq. 18 is fulfilled and convection is of no concern. But even in case of convection or some other flow where $u_{c\perp}$ cannot be neglected, it is still possible to determine $D_T$, as will be shown below.

The main effect of convection on the TDFRS signal stems from the velocity component of the flow along the x-axis, perpendicular to the fringes of the grating ($u_{c\perp}$ in Fig. 5b). Flow is easiest incorporated if we follow the line of arguments from section 3, but now in the fluid-fixed frame of reference (indicated by superscript $s$). Contrary to Section 3, the amplitude factors $T_q(t)$, $c_q(t)$, and $n_q(t)$ now become complex, to account for a phase shift relative to the optical excitation grating.

If $u_{c\perp}$ is sufficiently small, the retardation of the temperature grating can be neglected for simplicity. If we concentrate on the spatially periodic contribution only, the temperature grating in the fluid-fixed frame of reference is

$$T^s(x,t) = T_q^s \,e^{iqx} = \delta T \,e^{i\varphi(t)}\,e^{iqx} \tag{19}$$

The time dependent phase factor is given by

$$\varphi(t) = \frac{2\pi}{d} u_{c\perp} t \equiv \gamma t \;. \tag{20}$$

The concentration distribution in the fluid-fixed frame of reference now can be written as

$$c^s(x,t) = c_0 + c_q^s(t)\,e^{iqx} \tag{21}$$

with

$$c_q^s(t) = -q^2 D_T c_0 (1-c_0) \int_{-\infty}^{t} \delta T e^{i\varphi(t')}\,e^{-(t-t')/\tau} \tag{22}$$

Taking $\varGamma = Dq^2 = \tau^{-1}$, Eq. 22 is solved for a step excitation starting at $t = 0$ by

$$c_{\mathrm{q}}^{\mathrm{s}}(t) = -q^2 \tau D_{\mathrm{T}} c_0 (1 - c_0) \delta T \frac{\varGamma}{\varGamma + i\gamma} \left[ e^{i\gamma t} - \mathrm{e}^{-\varGamma t} \right] . \tag{23}$$

Following Eqs. 14 and 15, the amplitude factor of the resulting refractive index grating becomes now complex,

$$n^{\mathrm{s}}(x,t) - n_0 = n_{\mathrm{q}}^{\mathrm{s}}(t)\, \mathrm{e}^{iqx} = \left[ \left( \frac{\partial n}{\partial T} \right)_{p,c} T_{\mathrm{q}}^{\mathrm{s}}(t) + \left( \frac{\partial n}{\partial c} \right)_{p,T} c_{\mathrm{q}}^{\mathrm{s}}(t) \right] \mathrm{e}^{iqx} , \tag{24}$$

with $T_{\mathrm{q}}^{\mathrm{s}}(t)$ and $c_{\mathrm{q}}^{\mathrm{s}}(t)$ as defined in Eqs. 19 and 22, respectively. Since the retardation of the temperature grating has been neglected, it is stationary in the laboratory-fixed frame of reference. The refractive index grating (Eq. 24) is transformed to the laboratory-fixed frame of reference simply by multiplying with $\mathrm{e}^{-i\gamma t}$:

$$n(x,t) - n_0 = n_{\mathrm{q}}(t)\, \mathrm{e}^{iqx} = n_{\mathrm{q}}^{\mathrm{s}}(t)\, \mathrm{e}^{-i\gamma t}\, \mathrm{e}^{iqx} \tag{25}$$

It is convenient to normalize the refractive index grating amplitude to the amplitude of the instantaneous thermal contribution:

$$\tilde{n}_{\mathrm{q}}(t) = n_{\mathrm{q}}(t) \left[ \left( \frac{\partial n}{\partial T} \right)_{p,c} \delta T \right]^{-1} = 1 - \frac{A\varGamma}{\varGamma + i\gamma} \left( 1 - \mathrm{e}^{-(\varGamma + i\gamma)t} \right) \tag{26}$$

$A$ is the saturation amplitude of the concentration signal in case of $u_{\mathrm{c}\perp} = 0$ and has already been defined in Eq. 17.

The homodyne signal is not phase sensitive and, hence, the normalized homodyne diffraction efficiency is

$$\zeta_{\mathrm{hom}}(t) = |\tilde{n}_{\mathrm{q}}(t)|^2 . \tag{27}$$

Experimentally, the phase angle of the reference wave for heterodyne detection is adjusted to the fast temperature signal. Hence, the normalized heterodyne diffraction efficiency is given by

$$\zeta_{\mathrm{het}}(t) = Re(\tilde{n}_{\mathrm{q}}(t)) . \tag{28}$$

Technically, $\zeta_{\mathrm{het}}(t)$ is obtained according to Eq. 4 with subsequent normalization to the thermal signal. Note, however, that the phase angle $\varphi$ in Eqs. 2 and 4 now becomes time dependent, as can be seen by writing Eq. 26 in the form

$$\tilde{n}_{\mathrm{q}}(t) = |\tilde{n}_{\mathrm{q}}(t)|\, \mathrm{e}^{i\varphi(t)} \propto E_{\mathrm{s}}\, \mathrm{e}^{i\varphi} . \tag{29}$$

The proportionality corresponds to the respective term in Eq. 2.

Of special interest is the behavior of $\tilde{n}_{\mathrm{q}}(t)$ at short times,

$$\lim_{t \to 0} \tilde{n}_{\mathrm{q}}(t) = 1 - A\varGamma t , \tag{30}$$

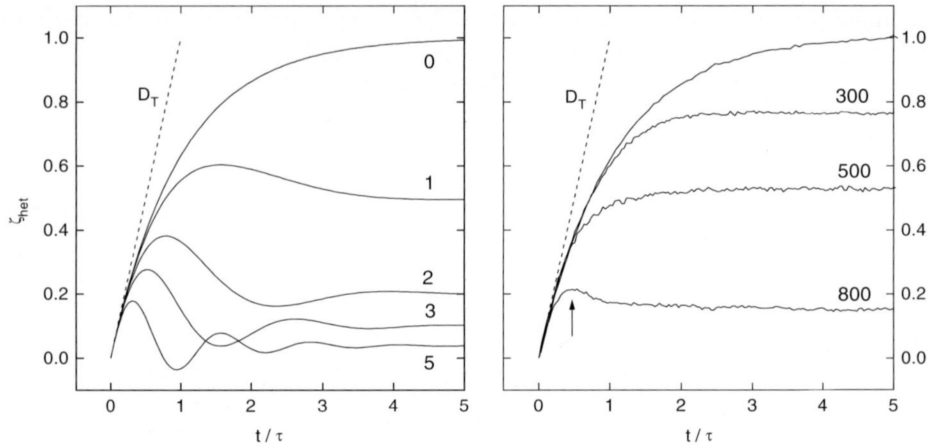


**Fig. 6.** Calculated (left) and measured (right, PS (48kg/mol) in toluene) concentration parts of the heterodyne signals in normalized representation. Numbers in the left plot are $\gamma/\Gamma$ and in the right plot laser power in mW. Convection increases with increasing laser power. The measurements have been made with thick cuvettes to enforce convection. From Ref. [23].

which is independent of $u_{c\perp}$. Hence, no matter whether there is convection or not, $D_T$ can always be determined from the initial slope $-A\Gamma = -A/\tau$ of the concentration signal by means of Eq. 17.

Fig. 6 shows calculated and measured concentration signals for different strengths of convection. A 1 mm thick cuvette, high laser power, and a high dye concentration have been used in order to enforce convection. The oscillations originate from the beating of the wave diffracted from the moving grating against the reference wave. They are less pronounced in the experiment, which is not sufficiently described by a single $u_{c\perp}$. For a more realistic model, a distribution of $u_{c\perp}$ across the illuminated spot would have to be taken into account. Nevertheless, the predicted decrease of the concentration signal with increasing convection and the initial overshooting, marked in Fig. 6 with an arrow, are observed in the experiment. Most important, however: the initial slope remains constant and allows for the determination of the correct $D_T$. Since it is difficult to account for convection in a quantitative way, $D$ and $S_T$ can, however, no longer be determined.

## 5 Measurements close to the critical point

The experimental determination of transport properties like thermodiffusion in a binary liquid mixture close to its critical point is a difficult task compared to other ordinary mixtures. In order to measure the transport coefficient, a gradient has to be applied, which disturbs the thermodynamic equilibrium. If one wants to study thermodiffusion in a system close to the critical point the applied temperature difference has to be as small as possible so that the system does not partially demix. As an additional complication of measurements close to the critical point the diffusion coefficient vanishes, which leads to the so-called *critical slowing down* of the system. Hence, one needs to allow for long equilibration times if the system dimensions are large. Furthermore, a system close to its critical point preserves concentration gradients once they have been formed [49].

The first qualitative measurements of the Soret coefficient in the critical binary mixture nitrobenzene/n-hexane were performed by Thomaes [50] in a flow cell . He could only demonstrate the divergence of the Soret coefficient $S_{\mathrm{T}}$, a calculation of the critical exponent was not possible. Later Giglio and Vendramini performed systematic measurements using an optical diffusion cell [51] . For the system aniline/cyclohexane they determined a critical exponent of $\varphi = 0.73$ for the divergence of $S_{\mathrm{T}} \propto (T - T_{\mathrm{c}})^{-\varphi}$, while the thermal diffusion coefficient $D_{\mathrm{T}}$ remaines constant with temperature (see also Ref. [52,53]).

Pohl [54] was the first who applied the holographic grating technique to a critical mixture. For the system 2,6-lutidine/water he showed for two temperatures below and two above $T_{\mathrm{c}}$ that the dynamical behavior changed. The intensity of the diffracted signal showed for temperatures in the one phase region below $T_{\mathrm{c}}$ a pure exponential decay, while for temperatures in the two phase region well above $T_{\mathrm{c}}$ the signal crossed over from exponential growth to decay. The experimental accuracy was not sufficient for a quantitative data analysis.

Due to the possibility to choose a very narrow fringe spacing of a few micrometers and because the temperature difference can be as low as a few micro-Kelvin, the TDFRS technique appears suitable for critical mixtures. First systematic measurements of the Soret coefficient using the TDFRS technique were recently performed for the binary mixture aniline/cyclohexane [25].

Theoretically, one expects a divergence of $S_{\mathrm{T}} = D_{\mathrm{T}}/D$. The mutual diffusion coefficient D varies with $t_{\mathrm{red}}^{\nu(1+z_\eta)}$ in the asymptotic range [53,55]. Here, $t_{\mathrm{red}} = (T - T_{\mathrm{c}})/T_{\mathrm{c}}$ denotes the reduced temperature and $\nu = 0.63$ and $z_\eta \approx 0.063$ the critical exponents for the correlation length and viscosity, respectively. According to theory, the thermal diffusion coefficient $D_{\mathrm{T}}$ shows no critical enhancement [53,55].

The first check for the reliability of measurement is to study the influence of the addition of dye. It is known that a third component modifies the critical concentration, critical temperature and the asymptotic critical behavior [56]. In order to adjust the optical density of toluene to OD=1 one needs 0.0035 wt % quinizarin. For the investigated system aniline/cyclohexane it was checked that the critical weight fraction $w_{\mathrm{c}} = 0.470 \pm 0.001$ and temperature $T_{\mathrm{c}} = 302.87$ K remained unchanged. The criticality of the sample was checked by the equal volume criterion 0.01 K below the critical temperature. According to Fisher and Scesny [56] the impurity does not lead to a change in the critical exponent, if the critical temperature remains unchanged.

The temperature stability during the measurements in the TDFRS set-up was 2 mK over 24 hours. In order to avoid convection, sample cells with a thickness of 0.2 mm were used. The cells were carefully sealed with a combination of

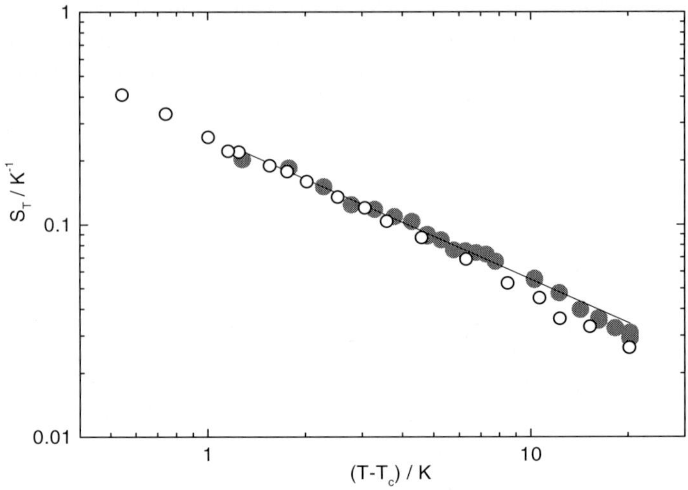


**Fig. 7.** Soret coefficient of aniline in cyclohexane (●) measured for the binary critical mixture aniline/cyclohexane versus the reduced temperature. The solid line represents the fit to the experimental data points. For comparison, we show the estimated values for $S_{\mathrm{T}}$ (○) using the displayed thermal diffusion factors $\alpha_{\mathrm{T}}$ in Fig. 2 of the paper by Giglio and Vendramini [51]

Teflon stoppers, Teflon tape and finally fixed by some glue, which did not come into direct contact with the sample. Nevertheless, although no weight loss was noticed during the measurement, the sample needed to be replaced by a fresh sample from a large critical batch after 2-3 days. The reason was a color change of the sample from pale yellow to yellow-red due to oxidation of the aniline.

Preliminary measurements in the strongly scattering samples close to the critical point showed that a reference wave caused by a scratch on the window of the sample cell is not sufficient for the heterodyne phase tracking. Strongly scattering samples require a local oscillator in form of a reference beam as explained in Fig. 4a. For measurements without this reference beam, the heterodyne phase tracking synchronizes to one of the slow critical fluctuations, which eventually leaves the scattering volume. This way the signal averages to zero over time. For this reason, all measurements were performed with a reference beam whose intensity was adjusted to the respective scattering intensity.

It was shown that the Soret coefficient diverges as $S_{\mathrm{T}} \propto (T - T_{\mathrm{c}})^{-\varphi}$ with $\varphi = 0.68 \pm 0.03$ [25]. The closest investigated temperature was roughly 1.2 K apart from the critical temperature. The results obtained very close to the critical point scatter over a large range. The most likely explanation for the poor reproducibility are concentration gradients, which do not resolve. Unfortunately, the small sample cells required for convection minimization allow no stirring in order to resolve the gradients. Presumably, this gradient are mainly caused by the global heating of the beam path during the measurement, which is of the or-

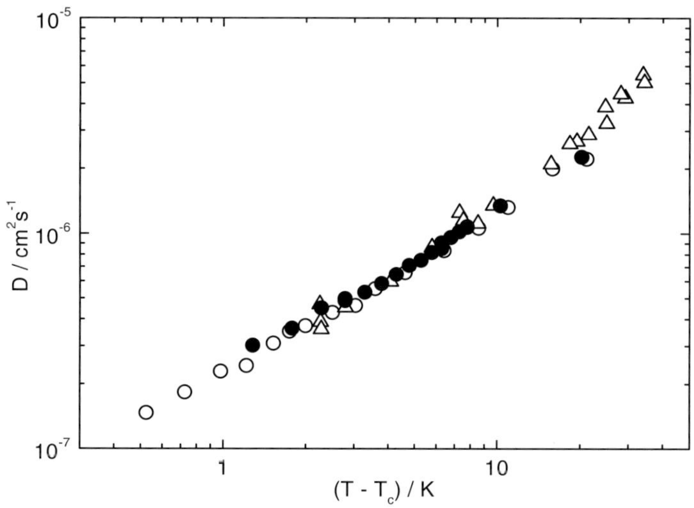


**Fig. 8.** Diffusion coefficient for the critical mixture aniline/cyclohexane plotted versus the temperature difference $T - T_c$. For comparison with the recent results ● [25] also former results determined by a light scattering experiment △ [57] and in a thermodiffusion cell ○ [51] are plotted.

der of 0.5 K and three order of magnitude larger than the temperature difference between the hot and cold stripes within the beam.

The Soret coefficient $S_T$ determined with the TDFRS technique agrees well with the $S_T$ values which we calculated from the thermal diffusion ratios $k_T$ determined by Giglio and Vendramini [51] by using the relation $S_T = k_T/(T \cdot c(1-c))$.

As already mentioned, the TDFRS method also allows for the simultaneous determination of the mutual diffusion coefficient $D$. Fig. 8 shows the results in comparison with the results obtained by Berge and Dubois [57] in a light scattering experiment and by Giglio and Vendramini [51] in an optical diffusion cell. In all experiments the agreement between the different measurement methods is good.

Figure 9 shows the thermal diffusion coefficient $D_T$ versus the temperature difference $T - T_c$. The thermal diffusion coefficient does not remain constant over the entire temperature range, but slightly decreases by 30 % when the critical temperature is approached.

The determination of the Soret coefficient $S_T$ with the TDFRS technique is so far limited to measurements at 1.2 K before the critical temperature is reached. This is still quite far away from the critical point, so we need to discuss, whether we are already in the asymptotic limit. This question was already raised and discussed for several properties and many systems 20 years ago by Levelt Sengers and Sengers [58]. In the following years up to now the theoreticians developed crossover theories and functions [59,52], which helped to distinguish between monotonic crossover and nonmonotonic crossover. The gas/liquid critical point

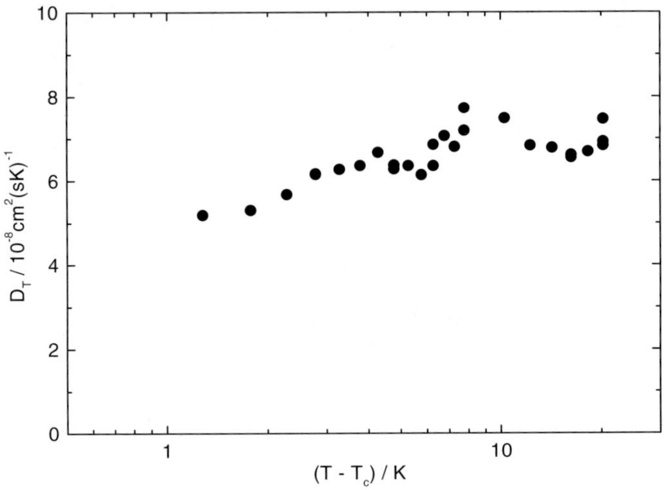


**Fig. 9.** Thermal diffusion coefficient of aniline in cyclohexane for the critical mixture aniline/cyclohexane plotted versus the temperature difference T-$T_c$.

is approached monotonically and the asymptotic behavior is not reached until the system is in the Milli Kelvin vicinity of the critical point. In contrast ionic liquid mixtures show a nonmonotonic sharp crossover, but the asymptotic exponent is reached several tenth parts of a degree away from the critical region [60]. In non-ionic solvent mixtures as aniline/cyclohexane the asymptotic region is often rather large of the order of several degrees. Far away from $T_c$ the system shows slight deviations from the asymptotic behavior. Folk and Moser [52] applied their crossover function to the experimental data of Giglio and Vendramini [51] and found good agreement with the experimental data over the entire temperature range.

There are serval reasons why TDFRS measurements very close to the critical point were not yet possible. First of all, the system scatters very strongly close to the critical point. Therefore, the thermal signal is just a very small contribution to the total signal which is mainly dominated by the concentration signal. For this reason, it becomes difficult to use the thermal signal as an internal measurement standard. It is possible to determine the thermal diffusivity $D_{th}$ far away from the critical temperature and fix it to its mean value for the analysis of the temperatures close to the critical point. Another complication close to the critical point is that the system rapidly forms concentration gradients. These gradients may not easily be removed without stirring of the sample. Once a concentration gradient has formed, it is impossible to resolve it completely. Therefore, one measures in an inhomogeneous system, which leads to erratic results.

It can be concluded that the TDFRS technique is suitable to study critical mixture. In order to measure very close to the critical point, it would be advantageous to use some kind of flow cell, which allows the continuous replacement of the sample from a large reservoir, which can be stirred. Much improvement is expected from sapphire windows, which, due to their high thermal conductivity, largely reduce the overall sample heating caused by the absorbed laser beam.

## 6 Summary and conclusions

We have discussed possibilities and limitations of the holographic grating technique TDFRS as a versatile tool for the study of the Ludwig-Soret effect in binary fluids. Due to the short diffusion lengths of a few micrometers, the equilibration times are only of the order of milliseconds to seconds. TDFRS allows for the simultaneous determination of the mutual diffusion coefficient $D$, the thermal diffusivity $D_{\mathrm{th}}$ and thermal diffusion coefficient $D_{\mathrm{T}}$. Recently, its potentials for polymer characterization have been discussed [22]. This contribution concentrates on two other aspects: the interplay between thermodiffusion and convection and measurements in critical binary mixtures as an example for strongly scattering samples.

Usually, relevant parameters like sample thickness, the amount of added dye, and laser power can be chosen such that convection within the sample does not disturb the measurements. Under unfavorable conditions with significant convection or constant flow of the fluid within a flow cell, the initial slope of the diffracted signal still leads to the correct thermal diffusion coefficient $D_{\mathrm{T}}$, but a correct determination of the mutual diffusion coefficient $D$ and the Soret coefficient $S_{\mathrm{T}}$ is no longer possible. The general experimental behavior can be illustrated by a simple model, which assumes heating in the sample volume illuminated by the laser and heat transport through the optical windows.

Due to the short diffusion length of a few micrometers, the TDFRS experiment is especially suitable for systems with slow diffusion dynamics, such as colloidal suspension and critical binary mixtures. For strongly scattering critical mixtures it is no longer advice able to derive the coherent reference wave for heterodyne detection from scattering on the cuvette surfaces. Better results are obtained by employing a separate reference wave which can be adjusted to the scattering strength of the sample. The experimental values for the mutual diffusion coefficient $D$ and the Soret coefficient $S_{\mathrm{T}}$ for a critical mixture of aniline/cyclohexane agree well with former data determined with a diffusion cell.

With the aid of special adjustments, as discussed here for the case of convection and strongly scattering samples, the technique has potential to investigate other systems that are difficult to handle because of their slow dynamics. As in the case of the study of a crystal-to-crystal photodimerization [61], an active stabilization may be necessary to study very slow processes as expected for e.g. polymer blends. Furthermore, the possibility to adjust the fringe spacing of the interference grating can be used to probe intrinsic length scales, like the correlation length in critical mixtures. These interesting modifications are subject of future research.

## References


1. C. Ludwig: Sitzber. Akad. Wiss. Wien Math.-naturw. Kl. **20**, 539 (1856).
2. C. Soret: Arch. Geneve **3**, 48 (1879).
3. G. Meyerhoff, K. Nachtigall: J. Polymer Sci. **57**, 227 (1962).
4. K. Nachtigall, G. Meyerhoff: Makromol. Chemie **33**, 85 (1959).
5. M. Giglio, A. Vendramini: Phys. Rev. Lett. **38**, 26 (1977).
6. M. Giglio, A. Vendramini: Phys. Rev. Lett. **34**, 561 (1975).
7. P. Kolodner, H. Williams, C. Moe: J. Chem. Phys. **88**, 6512 (1988).
8. K. J. Zhang, M. E. Briggs, R. W. Gammon, J. V. Sengers: J. Chem. Phys. **104**, 6881 (1996).
9. J. K. Platten, J. F. Dutrieux, G. Chavepeyer: same volume.
10. J. M. Harris: in *Analytical Applications of Lasers*, edited by E. H. Plepmeier (J. Wiley & Sons, Inc., ADDRESS, 1986), Chap. 13, p. 451.
11. M. Franko, C. D. Tran: J. Phys. Chem. **95**, 6688 (1991).
12. F. D. Hardcastle, J. M. Harris: Appl. Spectrosc. **40**, 606 (1986).
13. E. Freysz, A. Ponton, J. P. Delville, A. Ducasse: Opt. Commun. **78**, 436 (1990).
14. N. V. Tabiryan, W. L. Luo: Phys. Rev. A **57**, 4431 (1998).
15. N. V. Tabiryan, W. Luo: Phys. Rev. E **57**, 4431 (1998).
16. T. Du, W. Luo: Appl. Phys. Lett. **72**, 272 (1998).
17. W. Schaertl, C. Roos: Phys. Rev. E **90**, 2020 (1999).
18. R. Spill, W. Köhler, G. Lindenblatt, W. Schaertl: Phys. Rev. E **62**, 8361 (2000).
19. K. Thyagarajan, P. Lallemand: Opt. Commun. **26**, 54 (1978).
20. C. Allain, P. Lallemand: C. R. Acad. Sc. Paris, Ser. B **285**, 187 (1977).
21. D. W. Pohl: Phys. Rev. **77A**, 53 (1980).
22. W. Köhler, R. Schäfer: Adv. Polym. Sci. **151**, 1 (1999).
23. W. Köhler: in *Scattering in polymeric and colloidal systems*, edited by W. Brown and K. Mortensen (Gordon and Breach, Amsterdam, 2000), Chap. 2.
24. W. Köhler, B. Müller: J. Chem. Phys. **103**, 4367 (1995).
25. S. Wiegand, C. Rybarsch-Steinke: 2001, to be published in J.Chem. Phys.
26. P. Rossmanith, W. Köhler: Macromolecules **29**, 3203 (1996).
27. F. Renth *et al.*: Progr. Colloid Polym. Sci. **100**, 127 (1996).
28. J. C. Bacri *et al.*: Phys. Rev. Lett. **74**, 5032 (1995).
29. A. Bourdon, G. Demouchy, J. Bacri, E. Blums: presented at IMT4, Bayreuth, September 11-15, 2000.
30. I. Turek *et al.*: JMMM **201**, 167 (1999).
31. R. Schäfer, A. Becker, W. Köhler: Int. J. Thermophys. **20**, 1 (1999).
32. W. Köhler, C. Rosenauer, P. Rossmanith: Int. J. Thermophys. **16**, 11 (1995).
33. W. B. Li, P. N. Segre, R. W. Gammon, J. V. Sengers: Physica A **204**, 399 (1994).
34. O. Ecenarro *et al.*: J. Phys. Condens. Matter **2**, 2289 (1990).
35. M. M. Bou-Ali *et al.*: J. Phys. Condens. Matter **10**, 3321 (1998).
36. O. Ecenarro *et al.*: Macromolecules **27**, 4968 (1994).
37. M. E. Schimpf, J. C. Giddings: J. Polym. Sci., Polym. Phys. **B27**, 1317 (1989).
38. K. J. Zhang *et al.*: J. Chem. Phys. **111**, 2270 (1999).
39. R. Spill: Ph.D. Thesis, Univ. Bayreuth, 2000.
40. W. Köhler, P. Rossmanith: J. Phys. Chem. **99**, 5838 (1995).
41. J. D. G. Miles, P. D. Lamb, K. W. Rhee, J. C. S. Johnson: J. Phys. Chem. **87**, 4815 (1983).
42. J. Pinsl, M. Gehrtz, C. Bräuchle: J. Phys. Chem. **90**, 6754 (1986).
43. R. M. Odear, I. H. Park, K. Xia, J. C. S. Johnson: Rev. Sci. Instrum. **62**, 27 (1991).

44. W. B. Li *et al.*: J. Chem. Phys. **101**, 5058 (1994).
45. A. Becker, W. Köhler, B. Müller: Ber. Bunsenges. Phys. Chem. **99**, 600 (1995).
46. F. Bloisi *et al.*: Appl. Phys. **B44**, 103 (1987).
47. J. P. Garandet *et al.*: in *Materials and fluids under low gravity*, edited by L. Ratke, H. Walter, B. Feuerbacher (Springer, Berlin, 1996), p. 253.
48. S. V. Vaerenbergh, J. C. Legros: J. Phys. Chem. **102**, 4426 (1998).
49. S. Greer, T. E. Block, C. Knobler: Phys. Rev. Lett. **5**, 250 (1975).
50. G. Thomaes: J. Chem. Phys **25**, 32 (1956).
51. M. Giglio, A. Vendramini: Phys. Rev. Let. **34**, 561 (1975).
52. R. Folk, G. Moser: Phys. Rev. E **58**, 6246 (1998).
53. J. Luettmer-Strathmann: same volume.
54. D. Pohl: Phys. Lett. **77A**, 53 (1980).
55. J. Luettmer-Strathmann, J. Sengers: J. Chem. Phys. **104**, 3026 (1996).
56. M. Fisher, P. Scesney: Phys. Rev. A **2**, 825 (1970).
57. P. Berge, M. Dubois: Phys. Rev. Lett. **27**, 1125 (1971).
58. J. L. Sengers, J. Sengers: in *Perspectives in Statistical Physics*, edited by H. Raveche (North-Holland Publishing Company, Amsterdam, 1981).
59. M. Anisimov, A. Povodyrev, V. Kulikov, J. Sengers: Phys. Rev. Lett. **75**, 3146 (1995).
60. J. L. Sengers, A. Harvey, S. Wiegand: in *Equations of State for fluids and fluid mixtures*, edited by J. Sengers, R.F.Kayser, C.F.Peters, J. H.J. White (Elsevier, Amsterdam, 2001).
61. W. Köhler, K. Novak, V. Enkelmann: J. Chem. Phys. **101**, 10474 (1994).

## Supplementary glossary

| | |
|---|---|
| $\alpha_\lambda$ | absorption coefficient |
| $\Gamma$ | diffusion rate constant: $\Gamma = \tau^{-1}$ |
| $\gamma$ | angular velocity of phase shift |
| $\lambda_\mathrm{w}$ | writing wavelength |
| $\lambda_\mathrm{r}$ | readout wavelength |
| $\varphi$ | phase angle |
| $\tau$ | diffusion time constant |
| $\theta$ | angle of intersection |
| $\zeta_\mathrm{hom}(t), \zeta_\mathrm{het}(t)$ | homodyne, heterodyne diffraction efficiency |
| $A$ | normalized saturation amplitude of concentration signal |
| $c_0$ | initial concentration |
| $c_\mathrm{q}(t), c_\mathrm{q}^\mathrm{s}(t)$ | amplitude of concentration grating in laboratory- and fluid-fixed frame of reference |
| $d$ | fringe spacing of the grating |
| $E$ | electric field amplitude of light wave |
| $I$ | intensity of writing laser |
| $n$ | refractive index |
| $n_\mathrm{q}(t), n_\mathrm{q}^\mathrm{s}(t)$ | amplitude of refractive index grating in laboratory- and fluid-fixed frame of reference |
| $S_\mathrm{hom}, S_\mathrm{het}$ | homodyne, heterodyne signal |
| $S_\mathrm{m}(\varphi)$ | detected mixed signal |

| | |
|---|---|
| $T_\mathrm{q}(t), T^\mathrm{s}_\mathrm{q}(t)$ | amplitude of temperature grating in laboratory- and fluid-fixed frame of reference |
| $s$ | cuvette thickness |
| $u_\mathrm{c}$ | convective velocity |
| $T_\mathrm{c}$ | critical temperature |
| $\nu$ | critical exponent of the correlation length |
| $z_\eta$ | critical exponent of the viscosity |
| $\varphi'$ | critical exponent |
| $w_\mathrm{c}$ | critical weight fraction |
| $k_\mathrm{T}$ | thermal diffusion ratio: $k_\mathrm{T} = S_\mathrm{T} T c(1-c)$ |

# A Survey of the Thomaes Flow Cell Method for the Soret Coefficient

Guy Chavepeyer[1], Jean-François Dutrieux[1], Stéfan Van Vaerenbergh[2], and Jean-Claude Legros[2]

[1] Université de Mons-Hainaut, General Chemistry, Avenue du Champ de Mars, 24, B-7000 Mons, Belgium
[2] Université Libre de Bruxelles, Microgravity Research Center (MRC), Chemical Physics E.P. Dept., C.P. 165, Avenue F.D. Roosevelt, 50, B-1050 Bruxelles, Belgium

**Abstract.** The thermaldiffusion flow cell was imagined to measure small Soret coefficients but, even for normal values, perturbating phenomena were recorded and not explained. The major intriguing effect was the separation dependence in the basic Poiseuille flow, while the stability problems were partially understood. Recent studies report that natural convection generated by the inlet of the homogeneous mixture has no significant effect. More effective is the finite geometry of the outlet flow separator which provides a Poiseuille flow dependence of the vertical separation. Numerical simulations fit well existing experimental data for series of varied basic flows.

## 1 Introduction

The separation of the components of a mixture under a thermal constraint was reported by Ludwig (1856). The effect was formulated in term of thermaldiffusion by Soret (1880) and, till nowadays, a huge number of studies deal with the species separation for industrial benefit together with laboratory experiments performed to determine transport coefficients. In 1956, Thomaes imagined a Soret cell with an imposed Poiseuille flow able to extract significant volumes of enriched liquids. The basic idea consists in the mechanical splitting of the basic flow into two superposed outlet flows of modified average composition. The original theoretical approach considers a plane outlet domain and a point flow-divider. Turner (1961) proposed a refined theory, improving the perturbation-like approach of Thomaes but without change of the outlet region model. The seventies were the golden-age of Hydrodynamic Stability and Legros (1971) reported series of measurements oriented towards the stability study of binary mixtures in order to find systems with the heavier component migrating towards the hot wall. A separation dependence was recorded in function of the basic flow rate: the separation increased for high flow rates (despite of any stability effect). The opposite effect should be observed if the basic flow is such high that the transfert time of the liquid becomes of the same order of magnitude that the diffusion time constant. This "strange" effect was never explained and generally called "inlet-outlet and/or natural convection perturbating effect". In the present study we first recall the original theories. We then report recent computer simulations oriented towards the role of natural convection wich always takes place due to

W. Köhler and S. Wiegand (Eds.): LNP 584, pp. 211–232, 2002.

the inlet of homogeneous liquid, and finally we give a probable explanation of the separation dependence in the basic flow in term of the finite geometry of the outlet zone of the flow cell.

# 2 Flow cell basics

## 2.1 The early Thomaes theory

A flow cell is essentially made of two horizontal thermostated metal plates confining a thin fluid layer (Fig. 1). When heating from above, the upper plate is at temperature $T_{\mathrm{hot}}$ and the lower plate at temperature $T_{\mathrm{cold}}$. A given flow rate is imposed by maintaining a pressure difference between the inlet ($x = 0$) and the outlet ($x = L_{\mathrm{x}}$). A flow splitter device is built in the oulet zone to collect separately the upper and the lower parts of the fluid after species separation by thermodiffusion resulting from an imposed temperature difference between the planes $z = 0$ and $z = L_{\mathrm{z}}$.

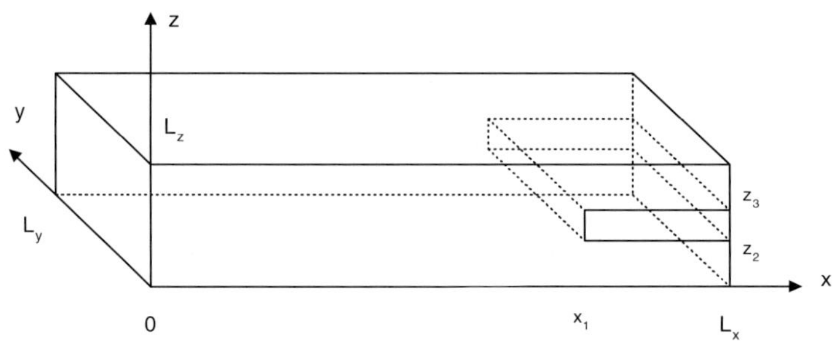


**Fig. 1.** Flow cell sketch

In his early paper, Thomaes[1] considers the 2D situation in the plane $y = 0$, with the flow splitter resumed to a single point ($L_{\mathrm{x}}$, $\frac{L_{\mathrm{z}}}{2}$ ). A plane Poiseuille flow is imposed along x and the velocity profile, resulting from an imposed horizontal pressure gradient:

$$U = U_{\mathrm{a}}(z - z_1)(z - z_4) \tag{1}$$

where the horizontal velocity profile depends on the vertical coordinate $z$ only. In (1) the amplitude $U_{\mathrm{a}} = \frac{\Delta p}{2\mu L_{\mathrm{x}}}$ with $\Delta p$ the imposed pressure difference along the length $L_{\mathrm{x}}$ of the cell and $\mu$ the dynamic viscosity; the lower and the upper limits of the cell are referenced as $z_1 = 0$ and $z_4 = L_{\mathrm{z}}$ respectively. The average velocity

$$\overline{U} = U_{\mathrm{a}} \frac{\int_{z_1}^{z_4} (z - z_1)(z - z_4) dz}{\int_{z_1}^{z_4} dz} \tag{2}$$

then becomes $\overline{U}=U_{\mathrm{a}}\frac{L_{\mathrm{z}}^2}{6}$. The maximum velocity $U_{\mathrm{m}}=U_{\mathrm{a}}\frac{L_{\mathrm{z}}^2}{4}$ and then $\overline{U}=\frac{2}{3}U_{\mathrm{m}}$.

A constant vertical temperature gradient is applied to the cell and, in equilibrium conditions, the thermodiffusion flux is zero

$$\frac{\partial c}{\partial z} + S_{\mathrm{T}} c_0 (1 - c_0) \frac{\partial T}{\partial z} = 0 \tag{3}$$

In (3), the mass fraction $c$ of the denser component of a binary liquid mixture is considered, with $c_0$ the mass fraction reference equal to the mass fraction of the homogeneous inlet mixture. At temperature reference $T_0 = \frac{T_{\mathrm{hot}} + T_{\mathrm{cold}}}{2}$, the Soret coefficient $S_{\mathrm{T}} = \frac{D_{\mathrm{T}}}{D}$, is defined as the ratio of the thermodiffusion coefficient $D_{\mathrm{T}}$ to the isothermal diffusion coefficient $D$.

The vertical separation of the species is small for moderate temperature differences and we assume the Boussinesq approximation. Over any height interval, the average mass fraction is defined as

$$\overline{c} = \frac{\int c(z) U(z) dz}{\int U(z) dz} \tag{4}$$

The flow splitter was considered [1] as punctual, say at vertical location $z = z_2 = z_3$ and from (4) the difference between the average mass fractions of the lower part $\overline{c}_{\mathrm{lower}}$ and the average mass fractions of the upper part $\overline{c}_{\mathrm{upper}}$

$$\Delta\overline{c} \equiv \overline{c}_{\mathrm{lower}} - \overline{c}_{\mathrm{upper}} = S_{\mathrm{T}} c_0 (1 - c_0) \Delta T \frac{3}{8} \tag{5}$$

with $\Delta T = T_{\mathrm{hot}} - T_{\mathrm{cold}}$ the imposed vertical temperature difference.

In the right-hand side of (5), the group of the parametric factors is nothing but the vertical separation resulting from pure Soret effect $\Delta c = c(z_1) - c(z_4)$. The $\frac{3}{8}$ factor comes simply from the convolution of the linear mass fraction with the parabolic velocity profile which reduces the contribution of the outer regions of the cell while promoting its bulk. Without flow (pure Soret effect) the factor $\frac{3}{8}$ becomes $\frac{1}{2}$, provided an averaging performed within two equal vertical domains and thus a reduction factor equal to 2 with respect to the pure Soret effect separation $\Delta c$.

The diffusion time constant, assuming a negligible contribution of the thermal diffusion as it should be for $D_{\mathrm{T}}$ of order of $10^{-12}$ $\mathrm{m^2 s^{-1} K^{-1}}$ and $\Delta T$ of order of a few degrees, read

$$\theta = \frac{L_{\mathrm{z}}^2}{\pi^2 D} \tag{6}$$

For $\overline{U}$, the average fluid velocity and $S = L_{\mathrm{y}} L_{\mathrm{z}}$ the cell section, the volume fluid flux

$$Q = \overline{U} S \tag{7}$$

and the transfer time of the fluid

$$\tau = \frac{L_{\mathrm{x}}}{\overline{U}} \equiv \frac{L_{\mathrm{x}} L_{\mathrm{y}} L_{\mathrm{z}}}{Q} \tag{8}$$

When the ratio

$$r = \frac{\tau}{\theta} \equiv \frac{\pi^2 L_y L_x D}{L_z Q} \tag{9}$$

is great enough, the inlet flow is able to reach the equilibrium separation due to thermodiffusion from an inlet at homogeneous composition.

After delay equal to $5\theta$, the equilibrium is approached within 1%. For a flow with a mean velocity $\overline{U}$, the fluid path x must be greater than $5\theta\,\overline{U}$. The original Thomaes cell [1] was designed with $L_x = 200$ mm, $L_y = 40$ mm and $L_z = 0.2$ mm and flow of order of 1 $\mu$l/s, corresponding to $\overline{U} \simeq 80$ $\mu$m/s. From (6), the equilibrium time $5\theta \simeq 40$ s and the transfer time of the fluid is equal to $\frac{L_x}{\overline{U}} \simeq 2500$ s, this is about 60 times the equilibrium time equal to $5\theta$. Such a design was build to handle systems with low diffusion coefficient, namely of order of $10^{-12}$ m$^2$s$^{-1}$ for which the equilibrium time is of order of the crossing time of the fluid.

Thomaes [1] studied the effect of the position of the splitting point along the $z$ direction and the result was that a splitter position away from midheight, simply modify the values of the mass fractions in $z = 0$ and in $z = L_z$, but without some important effect on the difference between these quantities. It was fairly found that an error of 33% on the vertical position generates but a 2% error on the mass fraction difference.

Shortly:

- From the velocity point of view, this is a 1D theory, with the horizontal velocity component (the only one considered) depending on $z$. The $x$ dimension appears as a dummy parameter enabling the setup of the vertical separation.
- The outlet zone is resumed to the plane $x = L_x$ and a dividing point separates the upper from the lower channel.
- The splitter position is not critical, given the former assumptions.
- Stability effect was ignored at the moment and the sign together the magnitude of the product $S_T \Delta T$ found no particular care.

## 2.2 Refinement by Turner

Starting in 1962, Turner [2] proposed a series of communications dealing with refinements of the Thomaes flow cell method. The main criticism addressed by Turner [2] to the Thomaes theory consists in its simple perturbation method nature.

Turner [2] assumed (as Thomaes [1]):

1. the flow is independent of the y direction
2. the velocity field is 1D along x and depends only of z
3. no inlet effect is considered on the velocity profile
4. a constant vertical temperature gradient is imposed
5. a Boussinesq fluid is considered

For a dilute mixture, the mass flux is equal to $\rho_0 DJ$ with

$$J = -\nabla c - S_{\mathrm{T}} c \nabla T \tag{10}$$

with $c(x, z)$ in steady condition. With (10), the mass balance becomes

$$D\left(\frac{\partial^2 c}{\partial x^2} + \frac{\partial^2 c}{\partial z^2}\right) + D_{\mathrm{T}} \frac{\partial T}{\partial z}\frac{\partial c}{\partial z} - U\frac{\partial c}{\partial x} = 0 \tag{11}$$

further simplified by dropping the $\frac{\partial^2 c}{\partial x^2}$ and $D_T \frac{\partial T}{\partial z}\frac{\partial c}{\partial z}$ terms in (11). New variables are introduced, $Z = \frac{2z}{L_{\mathrm{z}}} - 1$ and $X = \frac{Dx}{L_{\mathrm{z}}^2 U_{\mathrm{m}}}$ to build the velocity profile

$$U = \frac{3}{2}\overline{U}(1 - Z^2) \tag{12}$$

and (11) simplifies to

$$\frac{\partial^2 c}{\partial Z^2} - \frac{3}{8}(1 - Z^2)\frac{\partial c}{\partial X} = 0 \tag{13}$$

The splitting $c = c_\infty + c'$ between the asymptotic solution

$$c_\infty = c_0(1 - S_{\mathrm{T}} \frac{\partial T}{\partial Z} Z) \tag{14}$$

and the transient contribution $c'$, with the notation $c = c'$, gives a new equation of the exact form of equation (13) to be solved for convenient boundary conditions.

No analytic solution was found for (13) and a numeric solution was given from a finite differences algorithm. The separation between the two outlet streams was shown in function of $X$ up to 0.4, which corresponds to about 99% of the asymptotic solution. Owing to the structure of the $X$ and $Z$ variables, various values of $L_{\mathrm{z}}$, $S_{\mathrm{T}}$ and inlet flows can be considered and a universal curve is then presented for comparison with experimental data.

## 3 Experimental results

Historically, the group of Turner ([3]-[7]) gave the first reports of experimental results of an intensive use of the flow cell method. In [3] and [4], the need of a new cell building is explained, because the prime results were quite poor; the need of a calibration experiment is argued and the source of possible artefact are investigated in terms of inlet zone together with outlet zone perturbations due to the flow splitter setup. The cell is heated from above and the new cell size is $L_{\mathrm{z}} = 0.3$ mm, $L_{\mathrm{y}} = 5$ cm and $L_{\mathrm{x}} = 15$ cm and the operating flow rate range $Q = 2$ up to 200 cc/min. From (9), and $D = 5\ 10^{-10}$ m$^2$/s, the corresponding $r$ values are respectively equal to 35 and 0.35. For $Q = 15$ cc/min, $r = 4.6$ and $\overline{U} = 1700$ $\mu$m/s.

The 3 binary mixtures of the ternary system $CCl_4$–Benzene–Cyclohexane are studied in [5] and [6]. The Soret coefficients are found positive for $CCl_4$

with Benzene and with Cyclohexane but the reported values are generally of order of 50% higher than those obtained by other methods and other teams. In the Cyclohexane–Benzene system, the Soret coefficient if found negative for Cyclohexane mole fraction lower than 0.2, and positive for higher Cyclohexane content with higher data value than found in existing data file. No agreement of the experimental results with existing theories was found for the 3 binary mixtures of the ternary system.

In [7], Methanol is mixed with each of the 3 species studied before and given above. The $CCl_4$–Methanol system has large Soret coefficient for high alcohol content with a peak centered at the mole fraction 0.7 in $CCl_4$. The maximum $S_T$ value was found twice the values found by two other teams. In the Benzene–Methanol system, the Soret coefficient changes sign at about 0.6 Benzene mole fraction. The sign change is observed but quantitative fitting with other results is poor in the severe condition of strong $S_T$ dependence upon composition. The Cyclohexane–Methanol system is immiscible over the major part of the concentration range. Data are reported but without comparison with other data set and no further comment.

Generally, Turner et al. obtained Soret coefficient values significantly higher than data from other groups.

Intensive use of the flow cell was performed by Legros [8] in the Thomaes group in the Benzene–$CCl_4$, Benzene–Cyclohexane, Chlorobenzene–$CCl_4$ and Water–Ethanol systems. For the Benzene–$CCl_4$ the vertical separations were published [9] for various flow rates, for various temperature differences heating from above but also when heating from below in order to estimate the stability effect resulting from the mass contribution to the total vertical density gradient.

The cell geometry is $L_z = 0.2$ mm, $L_y = 5$ cm and $L_x = 20$ cm and the operating flow rate range $Q = 1$ up to 20 cc/h (a single one data at 80 cc/h). From (9), with $D = 5 \times 10^{-10}$ m$^2$/s, the corresponding $r$ values are 890 and 44 respectively. For $Q = 20$ cc/h, $\overline{U} = 560$ $\mu$m/s. When heating from above, in the Benzene–$CCl_4$ system, the denser component migrates towards the bottom side of the cell and no destabilizing effect can act. The derived $S_T$ values agree with results from literature for flows greater than 4 cc/h but, quoting Legros [9], "no interpretation can be given for the fact that separation grows for very small flows (rate less than 4 cc/h)" and also Platten and Legros [10], "The strange behavior when $\overline{U} \sim 1 - 2$ cc/h is still an enigma". For example, for $\Delta T = 8$ °C, $\Delta\bar{c} = 9.2 \times 10^{-3}$ at $Q = 0.5$ cc/h and $\Delta\bar{c} = 5.6\ 10^{-3}$ at $Q = 4$ cc/h with a gentle convergence of $\Delta\bar{c}$ for $Q$ up to 20 cc/h.

## 4 Analysis of additional effects

The prevailing feeling suggests that some significant effect(s) is (are) missing in the related theories of the flow cell method. Two possible major effects were studied:

1. the role of natural convection, always present in the flow cell because the homogeneous solution enters the cell at average temperature. Assuming stable

condition (heating from above and a positive Soret coefficient), the upper part of outlet region is lighter (or denser) than the bottom one because both the thermal expansion and the mass expansion effects contribute to modify the specific mass.

2. the geometry and position of the flow splitter. If the position of the divider was judged of no critical importance, the finite size of the dividing zone was not yet discussed and the generation of velocity components normal to the basic flow direction seems worth of interest.

The two next subsections deal with numerical simulations. A 2D centered finite difference method is applied to the balance laws for momentum, energy and mass fraction of the denser component of a binary mixture, in the Boussinesq approximation. The balance equations are written in the stream function $\Psi$ ($U = +\frac{\partial \Psi}{\partial z}$) and vorticity forms and $N_{\mathrm{x}}$ and $N_{\mathrm{z}}$ points are involved for $L_{\mathrm{x}}$ and $L_{\mathrm{z}}$ segmentation. The specific mass depends on temperature and composition

$$\rho = \rho_0[1 - \alpha(T - T_0) + \beta(c - c_0)] \tag{15}$$

with the reference specific mass $\rho_0$ at the reference temperature $T_0$ and mass fraction $c_0$ for the denser component; $\alpha$ and $\beta$ are the thermal and mass expansion coefficients.

## 4.1 Natural convection does exist

In [11], numerical simulations concern a 2D model of flow cell heated from above assuming a possible instability for negative values of $S_{\mathrm{T}}$. A model-fluid is considered, not a particular real mixture, but realistic fluids properties are taken, not too far from values that could apply, say, to alcohol-water or to organic-organic mixtures. The order of magnitude of the velocity amplitude of the natural convection should not be changed significantly when considering a particular mixture. The relevant parameters are:

| | | |
|---|---|---|
| $g$ | 9.81 | $\mathrm{ms^{-2}}$ |
| $L_{\mathrm{z}}$ | 0.3 | mm |
| $A$ | 1 up to 32 | unit |
| $\rho_0$ | 1000 | $\mathrm{kg\,m^{-3}}$ |
| $v$ | $10^{-6}$ | $\mathrm{m^2s^{-1}}$ |
| $D_{\mathrm{th}}$ | $1.435 \times 10^{-7}$ | $\mathrm{m^2s^{-1}}$ |
| $D$ | $10^{-9}$ | $\mathrm{m^2s^{-1}}$ |
| $D_{\mathrm{T}}$ | $10^{-11}$ | $\mathrm{m^2s^{-1}K^{-1}}$ |
| $\Delta T = T_{\mathrm{hot}} - T_{\mathrm{cold}}$ | 5 | K |
| $\alpha$ | $4 \times 10^{-4}$ | $\mathrm{K^{-1}}$ |
| $\beta$ | 0.25 | unit |
| $c_0$ | 0.5 | unit |

No-slip velocity conditions are imposed at the upper and lower limits which are kept at temperature $T_{\mathrm{hot}}$ and $T_{\mathrm{cold}}$ respectively, assuming a constant $\frac{\Delta T}{L_z}$ temperature gradient everywhere along x.

**No flow and fixed outlet composition.** Let be some experiment with the outlet mass fraction fixed by an hypothetical device, imposing a constant vertical composition gradient with the mass fraction $c$ in the range 0.50625 (top) down to 0.49375 (bottom), i.e. $\Delta c = c(L_{\mathrm{x}}, L_{\mathrm{z}}) - c(L_{\mathrm{x}}, 0) = 0.0125$ corresponding to $S_{\mathrm{T}} = -0.01\ \mathrm{K}^{-1}$. No forced inlet flow is imposed and the entrance vertical limit is kept at the reference mass fraction $c_0$. For an aspect ratio $A = 1$ , $N_{\mathrm{x}} = N_{\mathrm{z}} = 41$ grid points, the time and space numerical integration starts on, with an overall homogeneous composition $c_0$. A stationary state is reached, with composition governed by thermal diffusion as shown at Fig. 2. The greyscale interval indicates bright maximum value for any field.

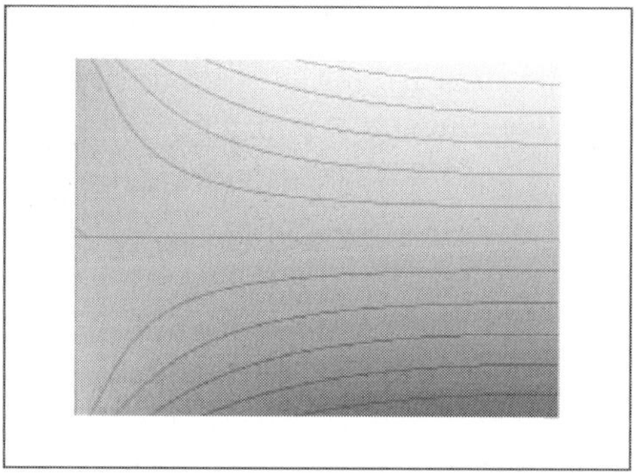

**Fig. 2.** Mass fraction without basic flow

The $c$ mapping of Fig. 2 indicates that the upper half cell is enriched by the denser component while the lower half part is depleted in the denser component, in accordance with the species conservation. In consequence, the average density in the upper (respectively lower) half-region is greater (respectively smaller) than the reference density $\rho_0$ , according to (15), because the mass expansion is dominant in comparison with thermal expansion.. Also, diffusion creates horizontal density gradients, mainly strong near the inlet zone. The composition remains unchanged in the plane $\frac{L_z}{2}$, and equal to $c_0$. The composition distribution of Fig. 2 and the resulting horizontal density gradients then contribute to generate natural convection as shown at Fig. 3. The stationary convection mode is characterized by two vortices. The convective regime is bicellular, made of two superposed cells rotating in opposite ways. The upper cell (respectively lower) rotates clockwise (respectively counter-clockwise) because the density of the outlet upper (respectively lower) zone is greater (respectively smaller) than the density $\rho_0$ at the inlet (fixed value in $x = 0$). This bicellular regime does not

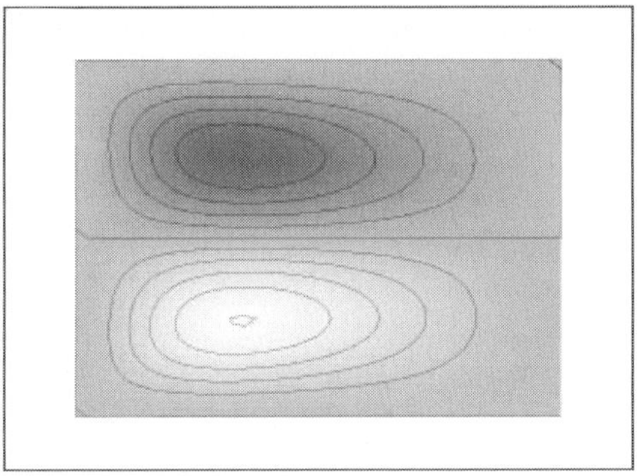

**Fig. 3.** Stream function without basic flow

enable any convective mass transport through the midheight plane $z = \frac{L_z}{2}$; the vortices are eccentric, shifted towards the strong composition gradients region.

By inverting the composition gradient at the right-hand side limit, the obtained stationary convective regime is unchanged, despite the opposite rotation of both convection cells.

Thus, even is the case of a negative value of $S_T$, the convective regime is due to the horizontal density gradient, despite the destabilizing vertical density gradient, too small to induce Rayleigh-Bénard convection.

The velocity amplitude of the natural convection is of order of 1 $\mu$m/s in both cases reported above. Here, nothing is to be said about the composition field because the mass fraction was imposed at the outlet limit.

**No flow and free outlet composition.** The composition was kept free at the outlet limit, assuming an achieved mass fraction profile along the $z$ axis: $\frac{\partial c}{\partial x} = 0$ in $x = L_x$. An incomplete Soret effect (in comparison with the former case where a pure Soret effect $\Delta c = |c(z_4) - c(z_1)| = 0.0125$ was imposed at the outlet) $\Delta c = 0.0116$ is obtained due to the small aspect ratio value, $A = 1$. The horizontal density gradients due to Soret effect and isothermal diffusion generate natural convection and a stable stationary state is still observed. The temperature and composition fields are practically unchanged in comparison with the former case. The velocity amplitude is equal to 0.8 $\mu$m/s, lower than the value equal to 1 $\mu$m/s cited above in reason of smaller horizontal density gradients resulting from an incomplete vertical separation.

It was verified that the velocity amplitude increases linearly for increasing Soret coefficient values, for $A = 1$. Figure 4 shows the vertical outlet separation $\Delta c$ (right axis), and the horizontal and vertical velocity amplitudes ($U$ and $V$ respectively at left axis) for various $S_T$ values. For $S_T$ up to 0.02 K$^{-1}$, $U$ and $V$ depends nearly linearly on $S_T$, no matter the sign of $S_T$. The $\Delta c$ dependence in function of $S_T$ is nearly linear. In comparison with the curve for an infinite cell (pure Soret, the main diagonal of the plot), an increasing lack of separation is detected at higher $S_T$ values due to the small aspect ratio $A = 1$. The curves

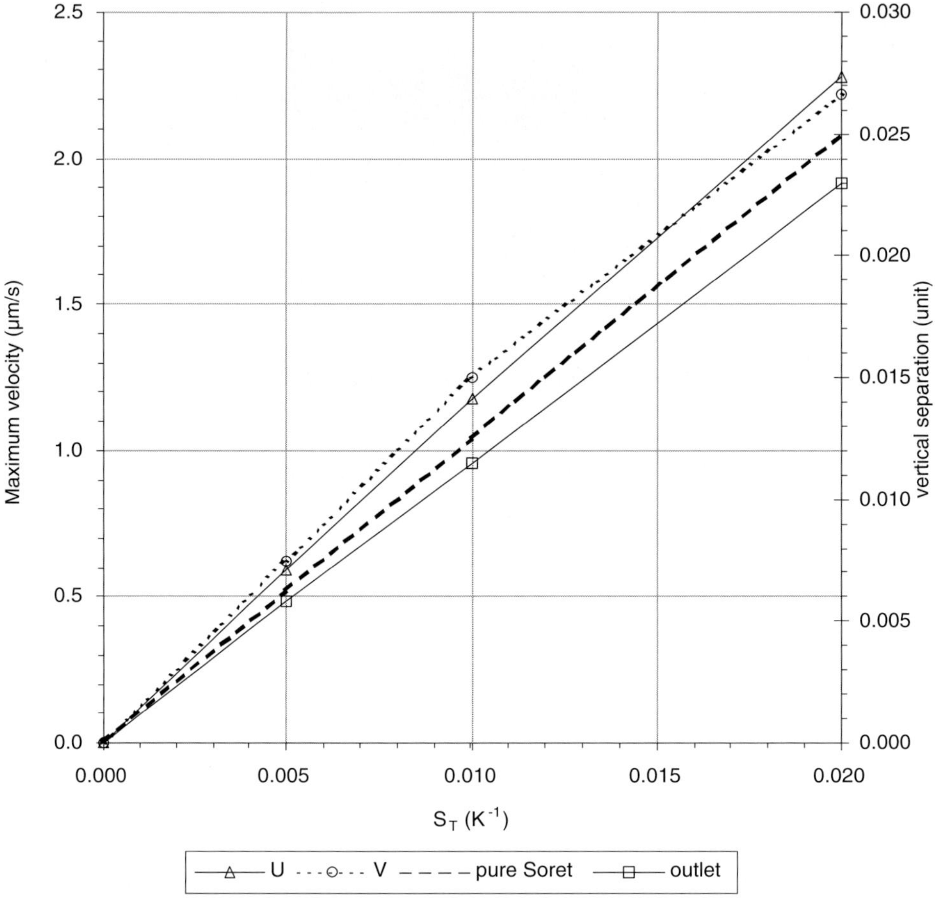


**Fig. 4.** Without Poiseuille flow, A = 1

for $U$ and $V$ are practically identical and the amplitude is of order of 1 $\mu$m/s for $|\Delta T S_{\mathrm{T}}| = 0.05$.

The identity of the curves $U$, $V$ and $\Delta c$ in function of $S_{\mathrm{T}}$ confirms clearly the effect of the mass contribution to the total density (15). Once again, instabilities was not observed for negative Soret coefficients.

**Basic flow and free outlet composition.** A Poiseuille flow was imposed [11] with a maximum velocity equal to $U = 300$ $\mu$m/s, for the properties of the fluid given above. Such a velocity value is currently operating in Thomaes flow cell experiments.

The prediction of the 2D numerical simulation is a stable stationary state with a "conductive" temperature field like in the former cases, owing to the small value of the thermal Péclet number $Pe_{\mathrm{T}}$, smaller than unity ($Pe_{\mathrm{T}} = \frac{UL_z}{\kappa} \sim 0.6$). The stream function and horizontal velocity component fields are shown at Fig. 5

**Fig. 5.** Stream function with Poiseuille flow, A = 1

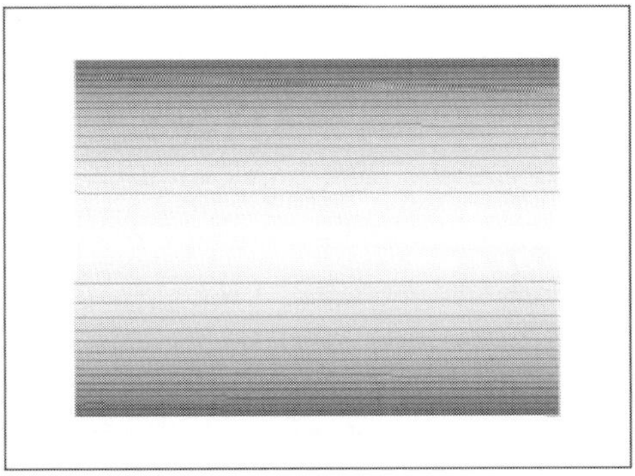

**Fig. 6.** Horizontal velocity component with Poiseuille flow, A = 1

and Fig. 6. At the resolution of the plots, nothing but the basic flow can be observed from these figures because the natural convection consists in a modulation of order of 0.1% of the basic flow.

The composition field (Fig. 7) is deeply modify by the basic flow, and the comparison with Fig. 2 enlightens the dragging effect of the Poiseuille flow.

The outlet separation $\Delta c = 0.0053$ instead of 0.0116 without Poiseuille flow. This is expected because the solutal Péclet number $Pe_\mathrm{s} = \frac{UL_\mathrm{z}}{D} \sim 90$ is much greater then unity.

The vertical velocity component (Fig. 8) and the corresponding buoyancy part of the stream function (Fig. 9) show two superposed convective cells. At Fig. 9, the centers of the vortices are located midway horizontally because the horizontal composition gradients (and the resulting density gradients) are smoothed by the dragging effect (see above the poor outlet separation achievement).

At both lateral side, the two opposite maximum vertical velocity are located nearly midway of the two half heigth call. The operating $S_\mathrm{T} = -0.01\ \mathrm{K}^{-1}$, and the upper (lower) convective cell rotates (counter-) clockwise. The vertical velocity component $V$ has no mixing effect because the vertical flow rate is very

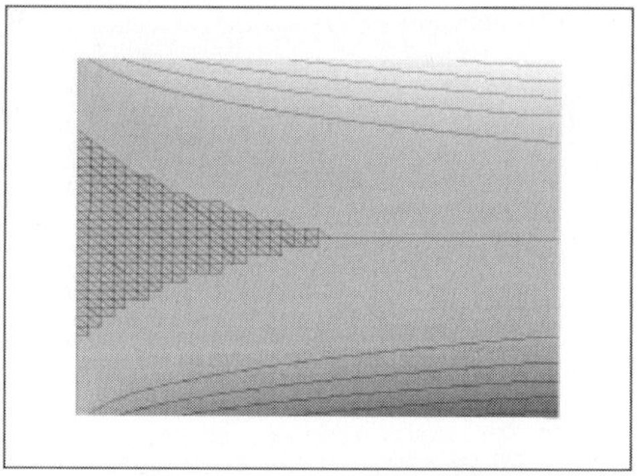

**Fig. 7.** Mass fraction with Poiseuille flow, A = 1

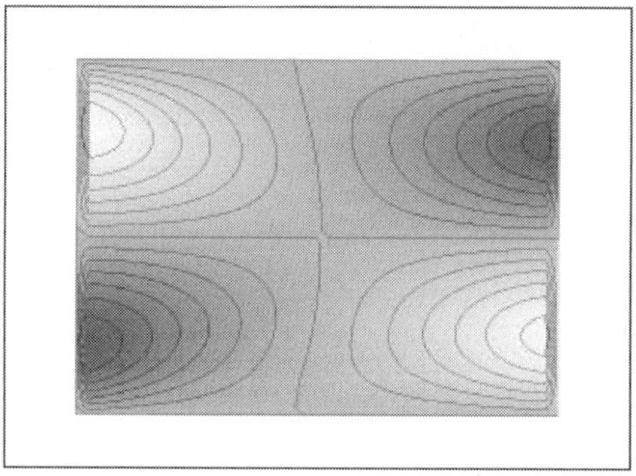

**Fig. 8.** Vertical velocity component with Poiseuille flow, A = 1

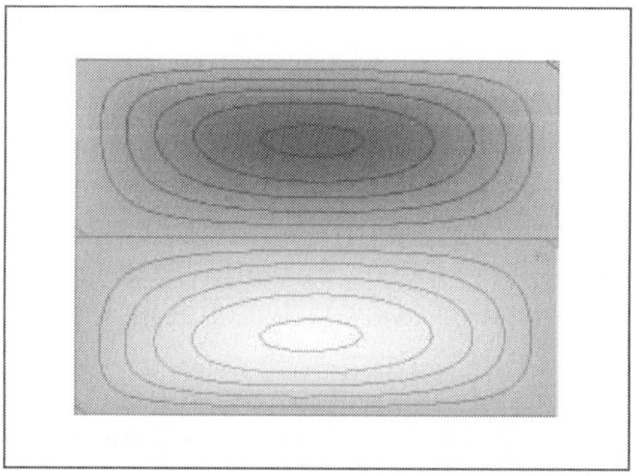

**Fig. 9.** Buoyancy stream function with Poiseuille flow, A = 1

low in comparison of the diffusion time constant. The solute Péclet number build with the half-height of the cell $Pe_s = \frac{V\frac{L_z}{2}}{D} \sim 0.15$, with $V = 1\,\mu\text{m/s}$ and $L_z = 0.3\,\text{mm}$.

Aspect ratios up to 8 were investigated in [11]. The series was completed and the results are given for $A$ up to 32 at Fig. 10. The graph shows an asymp-

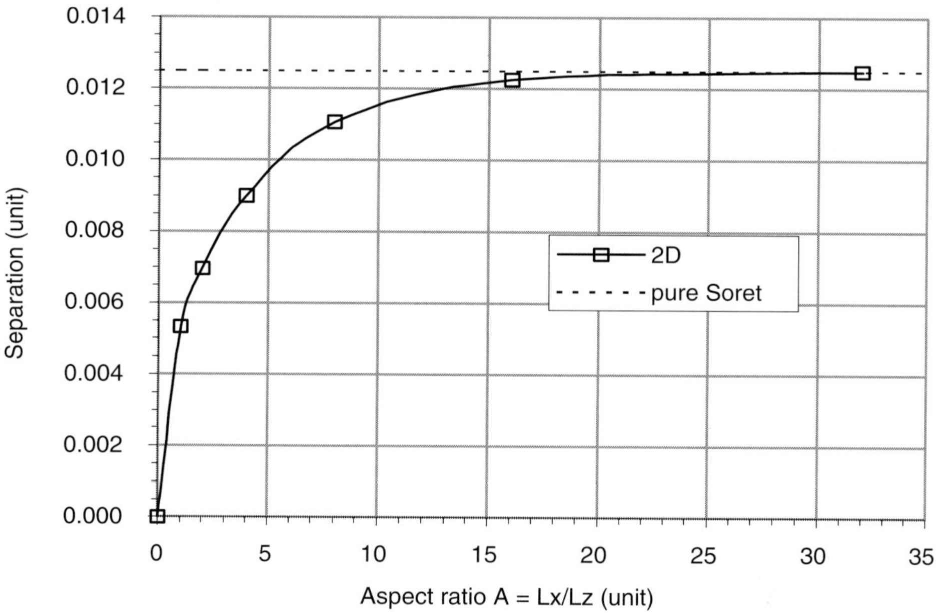


**Fig. 10.** With Poiseuille flow for a negative Soret coefficient

totic evolution of the outlet vertical separation in function of the aspect ratio A, with an evident convergence towards the "pure Soret effect" separation for an aspect ratio big enough. For $A = 32$, $\Delta c = 0.01225$, to be compared with the value 0.0125 for pure Soret effect. Even for negative Soret coefficient values (the denser component migrates upwards), no instability was observed and the convection mode remains characterized by two superposed convective cells rotating in opposite ways.

## 4.2 Utopian point flow separation

The assumption "point flow splitter" is always false when dealing with experimental devices. A razor blade (even fresh) has a front blade of, say, 20 $\mu$m size, and this is not negligible in comparison with the fluid path $L_z \sim 200$ $\mu$m of the flow cell.

Thomaes [1] and Turner [3] focused the theory on the diffusive aspect, taking the imposed Poiseuille flow just like a nice means to collect concentrated fluid. The precision of the position of the splitter was discussed, but never was argued the finite height of the blade. The flow separation was always considered like a mathematical convolution in the vertical outlet plane of the main channel, then limiting the system of interest just before the separation facility and resuming the splitter to a mathematical point. We study here the effect of a finite size splitter in the way of transverse velocity component generation.

**Simulation of an outlet finite domain.** According to Fig.1, in [12], 2D numerical simulations concern a flow cell with the finite size flow splitter limited by the planes $x = x_1$, $z = z_2$ and $z = z_3$. The Poiseuille flow is imposed from $x = 0$ with the velocity profile (1). For both outlet channels, the lower (index L) and the upper (index U) one, the velocity profiles have the form of (1), for no-slip boundary conditions at the limits of the flow splitter

$$U_{\mathrm{L}} = U_{\mathrm{aL}} z(z - z_2) \tag{16}$$

$$U_{\mathrm{U}} = U_{\mathrm{aU}}(z - z_3)(z - L_{\mathrm{z}}) \tag{17}$$

with $z_1 = 0$ and $z_4 = L_{\mathrm{z}}$. The corresponding stream function is imposed in $x = L_{\mathrm{x}}$, and a transient zone of length $L_{\mathrm{x}} - x_1$ is provided for field evolution (not only velocity, but also mass fraction).

The velocity amplitudes are linked to the inlet one through the conservation of the 2D volume flux $Q$ (eq. (7))

$$Q \equiv S\overline{U} = Q_{\mathrm{L}} + Q_{\mathrm{U}} \equiv S_{\mathrm{L}}\overline{U_{\mathrm{L}}} + S_{\mathrm{U}}\overline{U_{\mathrm{U}}} \tag{18}$$

with $\overline{U_{\mathrm{L}}}$ and $\overline{U_{\mathrm{U}}}$ the average velocities according to (2), and $S_{\mathrm{L}}$ and $S_{\mathrm{U}}$ the 2D channel sections. A relation between $Q_{\mathrm{L}}$ and $Q_{\mathrm{U}}$ is imposed by the extraction mode because the fluid can be either

1. pushed through the cell from the inlet by an imposed pressure difference, and the oulet fluxes are proportional to the 2D outlet sections $S_{\mathrm{L}} = z_2$ and $S_{\mathrm{U}} = L_{\mathrm{z}} - z_3$, with equal average velocities $\overline{U_{\mathrm{L}}} = \overline{U_{\mathrm{U}}}$ and $Q_{\mathrm{L}} \neq Q_{\mathrm{U}}$
2. pulled out by a depression imposed at the oulet. If equal fluxes are collected, i.e. $Q_{\mathrm{L}} = Q_{\mathrm{U}}$ equivalent to $z_2\overline{U_{\mathrm{L}}} = (L_{\mathrm{z}} - z_3)\overline{U_{\mathrm{U}}}$.

Selecting the suitable relationship among the two just given above, with (16), (17) and (18), the velocity amplitudes $U_{\mathrm{aL}}$ and $U_{\mathrm{aU}}$ can be easily computed in function of the given value of $U_{\mathrm{a}}$.

The initial conditions are generally:
– conductive temperature profile with heating from above with $\Delta T = 5\ °C$
– parabolic Poiseuille flows in each domain with respect to the corresponding channel sections and collection mode
– pure Soret separation in the inlet $x = 0$ plane and frozen mass fraction ($\frac{\partial c}{\partial x} = 0$) in $x = L_{\mathrm{x}}$

The fluid properties are identical to those involved in the "natural convection effect" paragraph and given above. The cell configuration is sketched at Fig. 1 and $L_{\mathrm{x}} = 0.45$ mm, $L_{\mathrm{z}} = 0.3$ mm, $x_1 = 0.3$ mm, $N_{\mathrm{x}} = 91$, $N_{\mathrm{z}} = 61$.

**A typical simulation.** For $S_{\mathrm{T}} = +0.01\ \mathrm{K}^{-1}$, the splitter blade of front height $z_3 - z_2 = 20\ \mu$m is centered in $\frac{L_{\mathrm{z}}}{2}$. The maximum velocity is equal to 300 $\mu$m/s and the fluid is pulled out. These conditions are very close to laboratory procedures, 3D excepted.

The stream function field at Fig. 11 shows the flow splitting by the blade of length equal to 0.15 mm. The curvature of the stream lines indicates a transient

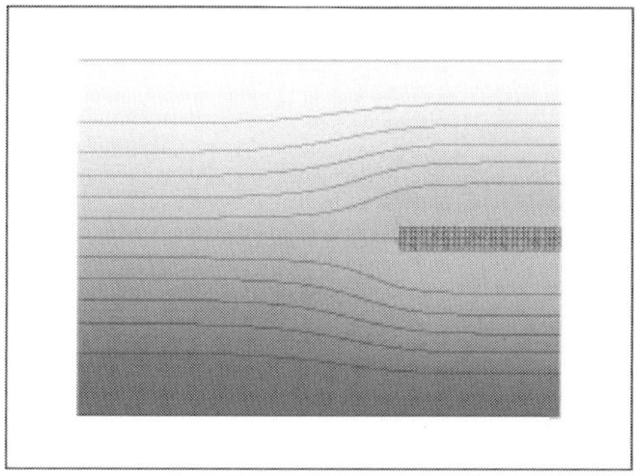

**Fig. 11.** Splitter 0.02 mm, stream function

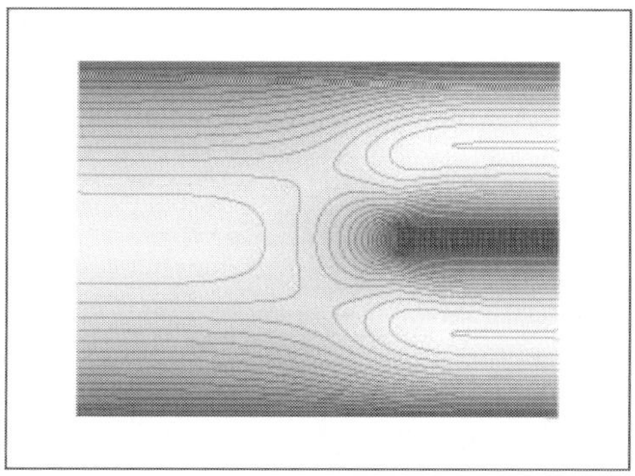

**Fig. 12.** Splitter 0.02 mm, horizontal velocity component

zone, with significant vertical velocity component in regions of horizontal extend of order of one half of the channels heights on each side of the obstacle front. About $\frac{1}{3}$ of the simulated domain is concerned by a significant 2D velocity field.

The horizontal velocity component field at Fig. 12 derives from Fig. 11 but the picture is more self-speaking about the transient zone and the greyscale indicates clearly the velocity amplitude increase in the outlet channels for extraction with two equal flow rates, and then equal velocity amplitudes for equal sections of the outlet channels. Of course, in the bulk region close to blade front, vanishing horizontal velocity corresponds to major vertical velocity domains. In Fig. 13, the mass fraction field c shows at first sight a trivial flow splitting. A close inspection near the splitter walls shows a transient zone suggesting a diffusion process. The $c_0$ line which marks the midplane $\frac{L_z}{2}$ in the inlet zone, is splitted into two lines, one in each outlet channels, and staying away from the blade limits. Thus, at a distance of about 50 $\mu$m behind the blade front, a stationary state is reached which is characterized by the upper part of the bottom channel less concentrated in the denser component than the lower part of the top channel. The vertical velocity field in Fig. 14 is the most spectacular to indicate the very localized effect of the blade front. Say 50% of the maximum vertical velocity amplitude acts

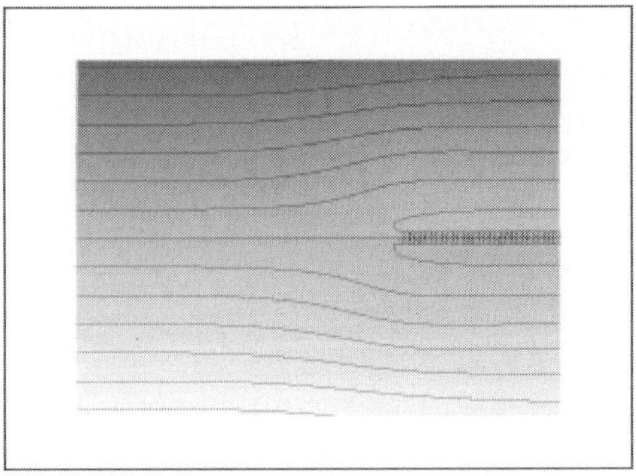

**Fig. 13.** Splitter 0.02 mm, mass fraction

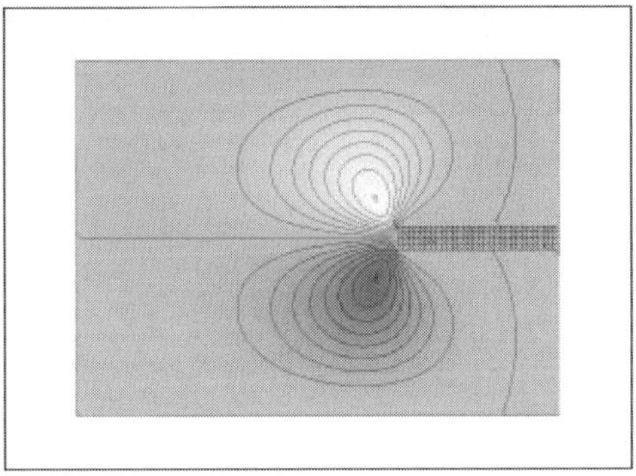

**Fig. 14.** Splitter 0.02 mm, vertical velocity component

(before the blade, evidently) in domains symmetrical with respect to the blade horizontal axis and of about 50 $\mu$m in diameter. The key point is the vertical velocity amplitude equal to 80 $\mu$m/s for a basic vertical velocity amplitude equal to 300 $\mu$m/s.

The solute Péclet number build with the half-height of the cell $Pe_\mathrm{s} = \frac{V \frac{L_z}{2}}{D} \sim 10$ and then, because $V_\mathrm{m}$ is intermediate between the contribution of natural convection (about 2 order of magnitude greater) and the amplitude of the basic flow (about 1 order of magnitude lower), an effect on the vertical separation could be observed for varied basic flow rate and proportional vertical velocity amplitude.

As expected, $V_\mathrm{m}$ is proportional to the flow rate as shown in Fig. 15 (squares). In Fig. 15, the 3 axis are in log scale because 3 decades are shown. The linear least square fitting (doted line) of $V_\mathrm{m}$ in function of $U_\mathrm{m}$ has a correlation coefficient equal to 1. The right axis shows the volume flux (triangles) corresponding to the basic $U_\mathrm{m}$ for an hypothetical $L_\mathrm{y} = 5$ cm. These data confirm the realistic flow conditions as used by Legros et al. [9]. The $V_\mathrm{m} = 0.2736 \times U_\mathrm{m}$ law is not universal, because the shape should depend on the size and geometry of the front blade, also on the blade position. For eccentric blade, the mirror symmetry of the $V$ field (see Fig. 14) is broken. That point was studied and it was found that

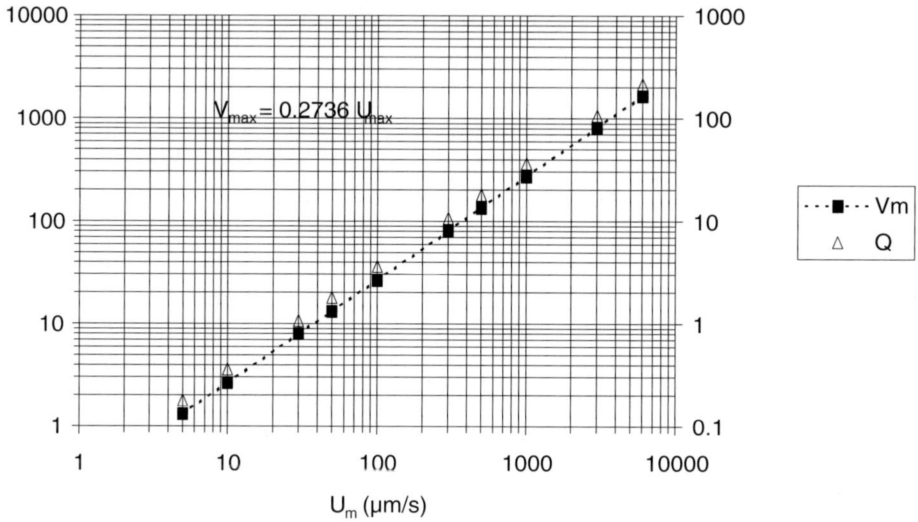


**Fig. 15.** Splitter 0.02 mm * 0.15 mm centered

this effect is of secondary importance owing to a possible significant effect on the outlet average vertical separation between the two channels.

The key-point is the vertical separation which was studied with respect to the prediction of Thomaes [1], equation (5), in a wide and realistic $U_m$ interval. Significant differences were found with the analytical prediction (5), and the differences depend on $U_m$, as shown in Fig.16. In Fig. 16, $\Delta c_{th} = \Delta c$ from eq. (5) and separations higher than $\Delta c_{th}$ are recorded which make negative the relative differences. First, let us consider the triangle and full line plot reporting results for the 0.15 mm long blade. For high rate flow, $U_m \geq 200\ \mu$m/s and $Q \geq 9$ cc/h (see Fig.16), the observed $\Delta c$ fit the analytical prediction $\Delta c_{th}$ within 5% and better for higher $U_m$. Within the $U_m$ interval 2 up to 200 $\mu$m/s ($0.1 < Q < 9$ cc/h) the data curve is S-shaped and converges towards an asymptotic overestimation against the analytical law of order of 40%.

The other quite similar curve (squares and doted line) corresponds to a model with an eccentric blade of the same front size (20 $\mu$m) but set in $z = \frac{1}{3}L_z$, and of zero length thus with the outlet zone resumed to a vertical segment of the $L_x$ plane. The differences between the two curves are meaningless and, in consequence, the precision of the position along the z axis together with the length of the flow splitter can be ignored about apparent average separation. The third curve of Fig. 16 (circles and full line) concerns positive ordinate values of the same variable but $\Delta c_{th}$ refers to the pure Soret separation, i.e. the equation (5) with the $\frac{3}{8}$ factor replaced by $\frac{1}{2}$. Simply, for low flow rates, the pure Soret effect law is the convenient relationship for separation versus Soret coefficient dependence. For increasing flow rate an S-shaped curve converges towards to an asymptotic depletion of order of 25%.

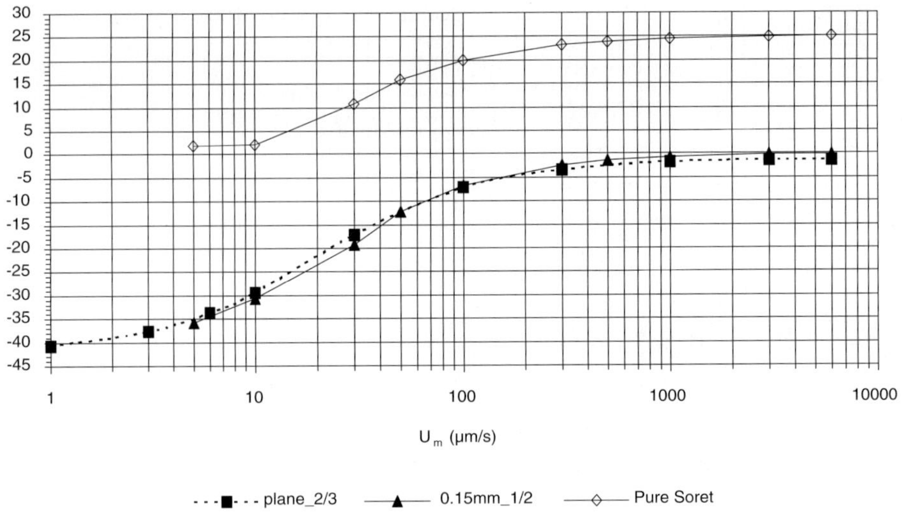


**Fig. 16.** Separation dependence in flow rate

For intermediate flow rates; $0.1 < Q < 9$ $cc/h$ i.e. $2 < U_{\mathrm{m}} < 200$ $\mu$m/s, no one of the two analytical laws can predict separation with an error of estimate less than 5%.

## 5 Relevance of dimensional effect

Qualitatively, the later paragraph explains the experimental separation in function of the flow rate (e.g. [9]). In a stable situation (heating from above with a positive Soret coefficient) increasing separation was observed with asymptotic value "for flows greater than 4 cc/h", quoting Legros.

Quantitative comparison is made for the $C_6H_6$-$CCl_4$ system, heating from above with various flow rates at temperature differences equal to 6, 8 and 10 °C. The 3 higher $\Delta T$ are selected because they correspond, a priori, to the more precise experimental separations. The 3 series involve together 25 data points which concern a unique Soret coefficient. For $U_{\mathrm{m}} > 300$ $\mu$m/s, the average Soret coefficient $S_{\mathrm{T}} = (7.8 \pm 0.6)$ $\mathrm{K}^{-1}$, i.e. a standard deviation equal to 8% computed from 14 data points. For each $\Delta T$, the relative difference $\frac{S_{\mathrm{T}} - S_{\mathrm{Tconv}}}{S_{\mathrm{Tconv}}}$ (in percent) with respect to the converged $S_{\mathrm{Tconv}}$ value of the Soret coefficient and (scaled) scattered plots are given in Fig.17, together with the best least square fit "fitErr/s" $= \sum_{\mathrm{i}=0}^{2} a_{\mathrm{i}} u^{\mathrm{i}}$ with $u = \log_{10}(U_{\mathrm{m}})$ and $(a_0, a_1, a_2) = (109.83, -66.50, 10.049)$.

In Fig. 17, the curves "$\frac{1}{2}$ gap" and "$\frac{2}{3}$ flat" are the corresponding curves of Fig. 16 for the actual situation and the full line "fit2D" $= \sum_{\mathrm{i}=0}^{2} a_{\mathrm{i}} u^{\mathrm{i}}$ is their least square fit with $(a_0, a_1, a_2) = (74.51, -49.05, 8.200)$. The differences between converged and actual $S_{\mathrm{T}}$ values are systematically higher for the experimental

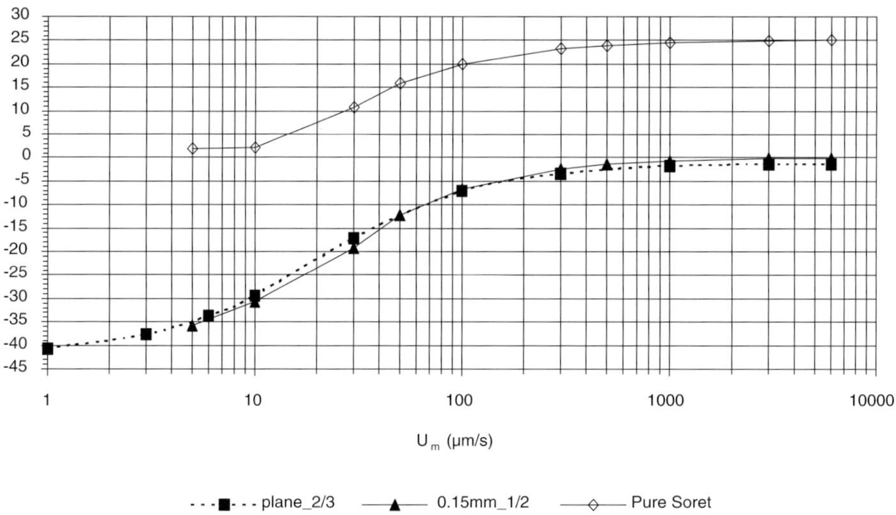


**Fig. 17.** Soret coefficient in function of the maximum basic velocity

(3D) data than for the 2D data from numerical simulation. However, the curves "fitErr/s" and "fit2D" are quasi identical (with the ordinate of Fig. 17 devoted to the crude "fit2D") with a standard deviation of order of 1%. A scaling factor $s = 1.614$ determined by least square fit (on this single one parameter) is sufficient to obtain a fair agreement between 2D simulations and the experimental series

$$\text{fit3D} = s \times \text{fit2D} \tag{19}$$

The insufficient correction estimated from a 2D model is justified by the Thomaes flow cell building with a significantly limited outlet section in the $y$ direction. Indeed, from the continuity equation, any decrease of both $x$ and $y$ velocity components must result in an increase of the vertical velocity component.

## 6 A few words on stability

Heating from above when $S_{\mathrm{T}} > 0$ (the denser component migrates towards the cold side, the bottom one) is a stable situation. For example, we observed no instability when studying natural convection through a 2D model even for negative Soret coefficient value. Nevertheless, for $S_{\mathrm{T}} < 0$, the existence of stationary instability is predicted and some experiments are reported (e.g. [10]) indicating a destabilizing effect.

When heating from below, i.e. a Bénard problem with Poiseuille flow, the flow stabilizes the system (e.g. [13], 10 references). A 3D theoretical study, by the local potential and normal mode analysis ([14], 26 references), showed the existence of longitudinal rolls for $S_{\mathrm{T}} > 0$ and the prevailing of transverse rolls for $S_T < 0$ without negating the stabilizing effect of the basic flow. In a pure fluid, the experimental proof was given ([15], 11 references) of the prevailing of transverse rolls, instead of longitudinal, for Reynolds numbers (depending on the Prandtl number) smaller than a threshold value. Recently, a numerical simulation study ([16], 34 references) indicates the effect of the inlet and outlet boundary conditions on the convection mode, again, without negating the stabilizing effect of Poiseuille flow.

From the beginning of this paragraph we deal with "normal mode" stability but other approaches exist interested in non-normal modes and from which a new lightening could be made on the Bénard-Poiseuille cell response, e.g. [17] (36 references) and [18] (63 references).

## 7 Conclusion

The asymptotic dependence of the Soret coefficient on the flow rate is well predicted by 2D simulation provided a scaling coefficient attributed to 3D effect when comparison is made with respect to experimental data from a flow cell device with limited outlet section. To definitely certify the role of transverse velocity field, 3D simulation is needed.

The negative Soret coefficient case was not closely studied yet from the 2D effect aspect, nor the heating from below situation. This will be done in a near future, hoping to explain experimental results (from Legros [9], for example) where the separation changes sign in function of the flow rate but, moreover, where the asymptotic Soret coefficient clearly depends on the temperature difference, indicating stability threshold points.

Natural convection is evidently always present because horizontal density gradients exist by the nature of the flow cell method. Nevertheless, it was shown that natural convection takes place with velocity amplitudes so small (of order of $1\mu$m/s) that the main velocity field, including the transverse contribution due to some possible outlet boundary effects, can not be significantly modified. Also, the composition distribution was not found perturbed by such a low velocity field, no matter the sign of the Soret coefficient, when heating from above.

### Acknowledgments

This paper presents results of the Belgian Programme on Interuniversity Poles of Attraction initiated by the Belgian State, Prime Minister's Office, Science Policy Programming. The scientific responsibility is assumed by its authors.

## References

1. G. Thomaes: J. de Chimie Physique, **50**, 407 (1957)
2. J.C.R. Turner: Chem. Eng. Sc., **17**, 95 (1962)
3. B.D. Butler, J.C.R. Turner: Trans. Faraday Soc., **62**, 3114 (1966)
4. B.D. Butler, J.C.R. Turner: Trans. Faraday Soc., **62**, 3121 (1966)

5. J.C.R. Turner, B.D. Butler, M.J. Story: Trans. Faraday Soc., **63**, 1906 (1967)
6. M.J. Story, J.C.R. Turner: Trans. Faraday Soc., **65**, 349 (1969)
7. M.J. Story, J.C.R. Turner: Trans. Faraday Soc., **65**, 1523 (1969)
8. J.C. Legros: Contribution à l'Etude de la Thermodiffusion et de la Stabilité Hydrodynamique des Systèmes Binaires. Ph.D. Thesis, Université Libre de Bruxelles (1971)
9. J.C. Legros, D. Rasse, G. Thomaes: Physica, **57**, 585 (1971)
10. J.K. Platten and J.C. Legros: *Convection in Liquids*, Springer-Verlag, Chap. X, pg 662 (1984)
11. G. Chavepeyer, S. Van Vaerenbergh, J.K. Platten, J.C. Legros, Y. Decroly: Entropie, **218**, 55 (1999)
12. G. Chavepeyer, J.-F. Dutrieux, S. Van Vaerenbergh, J.K. Platten: presented at IMT4, Bayreuth (2000)
13. C. Klapisz, M.-E. Weill: Physica, **71**, 106 (1974)
14. J.C. Legros, J.K. Platten: J. Non-Equilib., **2**, 211 (1977)
15. J.-M. Luijkx, J.K. Platten, J.C. Legros: Int. J. Heat Mass Transfer, **24**, 1287 (1981)
16. X. Nicolas, A. Mojtabi, J.K. Platten: Phys. Fluids, **9**, 337 (1997)
17. F. Waleffe: Phys. Fluids, **7**, 3060 (1995)
18. S.C. Reddy, P.J. Schmid, J.S. Baggett, D.S. Henningson: J. Fluid Mech., **365**, 269 (1998)


## Supplementary glossary

| | |
|---|---|
| $A$ | aspect ratio: $A = \frac{L_x}{L_z}$ |
| $g$ | gravity acceleration |
| $L$ | size of the cell |
| $N$ | grid points number |
| $p$ | pressure |
| $Pe$ | Péclet number |
| $Q$ | volume flux |
| $r$ | times ratio: $r = \frac{\tau}{\theta}$ |
| $S$ | cross section |
| $U$ | horizontal velocity component |
| $V$ | vertical velocity component |
| $\tau$ | transfer time |
| $\theta$ | diffusion time constant |
| $\Psi$ | stream function |
| $\mu$ | dynamic viscosity |
| $\nu$ | kinematic viscosity: $\nu = \frac{\mu}{\rho}$ |

### Subscripts

| | |
|---|---|
| $\infty$ | infinite |
| 0 | reference |
| $a$ | amplitude |
| $L$ | lower |
| $m$ | maximum |

$s$ solutal
$T$ thermal
$th$ theoretical
$U$ upper

## Superscripts

$\bar{\ }$ average value
$'$ transient

# Validity Limits of the FJO Thermogravitational Column Theory

Javier Valencia[1,2], Mohamed Mounir Bou-Ali[1,2], Oscar Ecenarro[1], José Antonio Madariaga[1], and Carlos María Santamaría[1]

[1] Departamento de Física Aplicada II, Universidad del País Vasco, Apdo. 644, 48080 Bilbao, Spain
[2] present address: Departamento de Ingeniería Mecánica, Energética y de Materiales, Universidad Pública de Navarra, Pamplona, Spain

**Abstract.** We have solved numerically in the formulation vorticity-stream function the equations that govern the separation process in a thermogravitational cell. The numerical values for the steady separation have been compared with the ones given by the Furry, Jones and Onsager theory for different values of the relevant parameters, namely the Grashof and Schmidt numbers and the aspect-ratio $A$. The obtained results show that with the exception of small regions near the cell ends where the vertical concentration gradient increases considerably, the FJO theory is accurate by more than 1% in the range $Gr\,Sc \lesssim 1000\,A$. Therefore, inside this range the standard formulation can be used with confidence to determine the Soret coefficient from steady separation measurements in a thermogravitational cell.

## 1 Introduction

The phenomenon of thermal diffusion, or Soret effect, results from the fact that a temperature gradient in a mixture gives rise to a concentration gradient. In a two component mixture, the denser component usually migrates to the cold region and a final steady state is reached when the ordinary diffusion and the thermal diffusion fluxes balance each other.

The separation produced by pure thermal diffusion is usually small. Larger separations are achieved in a thermogravitational column first discovered by Clusius and Dickel[1]. Essentially, the column consists of two vertical coaxial cylinders maintained at different temperatures (Fig. 1). The coupling of the horizontal Soret effect and the vertical convection currents give rise to a partial separation of the components of a mixture at the ends of the column.

The theoretical description of the thermogravitational cell was given by Furry, Jones and Onsager (FJO) [2,3]. These authors, starting from a simplified formulation of the fundamental hydrodynamic equations, arrived at a basic relation for the transport of one component through a column cross-section. At equilibrium, the condition of null transport allows one to determine the vertical separation in terms of the Soret coefficient of the mixture.

Because of the relatively large separations achieved in a thermogravitational column, many attempts have been made to determine the Soret coefficient from

W. Köhler and S. Wiegand (Eds.): LNP 584, pp. 233–249, 2002.

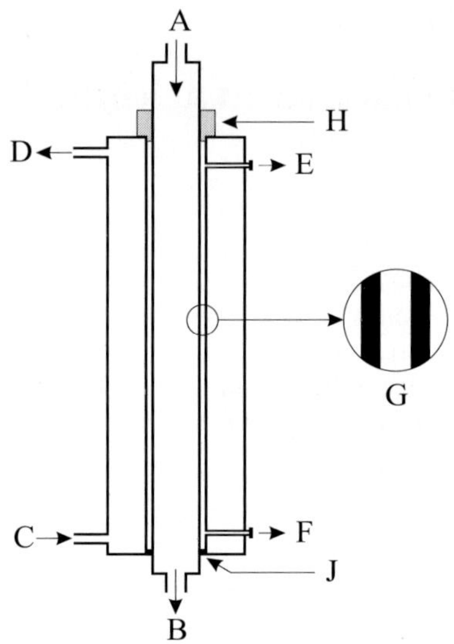


**Fig. 1.** A line drawing of the thermogravitational cell. A, B: the inlet and outlet of hot water; C, D: the connection to the circulating cooling water; E, F: sampling ports; G: the working space; J: centering ring, and H: a teflon stopper.

steady separation measurements, both in gaseous [4–8] and liquid mixtures [9–18]. We shall refer in this work to the case of binary liquid mixtures in which the separation does not depend on pressure as occurs in gases.

The reported values of the Soret coefficient in liquids generally reveal large deviations which are partly due to constructional faults [13] or to presence of reservoirs at the column ends[12]. These difficulties have been circumvented in the closed columns recently constructed by us. Using these columns, we have obtained reliable values of the Soret coefficient of different organic liquid systems even for the case of mixtures with negative values of this coefficient [15–18].

Another difficulty of the thermogravitational method concerns the validity of the FJO theory due to the simplifying assumptions made in its development. For instance, the dependence of density on composition in the buoyancy force is ignored. The influence of this dependence on the separation is known after DeGroot [19] as the *forgotten effect*. Several attempts have been made to include this effect in the theory concluding that it can be neglected at steady state in most liquid mixtures [11,20,21].

However, other of the FJO assumptions, such as the independence of the vertical composition gradient on the horizontal coordinate, have never been validated. To do so, we have solved numerically the full hydrodynamic equations comparing the results obtained with the ones given by the FJO theory. We have used the Boussinesq approximation as the temperature and concentration differences in a column to be used for determine the Soret coefficient are always small. Moreover, we shall consider a plane column instead a cylindrical one since cylindricity effects does not affect the vertical separation in columns utilized in liquids.

Finally, using the numerical results, we attempted to interpret the experimental results obtained for mixtures with negative Soret coefficient, in which an unstable adverse density gradient is established along the column.

## 2 The governing equations

The equation of motion and the balance equations of energy and mass that govern the separation process in a thermogravitational cell can be written in the Boussinesq approximation as:

$$\frac{\partial \mathbf{v}'}{\partial t'} + (\mathbf{v}' \cdot \nabla)\mathbf{v}' = -\nabla p' - Gr(T' - \psi\, c')\mathbf{k} + \nabla^2 \mathbf{v}' \tag{1}$$

$$\frac{\partial T'}{\partial t'} + (\mathbf{v}' \cdot \nabla)T' = \frac{1}{Pr}\nabla^2 T' \tag{2}$$

$$\frac{\partial c'}{\partial t'} + (\mathbf{v}' \cdot \nabla)c' = \frac{1}{Sc}(\nabla^2 c' + \nabla^2 T') \tag{3}$$

$$\nabla \cdot \mathbf{v}' = 0 \tag{4}$$

where the following dimensionless variables are used, respectively, for coordinates, $\mathbf{r}$, velocity, $\mathbf{v}$, time, $t$, and pressure, $p$:

$$\mathbf{r}' = \frac{\mathbf{r}}{d} \qquad \mathbf{v}' = \frac{\mathbf{v}d}{\nu} \qquad t' = \frac{t\nu}{d^2} \qquad p' = \frac{p}{\rho\nu^2}d^2$$

$d$ being the gap width, $\nu$ the kinematic viscosity and $\rho$ the density. The dimensionless temperature $T'$ is:

$$T' = \frac{T - T_0}{\Delta T}$$

with $T_0 = (T_1 + T_2)/2$ and $\Delta T = T_1 - T_2$, where $T_1$ and $T_2$ are, respectively, the hot and cold wall temperatures. The dimensionless concentration is defined by:

$$c' = \frac{c - c_0}{\overline{\Delta c}}$$

where $c_0$ is the initial concentration and $\overline{\Delta c} = S_{\mathrm{T}} c(1 - c)\Delta T$, with $S_{\mathrm{T}}$ the *Soret coefficient.* The dimesionless groups which appear in the above equations are:

- The *Grashof number*:

$$Gr = \frac{\alpha g \Delta T d^3}{\nu^2}$$

  where $\alpha$ is the coefficient of thermal expansivity and $g$ is the acceleration of gravity.
- The *Prandtl number*:

$$Pr = \frac{\nu}{D_{\mathrm{th}}}$$

  $D_{\mathrm{th}}$ being the thermal diffusivity.

- The *Schmidt number*:
$$Sc = \frac{\nu}{D}$$
with $D$ the ordinary diffusion coefficient
- The *buoyancy ratio*:
$$\psi = \frac{S_{\mathrm{T}}\beta c(1-c)}{\alpha}$$
where $\beta$ is the coefficient of variation of density with concentration.

As we are interested only in the case of small separations, we have considered the product $c(1-c)$ to be constant along the cell. For the concentration, we refer to the mass fraction of the denser component, and the Soret coefficient is taken to be positive when this component migrates to the colder regions.

In solving equations (1)–(4), we have considered the case of bidimensional flow in a plane cell (Fig. 2). The corresponding boundary conditions are written as:

$$\mathbf{v}' = \mathbf{0} \qquad \text{at } x' = 0;\ x' = 1;\ z' = 0 \text{ and } z' = A$$

$$T' = \begin{cases} +1/2 & \text{at } x' = 0 \\ -1/2 & \text{at } x' = 1 \end{cases}$$

$$\frac{\partial c'}{\partial x'} = -\frac{\partial T'}{\partial x'} \qquad \text{at } x' = 0 \text{ and } x' = 1$$

$$\frac{\partial c'}{\partial z'} = \frac{\partial T'}{\partial z'} = 0 \qquad \text{at } z' = 0 \text{ and } z' = A$$

where $A$ is the aspect-ratio, $A = L/d$, $L$ being the length of the cell. The last two equations correspond to the condition of zero mass flux at the vertical walls and adiabatic and impermeable horizontal walls at the cell ends.

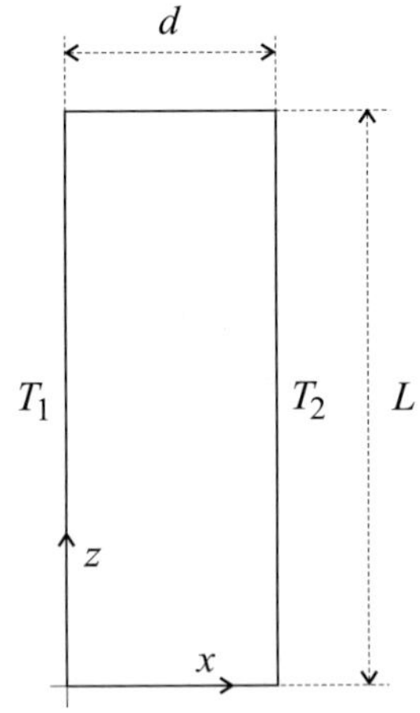


**Fig. 2.** Geometry and coordinates.

The initial conditions correspond to the fluid being at rest at uniform temperature, $T_0$, and at uniform concentration, $c_0$, respectively.

To solve the above equations, we have used the vorticity-stream function formulation. In terms of vorticity, $\zeta = \frac{\partial u'}{\partial z'} - \frac{\partial w'}{\partial x'}$, where $u$ and $w$ are the $x$ and $z$ components of the velocity. The equation of motion (1) is written as:

$$\frac{\partial \zeta}{\partial t'} + (\mathbf{v}' \cdot \nabla)\zeta = Gr\left(\frac{\partial T'}{\partial x'} - \psi \frac{\partial c'}{\partial x'}\right) + \nabla^2 \zeta \tag{5}$$

Defining the stream funtion $\varphi$ by:

$$u' = \frac{\partial \varphi}{\partial z'} \qquad w' = -\frac{\partial \varphi}{\partial x'} \tag{6}$$

the vorticity takes the form:

$$\zeta = \frac{\partial^2 \varphi}{\partial x'^2} + \frac{\partial^2 \varphi}{\partial z'^2} \tag{7}$$

The corresponding boundary conditions for $\zeta$ and $\varphi$ are: $\varphi = 0$ and $\zeta = \frac{d^2\varphi}{dn^2}$ at all surfaces, $n$ being the direction normal to the surface.

## 3 The FJO approximation

In the Furry, Jones and Onsager (FJO) approximation, the main quantity to be calculated is the transport through a column cross-section. In this respect, the following assumptions are used: i) The dependence of density on concentration is ignored. This means that the buoyancy ratio $\psi$ in equation (1) is $\psi = 0$; ii) A pure conductive regime in the column is assumed, and hence $T = T(x)$, $u = 0$, $w = w(x)$ and $c = c(x, z)$; iii) The dependence of the vertical concentration gradient with $x$ is neglected and therefore, $\partial^2 c/\partial x \partial z = 0$; iv) The vertical diffusion in calculating the transport is ignored and is taken into account by adding a diffusion term to the transport equation after it has been obtained.

By using the above assumptions, and considering steady state conditions, equations (1)–(4) reduce to:

$$\frac{d^3 w'}{dx'^3} - Gr = 0 \tag{8}$$

$$w' \frac{\partial c'}{\partial z'} = \frac{1}{Sc} \frac{\partial^2 c'}{\partial x'^2} \tag{9}$$

These equations, with the corresponding boundary conditions at the walls, and the condition of zero mass-flux through a column cross section, can be readily solved taking into account that the derivative $\partial c'/\partial z'$ is independent of $x'$ and can, thus, be considered as a parameter to be determined afterwards.

By using the solutions for $w'$ and $c'$, one can determine the convective transport of one component through a column cross section. By adding the contribution of vertical diffusion, one has for the total transport the expression:

$$q = Hc(1-c) - (K_\mathrm{c} + K_\mathrm{d})\frac{\partial c}{\partial z} \tag{10}$$

where $H$, $K_\mathrm{c}$ and $K_\mathrm{d}$ are the so-called column transport coefficients given by:

$$H = \frac{d^3 g\rho\alpha(\Delta T)^2 S_\mathrm{T}}{6!\,\nu} B \qquad K_\mathrm{c} = \frac{d^7 g^2 \rho\alpha^2(\Delta T)^2}{9!\,\nu^2 D} B \qquad K_\mathrm{d} = \rho DBd \tag{11}$$

$B$ being the breadth of the column.

The first term in equation (10) represents the contribution to transport due to the horizontal separation by the Soret effect. The last two terms represent remixing due to convective currents ($K_\mathrm{c}$) and diffusion ($K_\mathrm{d}$) along the tube.

At steady state, the transport $q$ is zero all along the tube. Setting $q = 0$, we have for the vertical concentration gradient:

$$\frac{\partial c}{\partial z} = \frac{Hc(1-c)}{K_\mathrm{c} + K_\mathrm{d}} \tag{12}$$

For purposes of comparison with the numerical results, it is convenient to write equation (12) in terms of the dimensionless variables used in equations (1)–(4). One obtains:

$$m_0 \equiv \frac{\partial c'}{\partial z'} = \frac{504}{Gr\,Sc} k \tag{13}$$

where $k$ is a correction factor due to the diffusion term in the transport equation given by:

$$k = \frac{K_\mathrm{c}}{K_\mathrm{d} + K_\mathrm{c}} = \frac{Gr^2 Sc^2}{9! + Gr^2 Sc^2} \tag{14}$$

Equations (13) and (14) present a maximum value $m_0 = 0.418$ at $Gr\,Sc = 602.4$. On the other hand, equation (14) shows that the correction factor $k$ approaches unity for $Gr^2 Sc^2 \geq 9!\,10^2$; that is, for $Gr\,Sc \geq 6000$. Inside this range, vertical diffusion can be neglected. For the typical values of the fluid properties: $\nu = 10^{-2}\ \mathrm{cm^2/s}$, $D = 10^{-5}\ \mathrm{cm^2/s}$, $\alpha = 10^{-3}\ \mathrm{K^{-1}}$ and a temperature difference of $\Delta T = 5$ K, this occurs for a gap width $d \geq 0.5$ mm. Therefore, for usual liquid columns with $d \gtrsim 1$ mm, the above condition for $Gr\,Sc$ is always fulfilled.

For $Gr\,Sc \geq 6000$, equation (13) gives for the dimensionless separation between the column ends:

$$\Delta c' = \frac{504\,A}{Gr\,Sc} \tag{15}$$

Taking into account that $\Delta c' = \Delta c / S_\mathrm{T} c(1-c)\Delta T$, one has for the steady separation:

$$\frac{\Delta c}{c(1-c)} = \frac{504L}{gd^4} \frac{S_\mathrm{T}\nu D}{\alpha} \tag{16}$$

which is the working equation to determine the Soret coefficient from steady separation measurements. As can be seen, the separation is independent of the temperature difference between the column walls and, therefore, it is not necessary to control the temperature with accuracy.

Finally, the solutions of equations (8) and (9) for $w'$ and $c'$ are, respectively:

$$w' = \frac{Gr}{12}\left(2x'^3 - 3x'^2 + x'\right) \tag{17}$$

$$\sigma = -\frac{7}{20}\left(12x'^5 - 30x'^4 + 20x'^3 - 1\right) + x' - \frac{1}{2} \tag{18}$$

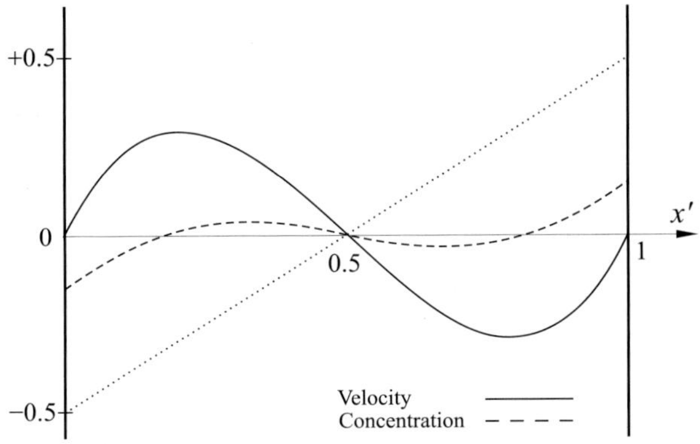


**Fig. 3.** Horizontal profiles of velocity and concentration.

where $\sigma = c'(x') - c'(0.5)$. Fig. 3 represents these equations with $w'$ in arbitrary units. The straight line in this figure corresponds to the balance between ordinary diffusion and thermal diffusion fluxes. This situation occurs at short times when convective remixing is still small. As can be seen, the solution for $\sigma$ indicates that, at steady state, the horizontal concentration differences are small.

The first assumption in the FJO theory can be removed, thus including in the formal treatment the dependence of density on composition in the buoyancy force and, therefore, the so–called *forgotten effect*. Then, the equation of motion writtes:

$$\frac{d^3 w'}{dx'^3} - Gr\left(1 + \psi \frac{\partial c'}{\partial x'}\right) = 0 \tag{19}$$

This equation and equation (9) can be solved as before for $w'$ and $c'$ [20,21]. Then, the condition of zero transport of one component through a column cross section allows one to determine the vertical composition gradient $\partial c'/\partial z'$ which we shall denote by $m$. The ratio between this value and the one given by equation (13) depends only on the buoyancy ratio $\psi$ and is a measure of the magnitude of the forgotten effect.

Fig. 4 represents the ratio $m/m_0$ as a function of $\psi$. As can be seen, the forgotten effect can be neglected at the steady state for $\psi \lesssim 2$ and, hence, for most liquid mixtures. However, for large values of $\psi$, as occurs for mixtures in the vicinity of critical conditions, the forgotten effect strongly amplifies the separation with respect to the values given by the FJO theory. Thermogravitational experiments and results for mixtures near critical conditions can be found in the literature [22].

Although all the above equations were derived for plane columns, they can be applied to cylindrical ones with small ratio of radii. In [23], we showed that the expansion of the separation in powers of $\ln(r_1/r_2)$, being $r_1$ and $r_2$ the outer and inner radii, leads up to quadratic terms in this quantity to:

$$\Delta c = \Delta c_0 \left[1 + 0.039 \ln^2(r_1/r_2)\right] \tag{20}$$

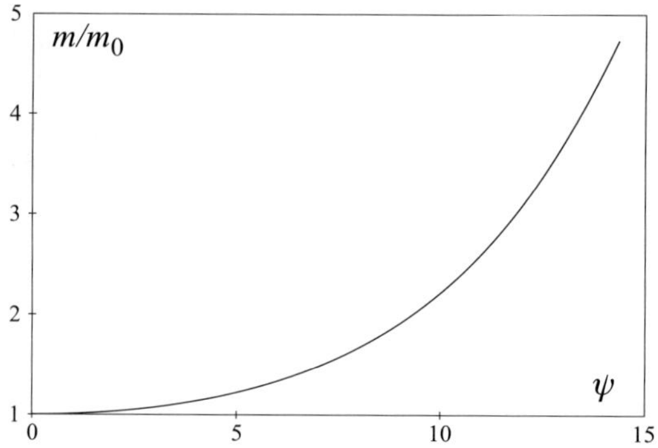


**Fig. 4.** The forgotten correction as a function of the buoyancy ratio.

where $\Delta c_0$ is the separation for the plane case. As can be seen, for $r_1/r_2 \lesssim 1.5$ the curvature effects can be neglected. In a liquid column, the gap width is of the order of 1 mm and hence, the ratio of radii is always small.

The FJO theory also describes the approach to equilibrium by introducing an additional assumption. It is assumed that the transport equation is valid at any time and coordinate $z'$. Therefore, the time evolution of the concentration is given by the continuity equation applied to the transport. Neglecting the $K_\mathrm{d}$ term, one obtains the following diffusion-type equation:

$$\frac{\partial c}{\partial t} = \frac{K_\mathrm{c}}{\mu}\frac{\partial^2 c}{\partial z^2} \tag{21}$$

where $\mu$ is the fluid mass per column unit length. The boundary conditions correspond to the annulation of the transport at the two column ends, and the initial condition is a uniform concentration along the cell.

The solution of equation (21) yields for the separation, $\Delta c$, between two points in the column:

$$\Delta c = \Delta c_\mathrm{e}\left(1 - a e^{-t/t_\mathrm{r}}\right) \tag{22}$$

where $\Delta c_\mathrm{e}$ is the separation at steady state and $a$ and $t_\mathrm{r}$ are given by:

$$a = \frac{4}{\pi^2}\frac{L}{\Delta z}\left(\cos\frac{\pi z_1}{L} - \cos\frac{\pi z_2}{L}\right) \tag{23}$$

$$t_\mathrm{r} = \frac{L^2\mu}{\pi^2 K_\mathrm{c}} = \frac{9\,!}{\pi^2}\frac{L^2\nu^2 D}{d^6 g^2\alpha^2(\Delta T)^2} \tag{24}$$

being $z_1$ and $z_2$ the coordinates of the lower and upper points considered and $\Delta z = z_2 - z_1$.

For $z_1$ and $z_2$ near the column ends, $a$ is of the order of unity. Therefore, $t_\mathrm{r}$ is the relaxation time of the separation process. At this time, about 63% of the equilibrium separation is reached. These equations have been used to determine the ordinary diffusion coefficient from non-steady separation measurements [13,24].

Equations (16) and (24) show that both the separation and the relaxation time increase with length and strongly decrease with gap width. It is therefore not possible to achieve large separations in short times. In the design of a column, it is convenient to choose a length and gap width sufficiently large for the relaxation time to be small and for the separation to be sufficiently large to be measured with precision.

By using equation (16), we have measured positive Soret coefficients of different organic liquid mixtures with a cylindrical column with gap width of 1.53 mm, length of 52 cm and distance between the sampling ports of 42 cm[15,17,18]. The obtained results, in particular for mixtures of toluene–hexane, are in excellent agreement with the ones recently measured by using two different methods [25,26]. As we shall see below, negative Soret coefficients can be also measured using the thermogravitational method.

For mixtures with positive Soret coefficient, the denser component migrates to the colder regions and enriches at the lower part of the column giving rise to a stable density stratification. Conversely, in the case of mixtures with negative Soret coefficient, is this component that becomes enriched in the upper part, giving rise to an adverse density gradient along the column.

For a quiescent isothermal fluid layer with an adverse density gradient, the critical value of the solutal Rayleigh number, defined by:

$$Ra = \frac{\beta g d^4 \Delta c}{\nu D L} \tag{25}$$

is, extrapolating tentatively the results of Ref. [27] for a rectangular cell to our cylindrical column, of $Ra_c \approx 0.12$. On the other hand, the Rayleigh number corresponding to the thermogravitational steady state is obtained by substituting in $Ra$ the value of $\Delta c$ given by equation (16). One obtains:

$$Ra = 504|\psi| \tag{26}$$

which for a value of, say, $\psi = -0.1$, gives $Ra = 50.4$, which is two orders of magnitude higher than the critical value for a quiescent layer. Despite this, in experiments performed with mixtures of water–ethanol [17], which present negative values of the Soret coefficient for concentrations below 30% in alcohol, we have found that for sufficiently large values of $\Delta T$ and, therefore, of the Grashof number, the thermogravitational steady state remains stable. In fact, we found that the critical Grashof number for stability is given by:

$$Gr_c \simeq 965|\psi| \tag{27}$$

For smaller values of $Gr$ than $Gr_c$, the separation increases with time untill a maximum value is reached and thereafter it decreases. On the contrary, for $Gr \gtrsim Gr_c$, the thermogravitational steady state is reached and it remains stable with time. Therefore, it seems that convection favours the stability of the adverse density gradient originated by thermogravitational effect.

These results show that it is possible to measure negative Soret coefficients using the thermogravitational method. In fact, we have measured the Soret coefficient of the binary systems formed by benzene and toluene with methanol and ethanol [18]. In particular, for the system benzene–methanol, many attempts have been made to determine the Soret coefficient as a function of the concentration by using different techniques. Our results are in good agreement with those of Tyrrell et al. [28] carefully measured in a thermodiffusion cell.

## 4 Numerical results

Equations (2), (3), (5) and (7) have been converted into finite difference equations using Taylor expansions for the derivatives. This conversion was performed by the alternating implicit method (ADI) and the resulting equations were solved with the Thomas algorithm [29].The Poisson equation for the stream function was solved in the same way using the false transient method. The numerical calculation scheme proceeds as follows: From the values of the variables $u'$, $\omega'$, $T'$, $c'$, $\zeta$ and $\varphi$ at time $t = 0$, the solutions of the temperature, concentration and vorticity equations give the values of these variables in the advanced time $\Delta t$. Then, the equation for stream function is solved using the fictitious time for iteration. Finally, from the advanced values of this function, the new values of $u'$, $w'$ and boundary vorticity are computed. The calculations proceed in time until steady state is attained.

A uniform mesh in the $x$ direction was chosen with a grid point number of 21. As the relevant variables changes slowly in the vertical direction except near the cell ends, a variable grid was taken in this direction with a grid growth factor such that the last spacing at the column ends would be of the order of $10^{-2}$. The number of grid points in the vertical direction was of 251. More refined grids reproduce the same results to within 1%. The time step used was about $10^{-4}$. For values of $\Delta t$ larger than about the square of the minimum value of the grid spacing, the solution diverges.

In the case $\psi = 0$, the velocity distribution does not depend on the concentration. On the other hand, the relaxation time for concentration is much higher than the corresponding time for temperature and velocity. Under these conditions, we operated by solving all the equations with a small time step ($\Delta t = 10^{-4}$) until steady state was attained for temperature and velocity. Following this, a larger time step was chosen ($\Delta t = 10^{-2}$) and only the concentration equation was computed. In this way, the computation time can be considerably reduced.

We checked the validity of equation (13) for the steady state vertical concentration gradient by using the numerical solutions corresponding to different values of the relevant parameters ($Gr$, $Sc$, $Pr$, $\psi$ and $A$). As the relaxation time for temperature is much smaller than the corresponding time for concentration, the influence of the Prandtl number can be always neglected. Thus, we performed the numerical calculations for a typical value of the Prandtl number: $Pr = 5$. On the other hand, as it has been shown [20,21], the buoyancy ratio $\psi$ can be neglected in describing the steady state. Therefore, here we took $\psi = 0$. For the aspect-ratio $A$, the values of 10, 25, 50, 100 and 275 were considered. Finally, the Grashof and Schmidt numbers were taken in the ranges of 10–2000 and 20–2500, respectively.

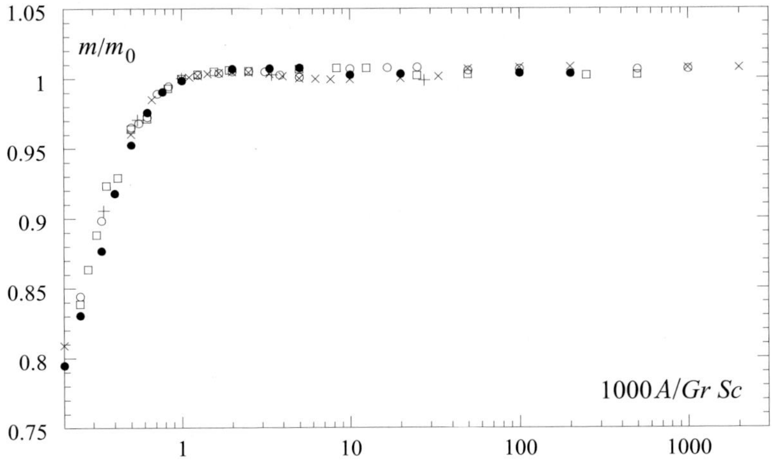


**Fig. 5.** Ratio between the numerical values of the vertical concentration gradient, $m$, and the corresponding FJO values, $m_0$, as a function of $A/Gr\,Sc$: • $A$=10, □ $A$=25, ◇ $A$=50, × $A$=100, + $A$=275

The numerical results show that the relevant parameter which determines the validity of the FJO theory is the group $A/Gr\,Sc$. Fig. 5 plots the ratio between the numerical values of the non-dimensional vertical concentration gradient $m$ and the corresponding values $m_0$ given by equation (13), as a function of $A/Gr\,Sc$. As can be seen, a universal correlation is obtained which shows that for values of this group in the range $1000A/Gr\,Sc \gtrsim 1$ the deviations of the FJO approximation are negligible. Outside this range, these deviations increase rapidly the FJO values being higher than the numerical ones. Therefore, the validity range of the FJO approximation is:

$$Gr\,Sc \lesssim 1000A \tag{28}$$

This condition can be written in a more practical form taking into account that for liquids is $\alpha \approx 10^{-3}$ K$^{-1}$ and, thus, $\alpha g \approx 1$ cm s$^{-2}$ K$^{-1}$. Then, one has:

$$\Delta T \frac{d^4}{L} \lesssim 10^3 \nu D \tag{29}$$

which is always fulfilled for sufficiently large columns. That means that the departures of the FJO theory are due to the influence of the column ends. For the case of ordinary liquid mixtures, is $\nu \approx 10^{-2}$ cm$^2$ s$^{-1}$, $D \approx 10^{-5}$ s$^{-1}$, and this condition writes:

$$\Delta T \frac{d^4}{L} \lesssim 10^{-4} \tag{30}$$

For the usual values of $\Delta T = 5$ K and for $d = 1.5 \times 10^{-1}$ cm, this expression requires the length of a column to be larger than 25 cm. We have used columns with gap width of 1.5 mm and, for security, with a length of 50 cm. Shorter columns can be used for a gap width of 1 mm, but the relaxation time, which is proportional to $d^{-6}$, increases considerably.

For mixtures with small values of $D$, as occurs in polymeric mixtures, in which $D \approx 10^{-7}$ cm$^2$s$^{-1}$, according to equation (29), larger and thinner columns with smaller temperature differences are to be used [14].

It can be readily shown that condition (28) can be interpreted in terms of the ratio between two characteristics times: the time one particle of fluid spends in going from one end to the other of the column and the relaxation time for horizontal diffusion. Indeed, according to equation (17), the mean velocity of the fluid is $\bar{w} = Gr\,\nu/192\,d$ and, thus, the transfer time is $t_{\mathrm{w}} = 192\,Ld/Gr\,\nu$. On the other hand, the relaxation time for diffusion is $t_{\mathrm{D}} = d^2/\pi^2 D$. The ratio between these two characteristic times gives $2000A/Gr\,Sc$. When this value is higher than 2, condition (28) is fulfilled. This time ratio was introduced by Chavepeyer et al. [30].

Fig. 6 represents the numerical values of the horizontal mean concentration as a function of the vertical coordinate, $z'$, for the case $A = 275$, $Gr = 32$ and $Sc = 1000$, which is inside the above range for $A/Gr\,Sc$. As can be seen, the concentration profile is linear throughout the $z'$ range, with the exception of

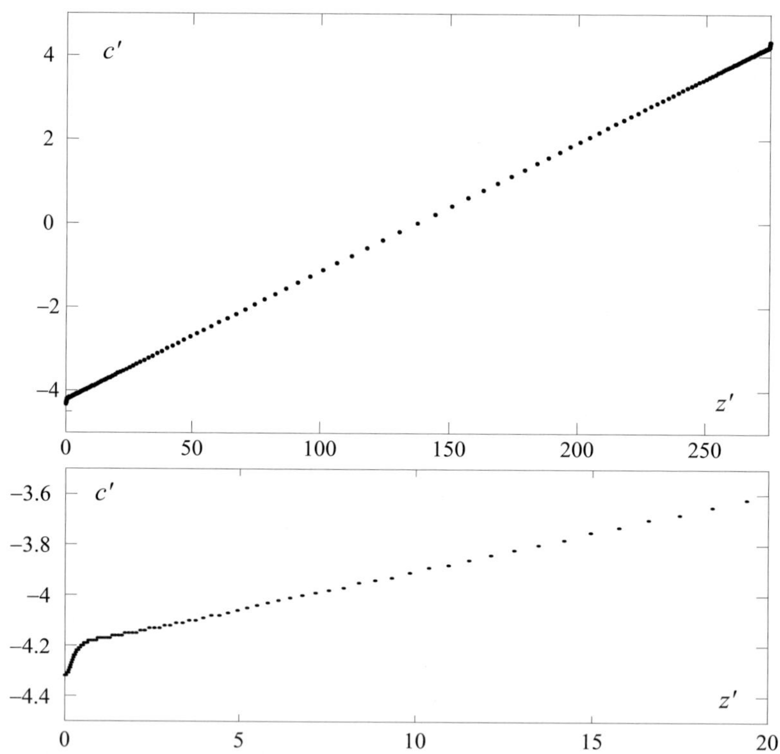


**Fig. 6.** Up: Numerical values of the horizontal mean concentration $c'$ as a function of the vertical coordinate $z'$; Down: Zoom on the range $0 \leq z' \leq 20$.

small regions near the column ends where the concentration gradient increases considerably due to the return currents in them. This effect must be taken into account when positioning the sampling ports in a column. The deviations from the FJO approximation found in a previous numerical work [31] were probably due to the same effect.

Fig. 7 shows the numerical values of $m$ as a function of the product $Gr\,Sc$ for different values of the aspect-ratio. The continuous line corresponds to equation (13). As can be seen, the FJO theory is accurate within 1% even at $Gr\,Sc < 6000$, where, as we have shown above, the diffusion term in the transport equation is relevant. The dotted line in this figure corresponds to the concentration gradient when this term is ignored.

Fig. 8 shows the numerical values obtained for the approach to equilibrium in terms of the relative separation $\Delta c/\Delta c_{\mathrm{e}}$ as a function of $t/t_{\mathrm{r}}$, where $\Delta c_{\mathrm{e}}$ is the steady state separation and $t_{\mathrm{r}}$ is the relaxation time as given by equation (24). The values of $Gr = 66$, $Sc = 504$, $A = 275$, and different negative values of the buoyancy ratio have been considered. The FJO approximation for the non-steady state given by equation (22) is represented by a solid line.

As can be seen, for $\psi = -0.2$ (Fig. 8-a), the FJO approximation describes fairly well the transient separation. Fig. 8-b corresponds to $\psi = -0.5$ and the deviations of this approximation are important due to the influence of the forgotten effect which, contrarily to what occurs at the steady state, seriously affects the approach to equilibrium.

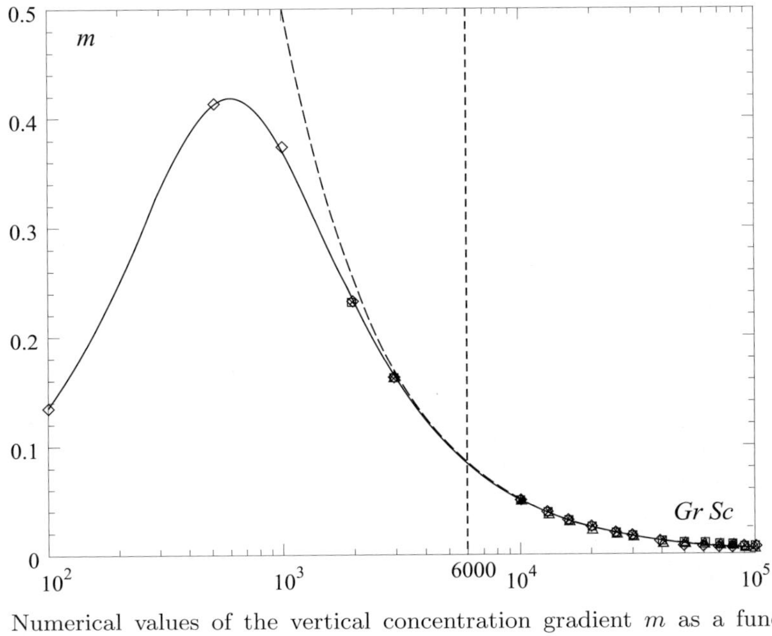


**Fig. 7.** Numerical values of the vertical concentration gradient $m$ as a function of the product $Gr\,Sc$ for different values of the aspect-ratio $A$: $\Diamond$ $A = 10$, $\times$ $A = 25$, $\Box$ $A = 50$, $\triangle$ $A = 100$, $\bullet$ $A = 275$

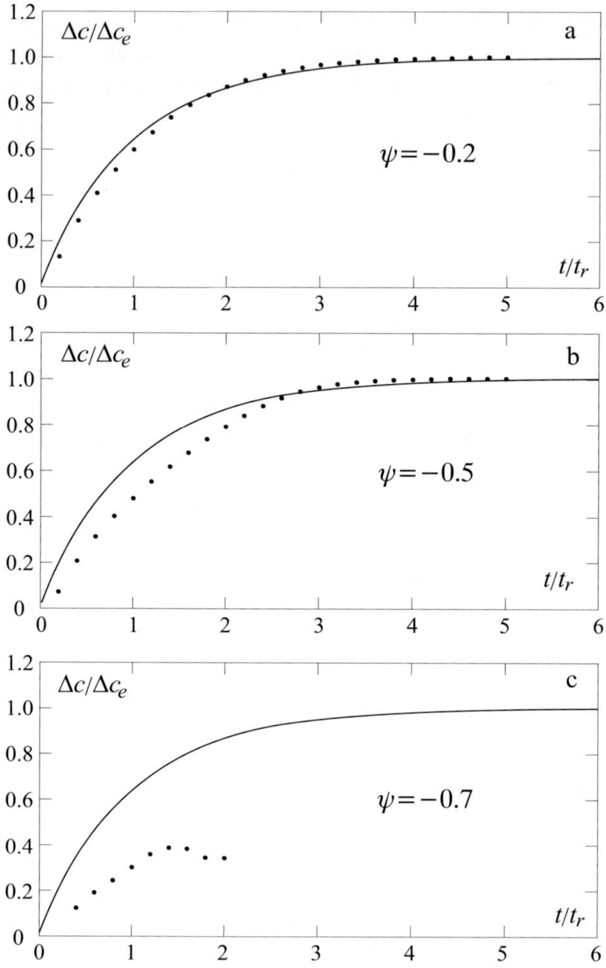


**Fig. 8.** The approach to equilibrium for negative values of the Soret coefficient: a) $\psi = -0.2$, b) $\psi = -0.5$, c) $\psi = -0.7$

Finally, Fig. 8-c corresponds to $\psi = -0.7$. As can be seen, the separation increases until a time of the order of $t_r$ and thereafter decreases following a behaviour similar to the experimental results for negative Soret coefficients we have described in Section 3. The stream function profile shows the presence of many rolls along the column which destroy the separation. However, for the $Gr$ number used, according to equation (26), the values of $|\psi|$ for the steady state to be stable are $|\psi| \lesssim 0.07$, and we have found stability for much higher values. Therefore, we can conclude that the 2D numerical solution considered cannot interpret the experimental facts for negative Soret coefficients. The most unstable modes will be perpendicular to the $xz$ plane considered in this work and, therefore, a 3D simulation or, alternatively, a 2D stability analysis for these modes is required.

## 5 Conclusions

We can conclude that the FJO formulation can be used with confidence to determine the Soret coefficient of a liquid mixture from column steady separation measurements, inside the range of temperature differences, gap width and length given by:

$$\Delta T \frac{d^4}{L} \lesssim 10^3 \nu D$$

where centimeter is to be used as unit length.

According to this condition, sufficiently small values of $\Delta T$ and $d$, and large values of $L$ are required. This favours the separation which is proportional to $L/d^4$, but increase the relaxation time proportional to $L^2/d^6(\Delta T)^2$. In the design of a column, a compromise between both effects must be reached.

The precision of the separation measurements is about 3%. In order to minimize errors in the value of the Soret coefficient, it is necessary to check the quality of the column geometry by measuring the separation as a function of the temperature gradient, for a mixture of well-known physical properties. If the separation results independent of this gradient within 3%, the column operates according to the theory. Then, the overall accuracy in the value of the thermal diffusion coefficient $D_{\mathrm{T}} = S_{\mathrm{T}} D$ can be estimated to be about 5%.

The principal advantages of the thermogravitational method are that the apparatus is easy to construct, no precise temperature control is needed and, lastly, the elementary separation by thermal diffusion can be considerably amplified. The disadvantage is that the diffusion coefficient is necessary to determine the Soret coefficient. However, we expect that a numerical and experimental analysis of the non–steady process would permit to determine both coefficients.

In order to analyse the experimental results about the stability of the adverse density gradient which appears in mixtures with a negative Soret coefficient, a three-dimensional numerical simulation or 2D stability analysis of the most unstable modes is required.

## References

1. K. Clusius, G. Dickel: Naturwissenschaften **26**, 456 (1938)
2. W. H. Furry, R. C. Jones, L. Onsager: Phys. Rev. **55**, 1083 (1939)
3. R.C. Jones, W. H. Furry: Rev. Mod. Phys. **18**, 151 (1946)
4. J. M. Savirón, M. A. Hidalgo, J. C. Yarza, D. González: J. Phys. A **4**, 101 (1971)
5. J. M. Savirón, C. M. Santamaría, J. A. Carrión, J. C. Yarza: J. Chem. Phys. **63**, 5318 (1975)
6. C. M. Santamaría, J. M. Savirón, J. C. Yarza: J. Chem. Phys. **3**, 1095 (1976)
7. W. M. Rutherford: J. Chem. Phys. **58**, 1613 (1973)
8. W. M. Rutherford: J. Chem. Phys. **59**, 6061 (1973)

9. H. J. V. Tyrrell: *Diffusion and Heat Flow in Liquids* (Butterworths, London, 1961)
10. F. H. Horne, R. J. Bearman: J. Chem. Phys. **46**, 4128 (1967)
11. F. H. Horne, R. J. Bearman: J. Chem. Phys. **49**, 2457 (1968)
12. D. J. Stanford, A. Beyerlein: J. Chem. Phys. **58**, 4338 (1973)
13. O. Ecenarro, J. A. Madariaga, J. Navarro, C. M. Santamaría, J. A. Carrión, J. M. Savirón: J. Phys., Condens. Matter **2**, 2289 (1990)
14. O. Ecenarro, J. A. Madariaga, J. Navarro, C. M. Santamaría, J. A. Carrión, J. M. Savirón: Macromolecules **27**, 4968 (1994)
15. M. M. Bou-Ali, O. Ecenarro, J. A. Madariaga, C. M. Santamaría, J. J. Valencia: J. Phys., Condens. Matter **10**, 3321 (1998)
16. M. M. Bou-Ali, O. Ecenarro, J. A. Madariaga, C. M. Santamaría, J. J. Valencia: J. Non-Equilib. Thermodyn. **24**, 228 (1999)
17. M. M. Bou-Ali, O. Ecenarro, J. A. Madariaga, C. M. Santamaría, J. J. Valencia: Phys. Rev. E **59**, 1250 (1999)
18. M. M. Bou-Ali, O. Ecenarro, J. A. Madariaga, C. M. Santamaría, J. J. Valencia: Phys. Rev. E **62**, 1420 (2000)
19. S. R. DeGroot, W. Hoogenstraaten, C. J. Gorter: Physica **9**, 923 (1942)
20. J. L. Navarro, J. A. Madariaga, J. M. Savirón: J. Phys. A, Math. Gen. **15**, 1683 (1982)
21. O. Ecenarro, J. A. Madariaga, C. M. Santamaría, M. M. Bou-Ali, J. Valencia: Entropie **198/199**, 71 (1996)
22. O. Ecenarro, J. A. Madariaga, J. L. Navarro, C. M. Santamaría, J. A. Carrión, J. M. Savirón: J. Phys., Condens. Matter **5**, 2289 (1993)
23. J. L. Navarro, J. A. Madariaga, J. M. Savirón: J. Phys. Soc. Japan **52**, 478 (1983)
24. O. Ecenarro, J. A. Madariaga, J. L. Navarro, C. M. Santamaría, J. A. Carrión, J. M. Savirón: J. Phys., Condens. Matter **2**, 2289 (1990)
25. K. J. Zhang, M. E. Briggs, R. W. Gammon, J. V. Sengers: J. Chem. Phys. **104**, 6881 (1996)
26. W. Köhler, B. Müller: J. Chem. Phys. **103**, 4367 (1995)
27. R. A. Wooding: J. Fluid Mech. **7**, 501 (1960)
28. G. Farsang, H. J. V. Tyrrell: J. Chem. Soc. A, 1839 (1969)
29. P. J. Roache: *Fundamentals of Computational Fluid Dynamics*, (Hermosa Publishers, Albuquerque (N.M., USA 1998))
30. G. Chavepeyer, J. K. Platten: Entropie **198/199**, 10 (1996)
31. G. Chavepeyer, J. K. Platten: Entropie **198/199**, 25 (1996)

## Supplementary glossary

| | |
|---|---|
| $A$ | aspect ratio |
| $B$ | breadth of the column |
| $d$ | gap width |
| $g$ | acceleration of gravity |
| $Gr$ | Grashof number |
| $H$, $K_c$, $K_d$ | transport column coefficients |
| $L$ | length of the cell |
| $m$ | vertical concentration gradient |
| $p$ | presure |
| $r_1$, $r_2$ | outer and inner radii |

| | |
|---|---|
| $t_r$, $t_D$ | relaxation times for separation and diffusion |
| $u$, $w$ | horizontal and vertical velocity components |
| $x$, $z$ | horizontal and vertical coordinates |
| $\mathbf{r}$ | position vector |
| $\mathbf{v}$ | velocity |
| $\mu$ | fluid mass per unit column length |
| $\nu$ | kinematic viscosity |
| $\varphi$ | stream function |
| $\zeta$ | vorticity |

# Determination of Thermodiffusion Parameters from Thermal Field-Flow Fractionation Retention Data

Michel Martin, Charles Van Batten, and Mauricio Hoyos

Ecole Supérieure de Physique et de Chimie Industrielles,
Laboratoire de Physique et Mécanique des Milieux Hétérogènes (UMR CNRS 7636),
10, rue Vauquelin, 75231 Paris Cedex 05, France

**Abstract.** Field-flow fractionation (FFF) encompasses a broad family of methods of separation and characterization of supramolecular species. They are all based on the application of a field force perpendicularly to the direction of the flow of a carrier liquid in a thin parallel plate channel. The principle of the method is described and the general relationships between retention time and analyte properties are presented. Then, the specific characteristics (such as range of applications, relaxation times, flow stability) of thermal field-flow fractionation (thermal FFF), which is the FFF method that is based on the application of a temperature gradient, are discussed. Thermal FFF basically relies on the Soret effect and the thermal FFF retention time is related to the Soret coefficient, $S_T$, of the polymer sample in the carrier liquid. However, the extraction of $S_T$ from retention time is rendered complicated by the temperature dependence of all relevant physico-chemical parameters (thermal conductivity, viscosity, density, and Soret coefficient itself). The various methods which have been developed to take into account some or all of these dependences are described. One of the most complete method, the so-called linear $\lambda$ method, is applied to the determination of the Soret coefficients of a series of polystyrene standards in ethylbenzene, and to their temperature dependence. The resulting data are further analyzed to obtain the molar mass dependence of $S_T$. This dependence can be expressed as a power law. It is found that the prefactor and the exponent of this power law both depend on temperature and have an optimum in the investigated temperature range. Furthermore, this exponent appears to be somewhat larger than the exponent of the power law relating the ordinary mass diffusion coefficient to the molar mass. Hence, depending on temperature, the thermodiffusion coefficient of polystyrene in ethylbenzene might slightly depend on molar mass.

## 1 Introduction

The chemical industry relies heavily on the efficiency of separation processes for fractionating mixtures of compounds or purifying a desired chemical species. Separation methods play a role of paramount importance in chemical analysis. Indeed, they allow separating complex sample mixtures into their individual components (or into groups of components of much lesser complexity than the initial sample) before analyzing them by means of another identification and/or quantification method of analysis. Furthermore, the behavior of the various sample components in the separator can provide qualitative information on these components.

W. Köhler and S. Wiegand (Eds.): LNP 584, pp. 250–284, 2002.

Chromatography is certainly the most important and most performant method of separation of chemical mixtures, at least at the analytical level. Chromatography is based on the differential distribution of the sample components between a carrier fluid and a retentive (usually stationary) phase. Some variants of the chromatographic method allow to fully separate several hundreds of components from each other. However, as one goes from relatively small molecular species to larger ones, the performance of chromatography becomes limited for a variety of reasons (excessive, sometimes irreversible, adsorption on the stationary phase; risk of shear degradation of long-chain molecules in the porous structure of a chromatographic column; clogging of the column,...).

It was in order to overcome these limitations that Giddings conceived, one-third of a century ago, a new separation method applicable to large chemical species and based on the coupling of concentration non-uniformities and flow non-uniformities, which he called field-flow fractionation (FFF) [1]. In a sense, chromatography is also based on the coupling between these two kinds of non-uniformities since the difference in concentration of a sample component between the two phases (carrier fluid and retentive phase) is associated with a difference in velocity of these two phases. However, in FFF, these two non-uniformities occur in a single, fluid phase, as a result of the application of a cross-field interacting with the sample components and of the shear flow.

Different kinds of external fields can be used for implementing FFF. They lead to as many different FFF methods, such as electrical FFF, magnetic FFF, etc. For a review of the principles and applications of these methods, see [2–5]. One of these methods uses a temperature gradient to induce a concentration non-uniformity. It is called thermal FFF and relies on the Soret effect. Using an appropriate model, it is possible from the experimental measurement of the residence time of a species in the thermal FFF separator to determine its Soret coefficient.

In the following, the general principle of FFF and the particular characteristics of thermal FFF are described. Then, the various methods for extracting thermodiffusion parameters are discussed according to their degree of sophistication. Finally, these methods are applied to synthetic polymers.

## 2 Principle of FFF

FFF can be implemented in a large variety of experimental configurations. However, the most common one is the thin ribbon-like parallel plate configuration depicted in Fig. 1. The channel has typical dimensions of 20-80 cm for the length, 1-2 cm for the breadth and 0.1-0.5 mm for the thickness. The extremities of the channel are tapered to allow connections to inlet and outlet tubes for carrier and sample introduction and evacuation. The carrier liquid flows at a constant rate in the channel. A small volume of sample to be analyzed is injected during a short time in the incoming carrier flow. The external field is applied, all along the channel and during the whole duration of the sample migration process, perpendicularly to the carrier flow, in the direction of the channel thickness. This

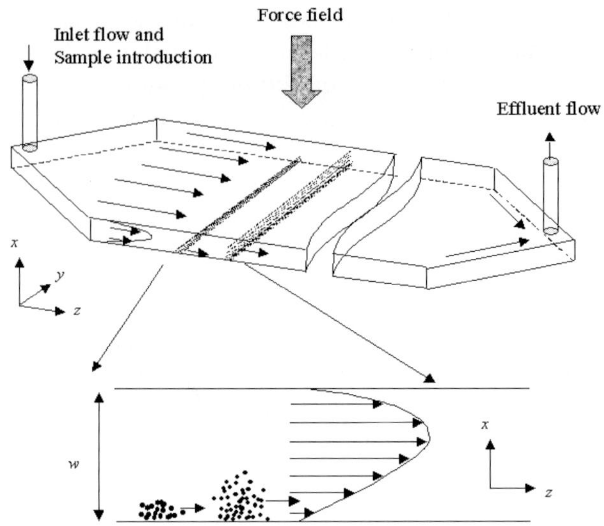


**Fig. 1.** Schematic diagram of the FFF channel and enlarged view of the channel thickness.

forces the sample components with which the field interacts to concentrate in the vicinity of one of the two main channel walls, the accumulation wall. The degree of compression of the cloud of molecules of a sample component near this wall depends on the field intensity and on the susceptibility of that component to the field. In the following, a sample component is called an *analyte*.

The viscous laminar flow of carrier liquid, induced by a pressure drop along the channel, leads to a velocity gradient and a Poiseuille-type profile of the axial velocity in the channel cross-section. This flow displaces the various analytes along the channel. As a result of the velocity profile, analytes which are highly compressed in the sluggish flow streamlines near the accumulation wall are displaced at a lower velocity than components which extend more significantly into the high velocity streamlines in the channel center. The latter components have then a shorter residence time in the channel than the former ones. At the channel outlet, the effluent flows through a low-volume detector providing a signal which is proportional to the concentration of the analytes in the carrier. The record of the detector signal versus time is called a *fractogram*. As the zone occupied by an analyte as its migrates along the channel somewhat spreads longitudinally due to a Taylor-type dispersion process, the signal resulting from the elution of a component is a peak, with a more or less Gaussian shape, whose mean residence time is called the *retention time*. This retention time depends on the analyte-field interaction parameter. Thus, the experimental measurement of the retention time allows to determine, in principle, this parameter, and eventually to characterize the analyte. An example of such a fractogram is given in Fig. 2.

In well optimized systems, the fractogram of a sample is made of a succession of well separated peaks, each peak corresponding to a given analyte. In other

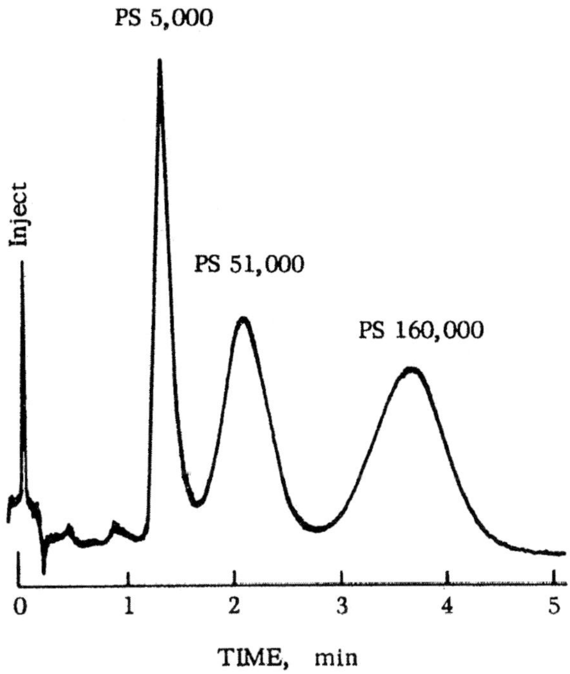


**Fig. 2.** Fractogram of a mixture of three polystyrene standards with a narrow molar mass distribution and weight-average molar masses of 5,000; 51,000; and 160,000 by thermal FFF. Carrier: ethylbenzene. Channel dimensions: 44.6 cm $\times$ 2.0 cm $\times$ 0.051 mm. Flow rate: 0.35 ml.min$^{-1}$. Temperature difference between the two channel walls: $\Delta T = 60$ K. Cold wall temperature: 306 K. Reprinted from Ref. 6, High-speed polymer separations by thermal field-flow fractionation, fig. 6, p. 434, with kind permission from Elsevier Science.

cases, e.g. for a polymer sample with a relatively large molar mass distribution, the fractogram is made of an envelope of the individual peaks of the various molar masses, as shown in Fig. 3. Then, the molar mass distribution can, in principle, be extracted from the fractogram.

Because of the flow, the system cannot be in equilibrium. However, in properly operated systems, an analyte zone migrates in near equilibrium conditions. Then, it can be shown that the retention time, $t_{\mathrm{R}}$, of an analyte which corresponds to the barycentre of the residence time distribution of that analyte, is related to the cross-sectional concentration distribution that the analyte would have at equilibrium if there was no carrier flow [8, 9]. This is an important finding as this allows to relate the retention time to the equilibrium parameters of the analyte and to characterize this analyte. Then, if $u(x, y)$ is the carrier velocity profile, and $c(x, y)$ the analyte equilibrium molar concentration, the average

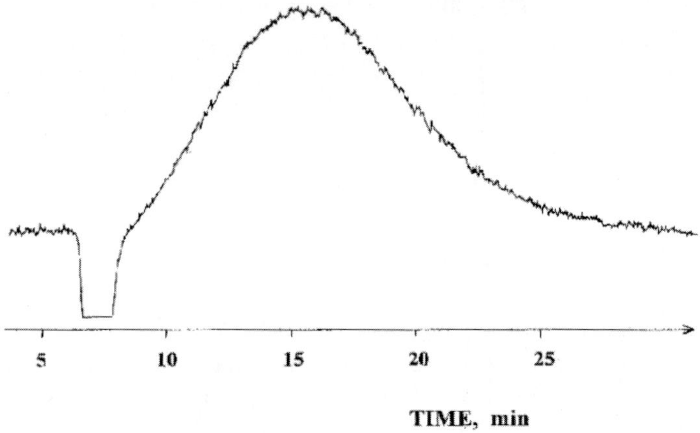


**Fig. 3.** Fractogram of a polymethylmethacrylate sample with a weight-average molar mass of 160,000 and a relatively large molar mass distribution by thermal FFF. Carrier: dimethylformamide. Channel dimensions: 42.6 cm × 2.05 cm × 0.1 mm. Flow rate: 0.2 ml.min$^{-1}$. Temperature difference between the two channel walls: $\Delta T = 34$ K. Cold wall temperature: 298 K. Reprinted from Ref. 7, Polymer analysis by thermal field-flow fractionation, fig. 4, p. 2297, with kind permission from American Chemical Society.

axial velocity of the analyte zone, $u_{\mathrm{zone}}$, is equal to

$$u_{\mathrm{zone}} = \frac{\int_{(x)} \int_{(y)} \frac{c(x,y)}{\langle c \rangle} u(x,y) dx dy}{\int_{(x)} \int_{(y)} dx dy} \tag{1}$$

$x$ and $y$ are the cross-sectional coordinates and $\langle c \rangle$ the cross-sectional average concentration

$$\langle c \rangle = \frac{\int_{(x)} \int_{(y)} c(x,y) dx dy}{\int_{(x)} \int_{(y)} dx dy} \tag{2}$$

The retention time is then equal to

$$t_{\mathrm{R}} = \frac{L}{u_{\mathrm{zone}}} \tag{3}$$

where $L$ is the channel length. Note that $L$ is conveniently defined as $V_{\mathrm{o}}/bw$, where $V_{\mathrm{o}}$, $b$ and $w$ are the channel geometrical volume, breadth and thickness, respectively.

The molar flux, $J_{\mathrm{x}}$, of an analyte in the field direction ($x$) in absence of flow arises from the field-induced convection and diffusion, and is expressed by

$$J_{\mathrm{x}} = U_{\mathrm{x}} c - D \frac{dc}{dx} = -\left| U_{\mathrm{x}} \right| c - D \frac{dc}{dx} \tag{4}$$

where $x$ is the distance from the accumulation wall (in the direction of the thickness), $U_{\mathrm{x}}$ is the velocity induced by the field on the analyte molecules in the $x$-direction, and $D$ the mutual diffusion coefficient of the analyte in the carrier. One notes that $U_{\mathrm{x}}$ is negative since the analyte migration occurs toward the accumulation wall in the negative $x$-direction. At equilibrium, the flux is zero at any distance from the wall and the concentration profile is given by

$$\frac{dc}{c} = -P\acute{e}\; ds = -\frac{ds}{\lambda_{\mathrm{F}}} \tag{5}$$

with

$$s = \frac{x}{w} \tag{6}$$

$s$ is a reduced distance varying between 0 (for the accumulation wall) and 1 (for the depletion wall). $P\acute{e}$ is the transversal Péclet number defined as

$$P\acute{e} = \frac{|U_{\mathrm{x}}|\, w}{D} = \frac{1}{\lambda_{\mathrm{F}}} \tag{7}$$

Its reciprocal, $\lambda_{\mathrm{F}}$, is called the *analyte-field interaction parameter* (it is frequenly used in the FFF literature where it is simply denoted $\lambda$). When $P\acute{e}$ or $\lambda_{\mathrm{F}}$ are constant, i.e. do not depend on the concentration and on the position in the channel, (5) becomes

$$c = c_{\mathrm{o}} \exp\left(-P\acute{e}\; s\right) = c_{\mathrm{o}} \exp\left(-s/\lambda_{\mathrm{F}}\right) \tag{8}$$

where $c_{\mathrm{o}}$ is the equilibrium analyte molar concentration at the accumulation wall. Hence $\lambda_{\mathrm{F}}$ appears as the characteristic dimensionless length of the exponential decrease of the equilibrium concentration profile from the accumulation wall.

Because of the generally large values of the cross-section aspect ratio in FFF instruments, $b/w$, the flow profile, in laminar conditions, can be considered as depending only on $x$. For an isoviscous carrier, it is given by

$$\frac{u}{\langle u \rangle} = 6\left(s - s^2\right) \tag{9}$$

where $\langle u \rangle$ is the average flow velocity, equal to $Q/(bw)$, where $Q$ is the carrier flow rate.

Combining (1)-(3), (5), (6), (8) and (9), one gets the expression of the retention time for an isoviscous flow and constant $P\acute{e}$

$$t_R = \frac{t_{\mathrm{o}}}{6} \frac{P\acute{e}}{\left[\coth\left(\frac{P\acute{e}}{2}\right) - \frac{2}{P\acute{e}}\right]} = \frac{t_{\mathrm{o}}}{6\lambda_F \left[\coth\left(\frac{1}{2\lambda_{\mathrm{F}}}\right) - 2\lambda_{\mathrm{F}}\right]} \tag{10}$$

where $t_{\mathrm{o}}$ is the mean residence time of the carrier in the channel, also called void time. Fig. 4 shows that the relative retention time, $t_{\mathrm{R}}/t_{\mathrm{o}}$, increases almost linearly with $P\acute{e}$ when $P\acute{e}$ is larger than about 5 (i.e. when $t_{\mathrm{R}}/t_{\mathrm{o}}$ larger than 1.36). Noting that $P\acute{e}$, defined by (7), is also equal to

$$P\acute{e} = \frac{|F_{\mathrm{x}}|\, w}{k_{\mathrm{B}} T} \tag{11}$$

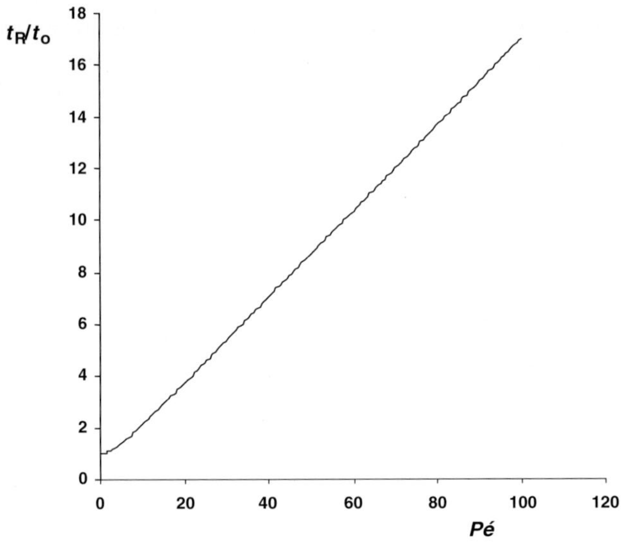


**Fig. 4.** Variation of the relative retention time, $t_R/t_o$, vs. *Pé* (from equation10).

where $F_x$ is the force induced by the field on an analyte molecule, $k_B$, the Boltzmann constant and $T$ the absolute temperature, $t_R$ appears linearly proportional to $F_x$ when *Pé* is large. Fig. 4 illustrates that $F_x$ can be easily determined, with a good accuracy, from the experimental measurement of the retention time, allowing to characterize the analyte.

## 3 Thermal FFF and the Soret Effect

In thermal FFF, the external field applied is a temperature gradient. The channel is similar to that shown on Fig. 1, the top plate is heated by electrical resistances and the bottom plate is cooled by a flow of tap water. This arrangement allows to control the steady temperature difference, $\Delta T$, between the plates from a few degrees to several tens of degrees, depending on the requirements for separation and on the boiling point of the carrier liquid (if needed, the boiling point can be increased by pressurizing the channel [10]). Because of the small thickness of the channel, high temperature gradients, typically in the range 500 - 10000 K/cm, are applied. The binary mixture made of the carrier liquid and a sample analyte can in these conditions experience a strong Soret effect and the analyte can become strongly compressed near the accumulation wall.

### 3.1 Mass flux equation in thermal FFF

In thermal FFF, (4) is generally expressed as [11-14]

$$J_{\mathrm{x}} = -Dc_{\mathrm{tot}}\left[\frac{dX}{dx} + S_{\mathrm{T}}X(1-X)\frac{dT}{dx}\right] \tag{12}$$

or

$$J_{\mathrm{x}} = -c_{\mathrm{tot}}\left[D\frac{dX}{dx} + D_{\mathrm{T}}X(1-X)\frac{dT}{dx}\right]$$

with

$$S_{\mathrm{T}} = \frac{D_{\mathrm{T}}}{D} \tag{13}$$

where $c_{\mathrm{tot}}$ is the total molar concentration of the solution, $X$ the analyte molar fraction, $S_{\mathrm{T}}$ the analyte Soret coefficient, and $D_{\mathrm{T}}$ the analyte thermodiffusion coefficient. Note that, here, $S_{\mathrm{T}}$ is positive if the wall at which the analyte accumulates is the cold one, negative in the opposite case. In steady state conditions ($J_{\mathrm{x}} = 0$), assuming in a first approximation that $S_{\mathrm{T}}$ does not significantly depend on $X$ and $T$, one gets the analyte molar fraction profile versus $T$ as

$$\frac{X}{X_{\mathrm{o}}} = \frac{\exp\left[-S_{\mathrm{T}}(T-T_{\mathrm{o}})\right]}{1 - X_{\mathrm{o}}\left\{1 - \exp\left[-S_{\mathrm{T}}(T-T_{\mathrm{o}})\right]\right\}} \tag{14}$$

where $X_{\mathrm{o}}$ is the molar fraction at the accumulation wall and $T_{\mathrm{o}}$ the temperature of this wall. Fig. 5 shows the variation of $X/X_{\mathrm{o}}$, versus $S_{\mathrm{T}}(T-T_{\mathrm{o}})$ for various values of $X_{\mathrm{o}}$. In the limit of dilute analyte solutions, $(1-X)$ in (12) becomes nearly equal to 1, and (14) becomes

$$\frac{X}{X_{\mathrm{o}}} = \exp\left[-S_{\mathrm{T}}(T-T_{\mathrm{o}})\right] \tag{15}$$

showing that, then, the analyte molar fraction decreases exponentially with temperature.

### 3.2 Validity of the diluted solution approximation in thermal FFF

Thermal FFF is an analytical separation method. Therefore, the analyte concentration in the sample is generally low. Still, because the analyte accumulates near one FFF channel wall, its concentration at this wall is larger than its average cross-sectional concentration. Since this latter decreases as the analyte zones progresses along the channel as a result of various dispersion processes, the highest analyte concentration is encountered at the accumulation wall at or near the channel inlet. However, due to the high molar mass of the polymeric analytes, the analyte molar fraction at this position is low even if its mass concentration is relatively large. In practice, the $X_{\mathrm{o}}$ value at the most concentrated position in the thermal FFF channel is always smaller than 0.01 and generally much smaller than this value. Fig. 5 shows that the molar fraction profile versus $T$ is then very

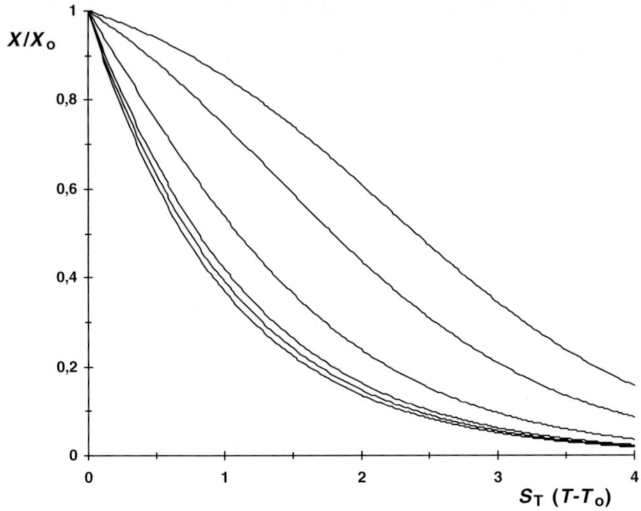


**Fig. 5.** Variation of the relative analyte molar fraction at equilibrium, $X/X_\mathrm{o}$, vs. $S_\mathrm{T}(T - T_\mathrm{o})$ for various values of $X_\mathrm{o}$, assuming $S_\mathrm{T}$ independent of $T$ and $X$. From lower to upper curve: $X_\mathrm{o} = 0.01, 0.1, 0.2, 0.5, 0.8, 0.9$.

similar to the exponential one. Consequently, it is justified to neglect the $(1-X)$ term in the flux equation 12.

Then, noting that the thermal expansion coefficient $\alpha$, defined as $-\frac{1}{\rho}d\rho/dT$, where $\rho$ is the solution mass density, is also equal to $-(1/c_\mathrm{tot})dc_\mathrm{tot}/dT$, and that $c = Xc_\mathrm{tot}$, the flux equation can be expressed as

$$J_\mathrm{x} = -D\left[\frac{dc}{dx} + (S_\mathrm{T} + \alpha)\, c\frac{dT}{dx}\right] \tag{16}$$

As the low $X$ approximation is valid, the mass density and the thermal expansion of the solution are, in practice, equal to those of the carrier liquid. Then, comparing (16) with (4) and combining with (5)-(7), one gets the expression of the Péclet number in thermal FFF

$$P\acute{e} = \frac{1}{\lambda_\mathrm{F}} = (S_\mathrm{T} + \alpha)\,\frac{dT}{ds} \tag{17}$$

As retention in thermal FFF depends on the Péclet number, (17) is the basic equation, which allows relating the Soret coefficient to the Péclet number.

### 3.3 Range of applications of thermal FFF

It has been seen above that, in general, FFF is applicable to species with Péclet numbers larger than about 5. This can be expressed as a limiting condition for $S_\mathrm{T}$ by noting, in first approximations, that the thermal expansion coefficient, for most carriers, is usually of the order of $10^{-3}$ K$^{-1}$ and can be neglected, and that $dT/ds$ is nearly equal to $\Delta T$. $Pé$ in (17) is then approximated by

$$Pé \approx S_{\mathrm{T}} \Delta T \tag{18}$$

This implies that thermal FFF is applicable to species for which the Soret coefficient exceeds

$$S_{\mathrm{T}} \geq \frac{5}{\Delta T} \tag{19}$$

$\Delta T$ does generally not exceed 50-100 K, unless the thermal FFF channel is pressurized in order to increase the boiling point of the carrier liquid [10]. Therefore, thermal FFF is generally applicable to species with $S_{\mathrm{T}}$ larger than about 0.05-0.1 K$^{-1}$. Because this value is larger than that usually encountered in mixtures of ordinary molecular liquids, thermal FFF is essentially applicable to species with a relatively large molar mass, such as macromolecules or polymers.

### 3.4 Relaxation toward steady state in thermal FFF

There are several characteristic times to be considered before the FFF process reaches its quasi-steady state. These three times, $t_{\mathrm{car},1}$, $t_{\mathrm{car},2}$, $t_{\mathrm{car},3}$, correspond to the establishments of, respectively, the fully developed flow velocity profile, the temperature profile, and the analyte concentration profile.

The first characteristic time corresponds to the development of the viscous boundary layer from the main channel walls and is equal to the time of diffusion of the vorticity across half of the channel thickness

$$t_{car,1} = \frac{w^2}{8\upsilon} = \frac{\rho w^2}{8\eta} \tag{20}$$

where $\upsilon$ and $\eta$ are, respectively, the carrier kinematic and dynamic viscosities.

The second characteristic time is that necessary for the steady-state temperature profile to be established in the carrier when the temperature of one surface is suddenly increased by $\Delta T$. In the FFF channel, this time is that of diffusion of heat across distance $w$

$$t_{car,2} = \frac{w^2}{2\beta_{\mathrm{T}}} = \frac{C_{\mathrm{p}} \rho w^2}{2\lambda_{\mathrm{T}}} \tag{21}$$

where $\beta_{\mathrm{T}}$ is the carrier thermal diffusivity, $C_{\mathrm{p}}$ its heat capacity at constant pressure, and $\lambda_{\mathrm{T}}$ its thermal conductivity.

When the sample enters the FFF channel, it is uniformly distributed across the FFF channel. It takes some time before it becomes exponentially distributed, according to (15), under the combined effects of diffusion and thermodiffusion. The evolution of the concentration of an analyte as a function of time and distance from the accumulation wall can be obtained by solving the convection-diffusion flux equation [3]. The result is then expressed as an infinite sum of terms exponentially decreasing with time. The characteristic time for the establishment of the steady concentration profile can be taken as the largest time constant, i.e.

$$t_{car,3} = \frac{w^2}{D\left(\pi^2 + \frac{Pé^2}{4}\right)} \tag{22}$$

In earlier thermodiffusion studies performed at low $Pé$ values, $t_{\mathrm{car},3}$ was approximated by $w^2/(\pi^2 D)$ [16]. However, since thermal FFF requires $Pé$ larger than 5, and is usually performed at $Pé$ much larger than $2\pi$, $t_{\mathrm{car},3}$ can be approximated by $4w^2/(Pé^2 D)$.

It is thus seen that these characteristic times are all proportional to the square of the channel thickness. They are thus much smaller in thermal FFF channels with typical thickness of 0.1 mm than in classical thermodiffusion cells with depths of the order of 1 or, even, 10 mm. Their ratios are essentially those of the corresponding diffusivities, momentum diffusivity, heat diffusivity and mass diffusivity, respectively, $\upsilon$, $\beta_{\mathrm{T}}$, and $D$. Since the analyte solution is, in practice, diluted, the two first coefficients are carrier liquid properties. Their ratio, the Prandtl number, $Pr$, is of the order of a few units for ordinary liquids (about 7 for water and toluene, 9 for ethylbenzene), so that the relaxation process for the establishment of the steady flow velocity profile is quite faster than that for the temperature profile, even if they are not completely independent of each other. Nevertheless, in typical thermal FFF conditions, for $w = 0.1$ mm and ethylbenzene as carrier liquid, $t_{\mathrm{car},1}$ and $t_{\mathrm{car},2}$ are respectively equal to 0.0016 s and 0.06 s, which is very short in comparison of the typical retention times (of the order of minutes or tens of minutes). In practice, the flow profile and temperature profile can be considered as instantaneously established when the carrier enters the FFF channel.

The establishment of the analyte concentration profile depends on the analyte diffusion coefficient in the carrier liquid, hence on the molar mass of the macromolecular analyte. Assuming that the thermodiffusion coefficient is independent of the molar mass, it can be shown from (18) and (22) that $t_{\mathrm{car},3}$ reaches a maximum value for $Pé = 2\pi$ and is then equal to

$$t_{\mathrm{car},3,\max} = \frac{w^2}{\pi D_{\mathrm{T}} \Delta T} \tag{23}$$

Then, $t_{\mathrm{R}}/t_{\mathrm{o}}$ is equal to 1.53 according to (10). Assuming that $D_{\mathrm{T}}$ is equal to $10^{-11}$ $\mathrm{m^2(sK)^{-1}}$, $w = 0.1$ mm and $\Delta T = 50$ K, $t_{\mathrm{car},3,\max}$ is equal to 6.4 s (then $D$ is equal to $8.0 \times 10^{-11}$ $\mathrm{m^2s^{-1}}$). If $\Delta T$ is selected such that the analyte of interest has a retention time 3 times larger than the void time $t_{\mathrm{o}}$, $t_{\mathrm{car},3}$ is then equal to 4.4 s. Obviously, the smaller $\Delta T$, the larger are $t_{\mathrm{car},3,\max}$ and $t_{\mathrm{car},3}$ at a given $t_{\mathrm{R}}/t_{\mathrm{o}}$ value. For instance if $\Delta T$ is only 20 K, $t_{\mathrm{car},3,\max}$ and $t_{\mathrm{car},3}$ at $t_{\mathrm{R}}/t_{\mathrm{o}} = 3$ become 16 s and 11 s, respectively. The characteristic time $t_{\mathrm{car},3}$ is seen to be much larger than $t_{\mathrm{car},1}$ or $t_{\mathrm{car},2}$. Its value is also usually quite lower than the analyte retention time. However, in some instances (low $D_{\mathrm{T}}$, high flow rate, relatively large channel thickness), it can reach a significant fraction of the retention time. In those instances, applying the retention model based on steady state concentration profile for determining thermodiffusion parameters from retention time measurements can lead to significant errors. It is then recommended to apply the stop flow procedure, i.e., to stop the carrier flow immediately after injecting the sample in the FFF channel for a time equal to $t_{\mathrm{car},3}$ (or, better, 1.5 $t_{\mathrm{car},3}$) before resuming it.

### 3.5 Flow stability in thermal FFF

The thermal FFF channel is generally oriented horizontally, the bottom plate being cooled and the top one heated, in order to avoid appearances of Rayleigh-Bénard convection instabilities. However, no change in retention time of polymers has been observed when the channel was heated from below. The Rayleigh number, $Ra$, defined as

$$Ra = \frac{\alpha g \varDelta T w^3}{\upsilon \beta_{\mathrm{T}}} = \frac{\alpha \rho^2 C_{\mathrm{p}} g \varDelta T w^3}{\eta \lambda_{\mathrm{T}}} \tag{24}$$

where $g$ is the Earth gravitational acceleration, is, in typical thermal FFF conditions, much lower than the critical value, $Ra_{\mathrm{crit}} = 1708$ [17], corresponding to the onset of flow instability convection rolls. Indeed, for ethylbenzene with $\varDelta T = 50$ K and $w = 0.1$ mm, $Ra$ is equal to 16.5.

## 4 Influence of the temperature dependence of the physico-chemical parameters in thermal FFF

By essence, any thermodiffusion experiment is performed in non-isothermal conditions. As all relevant physico-chemical parameters are temperature dependent, a thermodiffusion experiment is associated, not to single values, but to a range of values of these physico-chemical parameters. This seriously complicates the interpretation of thermodiffusion experimental measurements. Frequently, the temperature dependence of the physico-chemical parameters is neglected either because the temperature range over which the experiment is performed is narrow (this is definitely the case in forced Rayleigh scattering experiments [18]) or simply because of the theoretical intractability of this effect. It is then said that experiments are performed in Boussinesq conditions.

In thermal FFF, the temperature difference between the two plates may not be small. Ignoring the temperature dependence of the relevant physico-chemical parameters may lead to relatively significant errors in the determination of the thermodiffusion parameters or may invalidate conclusions derived from the comparison of the thermodiffusion parameters of various species. In the following, these dependences are discussed and methods used to account for them are presented. In fact, as discussed above, thermal FFF is generally performed in quite dilute solutions of the sample analytes in the carrier liquid. The values of some physico-chemical parameters of the analyte-carrier solution are then nearly equal to those of the pure carrier liquid. Their temperature dependence is thus the temperature dependence of the carrier liquid properties. They can more or less easily be taken into account. Other parameters, especially the Soret coefficient, depend on the analyte-carrier system and their temperature dependence is all the more complex to account for that it is a priori unknown. A model is then required to handle this problem.

### 4.1 Carrier thermal conductivity and temperature profile

The steady state temperature profile is obtained from the Fourier law

$$q = -\lambda_{\mathrm{T}} \frac{dT}{dx} \tag{25}$$

where $q$ is the heat flux, which is generally maintained constant in thermal FFF experiments. If the thermal conductivity, $\lambda_{\mathrm{T}}$, were constant, the temperature gradient would be constant and equal to $\Delta T/w$. In the moderate temperature range of a thermal FFF experiment, $\lambda_{\mathrm{T}}$ can be assumed to linearly vary with $T$

$$\lambda_{\mathrm{T}} = \lambda_{\mathrm{T_c}} \left[1 + \omega_{\mathrm{c}} \left(T - T_{\mathrm{c}}\right)\right] \tag{26}$$

where $T_{\mathrm{c}}$ is the cold wall temperature, $\lambda_{\mathrm{T_c}}$ the thermal conductivity at $T_{\mathrm{c}}$ and $\omega_{\mathrm{c}}$ the relative rate of variation of $\lambda_{\mathrm{T}}$, i.e. $(1/\lambda_{\mathrm{T_c}})d\lambda_{\mathrm{T}}/dT$. Then, from (25) and (26), for a constant heat flux, one gets

$$T = T_{\mathrm{c}} + \frac{1}{\omega_{\mathrm{c}}} \left[\sqrt{1 + 2\omega_{\mathrm{c}} \Delta T \left(1 + \frac{1}{2}\omega_{\mathrm{c}} \Delta T\right) s} - 1\right] \tag{27}$$

and

$$\frac{dT}{ds} = \Delta T \frac{1 + \frac{1}{2}\omega_{\mathrm{c}} \Delta T}{1 + \omega_{\mathrm{c}} \left(T - T_{\mathrm{c}}\right)} = \Delta T \frac{1 + \frac{1}{2}\omega_{\mathrm{c}} \Delta T}{\sqrt{1 + 2\omega_{\mathrm{c}} \Delta T \left(1 + \frac{1}{2}\omega_{\mathrm{c}} \Delta T\right) s}} \tag{28}$$

As $\omega_{\mathrm{c}}$ is generally negative, the temperature gradient at the cold wall is smaller than $\Delta T/w$, while the opposite is true at the hot wall. In Fig. 6, the temperature gradient profile is shown for typical thermal FFF conditions (ethylbenzene, $\Delta T = 40$ K). It is seen that the error made on the determination of $S_{\mathrm{T}}$ from $Pé$ using (17) by assuming a constant temperature gradient reaches about 4% near the walls.

### 4.2 Carrier viscosity and flow velocity profile

Equation 10, which relates the retention time to the Péclet number is based on the assumption of an isoviscous, hence isothermal, flow, for which the velocity profile is given by (9). In the case of non-isoviscous flow, the two-dimensional velocity profile is given by [3, 19]

$$\frac{u}{\langle u \rangle} = \frac{\int_0^s \frac{s}{\eta} ds - \frac{\int_0^1 \frac{s}{\eta} ds}{\int_0^1 \frac{ds}{\eta}} \int_0^s \frac{ds}{\eta}}{\int_0^1 \left(\int_0^s \frac{s}{\eta} ds - \frac{\int_0^1 \frac{s}{\eta} ds}{\int_0^1 \frac{ds}{\eta}} \int_0^s \frac{ds}{\eta}\right) ds} \tag{29}$$

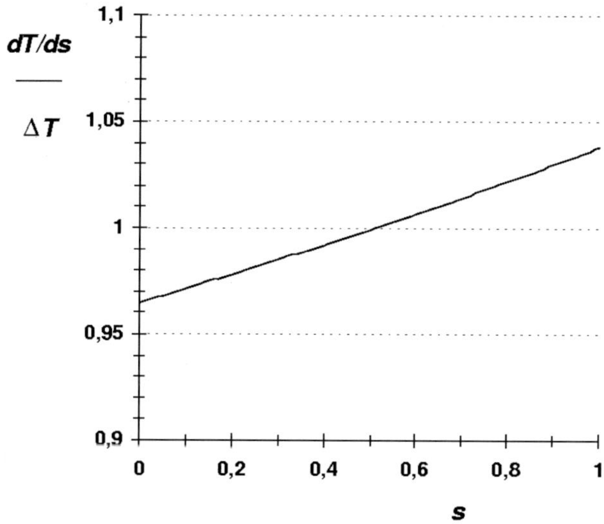


**Fig. 6.** Variation of the temperature gradient across channel thickness. Carrier: ethylbenzene, $\Delta T = 40$ K, $T_c = 293$ K.

The actual velocity profile differs from the parabolic Poiseuille profile. The behavior of the analyte is especially affected by their difference in the vicinity of the accumulation wall. Therefore, the extent of this difference is reflected by the flow distortion parameter, $\nu$, defined as the relative deviation of the velocity gradient from that of the Poiseuille profile at the accumulation wall, i.e.

$$\nu \equiv \frac{1}{6} \left. \frac{d\left(u/\langle u \rangle\right)}{ds} \right|_{s=0} - 1 \tag{30}$$

From (29) and (30), one gets

$$\nu = -1 - \frac{1}{6\eta_o} \frac{\int_0^1 \frac{s}{\eta} ds \Big/ \int_0^1 \frac{ds}{\eta}}{\int_0^1 \left( \int_0^s \frac{s}{\eta} ds - \int_0^1 \frac{s}{\eta} ds \int_0^s \frac{ds}{\eta} \Big/ \int_0^1 \frac{ds}{\eta} \right) ds} \tag{31}$$

where $\eta_o$ is the dynamic viscosity at the accumulation wall. The parameter $\nu$ can be estimated by different methods. For instance, one can use an Arrhenius-type relationship to obtain the temperature dependence of the viscosity

$$\eta = \eta_{\text{ref}} \exp \left[ B \left( \frac{1}{T} - \frac{1}{T_{\text{ref}}} \right) \right] \tag{32}$$

where $\eta_{\text{ref}}$ is the carrier viscosity at the reference temperature $T_{\text{ref}}$, and $B$ a constant characteristic of the carrier liquid, which is the activation energy for viscosity divided by $k_B$. Using (32) and the temperature profile (27) allows to obtain the viscosity profile, $\eta(s)$, then to perform the numerical computation of

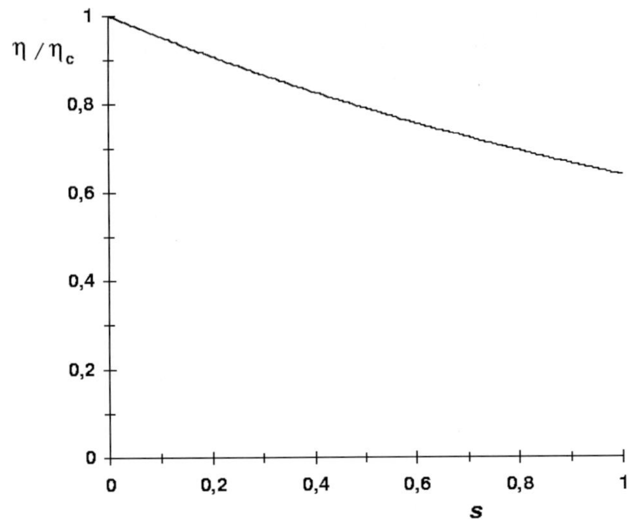


**Fig. 7.** Variation of carrier viscosity across channel thickness. Carrier: ethylbenzene, $\Delta T = 40$ K, $T_c = 293$ K.

$\nu$ using (31). The $\nu$ parameter depends on $T_c$ and $\Delta T$ as well as on the nature of the carrier liquid. From computations of $\nu$ using (31), an empirical relationship between $\nu$, $T_c$ and $\Delta T$ has been found, assuming that the accumulation wall is the cold one

$$\nu = (a_1 T_c + a_2)\,\Delta T + (a_3 T_c + a_4)\,\Delta T^2 + (a_5 T_c + a_6)\,\Delta T^3 \tag{33}$$

The $a_1 - a_6$ coefficients have been tabulated for various carrier liquids [20]. Fig. 7 shows the viscosity profile across the channel thickness for ethylbenzene with $\Delta T = 40$ K. The viscosity change is significant and reaches 36 % over this temperature interval. Since the viscosity is larger near the cold wall than elsewhere in the channel, the velocity near this wall is lower than it would be for an isoviscous carrier with the same flow rate, and $\nu$ is negative. In the present case, $\nu$ is equal to $-0.122$, indicating that the temperature dependence of the viscosity leads to a 12.2 % reduction in the slope of the relative velocity profile near the cold wall compared to the Poiseuille parabolic profile.

If $\eta$ is constant, then (29) reduces to (9) and obviously $\nu$ becomes equal to 0. Two approaches have been followed to take into account the variation of viscosity across the channel thickness. Both of them attempt to express that variation in a relatively simple, mathematically tractable form. The first one expresses the velocity profile as a third-degree polynomial in $s$. The second one expresses it as a fifth-degree polynomial in $s$.

**The third-degree polynomial velocity profile** Since the flow velocity must be equal to 0 at the contact of the channel walls and since the average carrier flow velocity is equal to $\langle u \rangle$, a third-degree polynomial velocity profile has a single degree of freedom. For convenience, one chooses to select $\nu$ as this adjustable parameter. The velocity profile becomes then [21]

$$\frac{u}{\langle u \rangle} = 6\left[(1+\nu)\,s - (1+3\nu)\,s^2 + 2\nu s^3\right] \tag{34}$$

The velocity profile described by (34) is not exactly equal to the true profile given by (29), but it becomes closer to this exact profile as one approaches the accumulation wall. The pertinence of using (34) is justified by the fact that for large *Pé* values, analytes move in close vicinity of the accumulation wall.

When the flow velocity profile is given by (34), the relationship between the retention time and *Pé*, for an exponential analyte concentration profile (8), is obtained by combining (1)-(3), (8) and (34) as [22]

$$\begin{aligned} t_{\mathrm{R}} &= \frac{t_{\mathrm{o}}}{6} \frac{Pé}{\left\{\left(1-\frac{6\nu}{Pé}\right)\left[\coth\left(\frac{Pé}{2}\right)-\frac{2}{Pé}\right]+\nu\right\}} \\ &= \frac{t_{\mathrm{o}}}{6\lambda_{\mathrm{F}}\left\{(1-6\nu\lambda_{\mathrm{F}})\left[\coth\left(\frac{1}{2\lambda_{\mathrm{F}}}\right)-2\lambda_{\mathrm{F}}\right]+\nu\right\}} \end{aligned} \tag{35}$$

The variation of $t_{\mathrm{R}}/t_{\mathrm{o}}$ vs. *Pé* for ethylbenzene with $\Delta T = 40$ K is shown on Fig. 8 and can be compared with that corresponding to a Poiseuille flow profile. The relative error made on the determination of *Pé* from retention time when neglecting the effect of non-constant viscosity is significant and is of the order of $\nu$, i.e. here about 12 %, at large *Pé*.

**The fifth-degree polynomial velocity profile** Alternatively, a third-degree polynomial expression of the carrier fluidity, $1/\eta$, versus $s$ can be obtained by appropriate fitting of either theoretical or experimental $\eta - T$ data together with the $T - s$ relation given by (27). This can be written as [23]

$$\frac{1}{\eta} = \sum_{\mathrm{i}=0}^{3} b_{\mathrm{i}} s^{\mathrm{i}} \tag{36}$$

where the coefficients $b_{\mathrm{i}}$ depends on $T_{\mathrm{c}}$, $\Delta T$ and $\omega_{\mathrm{c}}$. Note that $b_0 = 1/\eta_{\mathrm{o}}$. Inserting (36) in (29) and integrating leads to a fifth-degree polynomial expression of the velocity profile [23]

$$\frac{u}{\langle u \rangle} = \frac{\sum_{\mathrm{i}=1}^{5} h_{\mathrm{i}} s^{\mathrm{i}}}{\sum_{\mathrm{i}=1}^{5} \frac{h_{\mathrm{i}}}{\mathrm{i}+1}} = \sum_{\mathrm{i}=1}^{5} d_{\mathrm{i}} s^{\mathrm{i}} \tag{37}$$

with

$$h_i = \frac{1}{\mathrm{i}}\left(\theta b_{\mathrm{i}-1} - b_{\mathrm{i}-2}\right)$$

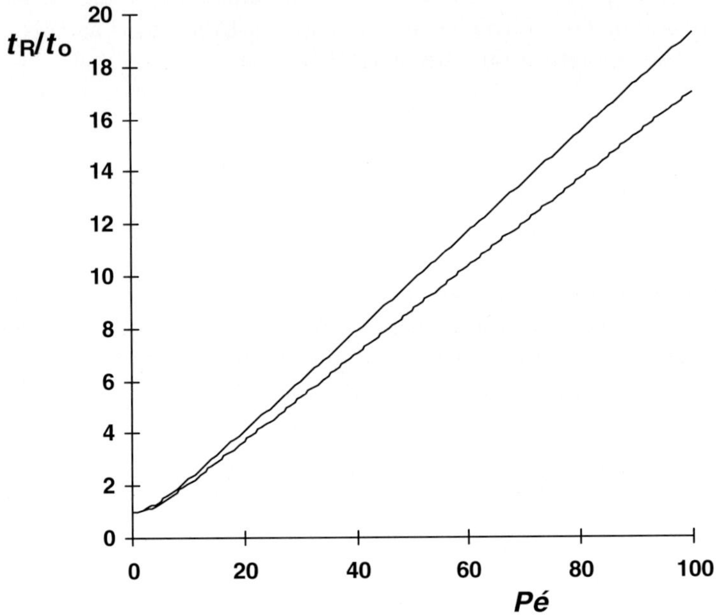


**Fig. 8.** Variation of the relative retention time, $t_R/t_o$, vs. *Pé* when taking into account the temperature dependence of viscosity, together with that of thermal conductivity (upper line) for exponential analyte concentration profiles Carrier: ethylbenzene, $\Delta T = 40$ K, $T_c = 293$ K. The lower line corresponds to the isoviscous case.

and

$$\theta = \sum_{i=0}^{3} \frac{b_i}{i+2} \Big/ \sum_{i=0}^{3} \frac{b_i}{i+1}$$

and

$$d_i = \frac{h_i}{\sum_{j=1}^{5} \frac{h_j}{j+1}}$$

in which one sets $b_{-1} = b_4 = 0$. Then, the $\nu$ parameter corresponding to this profile is obtained from (30) and (37) as

$$\nu = \frac{h_1}{6 \sum_{i=0}^{5} \frac{h_i}{i+1}} - 1 = \frac{d_1}{6} - 1 \tag{38}$$

The retention time corresponding to the fifth-degree polynomial flow velocity profile and the exponential analyte concentration profile is obtained by combining (1)-(3), (8) and (37) and integrating, which gives [25]

$$\begin{aligned} t_{\mathrm{R}} &= t_{\mathrm{o}} \frac{1-\exp(-P\acute{e})}{\sum\limits_{\mathrm{i}=1}^{5} \frac{d_{\mathrm{i}}\mathrm{i}!}{P\acute{e}^{\mathrm{i}}} - \exp(-P\acute{e}) \sum\limits_{\mathrm{i}=1}^{5} \frac{f_{\mathrm{i}}}{P\acute{e}^{\mathrm{i}}}} \\ &= t_{\mathrm{o}} \frac{1-\exp(-1/\lambda_{\mathrm{F}})}{\sum\limits_{\mathrm{i}=1}^{5} d_{\mathrm{i}}\mathrm{i}!\lambda_{\mathrm{F}}^{\mathrm{i}} - \exp(-1/\lambda_{\mathrm{F}}) \sum\limits_{\mathrm{i}=1}^{5} f_{\mathrm{i}}\lambda_{\mathrm{F}}^{\mathrm{i}}} \end{aligned} \tag{39}$$

with

$$f_{\mathrm{i}} = \sum_{\mathrm{j}=1}^{5} d_{\mathrm{j}} \frac{\mathrm{j}!}{(\mathrm{j}-\mathrm{i})!}$$

Alternatively, this can be written as

$$\begin{aligned} \frac{t_{\mathrm{o}}}{t_{\mathrm{R}}} &= \frac{1}{P\acute{e}} \left( K_1 - \frac{\exp(-P\acute{e})}{1-\exp(-P\acute{e})} K_2 \right) \\ &= \lambda_{\mathrm{F}} \left( K_1 - \frac{\exp(-1/\lambda_{\mathrm{F}})}{1-\exp(-1/\lambda_{\mathrm{F}})} K_2 \right) \end{aligned}$$

with

$$\begin{aligned} K_1 &= d_1 + \frac{2d_2}{P\acute{e}} + \frac{6d_3}{P\acute{e}^2} + \frac{24d_4}{P\acute{e}^3} + \frac{120d_5}{P\acute{e}^4} \\ &= d_1 + 2d_2\lambda_{\mathrm{F}} + 6d_3\lambda_{\mathrm{F}}^2 + 24d_4\lambda_{\mathrm{F}}^3 + 120d_5\lambda_{\mathrm{F}}^4 \end{aligned}$$

and

$$K_2 = 2d_2 + 3d_3\left(1+\frac{2}{P\acute{e}}\right) + 4d_4\left(1+\frac{3}{P\acute{e}}+\frac{6}{P\acute{e}^2}\right) + 5d_5\left(1+\frac{4}{P\acute{e}}+\frac{12}{P\acute{e}^2}+\frac{24}{P\acute{e}^3}\right)$$

or

$$K_2 = 2d_2 + 3d_3(1+2\lambda_{\mathrm{F}}) + 4d_4\left(1+3\lambda_{\mathrm{F}}+6\lambda_{\mathrm{F}}^2\right) + 5d_5\left(1+4\lambda_{\mathrm{F}}+12\lambda_{\mathrm{F}}^2+24\lambda_{\mathrm{F}}^3\right)$$

This retention time expression has been used for interpreting thermal FFF retention data [23, 26, 27]. It is more complicated than that obtained in (35) for a third-degree polynomial velocity profile, which has also been used [7, 22, 28, 29, 30], but it is expected to be more accurate. Note that (39), originally derived in [25] for a general polynomial velocity profile, has been incorrectly reported in [23] and [30].

### 4.3 Soret coefficient, Péclet number and analyte concentration profile

The analyte concentration profile is, according to (5) and (17), obtained by integration of

$$\frac{dc}{c} = -P\acute{e}ds = -(S_{\mathrm{T}}+\alpha)\frac{dT}{ds}ds \tag{40}$$

Because $S_\mathrm{T}$, $\alpha$ and $dT/ds$ are temperature dependent and thus depend on $s$, $P\acute{e}$ is not constant and (40) cannot be simply integrated as in (8). The $T$- and $s$-dependences of $dT/ds$ have been given in (28). Although $\alpha$ is small in comparison with $S_\mathrm{T}$, its dependence on $T$ can be accounted for by deriving it from that of $\rho$, itself derived from either experimental data or appropriate relationships [31]. However, it is far more complex to account for the dependence of $S_\mathrm{T}$ on $T$ because it is not a priori known, and one of the main objectives of a thermal FFF analysis is to determine $S_\mathrm{T}$. One is then forced to assume a model for the dependence of $S_\mathrm{T}$ or $P\acute{e}$ on $T$ or $s$. Four models have been proposed in the past.

**The constant $\lambda_\mathrm{F}$ model.** This is the model that most authors have used, sometimes unconsciously. It simply ignores the problem raised by the variation of $P\acute{e}$ across the channel thickness and relies on (8). In fact, according to (40), it amounts to state that $S_\mathrm{T}+\alpha$ varies with $T$ as does the reciprocal of $dT/ds$, which of course cannot reflect the true dependence of $S_\mathrm{T}$ on $T$, except maybe by accident for one specific analyte. One expects that such a model can only provide a rough estimate of $S_\mathrm{T}$.

**The quadratic $S_\mathrm{T}$ model.** This model [24] is based on the assumption of a quadratic variation of $S_\mathrm{T}$ (more exactly, in fact, of $S_\mathrm{T}+\alpha$) with $T$

$$S_\mathrm{T}+\alpha = e_0' + e_1'T + e_2'T^2 \tag{41}$$

From (40) and (41), $c$ is obtained as a function of $T$, assuming that the accumulation wall is the cold one, as

$$\frac{c}{c_\mathrm{o}} = \exp\left[-\left(e_0'+e_1'T_\mathrm{c}+e_2'T_\mathrm{c}^2\right)(T-T_\mathrm{c}) - \left(\frac{e_1'}{2}+e_2'T_\mathrm{c}\right)(T-T_\mathrm{c})^2 - \frac{e_2'}{3}(T-T_\mathrm{c})^3\right] \tag{42}$$

This expression can be converted into a concentration profile, $c(s)$, by combination with (27). The $e_0'-e_2'$ coefficients are not known. They have to be determined by adjustment to a set of experimental data.

**The hyperbolic $S_\mathrm{T}$ model.** An alternative model, based on the following hyperbolic expression, has been suggested to describe the variation of $S_\mathrm{T}$ with $T$ [32]

$$S_\mathrm{T}(\text{or } S_\mathrm{T}+\alpha) = \frac{m_1}{m_2+T} \tag{43}$$

As in the above case, the coefficients $m_1$ and $m_2$ have to be determined by means of experimental data. Combining with (40) and integrating gives, provided the cold wall is the accumulation wall

$$\frac{c}{c_\mathrm{o}} = \left(\frac{m_2+T_\mathrm{c}}{m_2+T}\right)^{m_1} \tag{44}$$

Again, use of (27) allows converting this expression in a concentration profile.

**The 'linear $\lambda_F$' model.** This model is based on the assumption that the effect of the temperature dependences of $S_T$, $\alpha$ and $dT/ds$ are such that $\lambda_F$ varies monotonously across the channel thickness at a nearly constant rate, or, at least, that it is so in the vicinity of the accumulation wall where most of the analyte is located, which, in terms of *Pé*, gives [33]

$$\frac{1}{Pé} = \frac{1}{Pé_o} + \delta s \tag{45}$$

or

$$\lambda_F = \lambda_{F,o} + \delta s$$

$Pé_o$ is the value of *Pé* at the accumulation wall where the $\lambda_F$ value is $\lambda_{F,o}$. $\delta$ is the difference in $1/Pé$ between the depletion and accumulation walls. Integrating (40) with (45) gives the concentration profile

$$\frac{c}{c_o} = (1 + Pé_o \delta s)^{-1/\delta} = \left(1 + \frac{\delta s}{\lambda_{F,o}}\right)^{-1/\delta} \tag{46}$$

This profile depends on two parameters, $Pé_o$ and $\delta$, which have to be determined by some appropriate means.

## 5 Methods of determination of Soret coefficients from thermal FFF retention data

The retention time in FFF depends, according to (1)-(3), on the velocity and concentration profiles. Because of the temperature dependences of the carrier thermal conductivity, carrier viscosity and analyte Soret coefficients, these profiles are somewhat complicated and the retrieval of the Soret coefficient, via the Péclet number, from retention time is a relatively complex task. Accordingly, several more or less sophisticated methods are been employed for this purpose. In fact, every $S_T$ determination relies on the selections of one of the various expressions of the velocity profiles among equations (9), (34) and (37), and of one expression for the analyte concentration profile generally among equations either (8), or (42) coupled with (27), or (44) coupled with (27), or (46). This gives a variety of possibilities of combinations. Some of the methods that have been used in the past are described below.

### 5.1 The crude method

The simplest method ignores all the complications arising from the temperature dependences of the relevant physico-chemical parameters. It is thus based on the parabolic flow profile (9) and the exponential analyte concentration profile (8). *Pé* is derived from the retention time, $t_R$, by means of (10) and $S_T$ is given, according to (18) as

$$S_T = \frac{Pé}{\Delta T} = \frac{1}{\lambda_F \Delta T} \tag{47}$$

which, in the high retention limit where the term in brackets in (10) is equal to 1, becomes

$$S_{\mathrm{T}} = \frac{6}{\Delta T}\frac{t_{\mathrm{R}}}{t_{\mathrm{o}}}$$

The resulting $S_{\mathrm{T}}$ value is only approximate but has been used to prove the validity of the thermal FFF method for $S_{\mathrm{T}}$ determinations in early thermal FFF experiments and/or to obtain rough estimates of $S_{\mathrm{T}}$ [34-36].

### 5.2 The constant $\lambda_{\mathrm{F}}$ methods

There are several methods which, like the above 'crude' method, are based on the constant $\lambda_{\mathrm{F}}$ assumption, but which, unlike the crude method, recognize the need to account for the influence of temperature [37, 38]. These methods of determination of $S_{\mathrm{T}}$ account for the non-linearity of the temperature profile and the deviation of the velocity profile from the parabolic shape. But they still assume an exponential concentration profile, hence on a constant value of $\lambda_{\mathrm{F}}$ or *Pé*. Most often, they are based on the third-degree or fifth-degree polynomial velocity profile and the value of *Pé* is obtained from $t_{\mathrm{R}}$ by means of (35) or (39), respectively. Then the Soret coefficient can still be obtained as above in (47), but most often it is derived from (17) as

$$S_{\mathrm{T}} = \frac{Pé}{dT/ds} - \alpha = \frac{1}{\lambda_{\mathrm{F}}\frac{dT}{ds}} - \alpha \qquad (48)$$

The question arises as to the computation of $dT/ds$, and, with lesser importance, of $\alpha$ since they are dependent on temperature. Since $dT/ds$, according to (28) depends on $T$, two approaches have been followed.

**The constant $\lambda_{\mathrm{F}}$-cold wall method.** Since the polymeric analyte is supposed to accumulate at the cold wall, one method associates $S_{\mathrm{T}}$ to the cold wall temperature [39]. This gives

$$S_{\mathrm{T}} = \frac{Pé}{\Delta T\left(1+\frac{1}{2}\omega_{\mathrm{c}}\Delta T\right)} - \alpha\left(T_{\mathrm{c}}\right) = \frac{1}{\lambda_{\mathrm{F}}\Delta T\left(1+\frac{1}{2}\omega_{\mathrm{c}}\Delta T\right)} - \alpha\left(T_{\mathrm{c}}\right) \qquad (49)$$

Generally, the $\alpha$ term is neglected.

**The constant $\lambda_{\mathrm{F}}$-center-of-gravity method.** The other approach takes into account the fact that analyte molecules sample different streamlines and thus different temperatures during their sojourn in the channel. Therefore, it associates $S_{\mathrm{T}}$ to the temperature of the center-of-gravity of the analyte cloud. Assuming an exponentially distributed concentration profile, the mean value of $s$, $s_{\mathrm{cg}}$, is given by

$$s_{\mathrm{cg}} = \frac{1}{Pé} - \frac{\exp(-Pé)}{1-\exp(-Pé)} = \lambda_{\mathrm{F}} - \frac{\exp(-1/\lambda_{\mathrm{F}})}{1-\exp(-1/\lambda_{\mathrm{F}})} \qquad (50)$$

Then replacing $s$ by $s_{\mathrm{cg}}$ in (27) and (28) provides the temperature gradient and the temperature $T_{\mathrm{cg}}$, respectively, at the center-of-gravity from which one gets

$$S_{\mathrm{T}} = \frac{P\acute{e}}{\left.\frac{dT}{ds}\right|_{T_{\mathrm{cg}}}} - \alpha\left(T_{\mathrm{cg}}\right) = \frac{1}{\lambda_{\mathrm{F}} \left.\frac{dT}{ds}\right|_{T_{\mathrm{cg}}}} - \alpha\left(T_{\mathrm{cg}}\right) \tag{51}$$

This method has been used in various studies [23, 29, 40]. The association of $S_{\mathrm{T}}$ with $T_{\mathrm{cg}}$ is made on an intuitive basis, but is arbitrary. It has been shown that it is not correct [33].

Clearly, these constant $\lambda_{\mathrm{F}}$ methods are not coherent and can only be approximate since they assume that $P\acute{e}$ (or $\lambda_{\mathrm{F}}$) is the same at all temperatures across the channel thickness and, simultaneously, recognize that it has to be associated with a specific position across this thickness.

### 5.3 The quadratic $S_{\mathrm{T}}$ method

According to (41), the variation of $S_{\mathrm{T}}$ across the channel thickness depends on three unknown parameters, the coefficients $e'_0$, $e'_1$, $e'_2$. Obviously, it is not possible from a single measurement of $t_{\mathrm{R}}$ to numerically evaluate these three parameters. Therefore, application of the quadratic $S_{\mathrm{T}}$ model requires that, at least, three independent $t_{\mathrm{R}}$ measurements, i.e. measurements of $t_{\mathrm{R}}$ in three different temperature conditions, be performed. It can be shown that the mean $S_{\mathrm{T}}$ in the thermal FFF channel is more sensitive to variations of $T_{\mathrm{c}}$ than to variations of $\Delta T$. Therefore the determinations for a given analyte are preferentially performed by varying $T_{\mathrm{c}}$ keeping $\Delta T$ constant. While at least three measurements are necessary, the experimental errors in the resulting determination of the $e'_{\mathrm{i}}$ coefficients are reduced by increasing the number of $t_{\mathrm{R}}$ measurements at different $T_{\mathrm{c}}$.

Fixing presupposed values for the $e'_{\mathrm{i}}$ coefficients, and using (27) and (28) for the temperature and thermal gradient profiles, (37) for the velocity profile, and (42) for the concentration profile, it is possible to numerically compute $t_{\mathrm{R}}/t_{\mathrm{o}}$ by means of (1)-(3) and to compare it with experimental retention data for each $T_{\mathrm{c}}$. If the gap between calculated and experimental data is too large, new values are assigned to the $e'_{\mathrm{i}}$ coefficients and the process is repeated until this gap is sufficiently narrow. In practice, this method has been applied by using a modified Simplex optimization procedure [24].

Provided the variation of $S_{\mathrm{T}}$ vs. $T$ is satisfyingly represented by a quadratic relationship (independent studies appear to validate this relationship), this approach is certainly, in principle, the most complete in order to account for the influence of temperature on all relevant physico-chemical parameters in thermal FFF. However, it has the drawback to require at least three (and, preferentially, much more than three) measurements in different experimental conditions and can therefore not be applied to single $t_{\mathrm{R}}$ determinations.

### 5.4 The linear $\lambda_{\mathrm{F}}$ method

This approach is based on the so-called linear $\lambda_{\mathrm{F}}$ model, described above, which assumes a linear variation of $1/P\acute{e}$ across the channel thickness. Then, applying a constant $P\acute{e}$ retention equation to an experimental retention time provides an apparent value of $P\acute{e}$, $P\acute{e}_{\mathrm{app}}$, which is the $P\acute{e}$ value of a hypothetical constant $P\acute{e}$, exponential concentration profile which would give the same retention time as the actual concentration profile in the same flow and temperature conditions. $P\acute{e}_{\mathrm{app}}$ must lie within the $P\acute{e}_0$ - $P\acute{e}_1$ range, where $P\acute{e}_1$ is the value of $P\acute{e}$ at the depletion wall. Let $s_{\mathrm{eq}}$ be that relative distance from the accumulation wall where $P\acute{e}_{\mathrm{app}}$ is equal to the actual value of $P\acute{e}$ in the varying $P\acute{e}$ profile. According to (45), $s_{\mathrm{eq}}$ is given by

$$s_{\mathrm{eq}} = \frac{1}{\delta}\left(\frac{1}{P\acute{e}_{\mathrm{app}}} - \frac{1}{P\acute{e}_{\mathrm{o}}}\right) = \frac{1}{\delta}\left(\lambda_{\mathrm{F,app}} - \lambda_{\mathrm{F,o}}\right) \tag{52}$$

We have found that, whatever $P\acute{e}_{\mathrm{o}}$, $\delta$, and $\nu$, there is a quasi-unique relationship between $s_{\mathrm{eq}}$ and $P\acute{e}_{\mathrm{app}}$, which can be expressed by [33]

$$s_{\mathrm{eq}} = \frac{2}{P\acute{e}_{\mathrm{app}}} - \frac{2.1365\,(1+2\nu)}{P\acute{e}^2_{\mathrm{app}}} - \frac{6.1678\,(2-\nu)}{P\acute{e}^3_{\mathrm{app}}} \tag{53}$$

or

$$s_{\mathrm{eq}} = 2\lambda_{\mathrm{F,app}} - 2.1365\,(1+2\nu)\,\lambda^2_{\mathrm{F,app}} - 6.1678\,(2-\nu)\,\lambda^3_{\mathrm{F,app}}$$

where $\nu$ is the flow distortion parameter. This equation should not be used for $P\acute{e}_{\mathrm{app}}$ lower than about 6.5. Therefore, from the value of $P\acute{e}_{\mathrm{app}}$ derived from $t_{\mathrm{R}}$, one gets $s_{\mathrm{eq}}$ by means of (53), then the temperature $T_{\mathrm{eq}}$ at this location using (27). Then, according to (17), the Soret coefficient at temperature $T_{\mathrm{eq}}$ is then obtained as

$$S_{\mathrm{T}}\left(T_{\mathrm{eq}}\right) = \frac{P\acute{e}_{\mathrm{app}}}{\left.\frac{dT}{ds}\right|_{T_{\mathrm{eq}}}} - \alpha\left(T_{\mathrm{eq}}\right) = \frac{1}{\lambda_{\mathrm{F,app}}\left.\frac{dT}{ds}\right|_{T_{\mathrm{eq}}}} - \alpha\left(T_{\mathrm{eq}}\right) \tag{54}$$

or, equivalently,

$$\begin{aligned} S_{\mathrm{T}}\left(T_{\mathrm{eq}}\right) &= \frac{P\acute{e}_{\mathrm{app}}\left[1+\omega_{\mathrm{c}}\left(T_{\mathrm{eq}}-T_{\mathrm{c}}\right)\right]}{\Delta T\left(1+\frac{1}{2}\omega_{\mathrm{c}}\Delta T\right)} - \alpha\left(T_{\mathrm{eq}}\right) \\ &= \frac{\left[1+\omega_{\mathrm{c}}\left(T_{\mathrm{eq}}-T_{\mathrm{c}}\right)\right]}{\lambda_{\mathrm{F,app}}\Delta T\left(1+\frac{1}{2}\omega_{\mathrm{c}}\Delta T\right)} - \alpha\left(T_{\mathrm{eq}}\right) \end{aligned} \tag{55}$$

It appears that, this method has the advantage over the quadratic $S_{\mathrm{T}}$ method described above, to be able, from a single retention time measurement, to provide one value of $P\acute{e}$, $P\acute{e}_{\mathrm{app}}$, at one location in the channel, $s_{\mathrm{eq}}$, which is transformed into one value of $S_{\mathrm{T}}$, at one temperature, $T_{\mathrm{eq}}$. Obviously, this does not unravel the full $P\acute{e}$ vs. $s$ profile, but this allows to state that the $P\acute{e}(s)$ curve goes to the ($P\acute{e}_{\mathrm{app}}$, $s_{\mathrm{eq}}$) point. The value of $s_{\mathrm{eq}}$, and consequently that of $T_{\mathrm{eq}}$, is not selectable at will. It derives from the particular $t_{\mathrm{R}}$ and its associated $P\acute{e}_{\mathrm{app}}$ by means of (52). In order to sample different $T_{eq}$ temperatures for determining the temperature dependence of $S_T$, measurements of $t_R$ have to be performed in various temperature conditions (change in $\Delta T$, or, preferably, in $T_c$). Such a determination has been performed for polymers, as described in the next section.

## 6 Application to the determination of the Soret coefficient of polystyrene in ethylbenzene

Thermal FFF has proved to be a particularly well-suited method for the separation of polymers and the determination of their molar mass distributions. The optimization of the method requires the knowledge of the relationship between retention time, hence Soret coefficient, and molar mass. Earlier thermal FFF investigations based either on the crude method or a constant $\lambda_F$ method have reached the conclusion that the thermodiffusion coefficient is, in a first approximation, independent of molar mass [29, 38, 40-42]. So the molar mass selectivity in thermal FFF arises essentially from the molar mass dependence of the ordinary mass diffusion coefficient. As more sophisticated methods of handling retention data are now available, as described above, it is interesting to revisit the relationship between the molar mass and the Soret coefficient derived from thermal FFF retention data in order to improve the accuracy of this relationship as well as that of the polymer characterization. For this purpose, we have performed retention measurements for polystyrene (PS) samples in ethylbenzene. PS was chosen, as standards of narrow molar mass distribution are available for various mean molar masses. Ethylbenzene was found a good solvent providing comparatively large retention times for PS [38].

### 6.1 Experimental

The thermal FFF channel was previously described [28]. It had a tip-to-tip length of 44.7 cm, a breadth of 2 cm and a thickness of 0.1 mm. Ethylbenzene (Prolabo, Paris, France) was used as carrier liquid and sample solvent. The carrier flow rate, provided by a syringe pump (Model A-99, Razel Scientific Instruments, Stamford, CT, USA), was selected in the range 0.02-0.04 ml.min$^{-1}$. Sample injection was made by means of a six-port homemade injection valve equipped with a 20 $\mu$l sample loop. The sample effluent was fed to a differential flow-through refractometer (Model R401, Waters Associates, Milford, MA, USA) by means of a Teflon connection tube. The detector signal was recorded on a computer using a data acquisition board (Model DAS-801, Keithley, Taunton, MA, USA) and mean (peak center-of-gravity) residence times were computed to get the experimental retention times, after removal of the time spent in the connection lines. Four PS samples (Polymer Laboratories, Church Stretton, Great-Britain) of nominal weight average molar masses 76000, 180000, 505000 and 1015000 were dissolved in the carrier liquid and injected at concentrations of 2 mg.ml$^{-1}$. In all experiments, the temperature drop, $\Delta T$, was fixed at 40 K. Experiments were performed at five different cold wall temperatures ($T_c$ = 293 K, 300 K, 307 K, 321 K and 335 K).

### 6.2 Data handling

For each sample molar mass, $M$, and $T_c$, experiments were repeated six times, so that a given retention time data is an average from six experimental fractograms. For each $T_c$, the $b_i$ coefficients of the fifth-degree polynomial velocity profile (37) were computed by least mean square regression of $1/\eta$ vs. $s$, according to (36) in which $\eta$ was computed from (32) and (27). For ethylbenzene, the input parameters were $B = 1095$ K, $d\lambda_T/dT = -2.33 \times 10^{-4}$ Wm$^{-1}$K$^{-2}$ and $\lambda_T = 0.132$ Wm$^{-1}$K$^{-1}$ at 293 K [20, 31]. A value of $Pé$ was derived from each $t_R$ value by means of (39). At this point, all the methods described above for the determination of $S_T$ can be applied provided that $\alpha$ can be calculated at the corresponding temperature. Using regression of $\rho$ vs. $T$ data computed according to the procedure recommended in [31], we found for ethylbenzene in the 273-400 K temperature range: $\alpha = 1.174 \times 10^{-3} - 3.079 \times 10^{-6} T + 8.414 \times 10^{-9} T^2$, with $T$ in K and $\alpha$ in K$^{-1}$. Note that it was previously shown that, for the PS - ethylbenzene system, the polymer migrates toward the cold region [21].

The linear $\lambda_F$ method is applied to these ($M$, $T_c$, $Pé$) sets of data because, among the methods accounting for temperature dependence on $S_T$, it is the most conveniently applicable to single data. Then, $Pé$ was considered as an apparent value, $Pé_{app}$, and the corresponding equivalent position, $s_{eq}$, was determined using (53) with $\nu$ given by (38). The equivalent temperature, $T_{eq}$, is calculated using (27) and the Soret coefficient, $S_T(T_{eq})$, is derived from (55). Only values of $t_R/t_o$ larger than 1.45 were taken into account.

### 6.3 Temperature dependences of the Soret coefficient and of the thermodiffusion coefficient

For each $M$, $Pé_{app}$ values at different $T_c$ are obtained. They are transformed into $S_T$ at different $T_{eq}$.

One can then plot, for each $M$, $S_T$ vs. $T_{eq}$. Such plots are shown in Fig. 9. $S_T$ is seen to decrease with increasing temperatures. Hence, thermal FFF retention and selectivity increase with decreasing cold wall temperatures.

Because, in FFF, the retention time reflects the steady state transversal concentration profile, thermal FFF can essentially give access to a steady state thermodiffusion parameter, i.e. to $S_T$, as shown in (17). Deriving the thermodiffusion coefficient, $D_T$, requires a numerical value of the ordinary diffusion coefficient $D$, which can be obtained either from appropriate experiment or from estimation. A correlation between $D$, $T$ and $M$ was previously suggested [29]

$$D = q_1 \frac{T}{M^{b'}} \exp\left(-\frac{q_2}{T}\right) \tag{56}$$

with, for PS in ethylbenzene, $q_1 = 8.421 \times 10^{-9}$, $q_2 = 1360$ and $b' = 0.552$, when $D$, $M$ and $T$ are expressed in m$^2$s$^{-1}$, g mol$^{-1}$ and K, respectively. The values of $D_T$, derived from $S_T$ vs. $T_{eq}$ data of Fig. 9, according to (13), as $S_T D$, are plotted vs. $T$ in Fig. 10. $D_T$ is seen to increase slightly, approximately linearly, with increasing $T$, which means that, on increasing $T$, $S_T$ decreases relatively less

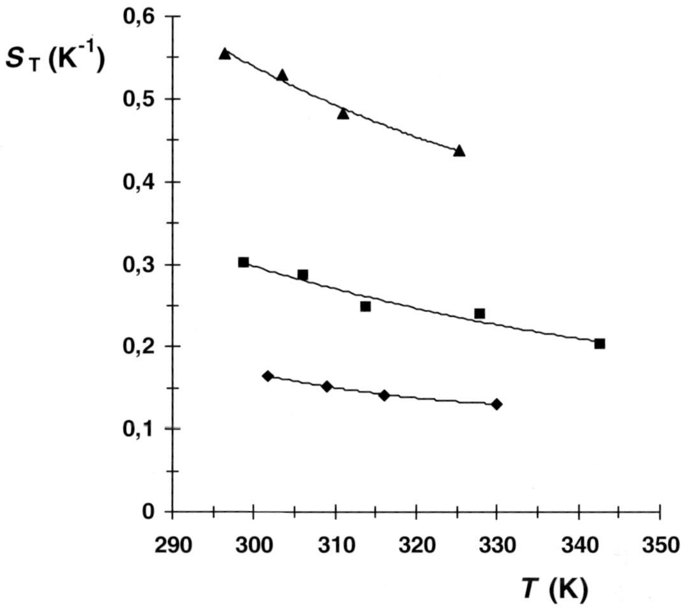


**Fig. 9.** Variation of the Soret coefficient of polystyrene in ethylbenzene vs. temperature for three PS molar masses. From lower to upper curve: $M = 76000$ (diamonds), 180000 (squares), 505000 (triangles). $S_T$ are derived, by means of the linear $\lambda$ method, from $t_R$ measured at different $T_c$.

than $D$ increases. Furthermore, Fig. 10 shows that $D_T$ depends slightly on PS molar mass, in contradiction with previous findings. This conclusion, however, depends on the validity of (56) to describe the $M$ and $T$ dependences of $D$ and on the accuracy with which the coefficients $q_1$, $q_2$ and $b'$ are determined.

### 6.4 Molar mass and temperature dependences of the Soret coefficient

Nevertheless, it is possible to investigate the dependence on $M$ of $S_T$ directly from thermal FFF retention data without the need of an expression for $D$. However, it is seen on Fig. 9, that, for one given $T_c$ temperature, the data points for various PS standards are not obtained at the same temperature and, therefore, cannot be straightforwardly compared. This is because the corresponding temperature, $T_{eq}$, varies with retention, according to (53). To be meaningful, the comparison of $S_T$ values for various $M$ must be made at a given temperature or to a series of given temperatures. The procedure followed for this purpose is described below.

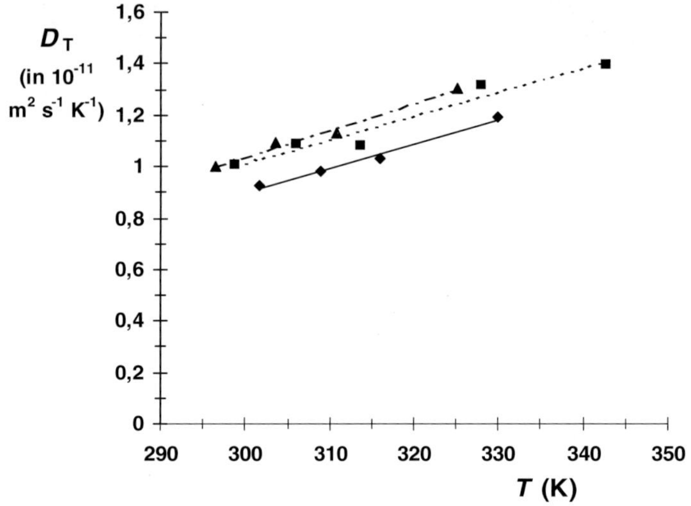


**Fig. 10.** Variation of the thermodiffusion coefficient of polystyrene in ethylbenzene vs. temperature, for three PS molar masses. $D_T$ is computed as $S_T D$ with $D$ given by (56). The curves are regression lines. From lower to upper lines: $M = 76000$ (solid line, diamonds), 180000 (dash line, squares), 505000 (dot-dash line, triangles).

It is seen, from Fig. 9, that the experimental data can be conveniently fitted according to a quadratic regression curve

$$S_T = e_0 + e_1 T + e_2 T^2 \tag{57}$$

The coefficients $e_{0-2}$ are determined, for each $M$, by least mean square regression of the experimental data. The corresponding regression curves are shown on Fig. 9. Then, for each $M$, the values of $S_T$ are calculated using (57) at nine reference temperatures, from 295 K to 335 K, every 5 K. At each reference temperature, a plot of $S_T$ vs. $M$ is performed. The curves are fitted according to a power law, as has been previously suggested for the dependence of $S_T$ on $M$ [40, 43-45]

$$S_T = S_0 M^n \tag{58}$$

where $S_0$ is the Soret coefficient of a hypothetical polymer extrapolated to a unit molar mass. The prefactor $S_0$ and the exponent $n$ are determined at each reference temperature by least mean square linear regression of the log $S_T$ vs. $\log M$ data. The correlation coefficients are, for all temperatures, larger than 0.997. The resulting data are plotted vs. the reference temperature on Figs. 11 and 12, respectively.

It is found that $S_0$ has a minimum and $n$ a maximum in the temperature range of interest. It was expected that the prefactor was temperature dependent, but it is noteworthy that the value of the exponent itself depends on temperature

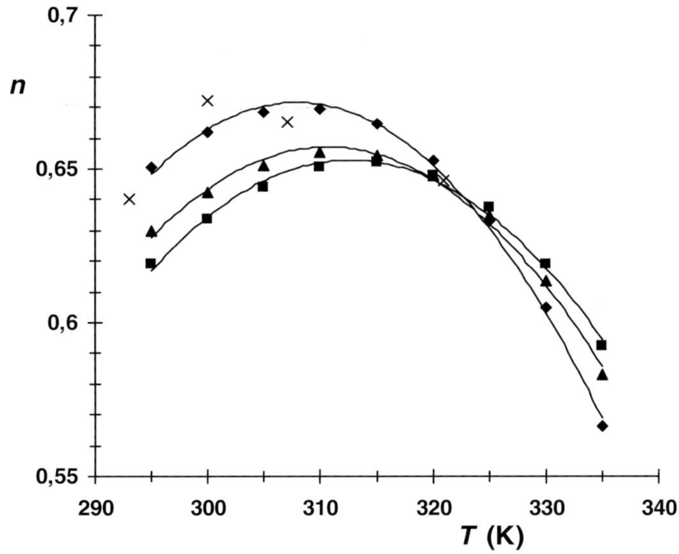


**Fig. 11.** Variation, vs. $T$, of the exponent, $n$, of the power law of $S_{\mathrm{T}}$ vs. $M$. From lower to upper curve in the 295-315 K range: linear $\lambda$ method (squares); constant $\lambda$ - center-of-gravity method (triangles); constant $\lambda$ - cold wall method (diamonds). The data obtained from the crude method (crosses) are plotted at the corresponding $T_{\mathrm{c}}$.

and varies from 0.62 to 0.59 with a maximum of 0.65 in the 295-335 K temperature range. These variations can reasonably well be represented by quadratic expressions

$$S_0 = s_{00} + s_{01}T + s_{02}T^2 \tag{59}$$

and

$$n = n_0 + n_1 T + n_2 T^2 \tag{60}$$

From least mean square regression, one gets, for PS in ethylbenzene, when $S_{\mathrm{T}}$, $M$ and $T$ are expressed, respectively, in $\mathrm{K}^{-1}$, $\mathrm{g\,mol}^{-1}$, and K: $s_{00} = +1.872 \times 10^{-2}$, $s_{01} = -1.181 \times 10^{-4}$, $s_{02} = +1.870 \times 10^{-7}$, $n_0 = -10.67$, $n_1 = +0.07246$, $n_2 = -0.0001159$. With (58-60), these coefficients allow calculating values of Soret coefficients and to compare them with those Soret coefficients obtained from experimental retention times by the linear $\lambda_{\mathrm{F}}$ method. The agreement between the calculated values and the "experimental" ones is quite satisfying as seen on Fig. 13. The mean absolute deviation from the first diagonal amounts to 2.6 %.

There has been discussion about the molar mass dependence of the thermodiffusion coefficient, $D_{\mathrm{T}}$. If this dependence can be expressed as a power law, as suggested [43], its exponent is, according to (13), (56) and (58), equal to $n - b'$. Since $n$ is found to depend on $T$, $n - b'$ is likely to also depend on $T$. Therefore, $D_{\mathrm{T}}$ may be independent of $M$ at some temperature, but not at another one.

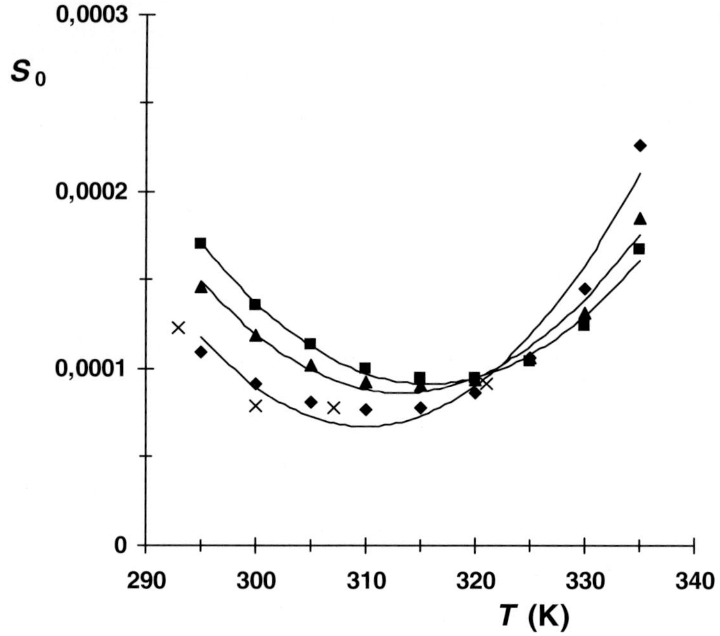


**Fig. 12.** Variation, vs. $T$, of the prefactor, $S_0$, of the power law of $S_T$ vs. $M$. From upper to lower curve in the 295-315 K range: linear $\lambda$ method (squares); constant $\lambda$ - center-of-gravity method (triangles); constant $\lambda$ - cold wall method (diamonds). The data obtained from the crude method (crosses) are plotted at the corresponding $T_c$.

This might in part explain the discrepancies found in the literature about the $D_T - M$ relationship for polymers. A detailed investigation requires an accurate expression of $D$ vs. $T$ and $M$. Nevertheless, since, for a polymer in a good solvent (such as, supposedly, PS in ethylbenzene), $b'$ is expected to be close to 0.6 [46], the value of $n - b'$ is likely to be lower than 0.1 and the molar mass dependence of $D_T$, at most, weak.

In order to check the susceptibility of the exponent $n$ to variations in experimental retention time measurements, two new sets of retention data were generated by assuming that $t_R/t_o$ are 2 % larger, for one set, and 2 % smaller, for the other set, than experimental values. Such relative errors are typical for $t_R/t_o$. As described above, $Pé_{app}$ and $S_T$ values at $T_{eq}$ were computed using (39), (53) and (55) and analyzed according to (57) and (58). The resulting dependences of $n$ on $T$ are shown in Fig. 14. The general trend of variations appears to be unchanged. Over the 295-335 K range, the mean relative change in $n$ for a 2 % change in $t_R/t_o$ amounts to 0.85 %. Therefore, a 2 % systematic error in $t_R/t_o$ does not affect significantly the value of $n$ and its temperature dependence.

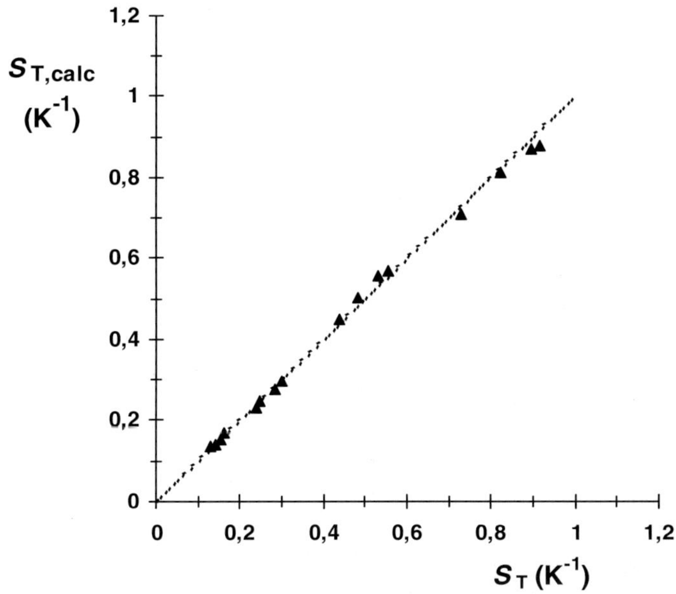


**Fig. 13.** Soret coefficients calculated according to (58-60) vs. Soret coefficients derived from retention data by the linear $\lambda$ method. The dash line represents the first diagonal.

### 6.5 Comparison with other thermal FFF data handling methods

The above method of investigation of the temperature and molar mass dependences of $S_T$ is based on the use of the linear $\lambda_F$ method to analyze the *Pé* values derived from experimental retention times by means of (39). The analysis of these *Pé* values can be made in a very similar way by either the constant $\lambda_F$ - cold wall method or by the constant $\lambda_F$ - center-of-gravity method. Then, instead of being associated to $T_{eq}$, $S_T$ is associated to $T_c$ or $T_{cg}$ by means of (49) or (50-51), respectively. The resulting dependences of $S_0$ and $n$ on $T$ are also shown on Figs. 11 and 12. Comparison with the crude method can be made by deriving $S_T$ from *Pé* using (47) for the various PS samples. The resulting $S_0$ and $n$ obtained using (58) are also plotted on Figs. 11 and 12 at the corresponding $T_c$.

Whatever the method employed, the trends are comparable. This similarity excludes the possibility that the optimums found for $S_0$ and $n$ at some temperature in the investigated range are an artifact of the data handling method. Such optimums have also been found for the PS-decalin system [47]. Variations of $n$ with $T$ observed from thermal FFF retention data using an approach based on the linear $\lambda_F$ method have recently been reported for various polymers in THF [48].

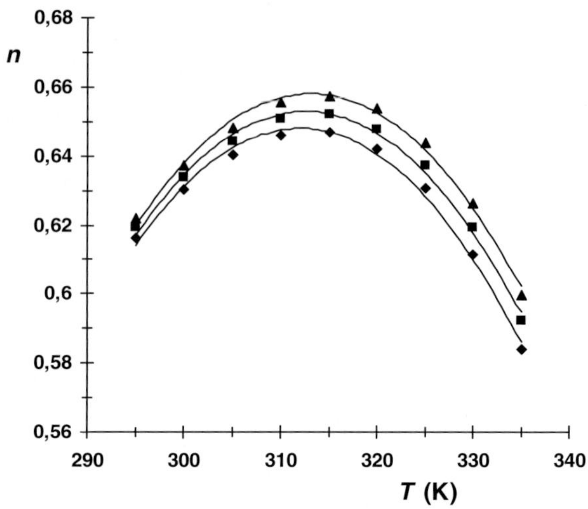


**Fig. 14.** Influence of a systematic error in $t_R/t_o$ on the temperature dependence of $n$ by the linear $\lambda$ method. Intermediate curve (squares): from experimental $t_R/t_o$ data. Upper curve (triangles): from $t_R/t_o$ data which are systematically 2% lower than experimental ones. Lower curve (diamonds): from $t_R/t_o$ data which are systematically 2% larger than experimental ones.

## 7 Conclusion

Thermal FFF appears to be a relatively fast, reliable and accurate method of determination of Soret coefficients of species with $S_T$ larger than about $5/\Delta T$. That the values of $S_T$ obtained by thermal FFF compare well with previously published data obtained by various thermodiffusion methods, such as the thermogravitational column, has been demonstrated soon after the implementation of the method [40]. But today, the development of the theory and of the methodology of thermal FFF has reached a level that can make it a method of reference for the determination of $S_T$ values. Mutual interactions of thermal FFF with the most advanced thermodiffusion methods applicable to polymers, such as forced Rayleigh scattering [49], should be greatly beneficial and lead to further progresses in this domain.

Because it takes into account the perturbations induced by temperature on the flow velocity profile and on the analyte transversal concentration profile, the linear $\lambda_F$ method is expected to provide more accurate $S_T$ and $M$ data than the methods based on an exponential concentration profile (constant $\lambda_F$ methods). It can therefore be adopted as the method of choice for interpreting retention data in thermal FFF.

As shown above, thermal FFF appears to be a convenient method for investigating the dependence of $S_T$ on molar mass and to unravel some relatively subtle effects such as optimum of $S_0$ and $n$ versus temperature. However, it must not be forgotten that the validity of such findings is heavily dependent on the accuracy with which average molar masses are determined. The availability of standards with very narrow polydispersity and accurate molar masses should lead to improvement in the $S_{\mathrm{T}} - M$ relationship.

Still, the various methods of handling thermal FFF retention data that have been described above are based on infinite dilution models. If it was shown that the sample molar fraction is, in all practical situations, sufficiently low for the $1 - X$ term to be negligible in the equation (12) defining $S_{\mathrm{T}}$, it remains to be verified that, in particular experimental conditions, the temperature, flow velocity and analyte concentration profiles are unaffected by the concentration of the injected sample. There are indications that, in some particular cases, the adjustment of sample concentration to satisfy detectability requirements does not allow the high dilution condition to be correctly fulfilled [50, 51]. In these cases, the interpretation of retention data becomes considerably more complex than in the infinite dilution situation.

The methods of determination of $S_{\mathrm{T}}$ presented above have been applied to the particular case of polystyrene in ethylbenzene. It should be noted that they are not restricted to macromolecular solutions, but should apply as well to supramolecular species for which the concentration profile is controlled by the balance between the diffusion flux and the convection flux induced by the temperature gradient, as shown in (12), such as gels [52], suspensions of colloidal particles [53-56] or dispersions of emulsion droplets [57]. The latter applications illustrate that there is a continuum in the spectrum of species for which a temperature gradient induces a concentration gradient. Thermal FFF is right at the place where the Soret effect meets thermophoresis. This justifies that the thermodiffusion coefficient $D_{\mathrm{T}}$ is also called the thermophoretic mobility, since, as shown by (12), like in other phoretic methods, it represents the field-induced velocity per unit value of the field, which is, in the present case, the temperature gradient.

## References


1. J.C. Giddings: Sep. Sci. **1**, 123 (1966)
2. J.C. Giddings: Science **260**, 1456 (1993)
3. M. Martin: 'Theory of field flow fractionation'. In: *Advances in Chromatography*, Vol. 39, eds. P.R. Brown, E. Grushka (Marcel Dekker, New York 1998) pp. 1-138
4. H. Cölfen, M. Antonietti: Adv. Polym. Sci. **150**, 67 (2000)
5. M. Schimpf, K. Caldwell, J.C. Giddings: *Field-Flow Fractionation Handbook* (Wiley-Interscience, New York 2000)
6. J.C. Giddings, M. Martin, M.N. Myers: J. Chromatogr. **158**, 419 (1978)
7. M. Martin, R. Reynaud: Anal. Chem. **52**, 2293 (1980)
8. J.C. Giddings: J. Chem. Phys. **49**, 81 (1968)
9. B.A. Buffham: J. Colloid Interface Sci. **181**, 490 (1996)
10. J.C. Giddings, L.K. Smith, M.N. Myers: Anal. Chem. **47**, 2389 (1975)
11. W. Jost: *Diffusion in Solids, Liquids, Gases* (Academic Press, New York 1960) p 523
12. H.J.V. Tyrrell: *Diffusion and Heat Flow in Liquids* (Butterworths, London 1961) p 44

13. S.R. De Groot, P. Mazur: *Non-Equilibrium Thermodynamics* (North-Holland, Amsterdam 1962) p 276
14. K.E. Grew: 'Thermal diffusion'. In: *Transport Phenomena in Fluids*, ed. H.J.M. Hanley (Marcel Dekker, New York 1969) p 367
15. M. Martin, C. Van Batten, M. Hoyos: Entropie **214**, 69 (1998)
16. S.R. De Groot: Physica **9**, 699 (1942)
17. A. Pelley, R.V. Southwell: Proc. Roy. Soc. (London) A **176**, 312 (1940)
18. W. Köhler: J. Chem. Phys. **98**, 660 (1993)
19. G. Westermann-Clark: Sep. Sci. Technol. **13**, 819 (1978)
20. J.E. Belgaied, M. Hoyos, M. Martin: J. Chromatogr. A **678**, 85 (1994)
21. J.C. Giddings, M. Martin, M.N. Myers: Sep. Sci. Technol. **14**, 611 (1979)
22. M. Martin, M.N. Myers, J.C. Giddings: J. Liq. Chromatogr. **2**, 147 (1979)
23. J.J. Gunderson, K.D. Caldwell, J.C. Giddings: Sep. Sci. Technol. **19**, 667 (1984). Corrigendum in [24]
24. A.C. Van Asten, H.F.M. Boelens, W.Th. Kok, H. Poppe, P.S. Williams, J.C. Giddings: Sep. Sci. Technol. **29**, 513 (1994)
25. M. Martin, J.C. Giddings: J. Phys. Chem. **85**, 727 (1981)
26. M.E. Schimpf, P.S. Williams, J.C. Giddings: J. Appl. Polym. Sci. **37**, 2059 (1989)
27. L. Pasti, S. Roccasalvo, F. Dondi, P. Reschiglian: J. Polym. Sci., Part B: Polym. Phys. **33**, 1225 (1995)
28. C. Van Batten, M. Hoyos, M. Martin: Chromatographia **45**, 121 (1997)
29. S.L. Brimhall, M.N. Myers, K.D. Caldwell, J.C. Giddings: J. Polym. Sci., Polym. Phys. Ed. **23**, 2443 (1985)
30. K.-H. Cho, Y.H. Park, S.J. Jeon, W.-S. Kim, D.W. Lee: J. Liq. Chromatogr. Relat. Technol. **20**, 2741 (1997)
31. R.C. Reid, J.M. Prausnitz, B.E. Poling: *The Properties of Gases & Liquids*, 4th edition (McGraw-Hill, New York 1987)
32. G.-H. Ko, R. Richards, M.E. Schimpf: Sep. Sci. Technol. **31**, 1035 (1996)
33. M. Martin, C. Van Batten, M. Hoyos: Anal. Chem. **69**, 1339 (1997)
34. Y.S. Gao, K.D. Caldwell, M.N. Myers, J.C. Giddings: Macromolecules **18**, 1272 (1985)
35. P.M. Shiundu, E.E. Remsen, J.C. Giddings: J. Appl. Polym. Sci. **60**, 1695 (1996)
36. S. Lee: J. Microcol. Sep. **9**, 281 (1997)
37. M.E. Hovingh, G.H. Thompson, J.C. Giddings: Anal. Chem. **42**, 195 (1970)
38. M.N. Myers, K.D. Caldwell, J.C. Giddings: Sep. Sci. **9**, 47 (1974)
39. M. Nguyen, R. Beckett: Sep. Sci. Technol. **31**, 291 (1996)
40. J.C. Giddings, K.D. Caldwell, M.N. Myers: Macromolecules **9**, 106 (1976)
41. J.J. Gunderson, J.C. Giddings: Macromolecules **19**, 2618 (1986)
42. M.E. Schimpf, J.C. Giddings: J. Polym. Sci., Part B: Polym. Phys. **27**, 1317 (1989)
43. G. Meyerhoff, B. Rauch: Makromol. Chem. **127**, 214 (1969)
44. F.S. Gaeta, A. Di Chiara, G. Perna: Il Nuovo Cimento **66**, 260 (1970)
45. J.C. Giddings: Anal. Chem. **66**, 2783 (1994)
46. P.-G. De Gennes: *Scaling Concepts in Polymer Physics* (Cornell University Press, Ithaca 1979)
47. D. Melucci, C. Contado, I. Mingozzi, M. Hoyos, M. Martin, F. Dondi: J. Liq. Chromatogr. Relat. Technol. **23**, 2067 (2000)
48. W. Cao, P.S. Williams, M.N. Myers, J.C. Giddings: Anal. Chem. **71**, 1597 (1999)
49. W. Köhler, R. Schäfer: Adv. Polym. Sci. **151**, 1 (2000)
50. P. Reschiglian, M. Martin, C. Contado, F. Dondi: Int. J. Polym. Anal. Charact. **3**, 107 (1997)

51. W.-J. Cao, M.N. Myers, P.S. Williams, J.C. Giddings: Int. J. Polym. Anal. Charact. **4**, 407 (1998)
52. S. Lee, A. Molnár: Macromolecules **28**, 6354 (1995)
53. G. Liu, J.C. Giddings: Chromatographia **34**, 483 (1992)
54. P.M. Shiundu, G. Liu, J.C. Giddings: Anal. Chem. **67**, 2705 (1995)
55. S.J. Jeon, M.E. Schimpf, A. Nyborg: Anal. Chem. **69**, 3442 (1997)
56. E.P.C. Mes, R. Tijssen, W.Th. Kok: J. Chromatogr. A **907**, 201 (2001)
57. C. Van Batten, M. Hoyos, M. Martin: to be published

## Supplementary glossary

| | |
|---|---|
| $a_{1\ldots6}$ | coefficients in (33) |
| $b$ | channel breadth |
| $b'$ | exponent of molar mass in (56) |
| $b_{0\ldots3}$ | coefficients in (36) |
| $B$ | constant characteristic of the carrier liquid in (32) |
| $c$ | analyte molar concentration at equilibrium |
| $c_{\mathrm{tot}}$ | total molar concentration of the solution |
| $\langle c \rangle$ | cross-sectional average analyte molar concentration |
| $C_{\mathrm{p}}$ | carrier heat capacity at constant pressure |
| $d_{1\ldots5}$ | coefficients in (37) |
| $e_{0-1-2}$ | coefficients in (57) |
| $e'_{0-1-2}$ | coefficients in (41) |
| $f_{1\ldots5}$ | coefficients in (39) |
| $g$ | Earth gravitational acceleration |
| $h_{1\ldots5}$ | coefficients in (37) |
| $J_{\mathrm{x}}$ | molar flux of the analyte in the field ($x$) direction |
| $K_{1-2}$ | intermediate term in (39) |
| $L$ | channel length |
| $m_{1-2}$ | coefficients in (43) |
| $M$ | molar mass of the polymer |
| $n$ | exponent of the $S_{\mathrm{T}} - M$ power law in (58) |
| $n_{0-1-2}$ | coefficients in (60) |
| $Pé$ | transversal Péclet number, defined by (7) |
| $Pé_{\mathrm{app}}$ | apparent transversal Péclet number obtained by applying a constant $Pé$ retention equation to experimental data |
| $Pé_{\mathrm{o}}$ | transversal Péclet number at the accumulation wall |
| $Pé_{1}$ | transversal Péclet number at the depletion wall |
| $Pr$ | Prandtl number |
| $q$ | heat flux |
| $q_{1-2}$ | coefficients in (56) |
| $Q$ | carrier flow rate |
| $Ra$ | Rayleigh number, defined by (24) |
| $Ra_{\mathrm{crit}}$ | critical Rayleigh number for onset of flow instabilities |
| $s$ | dimensionless coordinate along the thickness of the channel, defined by (6) |

$s_{00-01-02}$ coefficients in (59)
$S_0$ prefactor in (58)
$t_{\mathrm{car},1}$ characteristic time for establishment of the fully developed flow velocity profile
$t_{\mathrm{car},2}$ characteristic time for establishment of the steady-state temperature profile
$t_{\mathrm{car},3}$ characteristic time for establishment of the steady-state analyte concentration profile
$t_{\mathrm{car},3,\max}$ maximum value of characteristic time $t_{\mathrm{car},3}$
$t_{\mathrm{o}}$ void time
$t_{\mathrm{R}}$ analyte retention time
$T$ temperature
$T_{\mathrm{c}}$ cold wall temperature
$T_{\mathrm{ref}}$ reference temperature
$\Delta T$ (positive) temperature difference between channel plates
$u$ carrier flow velocity
$u_{\mathrm{zone}}$ average migration velocity of the analyte zone along the channel
$\langle u \rangle$ average carrier flow velocity
$U_{\mathrm{x}}$ velocity induced by the field on the analyte in the x-direction (field direction)
$V_{\mathrm{o}}$ channel geometrical volume
$w$ channel thickness
$x$ coordinate along the thickness of the channel (measured from the accumulation wall)
$X$ analyte molar fraction
$X_{\mathrm{o}}$ analyte molar fraction at the accumulation wall
$y$ coordinate along the breadth of the channel
$\alpha$ cubic thermal expansion coefficient of the analyte solution
$\beta_{\mathrm{T}}$ carrier thermal diffusivity
$\delta$ difference in $1/P\acute{e}$ values between depletion and accumulation walls in (45)
$\eta$ carrier dynamic viscosity
$\eta_{\mathrm{o}}$ carrier dynamic viscosity at the accumulation wall
$\eta_{\mathrm{ref}}$ carrier dynamic viscosity at the the reference temperature $T_{\mathrm{ref}}$
$\lambda_{\mathrm{F}}$ analyte-field interaction parameter, defined by (7)
$\lambda_{\mathrm{F,app}}$ apparent value of $\lambda_{\mathrm{F}}$ obtained using a constant $\lambda_{\mathrm{F}}$ retention equation to experimental data
$\lambda_{\mathrm{F,o}}$ value of $\lambda_{\mathrm{F}}$ at the accumulation wall
$\lambda_{\mathrm{T}}$ carrier thermal conductivity
$\lambda_{\mathrm{T_c}}$ carrier thermal conductivity at $T_{\mathrm{c}}$
$\nu$ flow distortion parameter, defined by (30)
$\upsilon$ carrier kinematic viscosity
$\omega_{\mathrm{c}}$ relative rate of variation of the carrier thermal conductivity with temperature

# Thermodiffusion of Polymer Solutions in Convectionless Cells

Martin E. Schimpf

Department of Chemistry, Boise State University, Boise ID 83725, USA

**Abstract.** This contribution reviews the study of polymer thermodiffusion in convectionless cells. First, the development of thermodiffusion theory is summarized. Next, the apparent independence of thermodiffusion and degree of polymerization in polymers is discussed. Thermodiffusion coefficients ($D_T$) measured by thermal field-flow fractionation on several polymer-solvent mixtures are then summarized. Polymers include a variety of different molecular weight samples of polystyrene, poly($\alpha$-methyl styrene) , polyisoprene , and polymethylmethacrylate, as well as copolymers of styrene and isoprene. Solvents include benzene, toluene, ethylbenzene, tetrahydrofuran, 2-butanone, ethyl acetate, and cyclohexane. Measured values of $D_T$ do not correlate with predictive models of thermodiffusion, but are found to correlate with polymer density and solvent viscosity. Finally, using a hydrodynamic model of polymer thermodiffusion in nonpolar solvents, an expression is derived that expresses $D_T$ in terms of physicochemical parameters that can be estimated or measured independently. These parameters include the thermal expansion coefficient, molar volume and viscosity of the solvent, the radius of a polymer repeat unit, and the Hamaker constants. The model predicts general trends in the thermodiffusion of polymethylmethacrylate and polystyrene in solvents where such parameters are available, but overestimates the values of $D_T$ by a factor of 2–4.

## 1 Introduction

Thermodiffusion is the underlying mechanism in thermal field-flow fractionation (thermal FFF), which is an analytical tool used to separate and characterize polymers by molecular weight and chemical composition [1]. Thus, our initial interest in thermodiffusion was connected to the development of thermal FFF methodology. However, as we came to appreciate the high precision with which we could measure Soret coefficients by thermal FFF compared to other methods, our research efforts broadened to include fundamental studies of polymer thermodiffusion. One of our primary goals in this work is to develop a working model of polymer thermodiffusion, that is, a model that relates thermodiffusion coefficients to fundamental physicochemical parameters of the polymer-solvent system. If we can predict thermodiffusion in polymer solutions, even qualitatively, then we can use thermal FFF more effectively to separate and characterize polymers.

In our studies, we have considered a variety of thermodiffusion models. Some of these models were developed specifically for polymers, but we believe that models developed for liquid mixtures are also relevant because the thermo-

W. Köhler and S. Wiegand (Eds.): LNP 584, pp. 285–309, 2002.

phoretic velocity of polymers is apparently independent of the degree of polymerization. In fact, it is this independence that makes thermodiffusion so effective for separating polymers. Thus, when we establish a temperature gradient in the confined space of a thermal FFF channel, polymer molecules migrate toward the cold wall at a velocity that is equal to that of its individual monomers. That velocity is defined by the thermodiffusion coefficient $D_{\mathrm{T}}$, which is the velocity per unit temperature gradient. The buildup in concentration is opposed by ordinary (mass) diffusion. In contrast to $D_T$, however, the opposing motion, as defined by the mass diffusion coefficient $D$, is much smaller for polymers compared to monomers or other small molecules; $D$ also varies with a polymer's degree of polymerization. The result is that polymers with molecular weights on the order of $10^4$ g/mol or greater can be concentrated within a relatively thin layer next to the cold wall of a thermal FFF channel. Furthermore, the thickness of that layer varies with molecular weight, not due to differences $D_{\mathrm{T}}$, but due to differences in $D$. By coupling the concentrated polymer zone in a thermal FFF channel to a differential velocity profile in an orthogonal direction, we can separate polymers based on differences in the ratio $D_{\mathrm{T}}/D$, that is, by their Soret coefficients. Further details of the thermal FFF channel and methodology are discussed below.

## 2 Models of thermodiffusion

### Kinetic and thermodynamic models

There are many theories that attempt to give physical insight into the phenomenon of thermodiffusion. The kinetic theories of Enskog [2] and Chapman [3] are remarkably successful at predicting thermodiffusion in simple gas mixtures, such as isotopic mixtures. However, such models fail in gases with more complex structures, and it is not surprising that attempts to extend such models to the liquid state also proved unsatisfactory [4–6]. When applied to liquids, kinetic approaches assume that the lattice model for liquids is valid and are based on the view that thermodiffusion proceeds by a series of activated transitions. The activation energy required for the transfer of a particle from its position in the quasicrystalline liquid lattice to a vacant site or "hole" is related to a heat of transfer.

Extending their work on mutual diffusion, Prager and Eyring [7] developed a kinetic model of thermodiffusion discussed in terms of Brownian motion. The model relates the Soret effect to derivatives in the jump frequencies with respect to temperature and concentration. It correctly predicts that in mixtures of components with widely varying coefficients of mass diffusion, such as a polymer solution, the component that has the lower diffusion coefficient will tend to accumulate at the cold wall. Unfortunately, the jump frequencies cannot be determined independently.

In a comprehensive treatise on irreversible thermodynamics presented in 1950 [8], de Groot argued that the analysis of thermodiffusion in liquids should be based on the thermodynamics of irreversible processes. Many researchers followed this lead, as the majority of liquid-state models developed over the next 10–15 years were based on that foundation.

Transport models that are rooted in irreversible thermodynamics utilize the first and second law together with equations that define entropy production and mass balance. Thus, internal heats that are transferred with mass are balanced against those associated with the temperature gradient. The entropy associated with such processes is defined as a sum of two terms. Each term is a product of a flux that characterizes the process, such as mass or heat flux, and a force that is related to the non-uniformity of the system, such as a concentration or temperature gradient. Because irreversible fluxes are unknown parameters, and so cannot be solved with given initial boundary conditions for the state of the system, equations of state and entropy balance are supplemented with phenomenological equations that relate fluxes and forces appearing in the entropy source equation. For a binary system of components a and b, the phenomenological equations for heat flux ($J_\mathrm{Q}$) and mass flux ($J_\mathrm{l}$) are

$$J_Q = -L_\mathrm{qq}(\mathrm{grad}\ T)/T^2 - (L_\mathrm{ql}\ \mu_\mathrm{a}/c_\mathrm{b}T)\mathrm{grad}\ c_\mathrm{a} \tag{1}$$

$$J_\mathrm{l} = -L_\mathrm{lq}(\mathrm{grad}\ T)/T^2 - (L_\mathrm{ll}\ \mu_\mathrm{a}/c_\mathrm{b}T)\mathrm{grad}\ c_\mathrm{a} \tag{2}$$

where $T$ is temperature and $\mu_\mathrm{a}$ is $d\mu/dc$ for component a, $\mu$ being the chemical potential and $c$ the concentration. Parameters $L_\mathrm{qq}$ and $L_\mathrm{ll}$ are the phenomenological coefficients for thermal conductivity and ordinary diffusion, respectively. $L_\mathrm{ql}$ and $L_\mathrm{lq}$ are the phenomenological coefficients for the Dufour and Soret effects, that is, the movement of heat in a concentration gradient and the movement of mass in a temperature gradient, respectively. In defining the equation of motion in terms of internal and external forces, an important assumption is made, which is perhaps the greatest drawback to using irreversible thermodynamics to model secondary phenomena such as the Soret and Dufour effects. The assumption is that the pressure tensor resulting from short-range interactions between particles is symmetric. In other words, the interactions are equal on all sides of the particle under consideration.

For thermodiffusion, we are interested in equation (2). De Groot eliminated $L_\mathrm{lq}$ from this equation by introducing a specific heat of transfer $Q$, which is the energy transported across a reference plane per mole of component when there is no temperature gradient and all other components are stationary. If we subtract the partial molar enthalpy of the component from $Q$, we are left with the heat that must be absorbed by the region in order to maintain a constant temperature and pressure, per mole of component diffusing out of the region, for an infinitesimal change in composition. This heat is termed the heat of transport $Q^*$ , and governs the gradients in the mole fraction of two components relative to the temperature gradient. The resulting expression for the Soret coefficient $S_\mathrm{T}$ is

$$S_\mathrm{T} = \frac{D_\mathrm{T}}{D} = \frac{d\ln X_\mathrm{a}}{dT} = \frac{Q_\mathrm{a}{}^*}{X_\mathrm{a}T\ \mathrm{grad}\ \mu} \approx \frac{Q_\mathrm{a}{}^*}{RT^2} \tag{3}$$

where $D_T$ is the thermodiffusion coefficient, $D$ is the ordinary diffusion coefficient, and $X_a$ and $Q_a{}^*$ are the mole fraction and heat of transport, respectively, of component $a$. For ideal solutions, the approximation $X_a \cdot (\text{grad}\ \mu) \approx RT$ is used to obtain the rightmost expression of $S_T$ in (3). For a binary liquid mixture of components $a$ and $b$, the Soret coefficient is defined by

$$S_T = \frac{d \ln(X_a/X_b)}{dT} = \frac{Q_a{}^* - Q_b{}^*}{RT^2} \tag{4}$$

Thermodynamic theory does not relate $Q^*$ to other characteristics of the components making up the system under consideration. Parameter $Q^*$ is not a function of state in the normal sense; it is defined with respect to a reference plane and its magnitude depends on the mechanism by which diffusion takes place, for example, on the height of an energy barrier. In fact, (3) and (4), which are due to de Groot, are similar in form to those derived in the purely kinetic approaches. The primary difference in the latter is its use of the difference in the energy associated with the detachment of a diffusing particle from its neighbors and that associated with creation of holes in the solvent, in place of $Q^*$. Another difference is the use of molar concentrations instead of mole fractions, which requires the thermal expansion to be considered. A detailed comparison of thermodynamic and kinetic approaches by Denbigh [9] outlines the relationship between equations obtained for the Soret coefficient using the two approaches. In this comparison, Denbigh argues that the energy terms used in kinetic approaches must be regarded as free energies, so that the substantive difference in the thermodynamic approach is the lack of an entropy term in $Q^*$. For binary liquid mixtures, however, we can equate the difference term in the kinetic approach with $Q_a^* - Q_b^*$, but only if we assume the entropies are equal and therefore cancel. Denbigh's approach is to begin with the model of Hiby and Wirtz, but replace activation energies with analogous thermodynamic energies for the detachment of a molecule from its neighbors and the creation or filling of a hole. Denbigh's model relies on the sum of configurational energies, one for each pair of nearest neighbors, to obtain the transport energies.

From our point of view, the practical differences between the various thermodynamic and kinetic approaches lie in the methods used to relate the "energy" terms, however defined, to physicochemical parameters. By practical we mean that only through equations that relate $S_T$ to independently measured parameters can the models be tested, and ultimately used to predict thermodiffusion a priori. Prigogine asserts [6] that the transport energies should be related to cohesive energy densities, which are directly measurable, so that differences in the cohesive energy densities could be used to qualitatively predict the Soret coefficient. Denbigh also asserts that differences in cohesive energy densities can be used in place of his energy parameters to qualitatively predict $S_T$ in liquid mixtures.

The approach of Bearman , Kirkwood, and Fixman [10] employs the methods of statistical mechanics used in the calculation of other transport coefficients in simple fluids. The method expresses $Q^*$ as the sum of two terms, $Q_a$ and $Q_b$, where the former involves averages over equilibrium ensembles and the latter arises from the perturbation induced in the equilibrium distribution function by the flow of heat and matter associated with the temperature gradient. Parameter $Q_\mathrm{a}$ is defined by

$$Q_\mathrm{a} = \left( \frac{V_\mathrm{s} V_\mathrm{p}}{V_\mathrm{s} + V_\mathrm{p}} \right) \left( \frac{L_\mathrm{p}}{V_\mathrm{p}} - \frac{L_\mathrm{s}}{V_\mathrm{s}} \right) \tag{5}$$

where $V$ and $L$ are the molar volume and latent heat of vaporization, respectively. The two components are designated by subscript $p$ (for the solute, which we take to be polymer) and $s$ (for the solvent). The perturbation term $Q_\mathrm{b}$ is more complex and contains an integral involving the radial distribution functions in the perturbed and unperturbed states, and the intermolecular potentials, which were not evaluated for any actual system. As a result, the theory cannot be used in its entirety for predicting values of $D_\mathrm{T}$, nor can it be tested experimentally. However, the utility of using the equilibrium term $Q_\mathrm{a}$ in predicting $D_\mathrm{T}$ values will be examined.

The kinetic model developed by Drickamer and coworkers [11,12] is based on the view of Wirtz [5] that thermal diffusion proceeds by a series of activated transitions. However, a shape term is introduced which allows more or less than one molecule to fill the hole left by a different type of molecule. The theory starts with Denbigh's equations but incorporates molar volumes to adjust for the relative sizes of the two types of molecules. Instead of using cohesive energy density to approximate differences in the transport energies, the model uses activation energies for viscous flow, $U$. The resulting equation for the Soret coefficient of a binary mixture of components 1 and 2 (indicated as subscripts) is

$$S_\mathrm{T} = \frac{M_1 V_2 + M_2 V_1}{2RT^2} \left[ \frac{U_2}{V_2} - \frac{U_1}{V_1} \right] \tag{6}$$

where $M$ is molecular weight and $V$ is partial molar volume. In extending the model to polymers [13], the polymer is considered to move in segments containing a characteristic number of chain atoms $N_C$. The resulting equation for $S_\mathrm{T}$ is

$$S_\mathrm{T} = \frac{(m + w')[U_\mathrm{S} - (V_\mathrm{s} \rho_\mathrm{P} / 52 N_C) U_\mathrm{P}]}{2RT^2 \,(1 - \phi_\mathrm{P})^2) \,[1 + 2m(1/2 - \mu')\phi_\mathrm{P} + m\phi_\mathrm{P}^2]} \tag{7}$$

where $\rho_\mathrm{P}$ is polymer density, $\phi_\mathrm{P}$ is the volume fraction of polymer, and $\mu'$ is the empirical polymer-solvent interaction constant. Parameter $m$ is the ratio of molar volumes of polymer and solvent, and $w'$ is the ratio of molecular weights of polymer and solvent. Parameters $U_\mathrm{S}$ and $U_\mathrm{P}$ are activation energies for the transfer of solvent and polymer segments, respectively.

Ham developed a Brownian motion model specifically for polymers. In this model, flow is broken up into three fluxes, the superposition of which also assumes that a lattice model is valid. The result is similar to that of Emery and Drickamer in predicting the Soret coefficient to depend on the difference in activation energies governing transition probabilities. The difference between the two theories lie primarily in the manner of taking into account the discrepancy in size between the polymer and solvent molecules, which in Ham's theory results in activation energies of diffusion being weighted by the diffusion constants. As a result the polymer contribution to thermodiffusion is negligible; the resulting equation for $D_{\mathrm{T}}$ is

$$D_{\mathrm{T}} = \frac{U_{\mathrm{P}} D_{\mathrm{P}} - U_{\mathrm{S}} D_{\mathrm{S}}}{RT^2} \approx -\frac{U_{\mathrm{S}} D_{\mathrm{S}}}{RT^2} \tag{8}$$

Here, $D_{\mathrm{P}}$ and $U_{\mathrm{P}}$ are the mass diffusion coefficient and activation energy for viscous flow, respectively, of the polymer, and $D_{\mathrm{S}}$ and $U_{\mathrm{S}}$ are the same parameters for the solvent. Ham neglects the flexibility of the polymer in asserting thermodiffusion should be independent of the polymer. In our empirical assessment of Ham's model below, this approximation will be relaxed in defining $D_{\mathrm{P}}$ and $U_{\mathrm{P}}$ for a polymer repeat unit (mer) rather than the polymer as a whole.

Khazanovich [14] uses a similar approach to Ham, but considers the flexibility of the polymer chains. He begins with the Smoluchowski equation for diffusion, which defines the probability density in configurational space as a function of the friction coefficient of moving polymer segments. In treating the frictional properties of the polymer in solution, the polymer chain is represented by the "pearl necklace" or Rouse model, which considers the polymer chain as a series of rigid beads connected by infinitely thin linkages. In contrast to Ham, Khazanovich concludes that thermodiffusion depends on the segmental mobility of the polymer chain, whereas the solvent can be ignored because in dilute solutions the solvent frame coincides with the laboratory frame. The resulting equation for $D_{\mathrm{T}}$ is

$$D_{\mathrm{T}} = \frac{U_{\mathrm{mer}} D_{\mathrm{mer}}}{RT^2} \tag{9}$$

Further analysis by Khazanovich indicates that a polymer segment consists of one mer, as indicated by the subscript "mer" in (9).

## Independence of thermodiffusion and degree of polymerization

The diffusion of a particle in solution results from a force on the particle due to random collisions with molecules of the fluid. The dependence of diffusion on particle size is a result of an additional frictional force on the particle, which opposes the thermal motion of the particle. For polymers, we may consider two limiting cases. In one case, the mers in the Rouse model are far enough apart that they act essentially independent of one another. This model is referred to as the "free-draining coil". If such a coil is forced to move through a viscous fluid, each bead will be subjected to a frictional resistance that is independent of other mers, and the effective friction coefficient $f$ of the coil as a whole will be proportional to the number of mers in the coil. This results in an ordinary diffusion coefficient $D$ that is inversely proportional to the chain length, which in turn is proportional to the molecular weight $M$, i.e., $D \propto M^{-1}$. Clearly, the free-draining coil is not a correct model for mass diffusion, since empirically we find that $D \propto M^{-(0.5-0.8)}$.

In the second limiting case, the solvent is effectively trapped within the coil, which is therefore treated as a rigid sphere with an "equivalent *hydrodynamic* radius" $R_h$. According to Stoke's law, the friction coefficient of this "hydrodynamically equivalent sphere" is given by

$$f = 6\pi\eta R_h \tag{10}$$

where $\eta$ is the solution viscosity. In a theta solvent, $R_h \propto M^{1/2}$, and Stoke's law predicts that $f \propto M^{1/2}$. In better solvents, where the molecular coil is more expanded, $f$ assumes a correspondingly higher value, with the limit $f \propto M^{0.58}$.

The criteria for applying these two limiting cases were discussed by Debye and Bueche [15]. They assert that the appropriate model to be used depends on the shielding factor, which is the ratio $R_h/L_f$, where $L_f$ is the characteristic depth at which the flow velocity has been reduced to $1/e$ of its value at the surface of the polymer-solvent sphere. A free-draining model is applicable when $L_f \gg R_h$, while $R_h \gg L_f$ corresponds to a situation where most of the sphere is shielded from outside flow, so that it may be thought of as a single unit with entrapped liquid. The value of parameter $L_f$ depends on the nature of the force field causing motion in the fluid. For ordinary diffusion, $L_f \ll R_h$ and the frictional properties affecting molecular weight dependence correspond to the non-draining model. In thermodiffusion, however, the forces are more intense and penetrate the coil to a much greater extent. The shielding factor approaches zero and the frictional properties affecting molecular weight dependence correspond to the free-draining model. Thus, the friction coefficient for motion induced by a temperature gradient depends on $M$. The velocity $U_{th}$ associated with this motion, which we call thermodiffusion, can be represented by $U_{th} = F_{th}/f$, where $F_{th}$ is the associated force. The force $F_{th}$ is equal to the momentum pressure ($P$) times the area presented by all of the individual mers in the pearl-necklace model. The area, in turn, is proportional to the degree of polymerization, or $M$. Thus, $F_{th} \propto M$, $f \propto M$ and $U_{th}$ is independent of $M$.

de Gennes used a different approach to explain the independence of $U_{th}$ from $M$ in polymers [16]. He started with formulations for heat flux in a concentration gradient, which arise from two factors, the friction of each mer in the pearl-necklace model and the flow of the solvent. With dilute polymers, the uniform flow of solvent does not produce any heat flux. In fact, formulations for heat flux indicate that it is the second derivative of the solvent velocity that is responsible for heat flux. Therefore, even if the velocity of the solvent has long-range Stokes-type (i.e., viscous) behavior, the second derivative is much more localized around the source of friction at each mer. As a result, the heat flux associated with mass flux (the so-called "Dufour effect") is independent of the number of mers. Based on the Onsager reciprocity relationship, the mass flux associated with heat flux must also be independent of the number of mers. Thus, each mer moves essentially independently in a temperature gradient.

### Gaeta's radiation pressure model

Gaeta proposes [17] that thermal excitations in liquids should be analogous to those in solids, and therefore consist of a superposition of collective vibrations composed largely of longitudinal waves. The velocity of such waves were shown by Brillouin scattering to be equal or slightly greater than the velocity of ultrasonic waves in the same medium, and to be relatively independent of frequency. Alterations in momentum flux that arise from the longitudinal waves lead to what is referred to as a radiation pressure. For solutes larger than the wavelength of the longitudinal wave, the pressure that develops at the interface between a solute and solvent, and the resulting force acting on the solute, will be directed along the temperature gradient. Furthermore, the force will be with or against the flow of thermal energy, depending on whether the momentum conductivity ($G = \lambda_{\mathrm{T,s}}/\nu$ where $\lambda_{\mathrm{T,s}}$ is the thermal conductivity of the liquid and $\nu$ is the velocity of sound in the material of interest) is greater for the solute or solvent. The resulting equation is

$$D_{\mathrm{T}} = Z_{\mathrm{A}} \left( \frac{\lambda_{\mathrm{T,s}}}{W_{\mathrm{s}}} - \frac{\lambda_{\mathrm{T,p}}}{W_{\mathrm{p}}} \right) \frac{\sigma}{f} \tag{11}$$

where $Z_{\mathrm{A}}$ is the acoustical impedance, a quantity related to the density and velocity of sound waves in the solute and solvent, $\lambda_{\mathrm{T,p}}$ and $\lambda_{\mathrm{T,s}}$ are the thermal conductivities of the solute and solvent, respectively, $W_{\mathrm{p}}$ and $W_{\mathrm{s}}$ are the corresponding velocities of sound in the two materials, $\sigma$ is the cross-sectional area of the solute molecule, and $f$ is the friction coefficient of the solute moving in the solvent.

## 3 Studies of polymer thermodiffusion by ThFFF

### Methodology

Field-flow fractionation (FFF) is a class of analytical separation techniques in which a potential gradient is applied across the thin dimension of a ribbon-shaped flow channel, as illustrated in Fig. 1. The length of a typical FFF channel is on the order of 40 cm, the breadth 1–2 cm, and the thickness, across which the field is applied, is on the order of 0.1 mm. Because of the channel's high aspect ratio, fluid forced through the channel flows in a laminar fashion, with a velocity profile $\nu(x)$ across the thin dimension ($x$) defined by

$$\nu(x) = \frac{\Delta P}{2L\eta} x(w - x) \tag{12}$$

where $\Delta P$ is the pressure drop along the channel length $L$, $\eta$ is the viscosity of the carrier liquid, and $w$ is the channel thickness. Thus, the velocity profile is parabolic across the thin dimension with reduced flow near the walls. A sample injected into the channel is forced to one wall by the applied field, where its velocity is reduced relative to the average velocity of fluid. The buildup in concentration at the accumulation wall is opposed by ordinary diffusion and a dynamic steady-state condition is reached in which the field-induced motion and back-diffusion are balanced. The result is a concentration profile defined by

$$c(x) = c_0 e^{-x|U|/D} \tag{13}$$

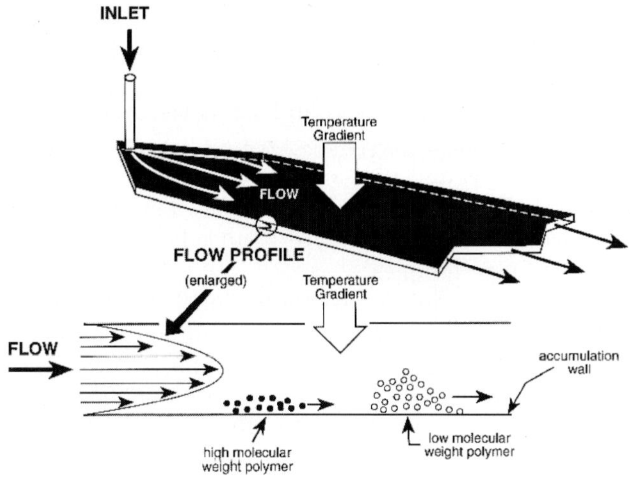


**Fig. 1.** The principles of separation by thermal FFF

where $c_0$ is the concentration at the accumulation wall, $U$ is the field-induced velocity, and $D$ is the ordinary diffusion coefficient. Analyte materials that interact more strongly with the field have a center of gravity closer to the wall and are therefore carried more slowly through the channel by the moving fluid, subsequently referred to as the carrier liquid. When a mixture is injected that contains components that interact differently with the field, the components are separated as they are carried through the channel.

In thermal FFF (ThFFF) the applied field is a temperature gradient and the distribution of analyte along the gradient is governed by its Soret coefficient. ThFFF is an effective tool for separating polymers based on differences in their degree of polymerization and chemical composition, both of which influence a polymer's Soret coefficient. In deriving equations that relate zone velocity in FFF to physicochemical parameters it is convenient to define a retention parameter as $\lambda_{\mathrm{F}}$ as follows:

$$\lambda_{\mathrm{F}} = \frac{D}{w|U|} \tag{14}$$

In ThFFF, the field-induced velocity is $U_{\mathrm{th}} = D_{\mathrm{T}}/(dT/dx)$, which can be substituted into (14) to obtain

$$\lambda_{\mathrm{F}} = \frac{D}{wD_{\mathrm{T}}(dT/dx)} \tag{15}$$

By considering the flow and concentration profiles defined in (12) and (13), the $\lambda_{\mathrm{F}}$-value of an analyte component can also be related to the volume $V_{\mathrm{r}}$ of carrier liquid that is required to flush it through the channel. If we neglect the temperature dependence of the various parameters involved, namely $U$ and $\eta$, we obtain

$$R = \frac{V^0}{V_r} = 6\lambda_F \left[\coth\left(\frac{1}{2\lambda_F}\right) - 2\lambda_F\right] \tag{16}$$

where $R_F$ is termed the retention ratio and $V^0$ is the geometric volume of the channel. Thus, retention parameter $\lambda_F$ serves as a link between $R_F$, which is measured in the laboratory, and physicochemical parameters such as the Soret coefficient ($S_T = D_T/D$) in the case of ThFFF. The ability to measure the physicochemical parameters of an analyte is an advantage that FFF has over chromatographic elution techniques.

The variation in temperature across the channel in ThFFF induces perturbations in the profiles of both sample concentration and parabolic flow. Martin and coworkers [18] developed corrections to (15) and (16) that account for such effects. Thus, one measures $R_F$ in the laboratory and calculates $S_T$ for a polymer-solvent system using the corrected forms of (15) and (16). If the corrected forms are not used, calculated values of $S_T$ can have errors of up to 6%, depending on experimental conditions [19]. Values of $D_T$ can be extracted from $S_T$-values provided an independent measure of $D$ is available. In our work, we have chosen to focus on $D_T$ rather than $S_T$ because the physicochemical parameters that govern $D$ are already well understood.

## Results

In our initial studies of thermodiffusion [20], the behavior of 30 different polystyrene samples were examined in ethylbenzene. Values of $D$ were measured by dynamic light scattering or calculated from sedimentation velocities, corrected for the temperature in the ThFFF channel, and combined with ThFFF retention data to obtain $D_T$ values. The samples varied in both molecular weight and branching configuration, and included linear, star, and comb-shaped polymers. The star-shaped samples included several different molecular weight examples each of 4-arm, 6-arm, and 12-arm configurations. The comb-shaped polymers had 28 arms attached to linear chains that varied in length between samples. $D_T$ was shown to be a constant for all these samples in ethylbenzene, regardless of molecular weight or branching configuration.

The independence of $D_T$ and molecular weight was confirmed in a subsequent study of 17 different polymer-solvent systems [21]. Figure 2 displays plots of $D_T$ versus molecular weight for several of these combinations. Although $D_T$ values vary with the composition of both polymer and solvent, no systematic influence of molecular weight is seen in this plot. Table 1 summarizes the molecular-weight-independent values of $D_T$ for the 17 polymer-solvent combinations. From three to five different molecular weights were examined for each combination. Table 1 reports the average values and their standard errors for each combination based on the different molecular weights. Error analysis indicates that the uncertainty in $D_T$ values comes primarily from uncertainties in $D$, which were determined by a secondary method in this case. When the corrected forms of (15) and (16) are used to calculate $S_T$ from ThFFF retention data, the relative uncertainties in $S_T$ are on the order of 2% or less.

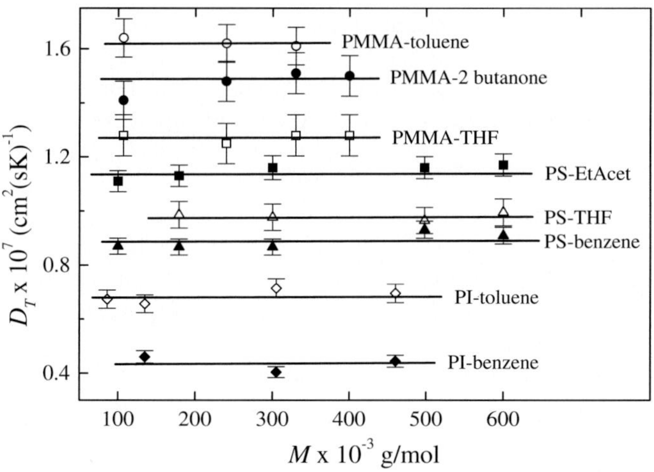


**Fig. 2.** Plot of $D_{\mathrm{T}}$ versus molecular weight $M$ for a representative set of polymer-solvent combinations, illustrating that $D_{\mathrm{T}}$ depends on both polymer and solvent but is essentially independent of $M$

**Table 1.** Summary of $D_{\mathrm{T}}$ Values x $10^8$ $(\mathrm{cm}^2(\mathrm{sK})^{-1})$ measured by ThFFF

| | PMMA | P$\alpha$MS | PS | PI |
|---|---|---|---|---|
| 2-Butanone | 14.8 (0.4) | | 13.9 (0.5) | |
| Ethyl Acetate | | | 11.6 (0.4) | |
| Tetrahydrofuran | 13.3 (0.9) | 12.7 (0.7) | 10.0 (0.5) | 5.7 (0.3) |
| Toluene | 16.3 (0.7) | 11.9 (0.5) | 10.3 (0.5) | 5.9 (0.3) |
| Ethylbenzene | | | 9.5 (0.4) | |
| Benzene | 13.7 (0.6) | 10.2 (0.4) | 8.9 (0.3) | 4.4 (03) |
| Cyclohexane | | | 6.6 (0.4) | |

An examination of the trends in $D_{\mathrm{T}}$ shows that in each solvent the polymer order for decreasing $D_{\mathrm{T}}$ values is the same: polymethylmethacrylate (PMMA) has the highest $D_{\mathrm{T}}$ value and polyisoprene (PI) has the lowest. This is the same order as the polymer densities. Within the individual polymer types the solvent order for increasing $D_{\mathrm{T}}$ values differs from one polymer to another. Polystyrene (PS) and poly($\alpha-$methyl styrene) (P$\alpha$MS) show similar solvent trends, but for PMMA the solvents toluene and benzene are displaced to relatively higher $D_{\mathrm{T}}$ values. For P$\alpha$MS and PI, changing from tetrahydrofuran (THF) to toluene has a markedly different effect. A two-way analysis of variance (ANOVA) with $D_{\mathrm{T}}$ as the dependent variable and solvent and polymer type as the independent variables indicates that $D_{\mathrm{T}}$ varies with both polymer and solvent [21]. Moreover, the two-way ANOVA indicates an interaction between polymer and solvent parameters with $> 99\%$ confidence. Thus, the influences of polymer and solvent cannot be considered separately in polymer thermodiffusion.

In solvent mixtures, the thermodiffusion of polymers is a linear function of solvent composition in some cases but not in others [22,23]. In studying this phenomenon, we converted our ThFFF channel into a Thomaes cell [25] in order to measure the partitioning of binary solvents in a temperature gradient. Our objective was to find a correlation between thermodiffusion of the solvent components and that of a polymer dissolved in the same mixture. We found that the denser component of a solvent mixture tends to partition to the cold wall, although there are exceptions, primarily with hydrogen-bonding solvents. We also found that thermodiffusion of polymers in solvent mixtures is enhanced beyond that expected from a linear dependence of $D_{\mathrm{T}}$ on solvent composition when the better solvent for the polymer partitions to the same (cold) wall as the polymer. When the better solvent partitions to the hot wall, thermodiffusion of the polymer is diminished from that expected from a linear dependence of $D_{\mathrm{T}}$ on solvent composition.

In its standard configuration, the ThFFF channel is oriented with the hot wall above the cold wall. Although we don't expect convection or gravitational effects in the thin FFF channel ($w = 0.1$ mm), we decided to examine the effect of channel orientation on polymer thermodiffusion because of the apparent correlation between thermodiffusion and either solvent density (in solvent mixtures) or polymer density [26]. If the correlation is an artifact of a separate density effect in operation, then we can expect measurements of thermodiffusion to be affected by channel orientation. The effect of channel orientation on polymer thermodiffusion in four homogeneous liquids and ten binary liquid mixtures was examined. In each case thermodiffusion of the polymer was not affected by channel orientation. Therefore, the correlation of thermodiffusion to density is real and not due to a separate density effect.

We also conducted a series of systematic studies on the thermodiffusion of copolymers in solution. The copolymers can be organized into groups whose members differ in one major aspect only. These differences include monomer ratios in random copolymers, block arrangements in block copolymers having both linear and star-shaped configurations, and arm numbers and lengths in star-shaped block copolymers. By examining the influence of these distinctive features on $D_{\mathrm{T}}$, two principal observations regarding the thermodiffusion of copolymers has been made. First, for random copolymers, $D_{\mathrm{T}}$ can be described by a weighted average of the $D_{\mathrm{T}}$ values of the corresponding homopolymers, where the weighting factors are the mole-fractions of each type of monomer unit (mer) in the copolymer [27,28]. Second, for block copolymers, thermodiffusion is a linear function of composition only when the solvent is equally good for all mers. When the solvent is better for one type of mer versus another, thermodiffusion of the copolymer is weighted more toward that of the better solvated mers. For example, Figure 3 plots $D_{\mathrm{T}}$ versus composition for several polystyrene-isoprene copolymers (PSI) in the solvents tetrahydrofuran and cyclohexane. In tetrahydrofuran, which is an equally good solvent for both polystyrene and polyisoprene, there is a linear dependence of $D_{\mathrm{T}}$ on copolymer composition. However, in cyclohexane, which is a better solvent for isoprene, values of $D_{\mathrm{T}}$ are closer to

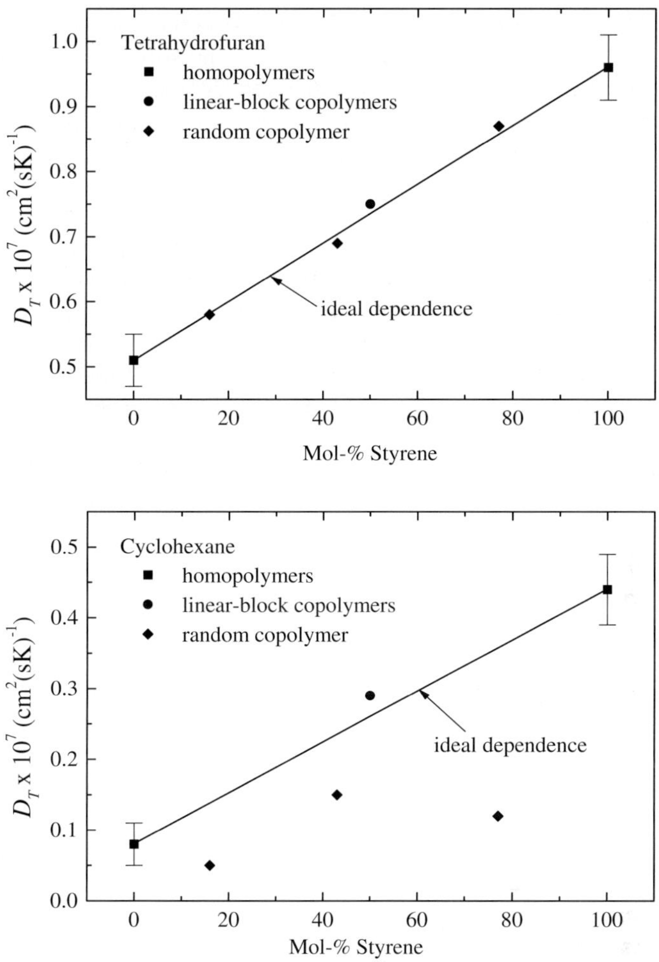


**Fig. 3.** Plots of $D_{\mathrm{T}}$ versus copolymer composition in copolymers of styrene and isoprene dissolved in tetrahydrofuran (top) and cyclohexane (bottom), illustrating that $D_{\mathrm{T}}$ values are biased toward that of the better solvated mers

that measured on polyisoprene homopolymers in the same solvent. In toluene, which is a better solvent for styrene, it was shown [27] that $D_{\mathrm{T}}$ values in PSI copolymers are more heavily weighted toward that of polystyrene.

## 4 Correlation of polymer thermodiffusion data

Thermodynamic approaches to thermodiffusion begin with (3), which can be rearranged to predict a linear dependence of $D_{\mathrm{T}}$ on $D/T^2$ in a given polymer-solvent system, $R$ and $Q^*$ being constants. In a paper by Brimhall et al. [29], thermodiffusion data obtained by ThFFF is reported on several different molecular weight polystyrene polymers dissolved in ethylbenzene and analyzed at temperatures in the range of 270–424 K. Using corrected forms of (15) and (16) that account for the temperature dependence of the solvent thermal conductivity and viscosity, we calculated Soret coefficients from their data. Corrections for the temperature dependence of $S_{\mathrm{T}}$ were not considered. The values of $S_{\mathrm{T}}$ were then combined with values of $D$, estimated using a semi-empirical relationship that expresses $D$ as a function of molecular weight and temperature, to obtain values of $D_{\mathrm{T}}$. Figure 4 plots the $D_{\mathrm{T}}$ values as a function of $D/T^2$ for this data. The correlation supports the general validity of the thermodynamic model. The slope of this plot is then related to the heat of transport for polystyrene in ethylbenzene. It should be noted that the thermodiffusion data represented in Fig. 4 was accumulated over a period of 10 years by several workers using different ThFFF channels, which were under development at the time. It is likely that a better correlation would emerge if the measurements were retaken using a single ThFFF channel.

As discussed above, we still need a way to estimate $Q^*$ from physicochemical parameters in order to predict values of $D_{\mathrm{T}}$ a priori. Prigogine asserted that $Q^*$ should be related to the cohesive energy density, which equals the square of the Hildebrand solubility parameter [30]. Solubility parameters of a variety of polymers and solvents are tabulated in the literature [31]. According to (4), $D_{\mathrm{T}}$ should correlate with the difference in the squares of the solubility parameters for the polymer and solvent, $\delta_{\mathrm{p}}^2 - \delta_{\mathrm{s}}^2$. Such a correlation is not obtained, as illustrated in Figure 5, which plots $D_{\mathrm{T}}$ versus $\delta_{\mathrm{p}}^2 - \delta_{\mathrm{s}}^2$ for polystyrene, polyisoprene, and polymethylmethacrylate in several solvents. In these and subsequent plots that incorporate the data summarized in Table 1, the effect of temperature is ignored because it varied by less than 2% in all of the measurements.

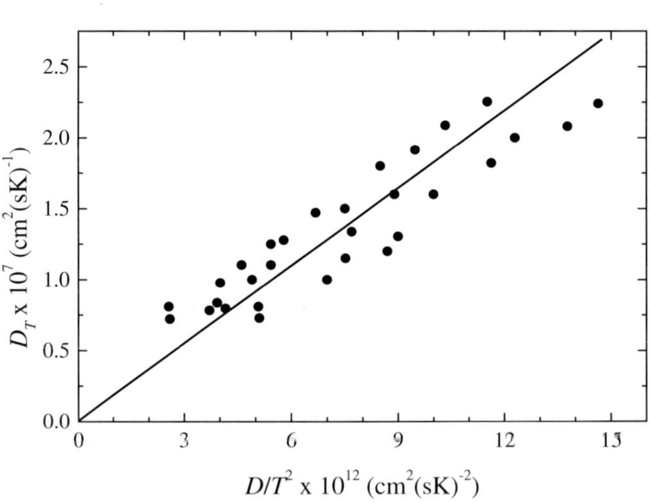


**Fig. 4.** Correlation of thermodiffusion with $D/T^2$, as predicted by the general thermodynamic approach

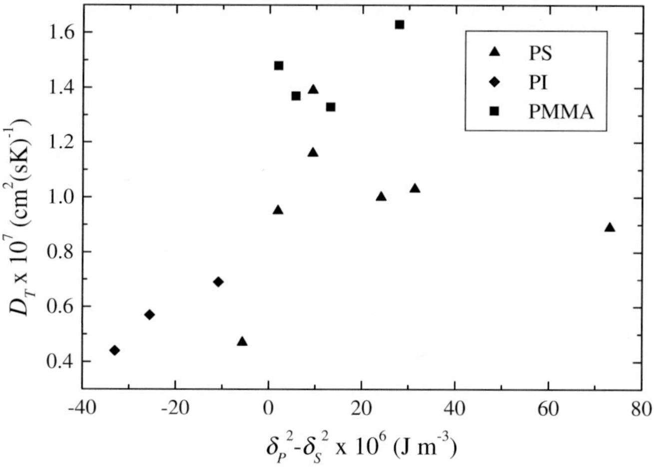


**Fig. 5.** Correlation of $D_{\mathrm{T}}$ with Prigogine's model of thermodiffusion

The model of Emery and Drickamer is similar in principle to that of Prigogine, except that they add a shape term and evaluate the non-ideality inherent in the chemical potential gradient, which includes a polymer-solvent interaction coefficient. The model is difficult to evaluate due to a lack of physicochemical parameters on polymer-solvent systems. However, polystyrene has been widely studied and well characterized. In evaluating the Emery–Drickamer model for polystyrene in the 7 solvents contained in Table 1, we obtained a correlation plot that is nearly identical to that obtained using the Prigogine model, as illustrated in Figure 5. Thus, the predictive capacity changes very little when the model is adjusted for differences in molecular shape and other non-idealities.

Ham's model, which is fundamentally kinetic in its design, predicts that $D_{\mathrm{T}}$ will correlate with differences in the products of the diffusion coefficient and activation energy for viscous flow for the polymer and solvent. If we use the diffusion coefficient of a mer ($D_{\mathrm{mer}}$) rather than that of the entire polymer, as Ham did, then the polymer terms in (8) cannot be ignored. However, $U_{\mathrm{p}}$ and $D_{\mathrm{p}}$ should be replaced with the corresponding values for a repeat unit, $U_{\mathrm{mer}}$ and $D_{\mathrm{mer}}$, respectively. Figure 6 plots $D_{\mathrm{T}}$ versus $D_{\mathrm{mer}}U_{\mathrm{mer}} - D_{\mathrm{s}}U_{\mathrm{s}}$ for polystyrene in several solvents. The resulting correlation is poor. Khazanovich's model, which is similar to Ham's without the solvent terms, correlates no better with measured values of $D_{\mathrm{T}}$, as illustrated in Figure 7.

The fundamental assumption in the kinetic models, which may be responsible for its failure to predict trends in polymer thermodiffusion is the lack of asymmetry in the transition probabilities, just as asymmetry in the interaction potentials is neglected in thermodynamic models. Asymmetry in the average distributions of collisions as a consequence of the temperature gradient forms the basis of the approach taken by Bearman, Kirkwood and Fixman. The model derives the nonequilibrium perturbation of the radial distribution function of the polymer and equates the resulting force with the thermodynamically modeled force to

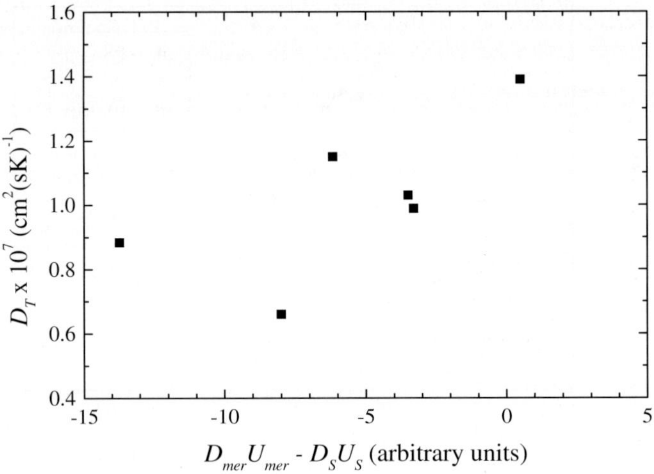


**Fig. 6.** Correlation of $D_{\mathrm{T}}$ with Ham's model of thermodiffusion

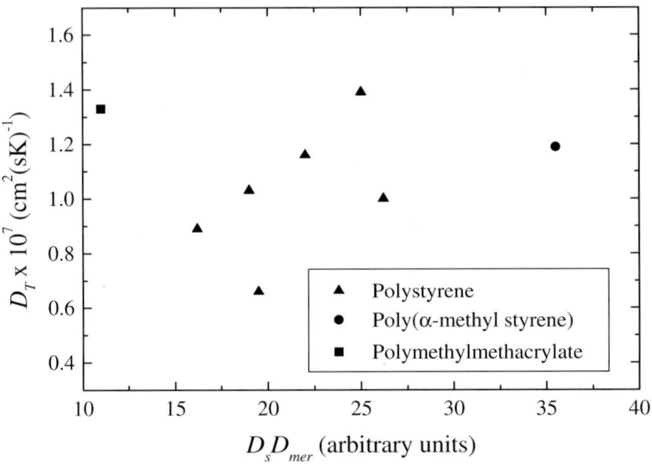


**Fig. 7.** Correlation of $D_{\mathrm{T}}$ with Khazanovich's model of thermodiffusion

obtain a statistical mechanical formula for thermodiffusion. Unfortunately, the perturbation term contains intermolecular forces that are neither available nor related to polymer-solvent parameters. The model also fails to consider flexibility in the polymer chain. Without the perturbation term, the model takes a similar form to that of Prigogine, with the addition of a term to account for differences in molecular size. As with the other models, trends in thermodiffusion data are not correctly predicted.

If we consider a polymer molecule as a single large solute molecule, then according to Gaeta's model (11), certain polymers satisfy the condition for migration to the hot wall, contrary to laboratory observations. Furthermore, $D_{\mathrm{T}}$ is predicted to increase directly with the surface area of the polymer, and therefore

**Table 2.** Polymer and Solvent Parameters Examined for Correlation to Thermodiffusion

| Polymer and Solvent | Solvent |
|---|---|
| density | boiling point |
| solubility parameter | reduced boiling point |
| sound velocity | surface tension |
| specific heat | temperature dependence |
| specific volume | of thermal conductivity |
| thermal conductivity | activation energy of |
| thermal expansion | viscous flow |

with molecular weight; this is also contrary to empirical data. In Gaeta's own analysis of his theory [32], he models polystyrene in cyclohexane as a long thin cylinder in evaluating its frictional resistance to thermodiffusion, and he uses the model to explain the dependence $S_{\mathrm{T}} \propto M^{1/2}$. Unfortunately, in order to obtain a stronger dependence of $S_{\mathrm{T}}$ on $M$, which is observed empirically in certain solvents, Gaeta asserts that free volume must decrease. Thus, the model is not consistent with the behavior of polystyrene in a good solvent, where the free volume increases, yet $S_{\mathrm{T}} \propto M^{0.6}$. If instead we consider the polymer molecule to consist of individual mers linked together, as in the free-draining limit of the Rouse model, then $D_{\mathrm{T}}$ can be expected to increase with the size of the mers rather than with the degree of polymerization. However, Gaeta asserts that in this case, where the individually affected molecules are small, (11) does not apply.

Failing to find a model for even a qualitative prediction of thermodiffusion in dilute polymer solutions, we turned to statistical methods in an attempt to find a relationship between the values of $D_{\mathrm{T}}$ summarized in Table 1 and commonly available polymer and solvent parameters. The list of parameters examined is given in Table 2. $D_{\mathrm{T}}$ was found to correlate most strongly with the density of the polymer ($\rho_{\mathrm{P}}$) and the solvent viscosity ($\eta$). Figure 8 illustrates the correlation of $D_{\mathrm{T}}$ with a combination of these two parameters. The correlation with solvent viscosity suggests that solvent flow plays a critical role. Furthermore, the fact that polymer thermodiffusion in solvent mixtures is enhanced when the better solvent partitions in the same direction suggests that polymer thermodiffusion is influenced by an interaction of the mers with solvent molecules as they flow past one another.

## 5 Hydrodynamic model of polymer thermodiffusion in nonpolar solvents

The basic difficulty with modeling thermodiffusion in liquids is the complicated interactions that occur between molecules. Yet it is clear that any model of polymer thermodiffusion in dilute solutions must consider such interactions. As

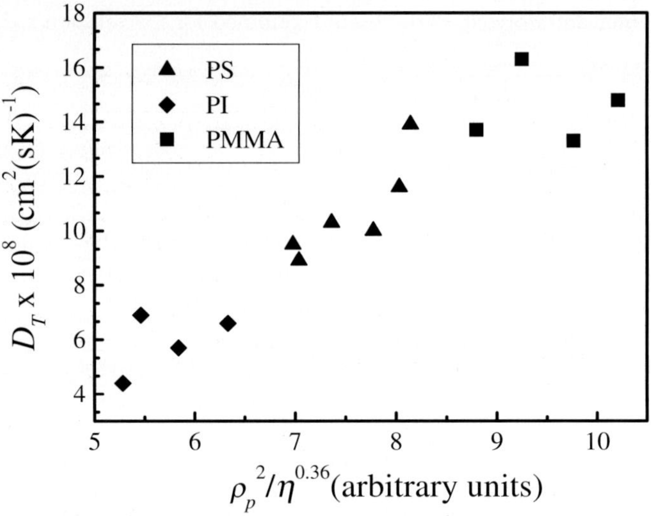


**Fig. 8.** Correlation of $D_{\mathrm{T}}$ with polymer density and solvent viscosity

stated above, models based on irreversible thermodynamics assume the pressure tensor resulting from short-range interactions is symmetric. In addition, transport energies are estimated by activation energies for molecular motion. Worse yet, activation energies for motion can't be put in terms of independently measurable parameters, so the best we can do is estimate them by activation energies for diffusion. Since this information is not readily available, we are tempted to use activation energies for viscous flow instead. Thermodynamically the latter two parameters differ by the heat of mixing. With all these assumptions and approximations, it is not surprising that thermodynamic and kinetic models of thermodiffusion fail to yield satisfactory results.

A hydrodynamic approach to modeling polymer thermodiffusion was recently proposed by Schimpf and Semenov [33]. The model begins with the assumption that thermodiffusion is independent of polymer molecular weight for the reasons discussed above. Consequently, the model considers the environment around, and consequent movement of, individual mers in the polymer chain. The underlying concept is that an osmotic pressure gradient is produced in solvent layers around the mers due to small changes in the concentration of solvent molecules (i.e., solvent density) as a result of the mer-solvent interaction energy. Thermodiffusion of polymers results from the temperature dependence of this osmotic pressure gradient. Using the general dependence of interaction potentials on the distance between interacting objects, an expression is obtained that relates polymer thermodiffusion to parameters that can be independently measured. Only an outline of the model is given here; a detailed description is contained in [33].

In nonpolar solvents, where ions are present in vanishing quantities, only dipole-dipole interactions play a role [34]. The potential of these interactions have a common dependence on the distance between dipoles and may be written in the following form:

$$\Phi(r) = -\frac{16A(r_{\rm m} \cdot r_0)^3}{9r^6} \tag{17}$$

Here $A$ is the Hamaker constant, which characterizes the potential energy of dipole-dipole interactions between mers of radius $r_{\rm m}$ and solvent molecules of radius $r_0$. Derivation of the thermodiffusion coefficient can be divided into three steps: 1) Calculation of the temperature distribution around the mer. 2) Derivation of the solvent molecule concentration and related osmotic pressure distribution around the mer due to mer-solvent interactions. 3) Derivation of the general expression for the solvent velocity profile around the mers and the resulting polymer thermophoretic velocity using (17) for the mer-solvent interaction potential.

The temperature distribution around the mer is given by the following equation:

$$T_e = T_0 + r_{\rm m} \cos \upsilon \left( \frac{r}{r_{\rm m}} + \frac{n-1}{n+2} \frac{r_{\rm m}^2}{r^2} \right) \nabla T \tag{18}$$

Here, $T_0$ is the temperature at the center of the particle, $\upsilon$ is the angle between the radius vector and the outer temperature gradient $\nabla T$, $r$ is the distance from its center, and $n$ is the ratio of mer to solvent thermal conductivity. Under steady-state conditions, the distribution of solvent molecules around the mer is defined by

$$\frac{d\Pi}{d\phi} \nabla \phi + \frac{6\phi}{\pi \cdot r_0^3} \nabla \Phi = 0 \tag{19}$$

where $\Pi$ is the excess osmotic pressure raised around the mer, $\Phi$ is the force potential given by (17), and $\phi$ is the volume fraction of the solvent molecules, which is related to the numeric concentration using a hard sphere model for the solvent molecules.

After introducing the small parameter $\delta = 1 - \phi$, perturbation theory is used to define the osmotic pressure resulting from small differences in the local concentration of solvent molecules:

$$\Pi = p - p_0 = c_{\rm m} kT \left[ -\left| p_0{}' \right| \cdot (\delta - \delta_0) + \frac{p_0{}''}{2} \cdot (\delta - \delta_0)^2 \right] \tag{20a}$$

Here $p$ is the local pressure and the subscript in $p_0$ and $\delta_0$ represent the respective values far from the mer surface, where the interaction potential is zero. Parameter $c_{\rm m}$ is the densest possible packing concentration for the solvent molecules and

$$p_0{}' = \frac{1}{c_{\rm m} kT} \left( \frac{\partial p}{\partial \delta} \right)_{\delta = \delta_0} \tag{20b}$$

$$p_0{}'' = \frac{1}{c_m kT} \left( \frac{\partial^2 p}{\partial \delta^2} \right)_{\delta = \delta_0} \tag{20c}$$

Equation (20) is a general expression that relates osmotic pressure to variations in the solvent molecule concentration without connecting it to a concrete model equation of state. Substituting (20) into (19) and solving for the perturbation term allows the excess osmotic pressure to be redefined as follows:

$$\Pi = c_{\mathrm{m}}\phi_0 \left[ -\Phi(r) + \frac{1}{2|p_0'|}\frac{\Phi^2(r)}{kT(r,\upsilon)} \right] \tag{21}$$

where only the first two powers of the small parameter were kept. By introducing the isothermal compressibility ($\beta_{\mathrm{T}}$) and thermal expansion coefficient ($\alpha$), the osmotic pressure and its gradient can be further redefined as

$$\Pi = -\frac{\Phi}{\nu_0} + \frac{\beta_{\mathrm{T}}\Phi^2}{2{\nu_0}^2} \tag{22}$$

$$\nabla\Pi = \frac{\alpha}{\nu_0}\left[\Phi + \left(\frac{\partial\beta_{\mathrm{T}}/\partial T}{2\alpha} - \beta_{\mathrm{T}}\right)\frac{\Phi^2}{\nu_0}\right]\nabla T \tag{23}$$

where $\nu_0$ is the volume occupied by a solvent molecule. Equation (23) represents the expression for the temperature-related osmotic pressure gradient, which is used to solve the hydrodynamic problem of solvent flow around the mers.

The flow velocity profile in the mer surface layer is defined by the Navier–Stokes equation, which is solved using the approach of Teubner [35]. In this approach, particle movement is related to the osmotic pressure gradient and fluid velocity distribution in the space around a particle moving with constant unit velocity $\boldsymbol{U_1}$. The velocity distribution, which satisfies the conditions $\boldsymbol{U_1}(r = r_{\mathrm{m}}) = \boldsymbol{u_0}$ and $\boldsymbol{U_1}(r = \infty) = 0$, is

$$\boldsymbol{U_1}(\boldsymbol{r}) = \frac{3}{4}\frac{r_{\mathrm{m}}}{r}[\boldsymbol{u_0} + \boldsymbol{n_0}\cdot(\boldsymbol{u_0}\cdot\boldsymbol{n_0})] + \frac{1}{4}\left(\frac{r_{\mathrm{m}}}{r}\right)^3[\boldsymbol{u_0} - 3\boldsymbol{n_0}\cdot(\boldsymbol{u_0}\cdot\boldsymbol{n_0})] \tag{24}$$

where $\boldsymbol{u_1}$ is the unit vector directed along the $\boldsymbol{U_1}$ vector. Instead of the volume force considered by Teubner, which is compensated for by the temperature dependence of the interaction potential, we combine the osmotic pressure gradient defined in (23) with the temperature distribution defined in (18). Under steady-state conditions, where the sum of the hydrodynamic friction force and thermophoretic force are zero, we obtain the following expression for thermophoretic velocity:

$$U_{\mathrm{T}} = \frac{1}{6\pi\eta r_{\mathrm{m}}}\int_0^{\pi}\sin\upsilon\cdot d\upsilon\int_{r_H}^{\infty}2\pi r\cdot dr\frac{d\Pi}{dT}(\boldsymbol{U_1}\nabla\boldsymbol{T}) \tag{25}$$

By substituting (17), (18), (23), and (24) into (25) and carrying out some simple but cumbersome integral calculations we obtain the following expression for the thermophoretic velocity of the mer, and therefore of the entire polymer chain:

$$\begin{aligned} U_{\mathrm{T}} = \frac{\alpha r_{\mathrm{m}}^2}{3\eta}\Bigg[ & -\left(1 + \frac{1}{28}\frac{n-1}{n+2}\right)\frac{A}{9\nu_0} \\ & + \frac{2\beta_{\mathrm{T}}}{5}\left(1 + \frac{5}{364}\frac{n-1}{n+2}\right)\left(\frac{\partial\ln\beta_{\mathrm{T}}/\partial T}{2\alpha} - 1\right)\left(\frac{16A}{9\nu_0}\right)^2\Bigg]\nabla T \end{aligned} \tag{26}$$

Evaluation of (26) shows that terms containing the ratio $n$ of thermal conductivities are negligible. Ignoring these terms, the thermodiffusion coefficient $D_{\mathrm{T}}$, which is defined as the thermophoretic velocity per unit temperature gradient, is expressed as

$$D_{\mathrm{T}} = \frac{\alpha r_{\mathrm{m}}^2}{3\eta}\left[-\frac{16A}{9\nu_0} + \frac{2\beta_{\mathrm{T}}}{5}\left(\frac{\partial \ln \beta_{\mathrm{T}}/\partial T}{2\alpha} - 1\right)\left(\frac{16A}{9\nu_0}\right)^2\right] \tag{27}$$

Equation (27) contains two parameters that characterize the mer, namely its radius $r_{\mathrm{m}}$ and its interaction potential with the solvent $A$. Several parameters characterize the solvent, including the thermal expansion coefficient $\alpha$, the isothermal compressibility $\beta_{\mathrm{T}}$ and its temperature dependence, molecular volume $\nu_0$, and the solvent viscosity $\eta$. In evaluating the model from empirical data, $r_{\mathrm{m}}$ and $\nu_0$ are approximated from the molecular mass and density using the hard sphere model. Values for molecular mass and density, as well as $\alpha$ and $\eta$ were obtained from a single reference [36]. Values for the isothermal compressibility are more difficult to obtain. Fortunately, the terms containing $\beta_{\mathrm{T}}$ and its temperature dependence in (27) can be ignored unless the Hamaker constant $A$ is very high, which simplifies (27) to the following form:

$$D_{\mathrm{T}} \approx -\frac{16\alpha r_{\mathrm{m}}^2}{27\eta}\left(\frac{A}{\nu_0}\right)^2 \tag{28}$$

There is no direct data on mer-solvent interaction parameters. Therefore, we use the following equation to calculate mer-solvent interaction constants $A_{\mathrm{m-s}}$ from monomer-monomer interaction constants $A_{\mathrm{m-m}}$ and solvent-solvent interaction constants $A_{\mathrm{s-s}}$ found in the literature:

$$A_{\mathrm{m-s}} = \sqrt{A_{\mathrm{m-m}} \cdot A_{\mathrm{s-s}}} \tag{29}$$

Equation 29 is based on a simple model of molecule-molecule interactions in the liquid state [37,38]. Interaction constants vary significantly among different sources in the literature for the same compounds, depending on how they are calculated. Therefore, we chose the most extensive source of values we could find [38] and used that source alone. The approach is based on that of Lifshitz [39] that relates $A$ to the surface tension, which can be measured from contact angles and the minimum equilibrium distance between molecules. The latter parameter has been shown to vary little among widely varying materials [38] but does not take into account Born repulsion. Nevertheless, this approach allows us to use $A-$values that have been calculated from the same model. Table 3 contains these values, along with a comparison of $D_{\mathrm{T}}$ values for polystyrene calculated using (28) and (29) versus those measured by ThFFF. The values agree within an order of magnitude but the calculated values are consistently higher by a factor of 2–4. The discrepancy is not unexpected considering the simplifications used in the calculations. Systematic errors can be expected to result from the assumption of hard spheres in the calculation of molecular radii. More important, the use of (29) to estimate interaction energies may be an oversimplification of the Lifshitz

**Table 3.** Hamaker constants and $D_T$-values calculated from (28) and (29), and $D_T$-values measured by ThFFF.

| Polymer | Solvent | A x $10^3$ (ergs) | $D_T$ x $10^8$ $(cm^2(sK)^{-1})$ Calculated | Measured |
|---|---|---|---|---|
| PS | | 6.58 | | |
| | cyclohexane | 4.69 | 14.4 | 6.6 |
| | ehtylbenzene | 5.43 | 17.5 | 9.5 |
| | toluene | 5.3 | 25.5 | 10.3 |
| | benzene | 5.37 | 28.7 | 8.9 |
| | tetrahydrofuran | 5.09 | 39 | 10 |
| | ethyl acetate | 4.44 | 44.2 | 11.6 |
| | 2-butanone | 4.57 | 41.7 | 13.9 |

model. It is significant, however, that the general order of $D_T$ values among the different solvents is the same, with cyclohexane having the lowest value, followed by the aromatic solvents, then the ether (tetrahydrofuran), and finally the solvents containing a carbonyl group. This consistency supports the general validity of the model.

Another important simplification of the model is the neglect of interactions between mers. Although the coupling of heat and mass flux is mainly localized around each mer, the interaction between mer and solvent will be affected by other mers. Such secondary affects are neglected in the current model but are essential components to refining the model, and will be the focus of a future publication.

Other assumptions that were examined and found to be satisfactory include:

1. Changes in solvent concentration due to mer-solvent interactions are negligible.
2. Changes in solvent pressure due to the presence of polymer are small but not insignificant ($< 10\%$), allowing the simplified form in (28) to be used.
3. Perturbation in the temperature distribution around mers due to differences in the thermal conductivities of the mer and solvent are negligible.

For a detailed discussion of these assumptions, see [33].

An important conclusion of our model is that thermodiffusion is highly sensitive to the size of the mer. It is most important, therefore, to have accurate values of the mer radius if the model is to be used to measure interaction constants from ThFFF measurements of thermodiffusion, or to estimate values of $D_T$ from estimates of interaction constants.

The model also indicates that dispersion forces alone may be responsible for the thermodiffusion behavior of polymers in organic solvents. In fact, dispersion forces in organic solvents are typically an order of magnitude greater than the sum of all intermolecular forces in more polar solvents such as water [37], which is consistent with the reduced thermodiffusion of polymers dissolved in such solvents. Although in principle the model does not preclude thermodiffusion of polymers to the hot wall, such behavior would require a negative interaction constant, in which case a polymer would not be dissolved by the solvent.

Finally, for polymers with high interaction constants, the quadratic term in (27) cannot be neglected. In such cases, the situation is more complex, especially for copolymers. Thus, the linear terms in (27) could be partially cancelled in a solvent that is good for one kind of mer and bad for the other. In that case, the quadratic term could dominate the thermodiffusion of the copolymer. Another complication with copolymers lies in the fact that mers of different chemical composition will move in the temperature gradient at different velocities. In such a case, where mers are no longer moving in concert, frictional drag from the slower moving mers will need to be considered.

## References

1. M. E. Schimpf: 'Thermal Field-Flow Fractionation'. In: Field-Flow Fractionation Handbook, ed. by M. E. Schimpf, K. D. Caldwell, J. C. Giddings (Wiley, New York 2000) pp. 239–256
2. D. Enskog: Physik Z. **12**, 533 (1911)
3. S. Chapman: Proc. Roy. Soc. Lond. **A93**, 1 (1916)
4. K. Wirtz: Naturwissenschaften **27**, 369 (1939)
5. K. Wirtz, J. W. Hiby: Z. Physik **44**, 369 (1943)
6. I. Prigogine, L. de Broukere, R. Amand: Physica **16**, 851 (1950)
7. S. Prager, H. Eyring: J. Chem. Phys. **21**, 1347 (1953)
8. S. R. de Groot: The Thermodynamics of Irreversible Processes (Interscience, New York, 1950)
9. K. G. Denbigh: Trans. Faraday Soc. **48**, 1 (1952)
10. R. J. Bearman, J. G. Kirkwood, M. Fixman: Advanc. Chem. Phys. **1**, 1 (1958)
11. W. M. Rutherford, H. G. Drickamer: J. Chem. Phys. **22**, 1157 (1954)
12. E. L. Doughergy, H. G. Drickamer: J. Phys. Chem. **59**, 443 (1955)
13. A. H. Emery, H. G. Drickamer: J. Chem. Phys. **23**, 2252 (1955)
14. T. N. Khazanovich: J. Polym. Sci. Part C **16**, 2463 (1967)
15. P. Debye, A. M. Bueche: High Polymer Physics (Chemical Publishing, New York, 1948) pp. 497–527
16. F. S. Gaeta: Phs. Rev. **182**, 289 (1969)
17. J. C. Giddings, 'The Field-Flow Fractionation Family: Underlying Principles. In: M. E. Schimpf, K. D. Caldwell, and J. C. Giddings, Eds., Field-Flow Fractionation Handbook (Wiley, New York, 2000) pp. 3–30.
18. M. Martin, C. Van Batten, M. Hoyos: Anal. Chem. **69**, 1339 (1997)
19. A. C. van Asten, G. F. M. Boelens, W. Th. Kok, P. S. Williams, J. C. Giddings: Sep. Sci. Technol. **29**, 513 (1994)
20. M. E. Schimpf, J. C. Giddings: Macromol. **20**, 1561 (1987)
21. M. E. Schimpf, J. C. Giddings: J. Polym. Sci., Polym. Phys. Ed. **17**, 1317 (1989)
22. R. Sisson, J. C. Giddings: Anal. Chem. **66**, 4043 (1994)
23. A. C. van Asten, W. Th. Kok, R. Tijssen, H. Poppe: J. Polym. Sci. B: Polym. Phys. **34**, 283 (1996)
24. C. A. Rue, M. E. Schimpf: Anal. Chem. **66**, 4054 (1994)

25. G. Thomaes: Physica **17**, 885 (1951)
26. J. Xu, C. A. Rue, M. E. Schimpf: J. Liq. Chrom. & Rel. Technol. **20**, 2703 (1997)
27. M. E. Schimpf, J. C. Giddings: J. Polym. Sci., Polym. Phys. Ed. **28**, 2673 (1990)
28. M. E. Schimpf, L. M. Wheeler, P. F. Romeo: ACS Symp. Ser. **521**, 63 (1992)
29. S. L. Brimhall, M. N. Myers, K. D. Caldwell, J. C. Giddings: J. Polym. Sci., Polym. Phys. **23**, 2443 (1985)
30. J. M. Prausnitz, Moleclar Thermodynamics of Fluid Phase Equilibria. (Prentice-Hall, Englewood Cliffs, 1969)
31. J. Brandrup, E. H. Immergut, ed.: Polymer Handbook (Wiley, New York, 1989)
32. F. S. Gaeta, G. Perna, G. Scala: J. Poly. Sci. Polym. Phys. Ed. **13**, 203 (1975)
33. M. E. Schimpf, S. M. Semenov: J. Phys. Chem. B **104**, 9935 (2000)
34. S. Ross, I. D. Morrison: Colloidal Systems and Interfaces. (Wiley, New York, 1988)
35. M. Teubner: J. Chem. Phys. **76**, 5564 (1982)
36. J. A. Riddik, W. B. Bunger, T. K. Sakano, eds.: Organic Solvents. Physical Properties and Methods of Purification (Wiley, New York, 1986)
37. K. S. Birdi, ed.: Handbook of Surface and Colloid Chemistry (CRC Press, New York, 1997)
38. J. van Oss: Interfacial Forces in Aqueous Media (Dekker, New York, 1994)
39. E. M. Lifshitz: Zh. Teor. Fiz. **29**, 94 (1955)

## Supplementary glossary

$A$ Hamaker constant
$c_{\mathrm{m}}$ solvent molecule packing concentration
$D$ mass diffusion coefficient
$D_{\mathrm{T}}$ thermodiffusion coefficient
$f$ friction coefficient
$J_{\mathrm{Q}}$ heat flux
$J_{\mathrm{l}}$ mass flux
$L$ FFF channel length
$L_{\mathrm{f}}$ characteristic depth
$L_{\mathrm{p}}$ solute latent heat of vaporization
$L_{\mathrm{s}}$ solvent latent heat of vaporization
$m$ ratio of molar volumes of polymer and solvent
$n$ ratio of mer to solvent thermal conductivity
$P$ pressure
$p$ local pressure near mer
$p_0$ pressure in bulk solvent
$Q^*$ heat of transport
$R$ gas constant
$R_{\mathrm{h}}$ hydrodynamic radius
$R_{\mathrm{F}}$ FFF retention ratio
$r$ radial distance
$r_0$ radius of solvent molecule
$r_{\mathrm{mer}}$ radius of mer
$T_0$ temperature at particle center
$T_{\mathrm{e}}$ temperature around mer

| | |
|---|---|
| $U$ | activation energy for viscous flow |
| $U_{\mathrm{T}}$ | thermophoretic velocity |
| $V$ | molar volume |
| $V^0$ | FFF channel volume |
| $V_{\mathrm{r}}$ | FFF retention volume |
| $W$ | velocity of sound |
| $w$ | FFF channel thickness |
| $w'$ | ratio of molecular weights of polymer and solvent |
| $X_{\mathrm{a}}$ | mole fraction of component a |
| $Z_{\mathrm{A}}$ | acoustical impedence |
| $\beta_{\mathrm{T}}$ | isothermal compressiblility |
| $\delta$ | $1 - \phi$ |
| $\delta\mathrm{p}$ | solubility parameter for polymer |
| $\delta\mathrm{s}$ | solubility parameter for solvent |
| $\eta$ | solution viscosity |
| $\Phi$ | interaction potenital |
| $\phi$ | volume fraction of solvent molecules |
| $\phi_{\mathrm{P}}$ | volume fraction of polymer |
| $\lambda_{\mathrm{F}}$ | FFF retention parameter |
| $\nu$ | solvent velocity |
| $\nu_0$ | volume occupied by a solvent molecule |
| $\mu'$ | empirical polymer-solvent interaction constant |
| $\Pi$ | osmotic pressure |
| $\sigma$ | cross-sectional area of molecule |

# Part III

# Convection and Porous Media

# Soret Effect and Free Convection: A Way to Measure Soret Coefficients

Jean-Karl Platten, Jean-François Dutrieux, and Guy Chavepeyer

University of Mons-Hainaut, General Chemistry,
Avenue du Champ de Mars, 24, B-7000 Mons, Belgium

**Abstract.** The measurement of Soret coefficients is discussed in the frame of natural convection, both for Rayleigh–Bénard and thermogravitational configurations. The Rayleigh–Bénard configuration, together with the measurement of the Hopf frequency, is recommended for negative values of Soret coefficients; the record of the thermogravitational transient vertical velocity, combined with the vertical separation at steady state, is a good way for an accurate determination of positive values of Soret coefficients.

## 1 Introduction

It has been said and written many times, including very recently [1] that "in many mixtures, accurate measurement of thermal diffusion factors is difficult to perform because of the obscuring effect of natural convection". This could have been the situation until ±1970, since before, the conditions for the onset of convection in horizontal binary liquid layers were not clearly established. An example that has been cited several times, is that of the benzene–methanol system. In order to illustrate the descrepancies between values reported by 6 different groups of researchers, we give in Table 1 numerical values of the Soret coefficient ($\frac{D_T}{D}$) of benzene at different compositions, a comparison that was already made in Ref. [2]. In Table 1, $x$ is the benzene mole fraction. If most of authors do predict a change of sign of the Soret coefficient of benzene, the mole fraction of benzene at which this change occurs ranges from 0.9 to 0.3! And at a given mole fraction, the Soret coefficient may differ by one order of magnitude between different authors. Before asserting that these descrepancies are really linked to obscuring effects of free convection, one should critically and minutely examine the different experimental conditions used by these six groups. Since this is not the object of this paper, we shall not do it here, but let us emphasize that, at the time these measurements were performed, i.e. before 1970, it was generally accepted that convection could not occur in systems heated from above, while in systems heated from below convection could only be initiated when the Rayleigh number Ra (a nondimensional measure of the temperature gradient) exceeds the critical value 1708, thought to be independent of the Soret effect.

However, it is well known [3] that convection can occur even if the system is heated from above and for extremely small values of the Rayleigh number. This is so-called "double diffusive convection": when two fields (here the temperature and the concentration field) diffuse at different rates (and this is the case since

W. Köhler and S. Wiegand (Eds.): LNP 584, pp. 313–333, 2002.

**Table 1.** $\frac{D_T}{D} \times 10^3$ $K^{-1}$: benzene–methanol.

| $x$ (mole fraction) | *Tichacek* (40 °C) | *Prigogine* (20 °C) | *Thomaes* (23.5 °C) |
|---|---|---|---|
| 0.9 | | 0 | −0.08 |
| 0.8 | −5.75 | +4.68 | −0.065 |
| 0.7 | | | −0.035 |
| 0.6 | | +6.62 | +0.01 |
| 0.5 | −0.479 | | +0.19 |
| 0.4 | | +6.55 | +0.25 |
| 0.3 | | | +0.27 |
| 0.2 | +2.56 | +5.87 | +0.29 |
| 0.1 | | | +0.31 |

| $x$ (mole fraction) | *Whitaker* (35 °C) | *Tyrrell* (25 °C) | *Turner* (25 °C) |
|---|---|---|---|
| 0.9 | | −5.17 | −9.5 |
| 0.8 | | −4.90 | −6.5 |
| 0.7 | | −3.86 | −3 |
| 0.6 | 0 | −2.89 | 0 |
| 0.5 | | −1.64 | +2 |
| 0.4 | +2.43 | −0.772 | +3.5 |
| 0.3 | | 0 | +4.4 |
| 0.2 | +2.99 | +0.772 | +4.2 |
| 0.1 | | +2.18 | +4.0 |

the thermal diffusivity has generally a very different magnitude from that of the ordinary diffusion coefficient) and if the slowest diffusing field is destabilizing, then the convection occurs. Therefore in our opinion, the old examples should not longer be invoked to assert that ground based experiments are inadequate to accurately measure a Soret coefficient and to propose experiments in microgravity to alleviate the difficulties with convection. Several ground based experiments did produce in the recent years values that agree with each other, in convection-free experiments [4], [5], [6]. We shall come back later on a comparison between these values. And even in microgravity, the effect of g-jitters is, to our knowledge, not yet clearly established. Finally even if an experiment in microgravity is convection-free during all the duration of the flight (sometimes several months [7]), after the recovery of the experimental cell on earth, the accuracy on the Soret coefficient greatly rests on the accuracy of concentration measurements by some appropriate techniques such as density measurements, index of refraction measurements, or gas chromatography. And when dealing with volatile products, this could be the mean reason of large errors. Therefore optical techniques like TDFRS (thermal diffusion forced Rayleigh scattering) as employed by e.g. Köhler [6] or Wiegand [8] could be superior to microgravity determinations of Soret coefficients.

Finally, instead of considering convection as a perturbation, i.e. as a foe in thermal diffusion experiments, one could use the coupling between convection and thermal diffusion as a natural way to determine Soret coefficients. This is the object of this review paper. Indeed the concentration gradient induced by thermal diffusion modifies the density gradient, since density depends on the local temperature and composition of the system. In the gravity field, this modification of the density gradient will change, more or less drastically, the convective patterns or the time dependence of some velocity amplitude. These modifications must be considered as an indirect, however not less precise, determination of the Soret coefficient. Of course the accuracy relies on the precise measurement of the velocity field, which was not possible 25 years ago. Today, we have at our disposal precise optical techniques to record velocity components, such as LDV (Laser–Doppler Velocimetry). Clearly old techniques to detect e.g. the onset of convection, such as Schmidt–Milverton plots that are based on global measurements of heat fluxes, or Nusselt numbers, are inappropriate but remain still used [7]; they failed to produce acceptable values of Soret coefficients. Another approach is the so-called thermogravitational technique in which the coupling between thermal diffusion and convection may enhance the molecular separation, making the determination of the Soret coefficient even easier, see e.g. [9]. In any case, one has to make use of a correct theoretical approach, modeling the precise experiment that is undertaken, including realistic boundary conditions. Use of unrealistic boundary conditions in the theory, as e.g. free and permeable (to matter) boundaries instead of rigid and impermeable conditions, results inevitably in a wrong value of the Soret coefficient. By "correct theoretical approach" we mean not necessarily an "analytical approach" which usually needs a lot of inadequate simplifications, since it is hard to solve analytically the coupled set of Navier–Stokes, energy and mass equations. More generally a "numerical approach", based on the numerical solution of the above mentioned equations by appropriate numerical techniques like finite differences, finite volume, or even Galerkin-type methods, is sufficient, provided that the convergence of these numerical simulations is achieved. This is generally not a problem with the use of today's efficient computer facilities.

We will show, in the body of this paper, that an almost exact theory (in the "numerical sense") together with LDV measurements, can provide Soret coefficients with a reasonnable accuracy.

## 2 Useful equations for systems heated from below

In this section, we consider an horizontal binary liquid layer heated from below. Hence, the temperature gradient is adverse, in the sense that cold and heavier fluid at the top rests on hot and lighter fluid at the bottom, thus leading to a stability problem and convection, if the temperature gradient, linked to the

Rayleigh number

$$Ra = \frac{g\alpha \Delta T h^3}{D_{\mathrm{th}}\nu} \tag{1}$$

exceeds a critical value. In (1) $\alpha$ is the expansion coefficient defined by the equation of state for the density $\rho$

$$\rho = \rho_0 \left[1 - \alpha \left(T - T_0\right)\right] \tag{2}$$

$D_{\mathrm{th}}$ is the thermal diffusivity $D_{\mathrm{th}} = \frac{\lambda_{\mathrm{T}}}{\rho c}$, ($\lambda_{\mathrm{T}}$: thermal conductivity, c: specific heat), $\nu$ is the kinematic viscosity $\nu = \frac{\eta}{\rho}$ ($\eta$: dynamic viscosity), $h$ is the thickness of the layer, and $\Delta T$ the applied temperature difference. The numerical value of the critical Rayleigh number depends on the boundary conditions. In the case of a pure fluid one has only to consider mechanical and thermal boundary conditions. In all what follows we will consider only perfect (on infinitely) conductive boundary conditions, with prescribed temperatures, not subjected to fluctuations. The boundaries, at least in the experimental case, are rigid plates on which all velocity components vanish ("no slip" boundary conditions). But before, we have to define our system of coordinates and the notations for the velocity components $u$, $v$, $w$(see Fig. 1)

The continuity equation:

$$\frac{\partial u}{\partial x} + \frac{\partial v}{\partial y} + \frac{\partial w}{\partial z} = 0 \tag{3}$$

together with the normal modes analysis [10], that considers periodic velocity components in both infinite horizontal directions, $x$ and $y$, i.e.

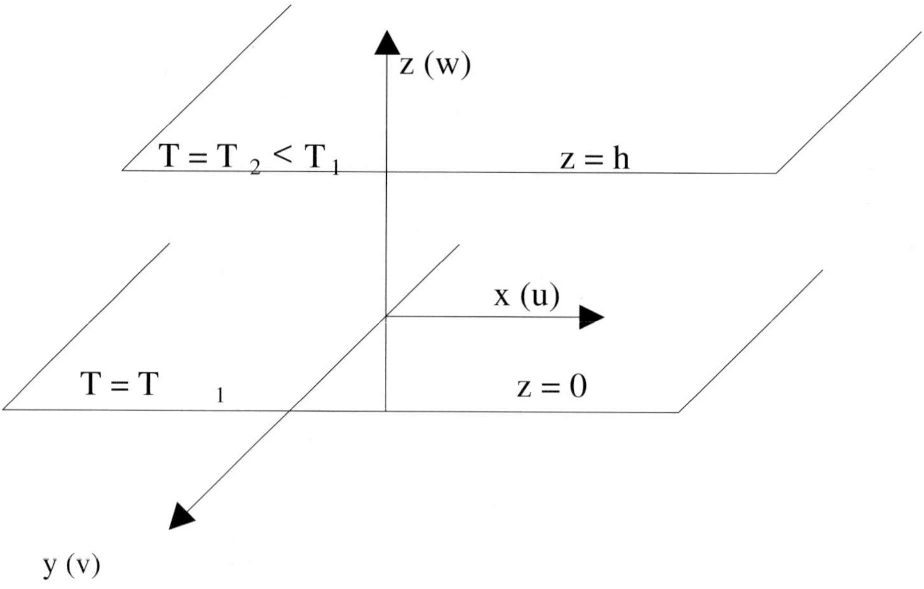


**Fig. 1.** Coordinates and boundary conditions

$$u(x,y,z) = U(z)e^{\mathrm{i}k_\mathrm{x}\mathrm{x}}e^{\mathrm{i}k_\mathrm{y}\mathrm{y}}$$
$$v(x,y,z) = V(z)e^{\mathrm{i}k_\mathrm{x}\mathrm{x}}e^{\mathrm{i}k_\mathrm{y}\mathrm{y}}$$
$$w(x,y,z) = W(z)e^{\mathrm{i}k_\mathrm{x}\mathrm{x}}e^{\mathrm{i}k_\mathrm{y}\mathrm{y}}$$

implies:

$$ik_\mathrm{x}U + ik_\mathrm{y}V + \frac{dW}{dz} = 0 \tag{4}$$

The no-slip boundary conditions

$$U = V = W = 0 \text{ at } z = 0, h \tag{5}$$

together with (4) imply:

$$W = \frac{dW}{dz} = 0 \text{ at } z = 0, h \tag{6}$$

One immediately recognizes the difficulty to find a solution $W(z)$ of the governing differential equations that vanishes at the boundaries together with its first derivative. Even if such functions do exist [10], they lead finally to a certain transcendental equation that has to be solved numerically in order to find the numerical value of the critical Rayleigh number:

$$Ra^{\mathrm{crit}} = 1707.762 \tag{7}$$

The case of less realistic conditions is much simpler to deal with. Suppose that the liquid layer is not bounded by rigid plates, but is "open" to air, or more correctly bounded by two fluids of zero viscosity. There is then no reason for the velocity components to vanish at the boundaries, except $w$ (or $W$) if the boundaries are considered to be underformed. The absence of shear stress do not imply an additional condition on $W$. However a careful inspection of the hydrodynamic equations (see e.g. [10] for more details) imposes that all even derivatives of $W$ are equal to zero at the boundaries:

$$W = 0 \text{ and } \frac{d^{2\mathrm{n}}W}{dz^{2\mathrm{n}}} = 0 \text{ at } z = 0, h \tag{8}$$

This is of course a very simple mathematical problem, since the solution reads simply:

$$W(z) \sim \sin(n\pi z) \tag{9}$$

and leads after a very simple calculation to:

$$Ra^{\mathrm{crit}} = \frac{27\pi^4}{4} \tag{10}$$

Let us come back to binary mixtures and Soret effect. In addition to thermal and mechanical conditions, one has also to impose boundary conditions on the concentration field, or more precisely on the mass fraction $c_1$ of component 1 taken here arbitrarily to be the densest component. Once again, realistic or non realistic conditions may be considered. In the non realistic case, one assumes that the the mass fraction $c_1$ is imposed at the boundaries and therefore cannot fluctuate. In other words the boundaries are considered to be permeable to matter, exactly as they are permeable to heat for a constant temperature. Thus for perfect conductive, stress free and permeable to matter boundary conditions, the stability problem is, once again, very simple to solve [11] since all the fields $T$, $\boldsymbol{V}$ and $c_1$ are of the form: $e^{\mathrm{i}k_\mathrm{x}\mathrm{x}}e^{\mathrm{i}k_\mathrm{y}\mathrm{y}}\sin(n\pi z)$ and the critical Rayleigh number is easily deduced analytically as a function of the relevant parameters of the problem: the Soret coefficient, the Schmidt number $Sc = \frac{\nu}{D}$ and the Prandtl number $Pr = \frac{\nu}{D_{\mathrm{th}}}$.

In the linear stability theory, the fluctuations around rest (i.e purely conductive state) behave with time according to $e^{\sigma\mathrm{t}}$. Here, two cases must be considered: the case of a monotonous increase in case of instability, or monotonous decrease in case of stability, i.e. $\sigma$ is real; the marginal stability is then characterized by $\sigma = 0$, and the critical Rayleigh number is given by:

$$\frac{Ra^{\mathrm{ex}}}{Ra_{(0)}} = (1 + \psi\frac{Pr + Sc}{Pr})^{-1} \tag{11}$$

where $Ra_{(0)}$ is the critical Rayleigh number without Soret effect, given by (10). $\psi$ is so-called separation ratio and represents the mass contribution divided by the temperature contribution to buoyancy forces, since now, instead of (2) we have:

$$\rho = \rho_0(1 - \alpha(T - T_0) + \beta(c_1 - c_1^0)) \text{ with } \beta > 0 \tag{12}$$

Thus

$$\psi = -\frac{\beta \Delta c_1}{\alpha \Delta T} = \frac{\beta}{\alpha}\frac{D_\mathrm{T}}{D}c_1^0(1 - c_1^0) \tag{13}$$

where $\frac{\Delta c_1}{\Delta T} = -\frac{D_\mathrm{T}}{D}c_1^0(1 - c_1^0)$ results from pure Soret effect. The superscript "ex" refers to the nomenclature of Chandrasekhar and means "exchange of stability" [10], or more simply "non oscillatory marginal stability".

The second case is an oscillatory increase (or decrease) of a given perturbation. That means that $\sigma$ is complex and the marginal stability is characterized by $\sigma = i\omega$, where $\omega$ is the frequency of oscillation. The critical Rayleigh number is then given by:

$$\frac{Ra^{\mathrm{over}}}{Ra_{(0)}} = \frac{(1 + Sc)(Pr + Sc)}{Sc^2}(1 + \psi\frac{Pr}{Pr + 1})^{-1} \tag{14}$$

Once again, the superscript "over" refers to the nomenclature of Chandrasekhar and means "overstability", i.e. oscillatory marginal stability. Sometimes such a bifurcation from the rest state to an oscillatory state, is called an Hopf bifurcation and $\omega$ the Hopf frequency. The Hopf frequency $\omega_\mathrm{H}$ is also a function of the parameters $\psi$, $Sc$ and $Pr$

$$\omega_\mathrm{H} = \frac{3\pi^2}{2}\sqrt{\frac{-\psi(1 + Sc)(Pr + Sc)}{Sc^2(1 + \frac{1}{Pr} + \psi)} - \frac{1}{Sc^2}} \tag{15}$$

Focusing only on liquid mixtures, it is well known that the thermal diffusivity is several orders of magnitude greater than mass diffusivity. It follows that the Prandtl number is several orders of magnitude smaller than the Schmidt number. Typically $Pr \approx 10$ (it is equal to 7 for pure water around 20°C) and $Sc \approx 1000$ to 2000. Therefore (14) and (15) can be simplified, since for example: $\frac{(1+Sc)(Pr+Sc)}{Sc^2} \sim 1$ within an error of less than 1%, thus irrelevant for comparison with experiments. For an Hopf bifurcation from the rest state, one has thus:

$$\frac{Ra^{\text{over}}}{Ra_{(0)}} = 1 - \frac{\psi}{1 + \frac{1}{Pr} + \psi} \tag{16}$$

$$\omega_H = \frac{3\pi^2}{2}\sqrt{\frac{-\psi}{1 + \frac{1}{Pr} + \psi}} \tag{17}$$

However these formulas are only valid for permeable boundaries and may not be used for comparison with experiments. For comparison with experiments, we have to use the proper boundary conditions, i.e. the liquid is bounded by rigid conducting plates (e.g. copper plates) implying not only conditions (6), an already difficult problem, but in addition, a zero mass flux across the boundaries,

$$-\rho D\frac{\partial c_1}{\partial z} - \rho D_{\mathrm{T}} c_1^0(1 - c_1^0)\frac{\partial T}{\partial z} = 0 \text{ at } z = 0, h \tag{18}$$

making the problem even more difficult. Nevertheless the problem of rigid boundary conditions has been solved by several techniques, see e.g. [12], [13], [14], [15], [16] and other references therein, and leads to a relation between critical values of the Rayleigh number and Hopf frequency and the Soret coefficient (or more precisely the separation ratio $\psi$). Of course, due to the numerical technique, this relation is usually in the form of tables: for many values of the relevant parameters Pr, Sc, $\psi$, one has tabulated values of the critical points. An example is extracted from [14] for $Pr = 10$ and $Sc = 1000$ (see Table 2). Such tables are very usefull because it helps in finding empirical relations between the critical Rayleigh number and the separation ratio. For example it can be verified that the results of Table 2 may be represented by (19) with a good accuracy:

$$\frac{Ra^{\text{over}}}{1708} = 1 - \frac{a\psi}{1 + \frac{1}{Pr} + \psi} \tag{19}$$

Equation (19) closely resembles (16) provided that $Ra_{(0)} = \frac{27\pi^4}{4}$ is replaced by 1708 (see (7) and (10)), together with the introduction of a totally empirical factor $a > 1$ that tells us that the relative increase of the critical point is faster for rigid boundaries than for permeable boundaries. Thus from numerical solutions for rigid boundaries and analytical solutions for permeable boundaries, we propose the following set of equations for an Hopf bifurcation:

$$\frac{Ra^{\text{over}}}{Ra_{(0)}} - 1 = -\frac{a\psi}{1 + \frac{1}{Pr} + \psi} \tag{20}$$

$$\omega_{\mathrm{H}} = \frac{3\pi^2}{2} b \sqrt{\frac{-\psi}{1 + \frac{1}{Pr} + \psi}} \tag{21}$$

together with

| | Permeable boundaries | Rigid boundaries |
|---|---|---|
| $Ra_{(0)}$ | $\frac{27\pi^4}{4}$ | 1708 |
| $a$ | 1 | 1.15 |
| $b$ | 1 | 1.43 |

This result shows that an Hopf bifurcation only exists for $\psi < 0$. This is not totally correct from a theoritical point of view since for an Hopf bifurcation to exist one has the condition $\psi < \psi^* < 0$, but $\psi^*$ is extremely small (e.g. $\psi^* \approx -\left(\frac{Pr}{Sc}\right)^2$ ) and thus from a practical point of view, (20)–(21) are sufficient for a comparison with experiments. To be complete, we should also add that (20) begins to show noticeable differences with numerical results for large absolute values of $\psi$ and should not be used for $\psi < -0.7$. But such cases are exceptional, as e.g. for LiI 1M in water [17]. A final remark concerns the empirical factor 1.15 in (20). This value has been recently proposed by Larre in his Ph.D. thesis [16] dealing with ternary systems, in a special "test" paragraph to binary systems. In the literature, this empirical factor appears for the first time with the value 1.05 instead of 1.15 [4]. This could simply be a misprint in the first publication and

**Table 2.** Variation of the critical Rayleigh number with $\Psi$ for rigid boundaries.

| $\Psi$ | $Ra^{\mathrm{cr}}$ |
|---|---|
| −0.00023 | 1720 |
| −0.00231 | 1729 |
| −0.01371 | 1750 |
| −0.02043 | 1762 |
| −0.04432 | 1805 |
| −0.06509 | 1844 |
| −0.08499 | 1882 |
| −0.1041 | 1921 |
| −0.1224 | 1960 |
| −0.1401 | 1999 |
| −0.1571 | 2037 |
| −0.1734 | 2076 |
| −0.1891 | 2115 |
| −0.3197 | 2502 |
| −0.4166 | 2881 |

reproduced many times. However, everybody agrees with the empirical factor 1.43 for the Hopf frequency.

Summarizing, the set (20)–(21) is sufficient for comparison with experiments and from the measured critical Rayleigh number or from the measured Hopf frequency, allows the determination of the separation factor $\psi$. If the two values of $\psi$ (the first determined from $Ra^{\text{over}}$, the second from $\omega_{\text{H}}$) do agree, this gives confidence in the procedure. Of course the price we have to pay to produce the desired values of $\frac{D_{\text{T}}}{D}$ is the knowledge of the thermal and mass expansion coefficients, see (12) and (13), but this is usually not a problem since one has only to measure in isothermal conditions the density of the mixture at different temperatures and compositions, and this may be achieved with an extraordinary accuracy since densities may be measured without difficulties with a $\pm 2\ 10^{-6}$ gcm$^{-3}$ precision, e.g. with a quartz U-tube vibrating densimeter.

The analog of (11) valid for $\psi > 0$, but modified for rigid boundaries could also be given. However, this is not very useful. The reason is evident: the separation ratio is multiplied by $\frac{Sc+Pr}{Pr}$, at least of the order of 100. Thus, for positive $\psi$ of the order 0.2 (a quite reasonable value) the decrease of the critical Rayleigh number is of the order 20. In other words, in an experiment in which the critical temperature difference would be of the order of 2 K in absence of the Soret effect, it drops to only 0.1 K due to the Soret effect and therefore the technique becomes less suitable, even if the temperature difference applied to the fluid layer is controled at the 0.01 K level. Moreover, since there is no oscillation, we cannot deduce two separated values of $\psi$ as in the case $\psi < 0$ and thus the consistency test of the technique is lost. Summarizing this paragraph, the precise detection of the onset of convection together with the Hopf frequency in non pertubative conditions (optical measurements by e.g. LDV) is a realistic way to measure accurately negative Soret coefficients. Moreover, if LDV is used, we may control the absence of any "obscuring convective perturbation" before the onset of free convection. Of course the LDV equipment must be dedicated to measure extremely small velocities, down to 5 $\mu$m s$^{-1}$ [18]. This is actually the case.

## 3 LDV experiments in systems with negative Soret coefficients

The experimental cell consists of two horizontal copper plates maintained at constant temperatures $T_1$ and $T_2$ with a 0.01 K control. Left and right plexiglass spacers impose the liquid layer depth. Front and back glass windows of good optical quality close the cell and allow optical access.

The imposed temperature difference $(T_2 - T_1)$ is gradually increased by steps of 0.1 K far from the critical point, reduced to 0.02 K and sometimes 0.01 K close to the critical point. Enough time is allowed between two successive temperatures increments for a well established steady state (typically five hours).

We give in Fig.2 a velocity–time record in the following conditions:

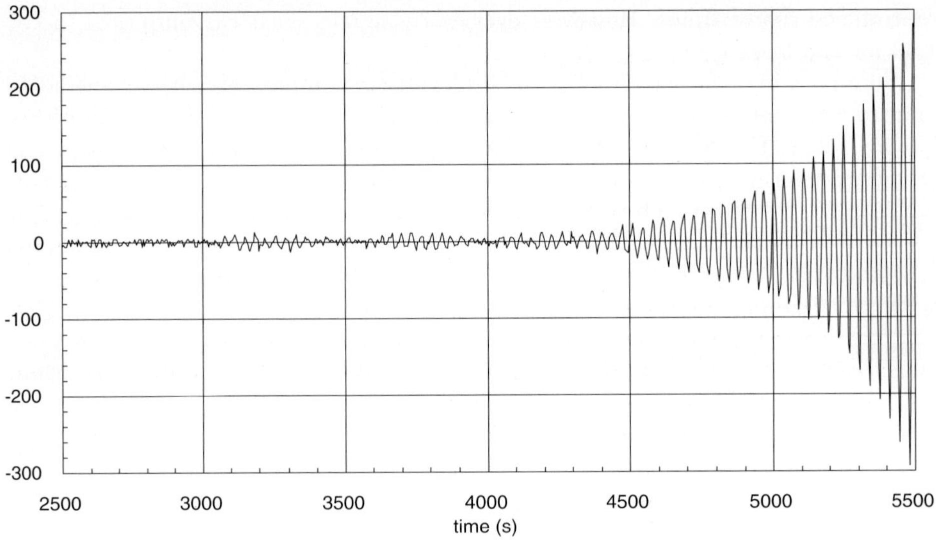


**Fig. 2.** Velocity ($\mu$m/s) versus time

| | |
|---|---|
| System | water–ethanol (8.01 wt % in ethanol) |
| Mean temperature | 25 $°C$ |
| Fluid depth | 2.99 mm |
| Initial temperature difference | $\Delta T = 4.79$ K |
| Final temperature difference | $\Delta T = 4.81$ K |
| An experimental point | each 2 sec. |

Thus at time $t = 0$ (not shown on Fig.2) the temperature difference is increased from $\Delta T = 4.79$ K to 4.81 K. One has to wait 3000 s before the onset of convection, the velocity amplitude being 5 $\mu$m/s, a very small value indeed, but more perceptible around 5000 s, before reaching saturation for $t > 5500$ sec (not shown on Fig.2). The Fourier transform of the time signal between 2900 s and 3900 s gives the dimensional frequency $f = 2.83 \times\ 10^{-2}$ Hz, corresponding to oscillations of 35.3 seconds period.

In order to use these two experimental informations ($\Delta T^{\text{crit}}$ and frequency) for the calculation of the Soret coefficient, one has to know a lot of physical parameters concerning the system, namely the kinematic viscosity and thermal diffusivity (needed to compute the Prandtl number and the Rayleigh number) but also the two expansion coefficients. The accuracy on the final value of the Soret coefficient is mainly determined by the accuracy on these thermophysical properties. They can be found in Tables [19] [20], or are measured in the laboratory. For the example given in details in this paragraph, called system B in Table 3, we have used:

$$\alpha = 2.97\ 10^{-4}\ \mathrm{K}^{-1}$$
$$\beta = 0.1561$$
$$\nu = 1.22 \times 10^{-2}\ \mathrm{cm}^2\mathrm{s}^{-1}$$
$$D_{\mathrm{th}} = 13.01 \times 10^{-4}\ \mathrm{cm}^2\mathrm{s}^{-1}$$
$$Pr = 9.38$$
$$Ra = 492 \times \Delta T$$

The experiment was duplicated, and we evaluate quite reasonnably the mean value of $\Delta T^{\mathrm{crit}} = 4.77$ K or $Ra^{\mathrm{crit}} = 2348$, or $r = Ra^{\mathrm{crit}}/1708 = 1.375$. This 37.5% increase of the critical point should allow the determination of $\psi$, using Eq. (20): $\psi = -0.272$ and thus $(\frac{D_{\mathrm{T}}}{D})_{\mathrm{water}} = -7.02\ 10^{-3}\ \mathrm{K}^{-1}$. Alternatively one could use the Hopf frequency. The time scale is $\frac{h^2}{D_{\mathrm{th}}} = 68.72$ sec and the reduced Hopf frequency is $\omega_{\mathrm{Hopf}} = 2\pi \times 2.81\ 10^{-2}\ \mathrm{Hz} \times 68.72\ \mathrm{sec} = 12.13$.

Using (21) we deduce: $\psi = -0.274$

**Table 3.** Systems: A: Water–ethanol (8.01 wt %) at 30 °$C$, B: Water–ethanol (8.01 wt %) at 25 °$C$, C: Water–ethanol (8.01 wt %) at 20 °$C$, D: Water–isopropanol (10.14 wt %) at 21 °$C$.

| | $Pr$ | $Ra^{\mathrm{over}}$ | $\omega_{\mathrm{H}}$ | $\psi$ from $Ra^{\mathrm{over}}$ | $\psi$ from $\omega_{\mathrm{H}}$ | diff ** | $\alpha \times 10^4\ \mathrm{K}^{-1}$ | $\beta$ |
|---|---|---|---|---|---|---|---|---|
| A | 8.11 | 2198 | 10.55 | –0.224 | –0.223 | < 1 % | 3.44 | 0.1609 |
| B | 9.38 | 2348 | 12.13 | –0.272 | –0.274 | < 1 % | 2.97 | 0.1561 |
| C | 10.93 | 2518 | 14.25 | –0.319 | –0.340 | 6 % | 2.45 | 0.1505 |
| D | 13.4 | 3218 | 16.81 | –0.467 | –0.416 | 11 % | 3.18 * | 0.1288 |

* From [20] at 15°$C$, 20°$C$, 30°$C$

** diff: difference between $\psi$ from $Ra^{\mathrm{over}}$ and $\psi$ from $\omega_{\mathrm{H}}$

| | $D_{\mathrm{T}}/D \times 10^3\ \mathrm{K}^{-1}$ | | | | |
|---|---|---|---|---|---|
| | present work | Kolodner (22) | Zhang [5] | Bou-Ali [23], interpolated | Legros [7,24] |
| A | –6.50 | –6.65 | – | – | – |
| B | –7.08 | –7.13 | –7.3 | –7.05 | – |
| C | –7.51 | –7.61 | – | – | – |
| D | –11.27 | – | – | – | from –2.2 to –21 |

Table 3, summarizes some studied cases, and the numerical values shown in this table are extracted from two Ph.D.thesis [21] [22]. The main information is the two values of $\psi$ deduced from the critical Rayleigh number and from the Hopf frequency. They sometimes agree better than 1% but usually they differ by 5 or 6%. It has been shown by Kolodner [4] that the value of $\psi$ deduced from the Hopf frequency is much more precise than that deduced from $Ra^{\mathrm{over}}$. In his paper, Kolodner reported two kinds of experiments:

*i)* A laser-beam-deflection technique; based on 60 determinations of $\frac{D_T}{D}$ at different mass fractions $c_2$ and temperatures $T(°C)$ the following empirical law is deduced for the water–ethanol system ($c_2$: mass fraction in ethanol)

$$10^3 \frac{D_T}{D} = (2.493 - \frac{T}{39.661})[1 - 3.190[\tanh(\frac{c_2 - 0.2387}{0.1668})^3]^{\frac{1}{3}}] \tag{22}$$

Let us note here that $\frac{D_T}{D}$ as reported by Kolodner is the Soret coefficient of ethanol, of opposite sign of that of water as reported by us. The rms deviation of the fitted curves from the data points is $\delta(\frac{D_T}{D}) = 0.12 \times 10^{-3}$ K$^{-1}$.

*ii)* Convection experiments (exactly as described here), but using a shadowgraphic technique to visualize the onset of convection and to measure the frequency of the linear traveling-wave oscillatory convective instability. The agreement between $\frac{D_T}{D}$ determined with the direct laser-beam-deflection technique is poor with that based on $Ra^{crit}$, but excellent with that based on the Hopf frequency.

Therefore in Table 3, we have used $\psi$ based on $\omega_H$, together with the two expansion coefficients $\alpha$ and $\beta$, to determine $\frac{D_T}{D}$, compared with the values given by Kolodner, (22). Remembering the rms deviation of $0.12 \times 10^{-3}$ K$^{-1}$, we conclude that for the water–ethanol system our values coincide with those of Kolodner. We also give an extrapolated value from results reported by Zhang et al. [5] who studied four different mass fractions in ethanol : 0.3912; 0.3302, 0.2204 and 0.1151. We extrapolated to $c_2 = 0.0801$. We also interpolate (graphically) between the results of Bou-Ali et al. [23] who studied a mass fraction in ethanol down to 0.0641 (Additional values are 0.1151; 0.2204 and 0.2587). In their paper, they report the "thermal diffusion factor $\alpha_T$" related to the Soret coefficient by $\frac{\alpha_T}{T} = \frac{D_T}{D}$.

We also add in Table 3 own results on the system water–isopropanol known to have a very negative Soret coefficient [24], but showing hudge descrepancies between values deduced from different techniques [7].

Thus in partial conclusion of the first part of this review paper, we may say that the coupling between Soret effect and convection (more precisely the onset of convection in a Rayleigh–Bénard configuration) can be used to determine negative Soret coefficients. The use of the Hopf frequency rather than the critical Rayleigh number, results in a more accurate value.

## 4 Useful equations for the free thermogravitational columns

In § 2, we have shown that when $\frac{D_T}{D} > 0$ or $\psi > 0$ (here $\frac{D_T}{D}$ is that of the densest component, our "reference" component) the lightest component migrates to the hot plate and the densest to the cold plate which is in the Rayleigh–Bénard configuration a destabilizing effect, leading to a drastic decrease of the critical Rayleigh number that cannot be used as a way to measure precisely the Soret coefficient. We have thus to imagine a different geometry in order to use the coupling between convection and Soret effect to determine positive Soret coefficients. Since in that case the lightest component goes to the hot plate, we consider a vertical slot (a liquid layer between two vertical rigid copper plates) submitted to an horizontal temperature gradient. In that case there is always convection and if the slot is of very large aspect ratio height/gap thickness, we may neglect horizontal velocity component at least far from the horizontal end walls: we suppose a parallel flow regime described by a vertical velocity $w$ solution of [25]

$$\nu \frac{d^3 w}{dx^3} = -\frac{g}{\rho_0} \frac{\partial \rho}{\partial x} \tag{23}$$

For a pure fluid, using (2) we have

$$\nu \frac{d^3 w}{dx^3} = g\alpha \frac{\partial T}{\partial x} = g\alpha \frac{\Delta T}{h} \tag{24}$$

supposing a pure diffusive regime for the temperature field. This leads after integration and the use of the no-slip boundary conditions at the walls, located at $x = 0$ and $x = h$ ($h$ is the gap between the walls) to

$$w = \frac{g\alpha \Delta T h^2}{6\nu} (\frac{x}{h})(\frac{x}{h} - \frac{1}{2})(\frac{x}{h} - 1) \tag{25}$$

with a velocity amplitude $w_{\mathrm{a}} = \frac{g\alpha \Delta T h^2}{6\nu} \frac{\sqrt{3}}{36}$

Let us now consider a mixture with positive Soret coefficient. Since the lightest component goes to the hot plate, the density there will be smaller and thus the velocity amplitude greater. In fact (23) is still correct (since it results from the momentum balance equation with the pressure eliminated) but we now use (12) for the density. Therefore (24) becomes:

$$\nu \frac{d^3 w}{dx^3} = g(\alpha \frac{\partial T}{\partial x} - \beta \frac{\partial c_1}{\partial x}) \tag{26}$$

Suppose for the time being also a pure diffusive regime for the concentration field

$$\frac{\partial c_1}{\partial x} = -\frac{D_{\mathrm{T}}}{D} c_1^0 (1 - c_1^0) \frac{\Delta T}{h} \tag{27}$$

Equation (26) becomes

$$\nu \frac{d^3 w}{dx^3} = g\alpha \frac{\Delta T}{h} [1 + \frac{\beta}{\alpha} \frac{D_{\mathrm{T}}}{D} c_1^0 (1 - c_1^0)] = g\alpha \frac{\Delta T}{h} (1 + \psi) \tag{28}$$

The integration of (28) leads to the same cubic profile (25) but with a modified velocity amplitude

$$w_{\mathrm{a}} (1 + \psi) \tag{29}$$

Thus, we propose to measure the velocity amplitude as a way to determine $\psi$ when it is positive. However the modified velocity amplitude (29) can only be transient in time, because we have supposed the diffusive regime for the concentration field, which is only true for small times. Indeed the mass conservation law for component 1

$$w\frac{\partial c_1}{\partial z} = D(\frac{\partial^2 c_1}{\partial x^2} + \frac{\partial^2 c_1}{\partial z^2}) \tag{30}$$

has two solutions. The first is the pure conductive regime with no vertical gradient produced by the velocity field ($\frac{\partial c_1}{\partial z} = 0$) and solution of

$$D\frac{\partial^2 c_1}{\partial x^2} = 0 \tag{31}$$

which leads to a linear mass fraction profile or a constant mass fraction gradient $\frac{\partial c_1}{\partial x}$ given by (27) due to boundary conditions (zero mass flux). But this solution prevails only for small times, since, due to the impervious conditions of the vertical plates, the lightest component that accumulates at the hot plate thanks to the Soret effect, is driven by the convection to the top of the slot, whereas the densest component is driven to the bottom, creating a vertical mass fraction gradient $\frac{\partial c_1}{\partial z}$. The objective of the Furry–Jones–Onsager theory is to find a relation between the difference in mass fraction at the top and at the bottom of the cell, and the parameters of the problem. Under particular circumstances (namely $h > 1$ mm, small values of the gradient $\frac{\partial c_1}{\partial z}$), the separation is given by [26]

$$c_1^{\mathrm{top}} - c_1^{\mathrm{bottom}} = 504\frac{\nu}{g\alpha}D_{\mathrm{T}}c_1^0(1-c_1^0)\frac{H}{h^4} \tag{32}$$

where $H$ is the cell height.
This relation is used by researchers working in free thermogravitational columns, e.g. the group of Bilbao [9], and shows that the vertical separation is independent of the imposed horizontal temperature gradient, and that only $D_{\mathrm{T}}$ can be measured, not the Soret coefficient itself! Thus once we have a constant vertical mass fraction gradient $\frac{\partial c_1}{\partial z}$ due to convection, (30) reduces to

$$w \times \mathrm{constant} = D\frac{\partial^2 c_1}{\partial x^2} \tag{33}$$

Integrating twice, we find for $c_1(x)$ a polynomial of fifth degree since $w(x)$ is cubical. This is the convective regime that prevails in the steady state. All this is confirmed by a direct numerical integration [26] of the full Navier–Stokes, energy and mass equations without any kind of simplifying assumption, starting with an initial condition that corresponds to the rest state and an homogeneous liquid layer (Fig.3).

For small times (part a of Fig.3) the isoconcentration lines are parallel and vertical, corresponding to an horizontal gradient $\frac{\partial c_1}{\partial x}$ and to the absence of any vertical gradient: $\frac{\partial c_1}{\partial z} = 0$. On the contrary in the steady state (part b of Fig.3) the isoconcentration lines show a vertical gradient $\frac{\partial c_1}{\partial z} \neq 0$. There is of course an horizontal gradient $\frac{\partial c_1}{\partial x}$ since $c_1(x)$ is of fifth degree, but that does not contribute to the buoyancy because the raising and the sinking liquid columns have the

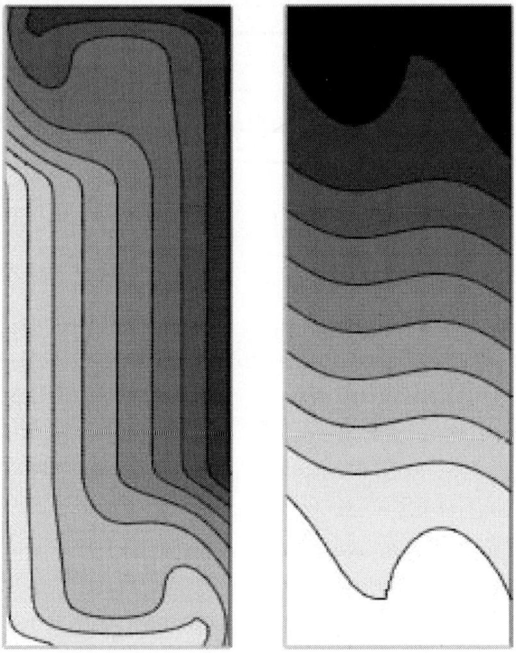

**Fig. 3.** early Soret effect (left) and vertical steady separation (right)

same mean concentration. Thus the enhanced velocity amplitude (29) linked to an enhanced buoyancy, due to a transient horizontal Soret induced mass fraction gradient, must itself be transient. In the steady state buoyancy is only due to the temperature gradient, and the usual velocity amplitude $w_{\mathrm{a}}$ is recovered. Thus if one records the velocity amplitude with time (Fig.4) there is an "overshoot" in the amplitude.

Curve (a) of Fig.4 corresponds to a pure fluid: after a very short time (the thermal relaxation time $\tau_{\mathrm{T}} \sim \frac{h^2}{D_{\mathrm{th}}}$) the steady state value is attained. Curve (b) is that of a fluid mixture in the diffusive regime (or in other words in a vertically infinite cell). After a short time $\tau_{\mathrm{T}}$, the temperature gradient creates an amplitude $w_{\mathrm{a}}$, that slowly increases with time due to the Soret effect, and reaches the saturation value $w_{\mathrm{a}}(1+\psi)$ after a time $\tau_{\mathrm{D}} \sim \frac{h^2}{D}$ characteristic for the establishment of a fully developped horizontal mass fraction gradient $\frac{\partial c_1}{\partial x}$ given by (27). But such a behaviour (b) is never observed since in any experiment the aspect ratio is finite, and the pure conductive regime will be replaced by the diffusive regime after some time in which a vertical mass fraction gradient $\frac{\partial c_1}{\partial z}$ is created and the horizontal one $\frac{\partial c_1}{\partial x}$ almost destroyed. Thus the behaviour of the velocity amplitude is given by curve (c) of Fig.4. There is an "overshoot" in the velocity amplitude related to a transient horizontal mass fraction gradient, and from this "overshoot" we should be able to deduce the separation ratio $\psi$. The question is to know whether the maximum velocity amplitude may reach

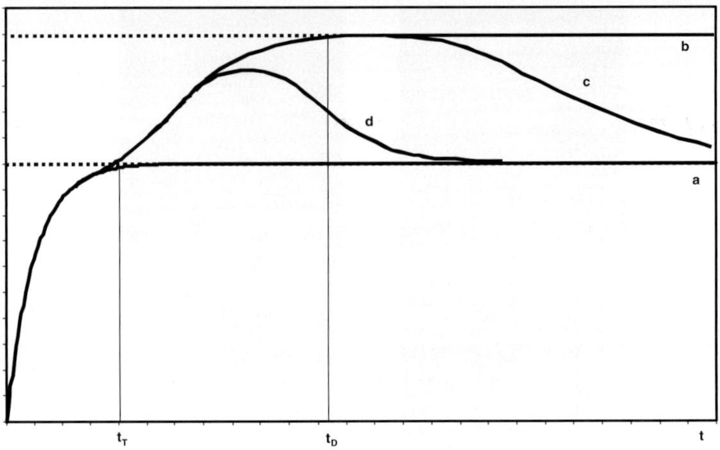


**Fig. 4.** Velocity amplitude versus time: (a) pure fluid, (b) infinite cell, (c) transient full mass effect, (d) partial mass effect

the expected value $w_a(1+\psi)$ as shown by curve(c), or not, like the behaviour shown in curve (d). Clearly the optimal operating conditions are:

- small relaxation time $\tau_D$ for diffusion (the gradient $\frac{\partial c_1}{\partial x}$ given by (27) is quickly established)
- small velocity and very high cell ($H \gg h$) to delay the convection regime

$$\tau_D \ll \frac{H}{w_a}$$

$$\frac{h^2}{\pi^2 D} \ll \frac{H}{\frac{g\alpha \Delta T h^2}{6\nu}\frac{\sqrt{3}}{36}}$$

or finally

$$\frac{GrSc}{A} \ll 1231 \ , \tag{34}$$

where Gr is the Grashof number $Gr = \frac{g\alpha \Delta T h^3}{\nu^2}$, A is the aspect ratio $A = \frac{H}{h}$.

For practical conditions like

$$H = 280 \text{ mm}$$
$$h = 1.5 \text{ mm}$$
$$\Delta T = 5 \text{ K}$$

and values of $\alpha$, $D$ and $\nu$ corresponding to water–ethanol or water–isopropanol mixtures, condition (34) is not totally fulfilled. For such conditions, the behaviour closely resembles that of curve (d) of Fig.4 and the maximum velocity amplitude recorded in an experiment must be multiplied by a correction factor $\gamma > 1$ to reach the optimal value $w_a(1+\psi)$, before deducing $\psi$ and next $\frac{D_T}{D}$. This correction factor is deduced as described below.

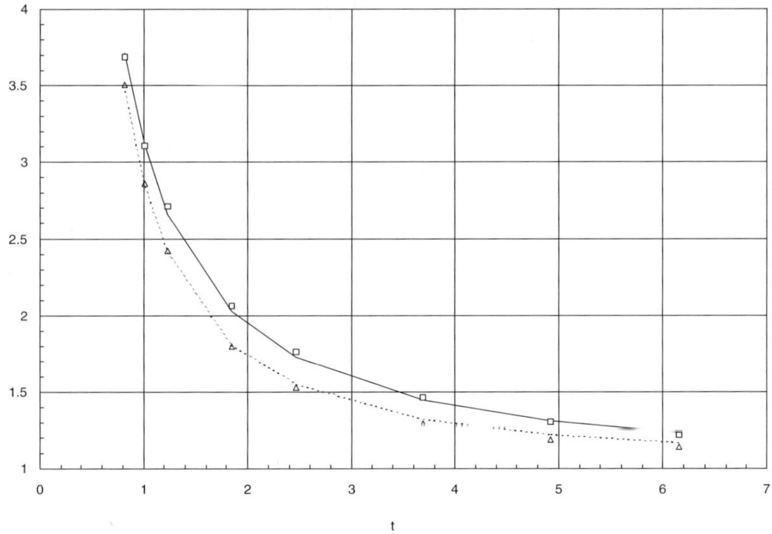


**Fig. 5.** Correction factor $\gamma$ in function of the time-ratio $\tau$: $\psi = 0.3$ (dot and triangle), $\psi = 1.5$ (square and full line)

For a given set of parameters ($H$, $h$, $\alpha$, $\beta$, $\nu$, $D_{\mathrm{T}}$, $D$, etc) the full Navier–Stokes, energy and mass equations are numerically integrated and thus we may follow the time evolution of each field, temperature, concentration and velocity. Therefore the velocity amplitude may be numerically recorded as a function of time. The maximum value that the velocity amplitude reaches is smaller than $w_{\mathrm{a}}(1+\psi)$, and their ratio defines the correction factor. Fig.5 shows this correction factor as a function of a parameter $\tau$ defined as

$$\tau = 1231 \frac{A}{GrSc} \tag{35}$$

representing the ratio of the time needed for a volume element to a travel from bottom to top of the cell (i.e. a distance $H$) at a velocity $w_{\mathrm{a}}$, divided by the time needed to reach horizontally a fully developped Soret separation ( $\tau_{\mathrm{D}} = \frac{h^2}{\pi^2 D}$).

Surprisingly this correction factor depends also on $\psi$. On Fig.5 the correction factor is displayed for two values of $\psi$ ($\psi = 0.3$ and $\psi = 1.5$). Based on 48 numerical simulations, the correction factor is fitted by a totally empirical law that represents quite satisfactorily Fig.5. Details may be found elsewhere [27].

## 5 LDV experiments in thermogravitational columns at $\psi > 0$

Up to now two sets of experiments have been undertaken. We now describe experiments on the system water–isopropanol (50 wt%).

$$\begin{aligned} \text{mean temp.} &= 22.5\ ^\circ\text{C} \\ \alpha &= 7.7\ 10^{-4}\ \text{K}^{-1} \\ \beta &= 0.256 \\ \nu &= 7.5\ 10^{-2}\ \text{cm}^2/\text{s} \\ h &= 1.61\ \text{mm} \\ H &\approx 300\ \text{mm} \\ D &\approx 5\ 10^{-6}\ \text{cm}^2/\text{s} \end{aligned}$$

Fig.6 shows the time evolution of the velocity amplitude during 4000 sec. At the steady state (2500 sec $< t <$ 4000 sec) $w_\text{a} = (196\pm5)\ \mu m/sec$. The maximum velocity, $w_\text{max} = 254\ \mu m/sec$, is found by a cubical fitting of the experimental points between 300 sec and 800 sec. The experiment is repeated three times and results are listed in the table below.

| Run | 1 | 2 | 3 | mean |
|---|---|---|---|---|
| $w_\text{max}$ | 249 | 254 | 249 | 250.67 |
| $w_\text{a}$ | 194 | 196 | 201 | 197 |
| $\frac{w_\text{max}}{w_\text{a}} - 1$ | 0.28 | 0.30 | 0.24 | $0.27 \pm 0.03$ |

If $w_\text{max}$ was really $w_\text{a}(1+\psi)$, i.e. in absence of any correction, then the deduced value of $\psi$ would be $\psi = 0.27 \pm 0.03$ (i.e. 11% standard deviation). For the present set of experiments, the correction factor is $\gamma = 1.65$ (from the empirical formula that fits results of Fig.5). Therefore the real separation factor is: $\psi = \gamma(\frac{w_\text{max}}{w_\text{a}} - 1) = 1.65 \times 0.27 = 0.45$. Knowing $\alpha$ and $\beta$, we deduce

Water–isopropanol (50 wt%): $\frac{D_\text{T}}{D} = (5.4 \pm 0.6) \times 10^{-3}\ \text{K}^{-1}$
in perfect agreement with the results of Poty et al.[24].

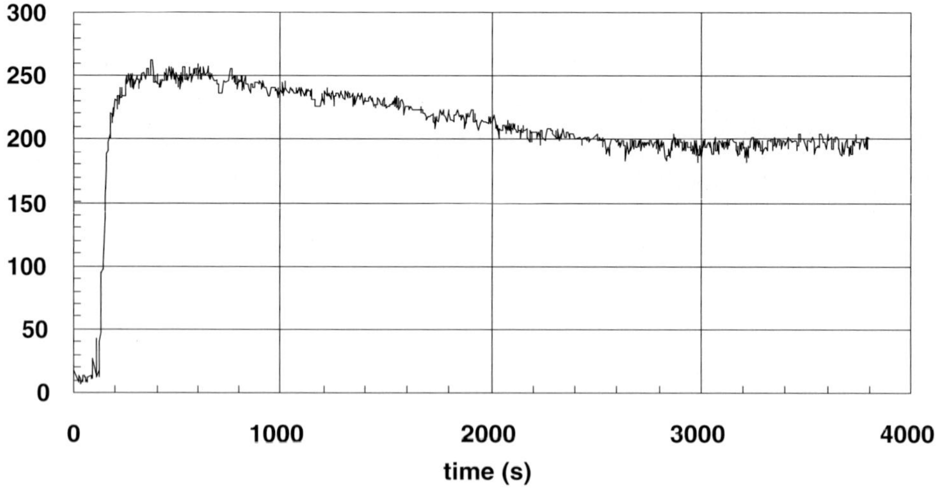


**Fig. 6.** Velocity amplitude $w_\text{a}(\mu m/sec)$ versus time

The second system, water–ethanol (ethanol 39.12 wt%) has been studied by at least four other research groups and is more appropriate for a validation of the present technique. Based on eight runs we have found a mean value of $\psi = 0.248$ and since $\alpha = 7.86 \times 10^{-4}$ K$^{-1}$ and $\beta = 0.217$ we have:

Water–ethanol (39.12 wt%): $\frac{D_\mathrm{T}}{D} = 3.70 \times 10^{-3}$ K$^{-1}$

This is not too bad compared to the results of Kolodner et al. [4], Zhang et al. [5], Bou-ali et al. [9], and Köhler et al. [6]

| | This work | [4] | [5] | [9] |
|---|---|---|---|---|
| $\frac{D_\mathrm{T}}{D} \times 10^3$ K$^{-1}$ | 3.70 | 3.38 | 3.21 | 3.22 |

The difference is at worst 14% (with Univ. of Maryland). To be complete, we should add that we have encountered experimental troubles (too long to explain here) and that therefore the standard deviation based on the eight runs is 24% which is far too much. These experiments will be repeated in the future, in a new cell with $H = 500$ mm, $h = 1.5$ mm in such a way that $\tau \sim 7$ and the correction factor $\gamma \sim 1.1$. This new cell is also dedicated for a sampling process at the steady state, and following (32) the steady state separation will give $D_\mathrm{T}$, whereas the transient velocity amplitude records give $\frac{D_\mathrm{T}}{D}$. Therefore as a byproduct we will measure the ordinary diffusion coefficient.

## 6 General conclusion

We have shown that, both for negative and positive values of the separation factor $\psi$, the Soret coupling with convection is a possible way to deduce the Soret coefficient $\frac{D_\mathrm{T}}{D}$, without any sampling process and any analysis of removed samples. Of course the price we have to pay is a precise measurement of the velocity field (in the present case by Laser–Doppler velocimetry) and of the two expansion coefficients $\alpha$ and $\beta$. The technique of detection of the onset of convection seems perfect for negative $\frac{D_\mathrm{T}}{D}$, mainly because the Hopf frequency, related to $\psi$ or $\frac{D_\mathrm{T}}{D}$ is rather easy to measure. Also spacings of 4 mm are optimal and do not cause any experimental difficulties.

The free thermogravitational configuration is more appropriate than the Rayleigh–Bénard configuration for the coupling with positive Soret coefficients, but ideally needs very small gaps ($\sim 1$ mm) and tall cells ($\sim 50$ cm ), a situation which is experimentally not so easy to achieve. And this could be responsible of a rather high standard deviation ($\sim 20\%$ ) even if the mean value of our results (based on eight successive runs) do not differ more than 14% with published values by other researchers. In any case, these deviations can be considered as satisfactory compared to techniques used before 1970 and give no confidence to what is written in the present literature concerning "the obscuring effect of natural convection".

**Acknowledgements**

This paper presents results of the Belgian Programme on Interuniversity Poles of Attraction initiated by the Belgian State, Prime Minister's Office, Science Policy Programming. The scientific responsibility is assumed by its authors.

## References

1. A. Firoozabadi, K. Ghorayeb and K. Shukla: AIChE Journal **46**, 892 (2000)
2. J.C. Legros,P. Goemare and J.K. Platten: Phys.Rev.A **32**, 1903 (1985)
3. R.Schechter, M. Velarde, J.K. Platten: Adv. Chem. Phys. **26**, 265 (1974)
4. P. Kolodner, H. Williams and C. Moe: J.Chem.Phys. **88**, 6512 (1988)
5. K.J. Zhang, M.E. Briggs, R.W. Gammon and J.V. Sengers: J. Chem. Phys. **104**, 6881 (1996)
6. W. Köhler and B. Müller: J. Chem. Phys. **103**, 4367 (1995)
7. S. Van Vaerenbergh, J.C. Legros: J. Phys. Chem. **102**, 4426 (1998)
8. S. Wiegand: Entropie **218**, 69 (1999) (Proceedings of IMT3)
9. M. M. Bou Ali, O.Ecenarro, J.A. Madariaga, C.M. Santamaria and J.J. Valencia: J. Phys: Condens. Matter **10**, 3321 (1998)
10. S. Chandrasekhar: Hydrodynamic and Hydromagnetic Stability, Oxford University Press (1968)
11. J.K. Platten: Bull. Acad. Roy. Belg., Cl. Sci. **57**, 669 (1971)
12. D.T.J. Hurle and E. Jakeman: J. Fluid Mech. **47**, 667 (1971)
13. J.C. Legros, J.K. Platten and P.G. Poty: Phys. Fluids **15**, 1383 (1972)
14. J.C. Legros, D. Longree, G. Chavepeyer, J.K. Platten: Physica **80(A)**, 76 (1975)
15. M.C. Cross and K. Kim: Phys. Rev. **A37**, 3909 and **A38**, 529 (1988)
16. J.Ph. Larre: Ph.D. Thesis, University of Toulouse (1999)
17. D.R. Caldwell: J. Phys. Chem. **79**, 1882 (1975)
18. J.K. Platten, D. Villers, O. Lhost: in "Laser Anemometry in Fluid Mechanics", Ed. by R.J. Adrian, T. Asanuma, D.F.G. Durao, F. Durst and J.H. Whitelaw, LADOAN **vol3**, 245 (1988)
19. N.B. Vargaftik: Handbook of Physical Properties of Liquids and Gases Pure Substances and Mixtures, Second Edition, Springer (1975)
20. International Critical Tables of Numerical Data, Physics, Chemistry and Technology, **vol 3**, Mac Graw–Hill (1928)
21. O. Lhost: Ph.D. Thesis, University of Mons (1990)
22. H. Touiri: Ph.D. Thesis, University of Mons (1992)
23. M.M. Bou Ali, O. Ecenarro, J.A. Madariaga, C.M. Santamaria, J.J. Valencia: Entropie **218**, 5 (1999), (Proceedings IMT3)
24. P. Poty, J.C. Legros and G. Thomaes: Z. Naturforsch **29a**, 1915 (1974)
25. J.K. Platten, J.C. Legros: Convection in Liquids, Springer Verlag (1984)
26. G. Chavepeyer, J.K. Platten: Entropie **198/199**, 25 (1996) (Proceedings of IMT2)
27. J.F. Dutrieux: Ph.D. Thesis, University of Mons, to appear

## Supplementary glossary

| | |
|---|---|
| $f$ | dimensional frequency in Hertz |
| $g$ | acceleration due to gravity |

| | |
|---|---|
| $Gr$ | Grashof number |
| $h$ | thickness of the liquid layer |
| $k_\mathrm{x}$, $k_\mathrm{y}$ | components of the wavenumber $\boldsymbol{k}$ |
| $u$, $v$, $w$ | velocity components |
| $U$, $V$, $W$ | amplitude of $u$, $v$, $w$ |
| $\Delta T$ | temperature difference across the layer |
| $\eta$ | dynamic viscosity |
| $\nu$ | kinematic viscosity |
| $\omega_\mathrm{H}$ | Hopf frequency (dimensionless) |

# Convective Patterns in Binary Fluid Mixtures with Positive Separation Ratios

Björn Huke and Manfred Lücke

Institut für Theoretische Physik, Universität des Saarlandes,
Postfach 151150, D-66041 Saarbrücken, Germany

**Abstract.** We summarize our findings about laterally periodic convection structures in binary mixtures in the Rayleigh–Bénard system for positive Soret effect. Stationary roll, square, and crossroll solutions and their stability are determined with a multi-mode Galerkin expansion. The oscillatory competition of squares and rolls in the form of crossroll oscillations is reviewed. They undergo a subharmonic bifurcation cascade where the oscillation period grows in integer steps as a consequence of an entrainment process.

## 1 Introduction

The Rayleigh–Bénard system is a prominent example for studying pattern formation in hydrodynamic systems driven away from equilibrium. In this system a fluid layer is confined between two extended plates perpendicular to the direction of gravity. A temperature difference between lower and upper plate is applied. Below a critical temperature difference a quiescent conductive state is established wherein the temperature varies linearly between the plates. But for larger differences buoyancy forces are strong enough to destabilize this conductive state and to start convection.

The pure fluid convection in the Rayleigh–Bénard system at moderate heating rates is experimentally and theoretically well investigated [1]. An extension of the problem that leads to more complex convection behavior is achieved by using binary mixtures such as ethanol-water instead. In binary mixtures the buoyancy forces are also influenced by concentration variations. The structural dynamics of the concentration distribution in mixtures results from an interplay between three competing mechanisms: nonlinear advection and mixing, weak solutal diffusion, and the Soret effect. The latter generates and sustains concentration gradients in linear response to local temperature gradients. Without Soret coupling measured by the separation ratio any concentration fluctuation diffuses away.

Convection in binary mixtures shows a rich spectrum of pattern formation behavior [1–3]. But the knowledge about these structures and in particular about their stability is more limited than for pure fluids. This is especially true for the case of a positive Soret effect where the fact that the novel structures are three-dimensional makes their numerical investigation more difficult.

A positive separation ratio implies that temperature differences drive the lighter component of the mixture into the direction of higher temperature. That

W. Köhler and S. Wiegand (Eds.): LNP 584, pp. 334–354, 2002.

means that the Soret effect enhances the buoyancy forces that result from the temperature dependence of the density. Convection is therefore established at temperature differences smaller than those needed for pure fluids. The interval of temperature differences that allows convection in a binary mixture but not yet in a pure fluid is called Soret region. Therein convection is driven only because of the presence of the solutal contribution to the buoyancy force. Above this range, i. e. in the Rayleigh region, the thermal part of the buoyancy is most important and the Soret effect is less effective for destabilizing the conductive state.

In the Soret region square-like convection patterns are often found as the primary stable form of convection whereas in pure fluids roll patterns are stable at onset. Such rolls can also be found in binary mixtures, especially at higher heating rates where convection is strong enough to reduce the concentration variations sufficiently by advective mixing. In between there exist two different kinds of crossroll structures: one is stationary and the other one is oscillatory. They are stable where neither rolls nor squares are [4–7]. In this article we summarize our findings about laterally periodic convection structures for positive separation ratios. In Sect. 2 we present the basic equations for binary mixture convection in the Rayleigh–Bénard system and the Galerkin technique we used to investigate the different convection structures numerically. In Sects. 3 – 5 we discuss the properties of rolls, squares, and crossrolls and their range of stability. We conclude in Sect. 6.

# 2 Numerical methods

## 2.1 System and basic equations

We consider a horizontal layer of a binary fluid mixture of thickness $d$ in a homogeneous gravitational field, $\boldsymbol{g} = -g\,\boldsymbol{e}_z$. A vertical temperature gradient is imposed by fixing the temperature

$$T = T_0 \pm \frac{\Delta T}{2} \quad \text{at} \quad z = \mp\frac{d}{2}\,, \tag{1}$$

e.g., via highly conducting plates in experiments. Here we consider the plates to be infinitely extended, rigid, and impermeable. $T_0$ is the mean temperature of the fluid layer.

In the conductive state a linear temperature profile

$$T_{\text{cond}}(z) = T_0 - \frac{\Delta T}{d} z \tag{2}$$

is established. If there is a Soret effect, the temperature gradients generate also a concentration gradient:

$$C_{\text{cond}}(z) = C_0 + S_{\text{T}} C_0 (1 - C_0) \frac{\Delta T}{d} z\,. \tag{3}$$

Here $C$ means the concentration of the lighter component with $S_{\text{T}}$ being its Soret coefficient. We will consider the case of negative $S_{\text{T}}$. In this case the lighter component of the mixture is driven into the direction of higher temperature. Concentration at the bottom is larger in the conductive state for positive $\Delta T$ therefore enhancing the density gradient and further destabilizing the fluid layer.

Convection is described in terms of the fields of $T$, $C$, velocity $\boldsymbol{u} = (u, v, w)$, total mass density $\varrho$, and pressure $P$. In the balance equations connecting these fields we scale lengths and positions by $d$, time by the vertical thermal diffusion time $d^2/D_{\mathrm{th}}$, temperature by $\nu D_{\mathrm{th}}/\alpha g d^3$, concentration by $\nu D_{\mathrm{th}}/\beta g d^3$, and pressure by $\varrho_0 D_{\mathrm{th}}^2/d^2$. Here $\varrho_0$ is the mean density, $D_{\mathrm{th}}$ the thermal diffusivity, $\nu$ the kinematic viscosity, and $\alpha = -(1/\rho)\partial\rho/\partial T$ and $\beta = -(1/\rho)\partial\rho/\partial C$ are thermal and solutal expansion coefficients, respectively. Using the Oberbeck–Boussinesq approximation the balance equations read [2,9]

$$\nabla \cdot \boldsymbol{u} = 0 \tag{4a}$$

$$(\partial_t + \boldsymbol{u} \cdot \nabla)\, \boldsymbol{u} = -\nabla p + Pr\, \left[ (\theta + c)\, \boldsymbol{e}_z + \nabla^2 \boldsymbol{u} \right] \tag{4b}$$

$$(\partial_t + \boldsymbol{u} \cdot \nabla)\, \theta = Ra\, w + \nabla^2 \theta \tag{4c}$$

$$(\partial_t + \boldsymbol{u} \cdot \nabla)\, c = Ra\, \psi\, w + Le\, \left( \nabla^2 c - \psi\, \nabla^2 \theta \right) . \tag{4d}$$

Here $\theta, c$, and $p$ are the reduced deviations of temperature, concentration, and pressure, respectively, from the conductive profiles.

The Lewis number $Le$ is the ratio of the concentration diffusivity $D$ to the thermal diffusivity $D_{\mathrm{th}}$, therefore measuring the velocity of concentration diffusion. The Prandtl number $Pr$ is the ratio of the momentum diffusivity $\nu$ and $D_{\mathrm{th}}$:

$$Le = \frac{D}{D_{\mathrm{th}}} \;;\; Pr = \frac{\nu}{D_{\mathrm{th}}} . \tag{5}$$

The Rayleigh number $Ra$ measures the thermal driving and the separation ratio $\psi$ measures the strength of the Soret coupling between temperature and concentration fields

$$Ra = \frac{\alpha g d^3 \Delta T}{\nu D_{\mathrm{th}}} \;;\; \psi = -\frac{\beta}{\alpha} S_{\mathrm{T}} C_0 (1 - C_0) . \tag{6}$$

We consider here $S_{\mathrm{T}} < 0$, i. e. $\Psi > 0$ for mixtures with $\alpha, \beta > 0$.

The first equation (4a) expresses the fact that the fluid is considered to be incompressible. (4b) – (4d) are the equations of motion for $\boldsymbol{u}$, $\theta$ and $c$. The left hand sides of these equations are the substantial time derivatives of the respective fields. The driving forces entering into the momentum balance equation (4b) are pressure gradients and the buoyancy caused by the temperature and concentration dependence of the density. The remaining term on the right hand side is the momentum diffusion term.

(4c) and (4d) contain also diffusion terms on the right hand side. The off-diagonal term $-Le\, \psi\, \nabla^2 \theta$ and the term $Ra\, \psi\, w$ in the concentration balance equation (4d) describe the action of the Soret effect, i.e., the generation of concentration currents and concentration gradients by temperature variations.

The Dufour effect, i.e., the driving of temperature currents by concentration variations is of interest only in gas mixtures [9]. But even there it is often small [10].

## 2.2 Galerkin method

We are concerned with three-dimensional convection patterns periodic in the $x$- and $y$-direction. To describe such patterns with wavenumbers $k_x$ and $k_y$ each field $X$ is expanded as

$$X(x, y, z; t) = \sum_{lmn} X_{lmn}(t) \mathrm{e}^{\mathrm{i}lk_x x} \mathrm{e}^{\mathrm{i}mk_y y} f_n(z) . \tag{7}$$

Here $l$ and $m$ are integers and the $f_n$ form a complete system of functions that fits the specific boundary condition for the field $X$ at the plates. To find suitable sets of functions $f_n$ we introduce some new fields. First, two scalar fields $\Phi$ and $\Psi$ are defined via

$$\boldsymbol{u} = \nabla \times \nabla \times \Phi \boldsymbol{e}_z + \nabla \times \Psi \boldsymbol{e}_z . \tag{8}$$

The structures we want to discuss do not show a horizontal mean flow for mirror symmetry reasons. Then, (8) is the most general expression that fulfills the incompressibility condition (4a) [11].

Second, instead of $c$ we use the field

$$\zeta = c - \psi\theta \tag{9}$$

that allows in a more convenient way to guarantee the impermeability of the horizontal boundaries: The diffusive part of the concentration current, driven by concentration gradients as well as by temperature gradients is given by $-Le\, \nabla(c - \psi\theta)$. At the impermeable plates the vertical component of this current vanishes which requires

$$0 = \partial_z (c - \psi\theta) = \partial_z \zeta \qquad \text{at } z = \pm 1/2 . \tag{10}$$

The advective concentration current vanishes at the plates because there $\mathbf{u} = 0$. The balance equation for $\zeta$ is obtained by combining (4c) and (4d).

The boundary conditions for the fields $\Phi, \Psi, \theta$, and $\zeta$ read

$$\Phi = \partial_z \Phi = \Psi = \theta = \partial_z \zeta = 0 \qquad \text{at } z = \pm 1/2 . \tag{11}$$

To expand the fields $\Psi$, $\theta, \zeta$, and $\Phi$ vertically we used different orthonormal sets $f_n(z)$ as follows

$$\Psi \text{ and } \theta : f_n(z) = \begin{cases} \sqrt{2}\cos(n\pi z) & n \text{ odd} \\ \sqrt{2}\sin(n\pi z) & n \text{ even} \end{cases} \tag{12a}$$

$$\zeta : f_n(z) = \begin{cases} 1 & n = 0 \\ \sqrt{2}\,\sin(n\pi z) & n \text{ odd} \\ \sqrt{2}\,\cos(n\pi z) & n \neq 0 \text{ even} \end{cases} \tag{12b}$$

$$\Phi : f_n(z) = \begin{cases} C_{\frac{n+1}{2}}(z) & n \text{ odd} \\ S_{\frac{n}{2}}(z) & n \text{ even} \end{cases} . \tag{12c}$$

Here $C_n$ and $S_n$ are Chandrasekhar–Reid functions [8].

The balance equations for the new fields are

$$\partial_t \Delta_2 \Psi = Pr \nabla^2 \Delta_2 \Psi + \{\nabla \times [(\boldsymbol{u} \cdot \nabla)\, \boldsymbol{u}]\}_z \tag{13a}$$

$$\begin{aligned} \partial_t \nabla^2 \Delta_2 \Phi &= Pr \left\{ \nabla^4 \Delta_2 \Phi - \Delta_2 \left[ (1 + \psi)\, \theta + \zeta \right] \right\} \\ &\quad - \{\nabla \times \nabla \times [(\boldsymbol{u} \cdot \nabla)\, \boldsymbol{u}]\}_z \end{aligned} \tag{13b}$$

$$(\partial_t + \boldsymbol{u} \cdot \nabla)\, \theta = -Ra\, \Delta_2 \Phi + \nabla^2 \theta \tag{13c}$$

$$(\partial_t + \boldsymbol{u} \cdot \nabla)\, \zeta = Le \nabla^2 \zeta - \psi \nabla^2 \theta\,. \tag{13d}$$

Here $\Delta_2 = \partial_x^2 + \partial_y^2$.

By inserting the ansatz (7) for each field into the balance equations and projecting them onto the basic functions one gets a nonlinear algebraic system of equations of the form

$$A_{\kappa\mu} \partial_t X_\mu = B_{\kappa\mu} X_\mu + C_{\kappa\mu\nu} X_\mu X_\nu\,. \tag{14}$$

For simplicity amplitudes are labelled here by a single Greek index and the summation convention is implied in (14) with $A_{\kappa\mu}$, $B_{\kappa\mu}$, and $C_{\kappa\mu\nu}$ being constant coefficients.

The number of modes has to be truncated to get a finite number of equations as discussed later on. For stationary convection structures the left hand side of (14) vanishes and the solution can be found using a multidimensional Newton method.

### 2.3 Stability analysis

To make a full stability analysis one has to check the stability of the patterns against perturbations with arbitrary wavevector $d\boldsymbol{e}_x + b\boldsymbol{e}_y$. To do so one has to introduce a Floquet term writing out the perturbation as

$$\delta X(x, y, z; t) = \mathrm{e}^{\mathrm{i}dx + \mathrm{i}by} \mathrm{e}^{st} \sum_{lmn} \delta X_{lmn} \mathrm{e}^{\mathrm{i}lk_x x + \mathrm{i}mk_y y} f_n(z)\,. \tag{15}$$

Such a perturbation is added to the known solution the stability of which is to be tested and inserted into the balance equations. After linearizing and projecting one gets a linear eigenvalue problem of the form

$$s A_{\kappa\mu} \delta X_\mu = B_{\kappa\mu} \delta X_\mu\,. \tag{16}$$

with constant coefficients $A_{\kappa\mu}$ and $B_{\kappa\mu}$. The aforementioned solution, i.e., the convective structure described by it is stable if every eigenvalue $s$ has a negative real part for every $d$ and $b$.

The symmetry of the convective pattern discussed above can under some circumstances be used to get separated classes of possible eigenvectors representing the perturbations. That means the eigenvalue problem can be reduced to finding the eigenvalues of two matrices of about half of the size. Because evaluating the eigenvalues of a matrix is a $O(N^3)$-process this always implies a reduction of the computation time.

### 2.4 Parameter space

In pure fluids convection in the Rayleigh–Bénard system starts at a critical Rayleigh number $Ra_c^0 = Ra_c(\psi = 0) = 1707.762$ and a wavenumber $k_c^0 = k_c(\psi = 0) = 3.117$ [8]. In binary mixtures with positive separation ratios that we are dealing with, the critical Rayleigh number is smaller, $Ra_c(\psi > 0) < Ra_c^0$, since the solutal contribution to the quiescent state's buoyancy force enhances the latter. Thus a smaller thermal driving, i.e., a smaller Rayleigh number suffices to reach the critical buoyancy force size for onset of convection. The critical wavenumber is also somewhat lower: $k_c(\psi > 0) < k_c^0$ [12]. When presenting our results we shall use the reduced Rayleigh number

$$r = \frac{Ra}{Ra_c^0} \,. \tag{17}$$

The convection above onset depends on the three parameters $\psi$, $Le$, and $Pr$. These parameters depend on the components of the mixture and the mean values of temperature and concentration. For alcohol-water for example, $\psi$ may assume values between $-0.6$ and $0.3$. Here the Lewis number $Le = 0.01$ or smaller whereas $Le \approx 1$ is typical for gas mixtures. A typical Prandtl number for liquid mixtures is $Pr = 10$, in gas mixtures $Pr = 0.1 ... 1$.

## 3 Rolls

In pure fluids, convection in the form of parallel rolls is a stable form of convection directly above onset for all values of $Pr$ [13], although they compete with spiral defect chaos at low Prandtl numbers where the latter structure has a larger basin of attraction [14]. In binary mixtures, they exist also as stable patterns above $Ra_c$ for all positive $\psi$ if $Le$ is large enough to equilibrate the concentration sufficiently. The required value of $Le$ depends mainly on the strength of the Soret effect but also on the Prandtl number. The rolls lose stability against squares at onset if $Le$ is too small, especially for strong Soret coupling [15].

### 3.1 Technical remarks

Calculating the fixed point solutions and analyzing their stability requires less numerical effort than for 3D structures. Only $X_{l0n}$-modes have to be taken into account and $\Psi \equiv 0$. Mirror symmetry in the lateral direction allows to set $X_{l0n} = X_{-l0n}$ so that the lateral functions $e^{\pm iklx}$ can be replaced by $\cos(klx)$. In addition the roll pattern is antisymmetric under reflection at the plane $z = 0$ combined with a translation by half a wavelength in $x$-direction. This mirror glide symmetry enforces half of the amplitudes to be zero, e. g. all amplitudes $\Phi_{l0n}$ where $l + n$ is an odd number.

To perform the stability analysis of rolls one determines the growth behavior of perturbations of the form

$$\delta X(x, y, z; t) = e^{idx+iby} e^{st} \sum_{ln} \delta X_{l0n} e^{ilkx} f_n(z) \,. \tag{18}$$

Because of the periodicity of the patterns in $x$ and its mirror symmetry it suffices to consider $d \in [0, k/2]$. In $y$-direction, however, all perturbation wavenumbers, say, $b \geq 0$ have to be investigated. For a discussion of the effect of mean flow perturbations, e. g., $\delta\Phi_{00n}$ we refer to [16].

The linear system of equations (18) always separates into two subsystems of perturbations $\delta X_{l0n}$ that belong to modes with amplitudes $X_{l0n}$ that are antisymmetric or symmetric under the mirror glide operation $(x, z) \to (x + \lambda/2, -z)$. E. g. all perturbations with amplitudes $\delta\Phi_{l0n}$ with even $l + n$ belong to one set, while the perturbations with odd $l + n$ belong to the other.

In special cases the system of equations can be separated even further. For $d = 0$ the perturbations can be divided into those that are symmetric and those that are antisymmetric under the operation $x \to -x$. Furthermore, if $b = 0$ then the perturbations contain either no or only $\delta\Psi$-amplitudes. This is also of practical interest, since some instabilities are most critical in these special cases.

The sets of modes taken into account in the Galerkin procedure were chosen as follows: We defined a maximal mode index $N$ and neglected all modes $X_{l0n}$ or $\delta X_{l0n}$ with $|l| + n > N_1$ for $\Phi$- and $\Psi$-fields and $|l| + n > N_2$ for $\theta$- and $\zeta$-fields. Here $N_2 = 2N_1$.

## 3.2 Numerical results

A quantitative description of the fixed point solutions and the stability behavior in binary mixtures requires in some cases sets of modes that are much larger than those needed for pure fluids. This is because the strong anharmonic narrow boundary layer behavior of the $\zeta$-field for small $Le$ and large $r$. For the most anharmonic roll structures at $r \approx 1.5$, $Le < 0.01$, and $\psi = 0.15$ that we have investigated expansions up to $N_2 = 40$ were needed. Although the velocity and temperature field are much smoother, a consistent description of the latter requires then also high $\theta$-modes as discussed in [17,18]. This is much more than for pure fluids, where truncations with $N \leq 8$ are sufficient to describe the stability behavior quantitatively even at large $r$.

### Roll solutions

We will discuss the fixed point solutions by focusing on the mode intensity $|\Phi_{101}|^2$ at $k = k_c^0$ as order parameter (see Fig. 1). The relation to the vertical velocity $w$ is $k^2\Phi_{101} = w_{101}$. It is useful to start with considering solutions for small $Le$. In this case, two different convection regimes can be distinguished. For $r < 1$, in the Soret regime, the applied temperature gradients cause convection only indirectly, by generating concentration and therefore density gradients via the Soret effect. The convection amplitudes are very small here. This changes in the Rayleigh regime, where the temperature gradients cause the necessary density gradients to develop convection also directly. $|\Phi_{101}|^2(r)$ strongly curves upward and $|\Phi_{101}|^2$ assumes values comparable to those of pure fluid convection, since concentration is more and more equilibrated.

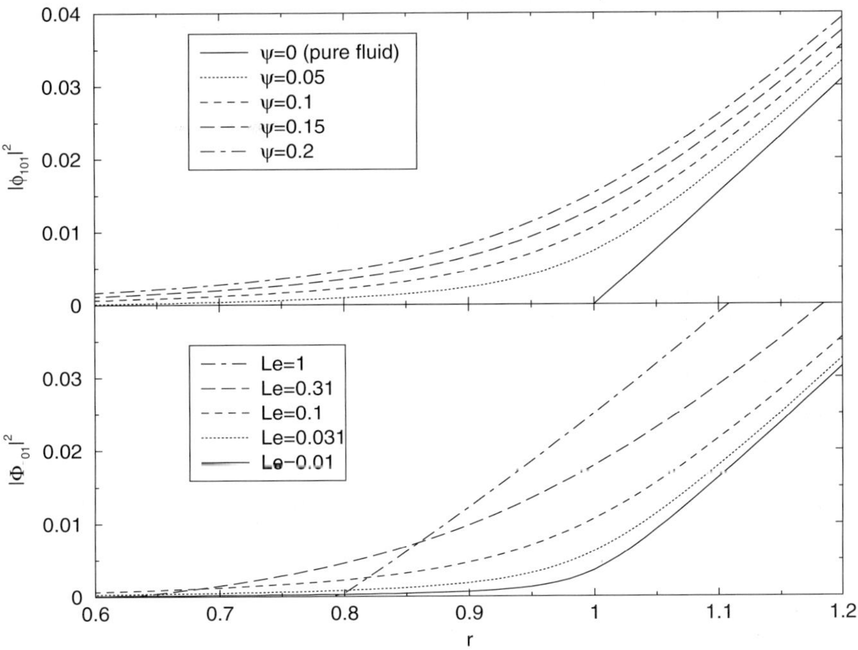


**Fig. 1.** Mode intensity $|\Phi_{101}|^2$ of roll solutions as a function of $r$ for $k = k_c^0$ and $Pr = 10$. Above: for $Le = 0.1$ and different $\psi$. Below: for $\psi = 0.1$ and different $Le$

The transition between Soret and Rayleigh region is especially sharp if $\psi$ is also small. It is also present for larger $Le$, but more smooth. It vanishes for $Le \approx 1$. The knowledge of these features will be helpful for a qualitative understanding of the stability behavior.

## Stability properties of rolls

The stability boundaries of roll patterns in pure fluids are known since the pioneering work of Busse and his coworkers [13]. At small Rayleigh numbers there exist five different instability mechanisms giving rise to five different stability boundaries that limit the region of stable rolls in the $(Ra, k, Pr)$-parameter space. At small Prandtl numbers the Eckhaus (EC), the skewed varicose (SV), and the oscillatory mechanism (OS) are the important instabilities [19]. At higher Prandtl numbers the zigzag (ZZ) and the crossroll (CR) mechanisms dominate [20]. Properties and symmetries of these perturbations are discussed in [21]. These five instabilities of roll patterns are the same as those that can be found in binary mixtures. A more detailed discussion of the symmetries and other features of these mechanisms can be found in [16].

The central qualitative difference between roll structures in pure fluids and binary mixtures concerning their stability is certainly their loss of stability in mixtures at onset in a wide range of parameter space. However, roll structures do in general still exist in mixtures as stable structures for these parameters,

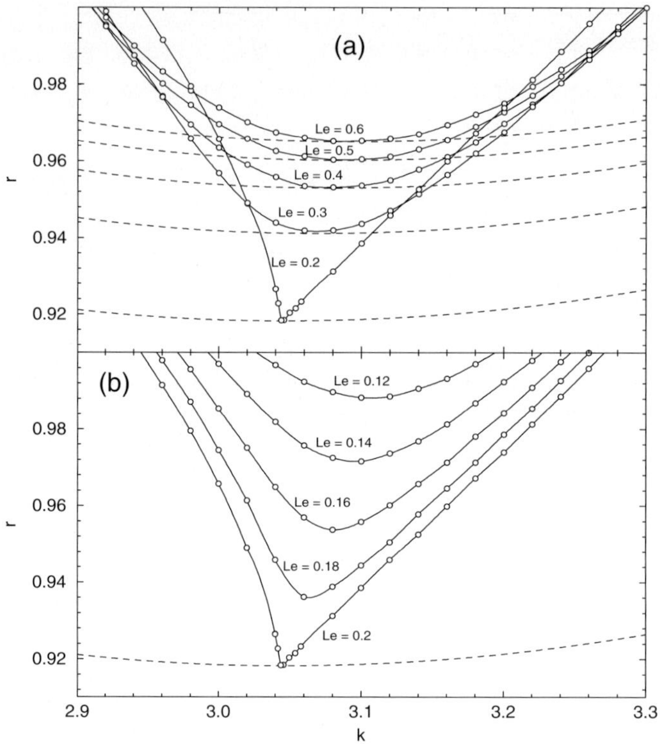


**Fig. 2.** CR instability boundaries (solid lines) of rolls in the Soret region for $Pr = 10$, $\psi = 0.01$ and several values of $Le$. Rolls are stable against CR-perturbations above the solid lines. (a): For $Le \geq 0.2$ the CR boundaries touch the neutral curve (dashed lines) in the critical point. (b): For $Le < 0.2$ the neutral curve goes further down (not shown) and is disconnected from the CR boundaries. Then rolls are not stable at the critical point anymore

not at onset, but at larger $r$. Figure 2 shows the CR instability boundaries of rolls in a parameter interval where an exchange of stability between rolls and squares at onset is predicted in [15]. One sees that the curvature of the CR boundary at the critical point diverges – a feature that follows also from cubic amplitude equations, see [16] – when this exchange occurs. For the parameters $Pr = 10, \psi = 0.01$ of Fig. 2 the exchange occurs at $Le = 0.2$. Decreasing $Le$ below this value the neutral stability curve (dashed line in Fig. 2) drops further down in $r$ (not shown in Fig. 2b) while the CR instability boundary disconnects from the neutral stability curve and moves up in $r$. Above the CR boundary rolls are still stable against CR perturbations.

The rolls could still be unstable there against other perturbations but we found the minimum of the CR boundary always to be the minimal Rayleigh number above which stable rolls exist. Only at $\psi = 0.15$, $Pr < 0.2$ and $Le < 0.015$ we found regions where rolls seem to be unstable everywhere in the $(k, r)$-plane. The two branches of the CR boundary meet again at higher $r$ near this region and limit an oval region of CR-stable rolls from below and also *above*. The oval region gets smaller by reducing $Le$ or $Pr$ until the region of stable rolls vanishes. The experimental observation of this behavior might be difficult because it occurs in a region of the parameter space that is not accessible by ordinary fluid mixtures.

We have calculated all stability boundaries at small $r$ for different values of $Le$, $\psi$, and $Pr$. Concerning the stability behavior of the roll structures one sees that in the Rayleigh region, $r > 1$, where the concentration field is nearly uniform the binary mixture behaves like a pure fluid. As for the fixed points, the transition between Soret and Rayleigh region is very sharp at small $Le$ and $\psi$. An example for such a behavior is given in Fig. 3. Here only the EC, CR, and ZZ boundaries are of importance. In the Rayleigh region of Fig. 3 the CR, ZZ, and EC boundaries of the mixture (full lines with circles, triangles, and squares, respectively) are lying close to the corresponding boundaries of the pure fluid (long-dashed lines). Note in particular the vase-like form of the EC boundary $r_{\mathrm{EC}}(k)$ and the dent in the ZZ boundary $r_{\mathrm{ZZ}}(k)$: Closer to onset ($r_c \simeq 0.6, k_c \simeq 2.6$ in Fig. 3), i.e., in the Soret regime $r_{\mathrm{EC}}(k)$ opens up parabolically and $r_{\mathrm{ZZ}}(k)$ comes out of the critical point linearly with *positive* slope. However, in the crossover range $r \sim 1$ between Soret and Rayleigh regime the curve $r_{\mathrm{EC}}(k)$ pinches inwards and develops a waist such as to follow in the Rayleigh

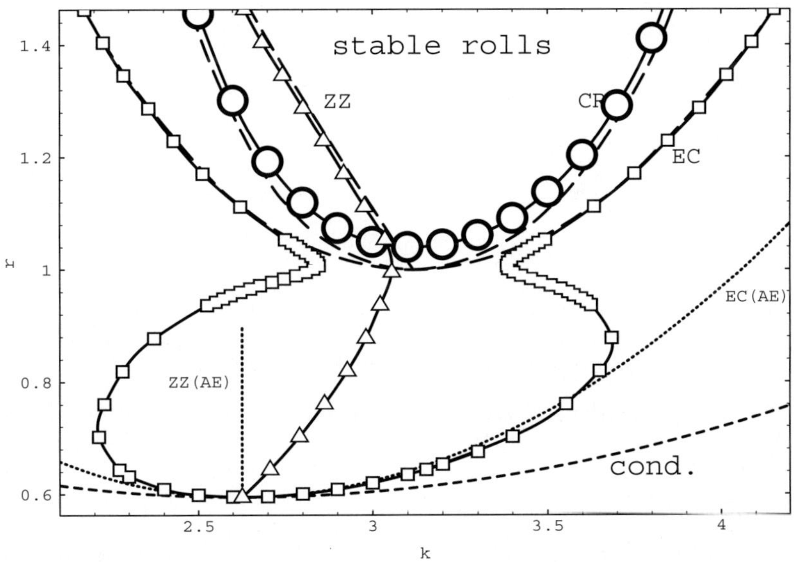


**Fig. 3.** Stability boundaries of rolls. Solid lines with open symbols refer to a mixture with $Pr = 7$, $\psi = 0.01$, and $Le = 0.025$. The corresponding boundaries in a pure fluid with $Pr = 7$ are the long dashed curves. Dotted curves labelled EC(AE) and ZZ(AE) are predictions of cubic amplitude equations for the EC and ZZ boundaries of the mixture, respectively. The conductive state is stable below the short–dashed curve labelled cond.

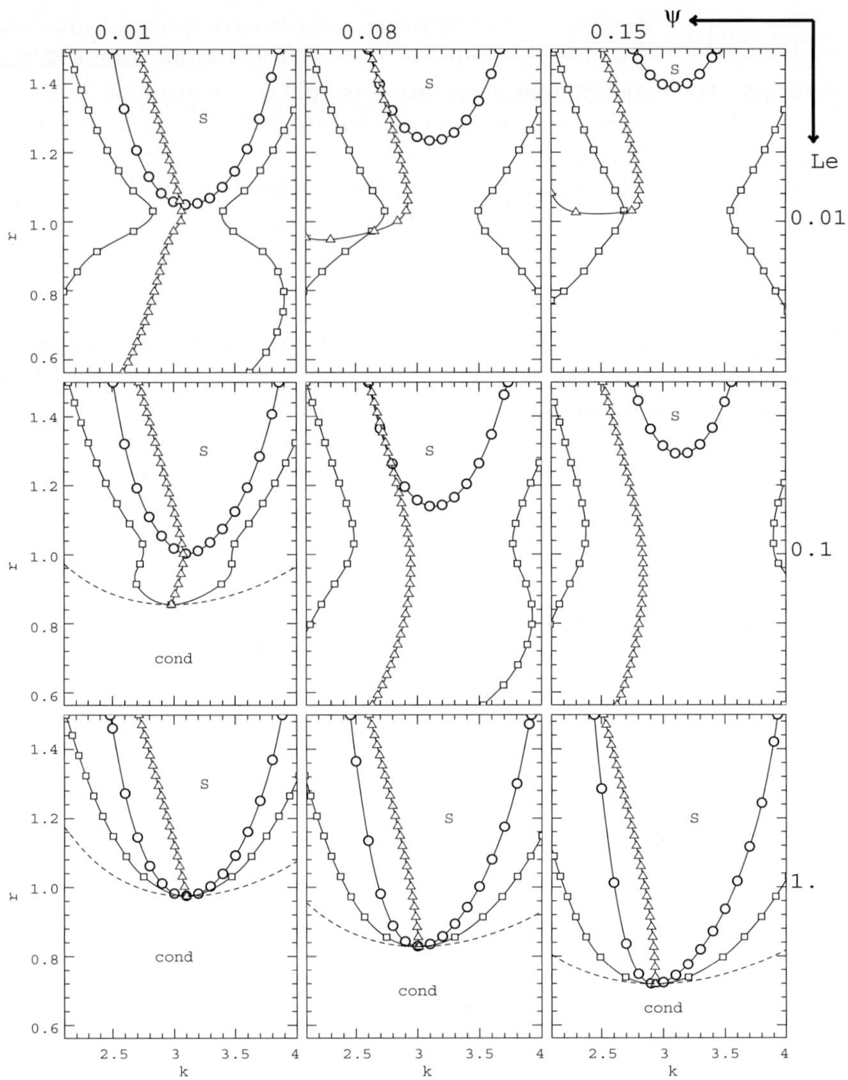


**Fig. 4.** Crossections of the stability balloon of rolls in the $k - r$ plane at $Pr = 10$. Open circles: CR, open squares: EC, open triangles: ZZ. S denotes the region of stable rolls

regime the parabolic shape of the EC curve of the pure fluid that starts out at $k_c^0 \simeq 3.1, r_c^0 = 1$. Similarly $r_{ZZ}(k)$ bends in the crossover range towards small $k$ to follow then the ZZ boundary of the pure fluid that shows *negative* slope.

Figure 4 show that this transition between Soret and Rayleigh region that causes the vase-like structure of the EC boundary and the sharp bend of the ZZ boundary is smoother for larger $Le$ or $\psi$. For $Le = 1$, the stability balloon does not qualitatively differ from its pure fluid counterpart.

# 4 Squares

Convection in square patterns is a fixed point solution that bifurcates out of the conductive state and that coexists with the roll solution. For parameter combinations where rolls are not stable at onset, square structures gain stability. They lose stability at higher $r$, roughly at the boundary between Soret and Rayleigh region.

## 4.1 Technical remarks

Squares have the same symmetry plane at $x = 0$ as rolls and an additional mirror plane at $y = 0$. Furthermore, squares show also a mirror glide symmetry. Here, however, the symmetry transformation consists of a reflection at the plane $z = 0$ combined with a translation by half a wavelength in $x$- *and* $y$-direction. To describe these three-dimensional patterns the $\Psi$-field cannot be neglected. We also mention that in contrast to the other fields $\Psi$ is odd in $x$ and $y$ and has positive parity under the mirror glide operation thereby reflecting the symmetries of the velocity field. A further reduction of the number of mode occurs since square patterns are invariant under rotation by 90° mapping the $x$- onto the $y$-directions and vice versa. Amplitudes like $\Phi_{lmn}$ and $\Phi_{mln}$ are the same. This is also true for $\theta$ and $\zeta$. Again $\Psi$ is different. Here $\Psi_{lmn} = -\Psi_{mln}$.

Because the amount of computational power needed to make a full stability analysis of these three-dimensional structures is too large, we will discuss perturbations only for periodic boundary conditions, i. e., $d = b = 0$. Here again a separation of perturbations is possible into those that change sign or not under the mirror glide operation. Furthermore, the stability problem is invariant under $x \to -x$ and $y \to -y$. Thus one can distinguish between perturbations that are even in $x$ and $y$, odd in $x$ and $y$, or even in $x$ and odd in $y$ (or equivalently odd in $x$ and even in $y$). If the perturbations have the same symmetry in both directions one can in the case of squares finally make use of a last symmetry property and separate between perturbations that are symmetric or antisymmetric under the exchange of the $x$- and $y$-direction.

The modes taken into account for the Galerkin procedure were chosen in a way analogous to the roll case: $X_{lmn}$ and $\delta X_{lmn}$ were neglected, if $|l| + |m| + n$ was larger than $N_1$ or $N_2$ respectively. Since square structures are stable mainly below $r = 1$ and the boundary layers are less narrow compared to the coexisting roll solutions (see below), values of $N_2 \leq 20$ were sufficient.

## 4.2 Numerical results

### Square solutions

The convection patterns of squares resemble linear superposition of two perpendicular sets of rolls. The intensities of the leading modes $\Phi^S_{101} = \Phi^S_{011}$ of the square solutions show qualitatively the same $r$-dependence as $\Phi^R_{101}$ for roll solutions but are always smaller. Arguments based on amplitude equations show

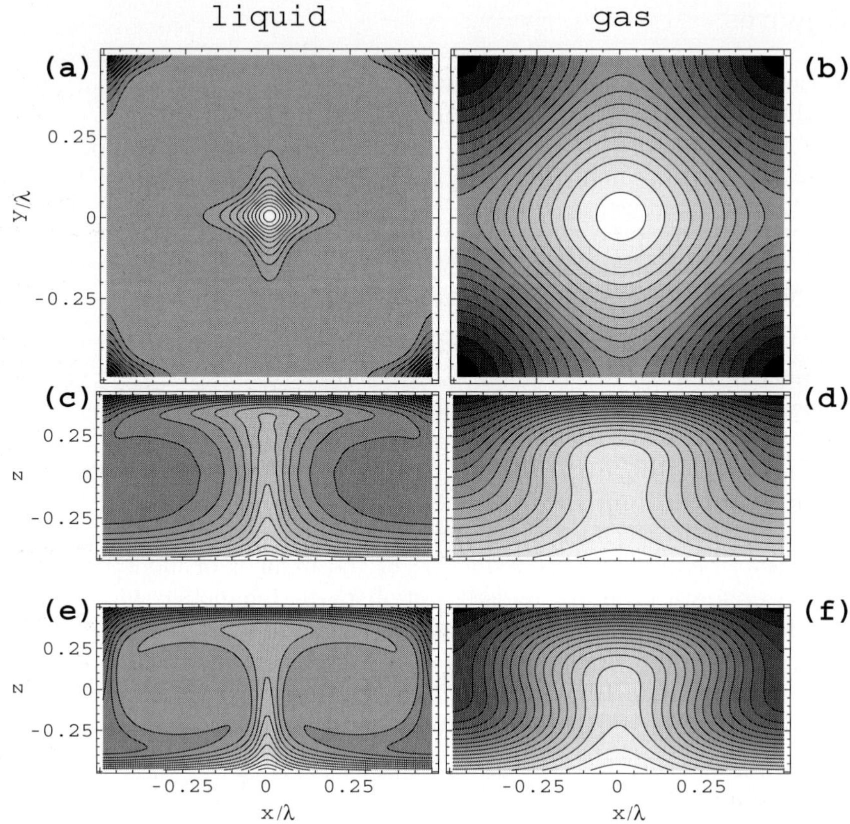


**Fig. 5.** Structural properties of square (a-d) and roll (e,f) convection for representative liquid parameters ($Le = 0.01$, $Pr = 10$, left column) and gas parameters ($Le = Pr = 1$, right column) at $r = 1, \psi = 0.15$. In (a,b) the concentration distribution of squares at mid height, $z = 0$, is shown. In (c-f) we show the concentration distribution in a vertical cross section at $y = 0$. The largest vertical upflow is at $x = y = 0$

however, that $|\Phi^{\mathrm{S}}_{101}|^2 + |\Phi^{\mathrm{S}}_{011}|^2 > |\Phi^{\mathrm{R}}_{101}|^2$ near onset if and only if squares are stable there. But this is no longer true at higher $r$ where the solutions are such that $|\Phi^{\mathrm{S}}_{101}|^2 + |\Phi^{\mathrm{S}}_{011}|^2 < |\Phi^{\mathrm{R}}_{101}|^2$. This change takes place before squares lose stability.

In Fig. 5 we show the concentration distribution of square convection for two parameter combinations that are representative for liquid and gas mixtures. This plot and the concentration field structure of rolls and squares in a vertical cross section shows a characteristic boundary layer and plume behavior at small $Le$. Such structures occur when advective mixing is large compared to diffusion in the bulk of the fluid. Consequently the boundary layers and plumes are more pronounced in rolls than in squares since the leading velocity amplitudes are greater for the former. Thus squares with their broader boundary layers are much smoother structures than rolls at the same parameters. The practically harmonic velocity and temperature fields are not shown.

**Stability properties of squares**

Performing the stability analysis of squares we had to restrict ourselves to the case $d = b = 0$ where the perturbations separate into different symmetry classes. Because both squares and rolls can be described as even in $x$ and $y$ and as mirror glide antisymmetric, one expects a perturbation that destabilizes the squares and favors the rolls to fulfill these symmetries, too. However such a perturbation should break the symmetry $x \leftrightarrow y$. We actually always found the most critical perturbation to fall into this symmetry class. Other perturbations that break the mirror symmetry in $x$- or $y$-direction are less critical.

Figure 6 shows typical examples for the stability region of squares. The left and right side of the stability boundaries should not be taken too serious – presumably square structures are destabilized earlier by instabilities with finite $b$ or $d$ that tune the wavenumber and that are not considered here.

Even if squares are stable at onset they always lose their stability against a roll pattern at higher $r$. Furthermore, for certain parameters there does also exist a band of $r$-values where neither squares nor rolls are stable. Therein crossrolls or oscillations appear as stable patterns.

## 5 Crossrolls

Crossrolls are 3D convection patterns that show the same symmetries as squares except the $x \leftrightarrow y$ symmetry. As squares, they can be qualitatively described as superpositions of perpendicular roll patterns. But in the case of crossrolls the amplitudes $\Phi_{101}$ and $\Phi_{011}$ of the $x$- and $y$-rolls are not the same.

Crossroll structures appear in two modifications: as stationary and as oscillatory crossrolls. For the latter, the amplitudes of $x$- and $y$-rolls are time dependent.

### 5.1 Technical remarks

The numerical investigations of the crossroll structures were performed as in the case of squares except that modes like $X_{lmn}$ and $X_{mln}$ could not be identified anymore due to the lack of the $x \leftrightarrow y$ symmetry. For $b = d = 0$ instabilities are either mirror glide symmetric or antisymmetric. They are either even or odd in the $x$- and the $y$-direction. We used expansions up to $N_2 = 20$ to investigate these structures. The time-dependent structures were investigated via direct numerical integration of the equations of motion for the modes.

### 5.2 Numerical results

Stationary crossrolls exist at parameters $Pr, L, \psi$ if squares are stable at onset. Then, they appear in the $r$-range were squares are already and rolls are still unstable. Oscillatory crossrolls exist only at small $Le$. The connection between the different structures can be extracted from the bifurcation diagram in Fig. 7.

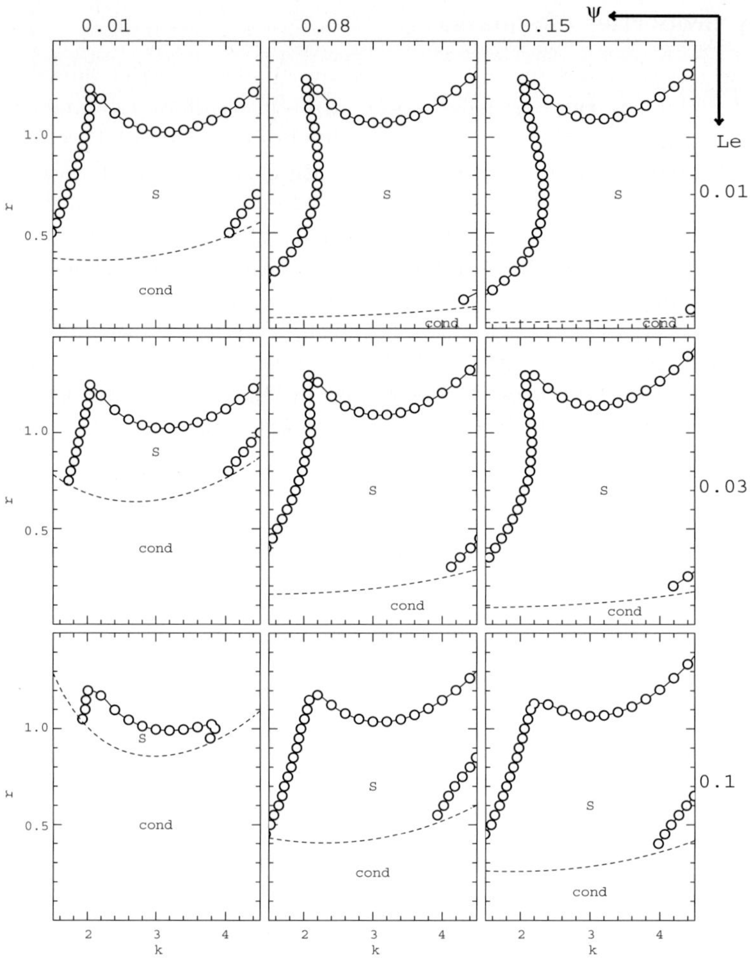


**Fig. 6.** Crossections of the stability balloon of squares in the $k - r$ plane for $Pr = 10$ obtained from a restricted stability analysis as explained in the text. S denotes the region of stable squares

For the parameters $Pr = 27$, $Le = 0.0047$, and $\psi = 0.23$ taken from [5], all four kinds of patterns exist. For the calculation of the fixed points a model with $N_2 = 10$ was used, the bifurcation diagram is therefore only qualitatively correct at higher $r$.

## Stationary crossrolls

We will first ignore the crossroll solutions of the oscillatory type and concentrate on the stationary patterns. At onset squares with $|\Phi_{101}|^2 = |\Phi_{011}|^2$ are the stable structure. The coexisting roll branch is unstable. At about $r = 1.11$ a new type of solutions bifurcates out of the square branch, the stationary crossrolls. In this

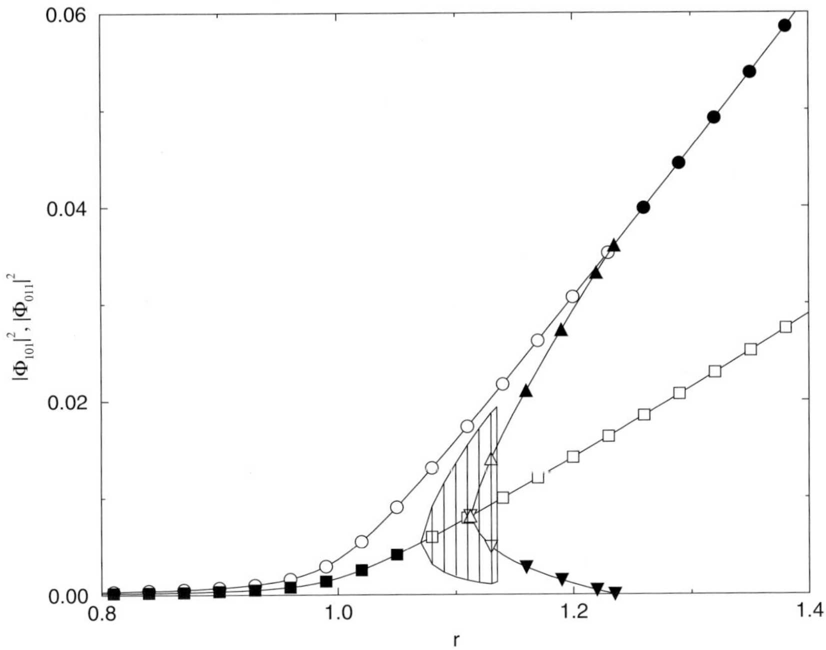


**Fig. 7.** Bifurcation diagram of $x$-roll intensity and $y$-roll intensity versus $r$ for squares (squares), rolls (circles), stationary (triangles), and oscillatory crossrolls. Filled (open) symbols denote stable (unstable) states. Full lines delimiting the hatched area mark maxima and minima of oscillations of $|\Phi_{101}|^2$ and $|\Phi_{011}|^2$. The parameters are [5] $\psi = 0.23$, $L = 0.0045$, $Pr = 27$, and $k = \pi$

$r$-range neither squares nor rolls are stable. For the crossroll structures the two important mode intensities are different, say $|\Phi_{101}|^2 > |\Phi_{011}|^2$. The difference between these intensities grows until $|\Phi_{011}|^2 = 0$ and $|\Phi_{101}|^2$ touches the roll branch at another bifurcation point at $r \approx 1.24$. The crossrolls transfers their stability to the rolls here.

The stationary crossrolls exist always inside an intermediate $r$-range if squares are stable at onset. The oscillatory type appears only at very small $Le$ as in Fig. 7. If these solutions are absent, the squares transfer their stability directly to the stationary crossroll branch, leading to a sequence squares – stationary crossrolls – rolls of stable structures for increasing $r$. In particular, the stability boundaries that limit the region of stable squares in Fig. 6 are all of the type leading to stationary rolls: oscillatory crossrolls do not exist for these parameters.

## Oscillatory crossrolls

But for the parameters in Fig. 7 the bifurcation behavior is more complicated. The squares lose their stability already at $r \approx 1.08$ before the stationary crossrolls emerge. At this point oscillatory crossrolls grow out of the square state in a

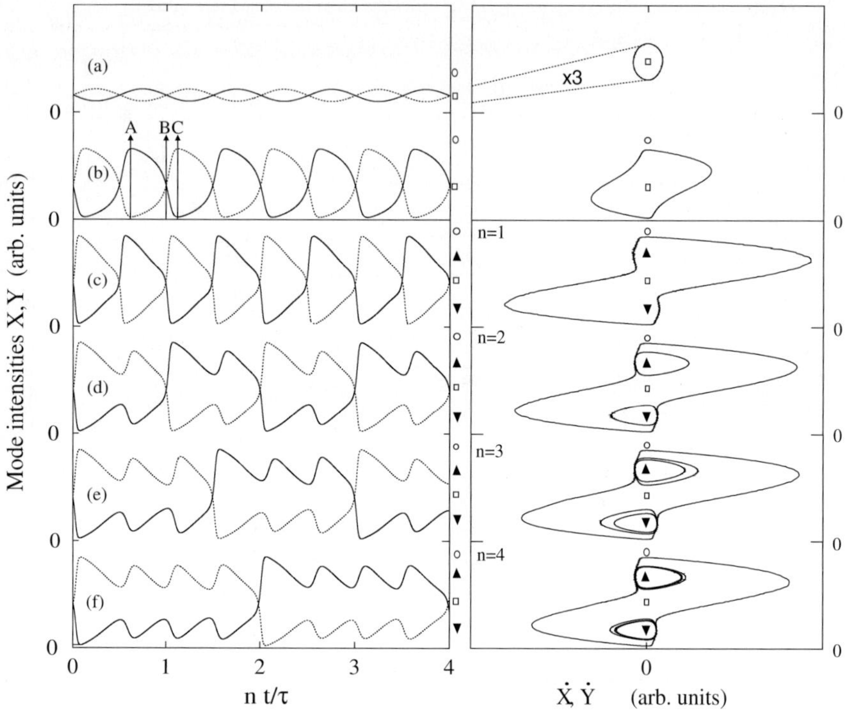


**Fig. 8.** Evolution of the oscillatory dynamics with increasing $r$ from top to bottom. Parameters as in Fig. 7. (a) and (b) are states located shortly above the Hopf threshold and right below the crossroll bifurcation, respectively. (c) – (f) are SC states. The right column contains phase space plots of $X \sim |\Phi_{101}|^2$ and $Y \sim |\Phi_{011}|^2$ versus their time derivatives for the intensities $X$, $Y$ shown in the left column with arbitrary common units versus reduced time $nt/\tau$. In the SC states $n$ is the number of windings around each of the stable crossroll fixed points (triangles). Squares and circles indicate unstable square and roll states, respectively

supercritical Hopf bifurcation. The stationary crossrolls that appear at $r = 1.11$ remain unstable until $r \approx 1.14$.

Fig. 8 shows for several Rayleigh numbers that $|\Phi_{101}|^2(t)$ and $|\Phi_{011}|^2(t)$ oscillate in opposite phase around a common mean value given by the unstable square state. The $x$-roll intensity of the pulsating pattern grows or decreases on cost of the $y$-roll intensity, however, without ever going to zero. Thus the two roll sets never die out completely or reverse their turning direction during the oscillations. Close to the Hopf bifurcation the oscillations are harmonic and of small amplitude (Fig. 8a). With increasing $r$ the frequency decreases roughly linearly [5]. Furthermore, and more importantly, with increasing amplitude the oscillation becomes more anharmonic and relaxational in Fig. 8b, c; see also Ref. [4,5]: While the system rapidly sweeps through the square state it spends more and more time in the vicinity of the roll state. The change from harmonic in Fig. 8a to strongly relaxational oscillations in Fig. 8c is also documented in the right column of Fig. 8. There we show phase space plots associated with the time histories of $|\Phi_{101}|^2$, $|\Phi_{011}|^2$ in the left column.

At larger $r$ the system gets attracted into one of the crossroll fixed points that have become stable shortly before oscillatory crossrolls cease to exist in Fig. 7. In this interval we have observed a novel subharmonic bifurcation cascade (SC) in which the crossroll attractors entrain the oscillations: First the crossroll attractors deform the phase trajectory (Fig. 8c). Then with increasing $r$ the oscillations execute an increasing number of windings around the crossroll states. In Figs. 8c – f the winding number around a crossroll fixed point increases from $n = 1$ to $n = 4$ and the period $\tau$ of the oscillations increases from $2\pi/\omega$ to $2\pi n/\omega$ in integer steps. This increase of the winding number in the SC continues beyond $n = 4$; we have found also $n = 5$. However, the control parameter interval $\delta r_n$ for an $n$-cycle becomes so narrow – $\delta r_n \cong (1.6, 0.6, 0.2)10^{-4}$ for $n = (2, 3, 4)$ – that our numerical resources were not sufficient to resolve the SC further. But we think that the SC is a robust, experimentally accessible phenomenon; Ref. [4] contains a hint for a 2-cycle. Beyond the SC interval in the system gets attracted into one of the crossroll fixed points. The transition between the oscillations and the crossroll state is slightly hysteretic. Upon reducing $r$ the system remains in the crossroll state until below the SC interval and then a transition to an $n = 1$ oscillation occurs. So in a small $r$-interval there is bistable coexistence of oscillations and stationary crossrolls.

## 6 Conclusion

We reviewed our findings of Rayleigh–Bénard convection in binary mixtures with positive Soret effect. In this case the lighter component of the mixture is driven into the direction of higher temperature. The enhanced density gradient leads to a convective instability at smaller temperature differences than in the pure fluid case. Convection at positive $\psi$ shows a rich variety of stable structures at moderate $r \leq 1.5$.

Using the Galerkin technique, we investigated two- and three-dimensional laterally periodic patterns as rolls, squares, and stationary and oscillatory crossrolls and determined their stability behavior. All these patterns are realized as stable convection structures somewhere in parameter space. The numerical examination of convection in liquid binary mixtures is in general much more difficult compared to pure fluids since small values of the Lewis number $Le \leq 0.01$ lead to a pronounced narrow boundary layer behavior of the concentration field. This is especially the case for rolls whereas squares are somewhat smoother. Nevertheless we had to restrict the stability analysis of the three-dimensional structures to perturbations with the same wave number.

Squares and rolls coexist as convective solutions above onset. The square symmetry requires that the main velocity modes are equal: $|\Phi_{101}| = |\Phi_{011}|$. For the two-dimensional rolls on the other hand one of these amplitudes, say $|\Phi_{011}|$ is zero. Squares are stable in the Soret region if $Le$ is sufficiently small. They transfer their stability to crossrolls at higher $r$. If $Le$ is not too small, only stationary crossrolls exist. As squares, they can be described as linear superpositions of roll structures. For them however, $|\Phi_{101}| > |\Phi_{011}|$ or vice versa. With increasing $r$ the smaller amplitude tends to zero until the crossroll branch ends on the now stable roll branch in the bifurcation diagram.

At very small $Le$ there exists also a branch of oscillatory crossrolls that emerges out of the square branch in a supercritical Hopf bifurcation at smaller $r$ than the branch of the stationary type. In the oscillatory crossrolls $|\Phi_{101}|^2(t)$ and $|\Phi_{011}|^2(t)$ oscillate in opposite phase around a common mean value given by the unstable square state. At higher $r$ the oscillatory crossrolls disappear in a subharmonic bifurcation cascade and the stationary ones gain stability.

At higher $Le$ roll structures are stable directly above onset and crossrolls structures do not appear. The analysis of the rolls shows that in the explored parameter range only the basic mechanisms of instability occur that are already known from the pure fluid, namely the Eckhaus, zigzag, crossroll, oscillatory, and skewed varicose mechanism. When the Soret region is small, the stability balloon of the mixtures resembles the Busse balloon for the pure fluid. However, at small $Le$ when the Soret region is large the situation is different. The fixed point solutions show a sharp transition between the two regimes. The convection amplitudes are very small in the Soret regime. But near $r = 1$ they increase strongly and become comparable to those of the pure fluid convection. The stability boundaries of roll convection show a similar transition here. In the Rayleigh region the boundaries are close to the boundaries of the pure fluid. But upon reducing the Rayleigh number the boundaries begin to deviate from their pure fluid counterparts. Near onset they finally agree with the predictions of the amplitude equations for the mixtures.

## References

1. M. C. Cross, P. C. Hohenberg: Rev. Mod. Phys. **49**, 581 (1993)
2. J. K. Platten, J. C. Legros: *Convection in Liquids.* (Springer, New York 1984)
3. M. Lücke, W. Barten, P. Büchel, C. Fütterer, St. Hollinger, Ch. Jung: 'Pattern formation in binary fluid convection and in systems with throughflow'. In: *Evolution of spontaneous structures in continuous systems* ed. by F. H. Busse, S. C. Müller. Lecture Notes in Physics **55**, 127. (Springer, Berlin, Heidelberg 1998)
4. P. Le Gal, A. Pocheau, V. Croquette: Phys. Rev. Lett. **54**, 2501 (1985)
5. E. Moses, V. Steinberg: Phys. Rev. Lett. **57**, 2018 (1986); Phys. Rev. A **43**, 707 (1991)
6. M. A. Dominguez-Lerma, G. Ahlers, D. S. Cannell: Phys. Rev. A **52**, 6159 (1995)
7. Ch. Jung, B. Huke, M. Lücke: Phys. Rev. Lett. **81**, 3651 (1998)
8. S. Chandrasekhar: *Hydrodynamic and hydromagnetic stability*, Appendix V (Dover, 1981)
9. W. Hort, S. Linz, M. Lücke: Phys. Rev. A **45**, 3737 (1992)
10. J. L. Liu, G. Ahlers: Phys. Rev. E **55**, 6950 (1997)
11. R. M. Clever, F. H. Busse: J. Fluid Mech. **198**, 345 (1989)
12. E. Knobloch, D. R. Moore: Phys. Rev. A **37**, 860 (1988)
13. F. H. Busse: Rep. Prog. Phys. **41**, 1929 (1978)
14. J. L. Liu, G. Ahlers: Phys. Rev. Lett. **77**, 3126 (1996)

15. T. Clune, E. Knobloch: Phys. Rev. A **44**, 8084 (1991)
16. B. Huke, M. Lücke, P. Büchel, Ch. Jung: J. Fluid Mech. **408**, 121 (2000)
17. St. Hollinger, M. Lücke: Phys. Rev. E **57**, 4238 (1998)
18. St. Hollinger: Theorie der ausgedehnten stationären und wandernden Konvektion in binären Fluidmischungen. PhD Thesis, Universität des Saarlandes, Saarbrücken (1996)
19. R. M. Clever, F. H. Busse: Phys. Fluids A **2**, 334 (1990)
20. F. H. Busse: Journal of Mathematics and Physics **46**, 140 (1967)
21. E. W. Bolton, F. H. Busse, R. M. Clever: J. Fluid Mech. **164**, 469 (1985)

## Supplementary glossary

| | |
|---|---|
| $b$ | wave vector deviation in $y$ |
| $c$ | concentration deviation |
| $C$ | concentration |
| $C_0$ | mean concentration |
| $C_{\rm cond}$ | concentration in conductive state |
| CR | crossroll instability |
| $d$ | wave vector deviation in $x$, also plate distance |
| $\mathbf{e}_x, \mathbf{e}_y, \mathbf{e}_z$ | unit vectors |
| EC | Eckhaus instability |
| $\boldsymbol{g}$ | gravitational field |
| $k$ | wave number |
| $k_x$ | wave number in x |
| $k_y$ | wave number in y |
| $k_{\rm c}$ | critical wave number |
| $k_{\rm c}^0$ | critical wave number for pure fluid |
| OS | oscillatory instability |
| $p$ | pressure deviation |
| $P$ | pressure |
| $r$ | reduced Rayleigh number |
| $Ra_{\rm c}$ | critical Rayleigh number |
| $Ra_{\rm c}^0$ | critical Rayleigh number for pure fluid |
| $s$ | growth rate |
| SV | skeved varicose instability |
| $T_0$ | mean temperature |
| $T_{\rm cond}$ | temperature in conductive state |
| $\Delta T$ | temperature difference |
| $u$ | velocity, x component |
| $v$ | velocity, y component |
| $w$ | velocity, z component |
| $X$ | arbitrary field |
| $X_{lmn}$ | field mode |
| $\delta X$ | perturbation in field $X$ |
| $\delta X_{lmn}$ | perturbation mode |
| ZZ | zigzag instability |

| | |
|---|---|
| $\zeta$ | $c - \psi\theta$ |
| $\theta$ | temperature deviation |
| $\nu$ | kinematic viscosity |
| $\rho_0$ | mean density |
| $\Phi$ | velocity potential |
| $\tau$ | period of oscillations |
| $\Psi$ | velocity potential |
| $\omega$ | frequency of oscillations |

# Convective Instability of Magnetized Ferrofluids: Influence of Magnetophoresis and Soret Effect

Mark I. Shliomis

Department of Mechanical Engineering, Ben-Gurion University of the Negev, P.O.B. 653, Beer-Sheva 84105, Israel

**Abstract.** Convective instability in a ferrofluid layer heated from below or from above in the presence of a uniform vertical magnetic field is investigated theoretically. Convection is caused by a magnetic mechanism based on the temperature and concentration dependence of magnetization. An imposed temperature gradient establishes (by the *Soret effect*) a concentration gradient of magnetic particles of which the ferrofluid is composed. Both these gradients cause a spatial variation in magnetization, which induces a gradient of magnetic field intensity within the fluid layer. The field gradient induces in its turn an additional redistribution of magnetic grains due to *magnetophoresis*. Resulting self-consistent magnetic force tries to mix the fluid. A linear stability analysis predicts *oscillatory instability* in a certain region of the magnetic field strength and the fluid parameters. The instability owes magneto- and thermophoresis its origin: were the particle diffusion not operative, then only stationary instability would occur. A discovery of predicted convective oscillations is expected in ferrofluid layers about 1 mm thick, where the buoyancy mechanism is negligible and the characteristic diffusion time is not too long.

## 1 Introduction

Magnetic colloids (so-called *ferrofluids* [1,2]) represent very interesting object for studies of convection. Even in the absence of the gravity, mechanical equilibrium of a nonisothermal ferrofluid in a magnetic field $H$ is in general impossible. At the basis of the mechanism of *thermomagnetic* convection [1-5] lies the temperature dependence of the magnetization $M$: under otherwise equal conditions, the colder elements of the fluid are more strongly magnetized, and therefore they are subject to a larger magnetic force in the direction of $\nabla H$. The gradients of the magnetic field intensity here play the same role as does the gravitational field $\mathbf{g}$ in the mechanism of ordinary thermogravitational convection, whilst the temperature dependence of the magnetization, $M(T)$, plays the role of $T$-dependence of the fluid density $\rho$.

Interestingly, thermomagnetic convection can arise even in a *uniform* applied magnetic field [4]. Let a ferrofluid confined between two horizontal planes be in a uniform vertical external field $\mathbf{H}_e = (0, 0, H_e)$ at the temperature $T(z)$. Then the dependence $M(T)$ leads to the result that the field *inside* the fluid, $\mathbf{H} = (0, 0, H)$, appears to be *nonuniform*. Indeed, from the Maxwell equation $\operatorname{div} \mathbf{B} = 0$ and the definition of magnetic induction $\mathbf{B} = \mathbf{H} + 4\pi\mathbf{M}$ it follows

$$\frac{\mathrm{d}H}{\mathrm{d}z} = -4\pi \frac{\mathrm{d}M}{\mathrm{d}z}, \tag{1}$$

W. Köhler and S. Wiegand (Eds.): LNP 584, pp. 355–371, 2002.

that yields

$$\frac{\mathrm{d}H}{\mathrm{d}z} = -\frac{4\pi}{\hat{\mu}} \left(\frac{\partial M}{\partial T}\right)_{\mathrm{H}} \frac{\mathrm{d}T}{\mathrm{d}z}, \tag{2}$$

where $\hat{\mu} = 1{+}4\pi(\partial M/\partial H)$ is the differential magnetic permeability. Equation (1) shows that the induced gradient of magnetic field is directed always *opposite* to the gradient of magnetization. It results in unstable magnetization stratification: the magnetic force $\mathbf{F} = M\nabla H$ tries to mix ferrofluid. According to eq. (2), this force is proportional to the imposed temperature gradient. Therefore, when both the magnetic and buoyancy mechanisms are operative, the first of two predominates over the second one at the onset of convection in a *thin enough* fluid layers, where the critical temperature gradient is sufficiently large. One can show [1] that the condition

$$\left|\left(\frac{\partial M}{\partial T}\right)_{\mathrm{H}} \frac{\mathrm{d}H}{\mathrm{d}z}\right| \gg \left|\left(\frac{\partial \rho}{\partial T}\right)_{\mathrm{p}} g\right|,$$

which permits neglect the buoyancy mechanism of convection, is reduced to

$$d^4 \ll 4\pi Ra_{\mathrm{c}}\eta\kappa \left[\frac{(\partial M/\partial T)_{\mathrm{H}}}{g(\partial \rho/\partial T)_{\mathrm{p}}}\right]^2, \tag{3}$$

where $d$ is the layer thickness, $Ra_{\mathrm{c}} \sim 10^3$ the critical Rayleigh number ($Ra_{\mathrm{c}} = 1708$ in the case of rigid boundaries and $Ra_{\mathrm{c}} = 27\pi^4/4 \approx 657.5$ for free boundaries), $\eta$ and $\kappa$ the fluid viscosity and thermodiffusivity. For commonly used magnetite ferrofluids on the base of water or kerosene the pyromagnetic coefficient $-(\partial M/\partial T)_{\mathrm{H}}$ in the field $H \sim 200 - 500\,\mathrm{Oe}$ is about $0.5\phi$ G/K where $\phi$ is the volume fraction of magnetic grains. Then for $\phi \sim 0.1,\;\; \eta \sim 10^{-2}\mathrm{Ps},\;\; \kappa \sim 10^{-3}\,\mathrm{cm^2/s}$, and $-(\partial\rho/\partial T)_{\mathrm{p}} \sim 5 \times 10^{-4}\,\mathrm{g\,cm^{-3}\,K^{-1}}$, the inequality (3) holds true down to $d = 1\,\mathrm{mm}$. The same conclusion has been done by Finlayson [4] who first studied theoretically convective instability of a ferrofluid layer heated from below in the presence of a uniform vertical magnetic field.

It is important to notice that in [4] and posterior theoretical works [5–8] ferrofluids were treated as pure (i.e., single-component) magnetized fluids. Such a model assumes a fixed uniform concentration of magnetic grains, $\phi = \mathrm{const}$. Ignoring in that way *diffusion effects*, authors of [4-8] came to conclusion that only *stationary* convective instability is possible in ferrofluids. As it pointed out already in the abstract of Ref. [4], "oscillatory instability cannot occur". But a real ferrofluid should be treated as a *binary mixture* with allowance for concentration dependence of its magnetization. An externally imposed temperature gradient induces in a ferrofluid layer a concentration gradient owing to the *Soret effect*. Both these gradients cause a spatial variation in magnetization $M(\phi, T, H)$ that leads in turn to appearance of a magnetic field gradient within the fluid layer. Thus there arise magnetic forces, which try to mix the ferrofluid and, in addition, they induce a certain redistribution of magnetic grains due to *magnetophoresis.* As it will be shown below, just the magnetophoresis together with the thermophoresis (the Soret effect) give rise to *oscillatory instability* in a certain region of the magnetic field intensity and the fluid parameters. Appropriate experimental conditions to observe predicted instability take place in a thin ferrofluid layer, when the ordinary (buoyancy) mechanism of convection is negligible and the characteristic diffusion time, $\tau_{\rm D} = d^2/\pi^2 D$, required to settle a steady state, is not too long (here $D$ is the diffusion coefficient of magnetic particles in the basic liquid). Recent measurements [9] yielded $D = 2.7\times 10^{-7}\,\mathrm{cm}^2/\mathrm{s}$ in water and cyclohexane. (This magnitude of $D$ agrees well with a theoretical estimate by the Einstein's formula for the diffusion coefficient, $D = k_{\rm B}T/6\pi\eta a$, where $2a \simeq 10\,\mathrm{nm}$ is the diameter of a magnetic grain). According to above formula for $\tau_{\rm D}$, the concentration equilibrium in a ferrofluid layer of the thickness $d = 1\,\mathrm{mm}$ is reached after one hour.

In the next Section we obtain the description of the magnetophoresis and Soret effect in a ferrofluid assuming that the fluid behaves with respect to an external magnetic field like a Langevin's paramagnet [1, 10]. In Secs. 3 and 4 we derive the complete set of modified linear equations for perturbations and settle the boundary conditions for them. The eigenvalue problem is solved in Sec. 5 for the Bénard configuration. On the base of the solution, oscillatory instability is predicted to occur for a wide range of magnitudes of magnetic field, Soret coefficient, and other fluid parameters.

Owing to the small magnitude of $D$ (see above), there are possible two scenarios of the beginnings of convection. Depending on the imposed heating rate, the ferrofluid may behave either as a binary mixture or like a pure fluid. Both the possibilities are discussed in Sec. 6.

## 2 Magnetophoresis and Soret effect

In a nonuniform magnetic field there acts on each particle a force $\mathbf{f} = (\mathbf{m}\nabla)\mathbf{H}$. Orientation of the particle magnetic moment $\mathbf{m}$ in the direction of $\mathbf{H}$ is impeded by thermal motion. As a result, the mean value of $\mathbf{m}$ appears to be equal to

$$\langle \mathbf{m} \rangle = m\mathcal{L}\left(\frac{mH}{k_{\rm B}T}\right)\mathbf{e}\,, \quad \mathbf{e} = \frac{\mathbf{H}}{H}\,, \quad \mathcal{L}(\xi) = \coth\xi - \xi^{-1}\,. \tag{4}$$

With allowance for the identity $\mathbf{H}\times\mathrm{rot}\mathbf{H} = \frac{1}{2}\nabla H^2 - (\mathbf{H}\nabla)\mathbf{H}$ and the condition of absence of electric currents in nonconducting ferrofluids, $\mathrm{rot}\,\mathbf{H} = \mathbf{0}$, formula (4) enable us to write the mean magnetophoretic force $\langle \mathbf{f} \rangle = m\,\mathcal{L}(\xi)\nabla H$. This force and the Stokes drag coefficient, $6\pi\eta a$, for a sphere of the radius $a$ determine the regular component of the Brownian velocity of the particle with respect to the liquid,

$$\mathbf{u} = \frac{m}{6\pi\eta a}\mathcal{L}(\xi)\nabla H. \tag{5}$$

The volume density of the diffusion flux of the matter in a ferrofluid may be written

$$\mathbf{j} = \phi\mathbf{u} - D(\nabla\phi + S_{\rm T}\nabla T), \tag{6}$$

where $S_{\rm T}$ is the Soret coefficient. Substituting (5) in (6) and using the Einstein's formula for the diffusion coefficient, we obtain

$$\mathbf{j} = -D[\nabla\phi + S_{\rm T}\nabla T - (\phi/H)\xi\mathcal{L}(\xi)\nabla H]. \tag{7}$$

Thus the matter flux is provided by the "diffusiophoresis" ($\propto \nabla\phi$; this term was proposed by Derjaguin et al. [11]), thermophoresis ($\propto \nabla T$), and magnetophoresis ($\propto \nabla H$). If, however, confined horizontal planes are impervious and an applied magnetic field is uniform, then gradients of $\phi$ and $H$ can be caused only by an imposed temperature gradient. Therefore in the steady state ($\mathbf{j} = 0$) the values of $\nabla\phi$ and $\nabla H$ are expressed via $\nabla T$. As it is seen from (7), the equilibrium equation $j_{\rm z} = 0$ reads

$$\frac{{\rm d}\phi}{{\rm d}z} = -S_{\rm T}\frac{{\rm d}T}{{\rm d}z} + \phi\xi\mathcal{L}\,\frac{{\rm d}\ln H}{{\rm d}z}\,. \tag{8}$$

Substituting in (1) $M = \phi M_{\rm s}\mathcal{L}(mH/k_{\rm B}T)$ where $M_{\rm s}$ is the saturation magnetization of the particle material, we find

$$\hat{\mu}\frac{{\rm d}H}{{\rm d}z} = -4\pi\left(\frac{{\rm d}\ln\phi}{{\rm d}z} - \frac{\xi\mathcal{L}'}{\mathcal{L}}\frac{{\rm d}\ln T}{{\rm d}z}\right). \tag{9}$$

Eliminating now the concentration gradient from (8) and (9), we arrive at

$$\frac{{\rm d}H}{{\rm d}z} = \frac{4\pi M}{\sigma T}\left(\Psi + \frac{\xi\mathcal{L}'}{\mathcal{L}}\right)\frac{{\rm d}T}{{\rm d}z}\,, \tag{10}$$

where $\sigma = \hat{\mu} + 12\pi\chi_0\mathcal{L}^2(\xi) = 1 + 12\pi\chi_0(\mathcal{L}' + \mathcal{L}^2)$, $\chi_0 = \phi M_{\rm s}m/3k_{\rm B}T$ the initial Langevin magnetic susceptibility, and $\Psi = (T/\phi)S_{\rm T}$ plays the role of the *separation ratio* in the mechanism of thermomagnetic convection. Substituting ${\rm d}H/{\rm d}z$ from (10) into (8) we obtain, finally,

$$\frac{{\rm d}\phi}{{\rm d}z} = -\frac{\hat{\mu}\phi}{\sigma T}\left[\Psi - \frac{(\hat{\mu}-1)}{\hat{\mu}}\,\xi\mathcal{L}\right]\frac{{\rm d}T}{{\rm d}z}\,. \tag{11}$$

One can rewrite the latter in the form

$$\frac{{\rm d}\phi}{{\rm d}z} = -\frac{\phi}{T}\Psi_{\rm H}\frac{{\rm d}T}{{\rm d}z}\,, \tag{12}$$

introducing the field-dependent separation ratio

$$\Psi_{\rm H} = \frac{\hat{\mu}}{\sigma}\left[\Psi - \frac{(\hat{\mu}-1)}{\hat{\mu}}\,\xi\mathcal{L}\right]. \tag{13}$$

Note that $\Psi_{\rm H}$ differs from zero even if $\Psi = 0$, i.e., in the absence of the Soret effect. The *non-Soret* $\phi - T$ coupling originates from the magnetophoresis, which is manifested as a negative separation ratio:

$$\Psi_{\rm H}|_{\Psi=0} = -(\hat{\mu}-1)\,\xi\mathcal{L}/\sigma < 0. \tag{14}$$

Actually, the induced gradient of magnetic field (10) is directed along the temperature gradient, thus the hotter fluid (and hence the stronger magnetic field) – the higher concentration of magnetic grains. By the same manner it behaves a nonmagnetic binary mixture with a negative separation ratio. The function defined by Eq. (14) is proportional to $\xi^2$ at $\xi \ll 1$ and inversely proportional to $\xi$ at $\xi \gg 1$. For a ferrofluid with the initial magnetic permeability $\mu_0 = 1 + 4\pi\chi_0 = 3$ this function has a distinct minimum, $\Psi_{\mathrm{H}}|_{\Psi=0} = -0.303$, at $\xi = 2.30$.

## 3 Equations for stability analysis

The general expression for the volume density of magnetic force acting a nonconducting and incompressible ferrofluid may be written [12]

$$\mathbf{F} = \nabla \left\{ \frac{H^2}{2} \left[ \rho \left( \frac{\partial \chi}{\partial \rho} \right)_{\mathrm{T,H}} - \chi \right] \right\} + M \nabla H, \tag{15}$$

where $\chi = M/H$ stands for the magnetic susceptibility. The first term in (15) is known as the *magnetostrictive force* and the second one called the *Kelvin force*. Under the substitution $\mathbf{F}$ in the equation of fluid motion, the magnetostrictive force which originates from a potential is included in the pressure gradient, $\nabla p$, so this force is equilibrated *automatically*. Thus it remains only the Kelvin force, which in equilibrium should be also balanced at each point by the pressure gradient: $\nabla p = M \nabla H$. On applying the operation rot to this equation, we obtain a necessary condition for equilibrium,

$$\left( \frac{\partial M}{\partial T} \nabla T + \frac{\partial M}{\partial \phi} \nabla \phi \right) \times \nabla H = 0. \tag{16}$$

In the case under consideration, all three gradients contained in (16) are parallel to each other (they have only $z$-components), so the condition (16) is satisfied. Thus, the equilibrium is possible, and we set about a study of its stability.

Small perturbations of a standing mode will be characterized by velocity $\mathbf{v}$, pressure $p$, temperature $\theta$, concentration $\varphi$, and magnetic field $\mathbf{h}$. Then the perturbation $\mathbf{F}'$ of the magnetic force density $\mathbf{F} = M \nabla H$ is expressed via $\theta$, $\varphi$ and $\mathbf{h}$:

$$\mathbf{F}' = -h_{\mathrm{z}} \nabla M + \left( \frac{\partial M}{\partial T} \theta + \frac{\partial M}{\partial \phi} \varphi + \frac{\partial M}{\partial H} h_{\mathrm{z}} \right) \nabla H. \tag{17}$$

[We have omitted here the term $\nabla(M h_{\mathrm{z}})$. Later on it is assumed to be included in the gradient of the pressure perturbation, $\nabla p$, see Eq. (20)]. In the equilibrium $\nabla M$ and $\nabla H$ are directed along the $z$ axis; so, using (1) we obtain from (17) the only component of the vector $\mathbf{F}'$:

$$F'_{\mathrm{z}} = \left( \frac{\hat{\mu}}{4\pi} h_{\mathrm{z}} + \frac{\partial M}{\partial T} \theta + \frac{\partial M}{\partial \phi} \varphi \right) \frac{\mathrm{d}H}{\mathrm{d}z}. \tag{18}$$

Substituting in the formula $M = \phi M_s \mathcal{L}(\xi)$, $\xi = mH/k_B T$, and $\mathrm{d}H/\mathrm{d}z$ from (10), we get

$$F_z' = \frac{M}{\sigma T}\left(\Psi + \frac{\xi\mathcal{L}'}{\mathcal{L}}\right)\left[\hat{\mu}h_z + 4\pi M\left(\frac{\varphi}{\phi} - \frac{\xi\mathcal{L}'}{\mathcal{L}}\frac{\theta}{T}\right)\right]\frac{\mathrm{d}T}{\mathrm{d}z}. \quad (19)$$

Thus, the linear equation of fluid motion takes the form

$$\rho\frac{\partial \mathbf{v}}{\partial t} = -\nabla p + \eta\nabla^2\mathbf{v} + \frac{M\Delta T}{\sigma T d}\left(\Psi + \frac{\xi\mathcal{L}'}{\mathcal{L}}\right)\left[4\pi M\left(\frac{\varphi}{\phi} - \frac{\xi\mathcal{L}'}{\mathcal{L}}\frac{\theta}{T}\right) + \hat{\mu}\frac{\partial \Phi}{\partial z}\right]\mathbf{e}. \quad (20)$$

We have substituted in (20) the magnitude of the stationary temperature gradient, $\mathrm{d}T/\mathrm{d}z = \Delta T/d$, where $\Delta T$ is the temperature difference between confined surfaces of the layer. In (20) $\mathbf{e}$ is the unit vector along the $z$ axis, and $\Phi$ the potential of magnetic field perturbations, $\mathbf{h} = \nabla\Phi$. To derive an equation for $\Phi$ one should substitute the perturbation of magnetic induction,

$$\mathbf{b} = \mathbf{h} + 4\pi\frac{M}{H}(\mathbf{h} - h_z\mathbf{e}) + 4\pi\left(\frac{\partial M}{\partial T}\theta + \frac{\partial M}{\partial \phi}\varphi + \frac{\partial M}{\partial H}h_z\right)\mathbf{e},$$

in the Maxwell equation $\operatorname{div}\mathbf{b} = 0$. It yields

$$\mu\nabla^2\Phi - (\mu - \hat{\mu})\frac{\partial^2\Phi}{\partial z^2} = -4\pi M\left(\frac{1}{\phi}\frac{\partial\varphi}{\partial z} - \frac{\xi\mathcal{L}'}{T\mathcal{L}}\frac{\partial\theta}{\partial z}\right), \quad (21)$$

where $\mu = 1 + 4\pi(M/H) = 1 + 12\pi\chi_0\mathcal{L}(\xi)/\xi$.

Equation (7) under the replacement $\phi \to \phi + \varphi$, $T \to T + \theta$, and $H \to H + \partial\Phi/\partial z$ determines the matter flux perturbation. So, making allowance for Eq. (12), we obtain the diffusion equation

$$\frac{\partial\varphi}{\partial t} = D\nabla^2\left(\varphi + \frac{\phi}{T}\Psi\theta - \xi\mathcal{L}\frac{\phi}{H}\frac{\partial\Phi}{\partial z}\right) + \frac{\phi\Delta T}{Td}\Psi_H(\mathbf{ve}). \quad (22)$$

Equations (20)-(22) together with the heat conduction equation,

$$\frac{\partial\theta}{\partial t} = \kappa\nabla^2\theta - (\Delta T/d)(\mathbf{ve}), \quad (23)$$

and the condition of incompressibility, $\operatorname{div}\mathbf{v} = 0$, form the complete set of equations which determine the onset of convection in a magnetized ferrofluid.

Pass over to dimensionless variables by choosing a unit of length $d$, time $d^2/\kappa$, velocity $\kappa/d$, temperature $\Delta T$, concentration $(\Delta T/T)\phi$, and field potential $4\pi M(\Delta T/T)d$. Then the above equations take the form

$$Pr^{-1}\frac{\partial\mathbf{v}}{\partial t} = -\nabla p + \nabla^2\mathbf{v} + Rm\,G\left[\Omega - \left(\Psi + \frac{\xi\mathcal{L}'}{\mathcal{L}}\right)\theta + \sigma \mathrm{D}\Phi\right]\mathbf{e}, \quad (24a)$$

$$\frac{\partial \theta}{\partial t} = \nabla^2 \theta - (\mathbf{ve}), \qquad \operatorname{div} \mathbf{v} = 0, \tag{24b}$$

$$\frac{\partial \Omega}{\partial t} = Le \nabla^2 \Omega + \Psi \frac{\partial \theta}{\partial t} - (\sigma - \hat{\mu})\, \mathrm{D} \frac{\partial \Phi}{\partial t} + \Psi_{\mathrm{H}}\, (\mathbf{ve}), \tag{24c}$$

$$\mu \nabla^2 \Phi + (\sigma - \mu)\, \mathrm{D}^2 \Phi = -\mathrm{D} \left[ \Omega - \left( \Psi + \frac{\xi \mathcal{L}'}{\mathcal{L}} \right) \theta \right], \qquad \nabla^2 \Phi_{\mathrm{e}} = 0\,. \tag{24d, e}$$

Here $Pr = \eta / \rho\kappa$ is the Prandtl number, $Le = D/\kappa$ the Lewis number, $\mathrm{D} \equiv \mathrm{d}/\,\mathrm{d}z$,

$$Rm = \frac{[\phi M_{\mathrm{s}}\, \Delta T\, d]^2}{\eta \kappa\, T^2}, \qquad G = \frac{4\pi \mathcal{L}^2}{\sigma} \left( \Psi + \frac{\xi \mathcal{L}'}{\mathcal{L}} \right), \tag{25a, b}$$

where $Rm$ we shall call the magnetic Rayleigh number, $\Phi_{\mathrm{e}}$ is the potential of field perturbations outside the fluid layer, and $\Omega = \varphi + \Psi\theta - (\sigma - \hat{\mu})\, \mathrm{D}\Phi$ is introduced as an independent variable instead of the concentration perturbation $\varphi$. It is worth to note that $\nabla \Omega$ has the physical meaning of the matter flux perturbation.

Studying the onset of convection in the Bénard configuration, one should put all perturbations be dependent on time and horizontal coordinates as $\exp[\lambda t + \imath(k_{\mathrm{x}} x + k_{\mathrm{y}} y)]$. Then from Eqs. (24) it follows the system of equations for $z$-dependent amplitudes of $z$-component of velocity $W = (\mathbf{ve})$, temperature $\theta$, matter flux potential $\Omega$, and magnetic field potentials $\Phi$ and $\Phi_{\mathrm{e}}$:

$$(\mathrm{D}^2 - k^2)(\mathrm{D}^2 - k^2 - \lambda/Pr)W - k^2 Rm\, G\, [\Omega - (\Psi + \xi\mathcal{L}'/\mathcal{L})\theta + \sigma \mathrm{D}\Phi] = 0\,, \tag{26a}$$

$$(\mathrm{D}^2 - k^2 - \lambda)\theta - W = 0\,, \tag{26b}$$

$$Le(\mathrm{D}^2 - k^2 - \lambda/Le)\Omega + \lambda\Psi\theta - (\sigma - \hat{\mu})\lambda \mathrm{D}\Phi + \Psi_{\mathrm{H}} W = 0\,, \tag{26c}$$

$$(\sigma \mathrm{D}^2 - \mu k^2)\Phi = -\mathrm{D}[\Omega - (\Psi + \xi\mathcal{L}'/\mathcal{L})\theta], \qquad (\mathrm{D}^2 - k^2)\Phi_{\mathrm{e}} = 0\,. \tag{26d, e}$$

## 4 Boundary conditions

To completely specify solutions of Eqs. (26), one needs ten boundary conditions at confined surfaces with the addition of two conditions on $\Phi_{\mathrm{e}}$ far from the layer.

For the case of rigid boundary surfaces whose heat conductivity is very high in comparison with the enclosed ferrofluid, the boundary conditions on velocity and temperature read:

$$W = \mathrm{D}W = \theta = 0 \qquad \text{at} \quad z = \pm 1/2\,. \tag{27}$$

It is natural to assume such boundaries to be at the same time completely impervious. The latter condition is expressed by setting the matter flux perturbation equal to zero,

$$\mathrm{D}\Omega = 0 \qquad \text{at} \quad z = \pm 1/2\,. \tag{28}$$

Boundary conditions on the magnetic potential $\Phi$ are pretty complicated by the fact that $\Phi$ induces outside the layer a magnetic potential $\Phi_{\mathrm{e}}$ governed by Eq.

(26e). This equation have the solution $\Phi_{\mathrm{e}} \propto \exp(\pm kz)$ which must decay far from the layer and be valid everywhere outside the layer including the layer boundaries too. Hence we come to the relations

$$\mathrm{D}\Phi_{\mathrm{e}} = -k\Phi_e \quad \text{at} \quad z = 1/2\,; \qquad \mathrm{D}\Phi_{\mathrm{e}} = k\Phi_e \quad \text{at} \quad z = -1/2\,. \tag{29}$$

Boundary conditions of continuity of the tangential components of magnetic field and the normal component of magnetic induction on the layer boundaries have the form

$$\partial\Phi/\partial x = \partial\Phi_e/\partial x\,, \qquad \partial\Phi/\partial y = \partial\Phi_e/\partial y\,, \tag{30}$$

and

$$\mathrm{D}\Phi + 4\pi\left(\frac{\partial M}{\partial\phi}\varphi + \frac{\partial M}{\partial T}\theta + \frac{\partial M}{\partial H}\mathrm{D}\Phi\right) = \mathrm{D}\Phi_{\mathrm{e}}\,. \tag{31}$$

The conditions (30) are satisfied if

$$\Phi = \Phi_{\mathrm{e}} \quad \text{at} \quad z = \pm 1/2\,, \tag{32}$$

whereas Eq. (31) on the substitution $M = M(\phi, T, \xi)$ takes the dimensionless form

$$\sigma\mathrm{D}\Phi + \Omega - \left(\Psi + \frac{\xi\mathcal{L}'}{\mathcal{L}}\right)\theta = \mathrm{D}\Phi_{\mathrm{e}} \quad \text{at} \quad z = \pm 1/2\,. \tag{33}$$

Relations (29) and (32) permit to write down: $\mathrm{D}\Phi_{\mathrm{e}} = \mp k\Phi$ at $z = \pm 1/2$, that gives us an opportunity to eliminate $\mathrm{D}\Phi_{\mathrm{e}}$ from (33). Taking into account the boundary conditions on the temperature (27), we obtain, finally, the closed boundary conditions on $\Phi$:

$$\sigma\mathrm{D}\Phi + k\Phi = -\Omega \quad \text{at} \quad z = 1/2\,; \qquad \sigma\mathrm{D}\Phi - k\Phi = -\Omega \quad \text{at} \quad z = -1/2\,. \tag{34}$$

Thus, there is no necessity at all to find the magnetic potential outside the layer. The relations (27), (28), and (34) represent the complete set of the rather realistic boundary conditions for the set of equations (26a)–(26d).

Below, however, we consider the case of idealized boundary conditions

$$W = \mathrm{D}^2 W = \theta = \Omega = \mathrm{D}\Phi = 0 \quad \text{at} \quad z = \pm 1/2\,, \tag{35}$$

which would be true for free, perfectly heat conducting and pervious boundaries made of a superconductor. In the latter case, magnetic field perturbations do not penetrate into the bulk of constrained horizontal plates, so that $\mathrm{D}\Phi_{\mathrm{e}} = 0$. Thus, as it follows from (33) with allowance for the boundary conditions (35) on $\theta$ and $\Omega$, the value $\mathrm{D}\Phi$ at $z = \pm 1/2$ appears to be zero too.

## 5 Exact solution for free boundaries

In the case of boundary conditions (35) there exists an analytical solution of Eqs. (26a) – –(26d). Satisfying the boundary conditions and the equations by means of

$$W \propto \theta \propto \Omega \propto \cos \pi z\,, \qquad \Phi \propto \sin \pi z\,,$$

we arrive at the characteristic equation for increments $\lambda$,

$$(\pi^2+k^2+\lambda)(\pi^2+k^2+\lambda/Pr)(\pi^2+k^2)[\lambda(\alpha\pi^2+k^2)+Le(\beta\pi^2+k^2)(\pi^2+k^2)]$$
$$-k^4 Rm\, G\,[\lambda(\xi\mathcal{L}'/\mathcal{L})+(\pi^2+k^2+\lambda)\Psi_{\rm H}+Le(\pi^2+k^2)(\Psi+\xi\mathcal{L}'/\mathcal{L})]=0\,, \quad (36)$$

where $\alpha=\hat{\mu}/\mu$ and $\beta=\sigma/\mu$. For stationary convective instability ($\lambda=0$) the magnetic Rayleigh number is given by

$$Rm^{\rm st}=\frac{Le\,(\beta\pi^2+k^2)(\pi^2+k^2)^3}{k^4G\,[\Psi_{\rm H}+Le\,(\Psi+\xi\mathcal{L}'/\mathcal{L})]}\,.$$

As it mentioned above, magnetic colloids are characterized by extremely small Lewis numbers (typically $Le\sim10^{-4}$). Therefore, using the definitions (13) and (25b), one may rewrite the last equation in the form

$$Rm^{\rm st}=\frac{Le\,\sigma^2\,(\beta\pi^2+k^2)\,(\pi^2+k^2)^3}{4\pi\hat{\mu}k^4\mathcal{L}^2\,(\Psi+\xi\mathcal{L}'/\mathcal{L})[\Psi-(\hat{\mu}-1)\xi\mathcal{L}/\hat{\mu}]}\,. \quad (37)$$

The magnetic Rayleigh number according to its definition (26a) is always *positive.* Therefore, as it seen from (37), stationary instability sets in only *outside* the interval

$$-\frac{\xi\mathcal{L}'(\xi)}{\mathcal{L}(\xi)}<\Psi<\frac{\hat{\mu}-1}{\hat{\mu}}\,\xi\mathcal{L}(\xi)\,, \quad (38)$$

i.e., either for a large enough positive separation ratio or for a sufficiently small negative $\Psi$. At the same time, as it will be shown below, *inside* the interval it can occur only *oscillatory* convective instability. In particular, the oscillatory instability is only possible in the absence of the Soret effect, i.e., when $\Psi=0$.

Omitting in (36) small terms with the coefficient $Le$, we obtain the cubic equation for $\lambda$:

$$(\alpha\pi^2+k^2)(\pi^2+k^2)\lambda^3+(Pr+1)(\alpha\pi^2+k^2)(\pi^2+k^2)^2\lambda^2$$
$$+[(\alpha\pi^2+k^2)(\pi^2+k^2)^3-k^4Rm\,G\,(\Psi_{\rm H}+\xi\mathcal{L}'/\mathcal{L})]Pr\lambda$$
$$=k^4(\pi^2+k^2)Pr\,Rm\,G\,\Psi_{\rm H}\,. \quad (39)$$

Substituting in this equation $\lambda=\mathrm{i}\omega$ we arrive at two relations which determine the magnetic Rayleigh number for oscillatory instability,

$$Rm^{\rm osc}=\frac{(Pr+1)\,\sigma^2\,(\alpha\pi^2+k^2)\,(\pi^2+k^2)^3}{4\pi\hat{\mu}Pr\,k^4\mathcal{L}^2(\Psi+\xi\mathcal{L}'/\mathcal{L})[\Psi+(\xi\mathcal{L}'/\mathcal{L})(1+\sigma/\hat{\mu}\,Pr)]}\,, \quad (40)$$

and the frequency of neutral convective oscillations,

$$\omega^2=-\frac{[\Psi-(\hat{\mu}-1)\xi\mathcal{L}/\hat{\mu}]\,(\pi^2+k^2)^2}{\Psi+(\xi\mathcal{L}'/\mathcal{L})(1+\sigma/\hat{\mu}\,Pr)}\,. \quad (41)$$

Since $Rm^{\rm osc}$ and $\omega^2$ must be both positive, the oscillatory instability can occur only if $\Psi$ lies within the interval (38). The corridor of oscillatory instability for

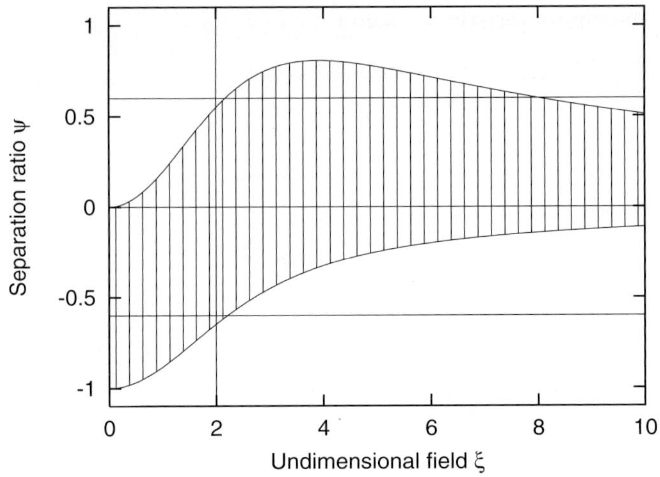


**Fig. 1.** The corridor of oscillatory instability (*dashed*) in $(\xi\,,\Psi)$ plane. *Straight lines* correspond to $\Psi = 0$, $\Psi = -0.6$, $\Psi = +0.6$ and $\xi = 2$. Cross sections of the surface $Rm_\mathrm{c}(\xi, \Psi)$ along these lines are presented below in Figs. 2–5 respectively.

a ferrofluid with the initial magnetic permeability $\mu_0 = 3$ is plotted in Fig. 1 for a wide enough range of the Langevin parameter $\xi$. On the upper border of the corridor, the frequency of convective oscillations turns into zero, whereas on its lower border the critical Rayleigh number turns into infinity. In the absence of magnetophoresis the corridor grows narrow so that its upper border passes along the line $\Psi = 0$. In this case the oscillatory instability would be possible owing to the Soret effect only for $\Psi < 0$. At last, if neither magnetophoresis nor Soret effect are in operation (i.e., *no particle diffusion*), then the cubic equation (39) for $\lambda$ is reduced to the square one,

$$\hat{\mu}\,(\alpha\pi^2 + k^2)\,(\pi^2 + k^2)\,[\lambda^2 + (Pr + 1)\,(\pi^2 + k^2)\,\lambda + Pr\,(\pi^2 + k^2)^2] = 4\pi Pr\, k^4\, Rm\,(\xi\mathcal{L}')^2. \tag{42}$$

This equation has no solutions of kind of $\lambda = \mathrm{i}\omega$. Thus, if there is no diffusion, then there is no oscillatory instability as well. For stationary instability Eq. (42) gives

$$Rm^{(0)} = \frac{\hat{\mu}\,(\alpha\pi^2 + k^2)\,(\pi^2 + k^2)^3}{4\pi k^4(\xi\mathcal{L}')^2}\,. \tag{43}$$

Below we shall consider $Rm$ as a function of $\xi$ and $\Psi$. Any pair of fixed magnitudes of the parameters determines a neutral stability curve in the $(Rm\,,\ k)$ plane. The minima of these curves, $Rm_\mathrm{c}$, and the critical circular frequency of convective oscillations, $f = \omega_\mathrm{c}/2\pi$, are presented in Figs. 2–5. Figures 2–4 show $Rm_\mathrm{c}$ and $f$ as functions of $\xi$ for three magnitudes of $\Psi$ marked in Fig. 1 by horizontal lines $\Psi = 0$, $\Psi = -0.6$, and $\Psi = +0.6$. Figure 5 represents a stability diagram in the plane $(Rm_\mathrm{c},\ \Psi)$ for the magnitude $\xi = 2$ marked in Fig. 1 by the vertical line.

## 6 Discussion of results and conclusion

At first, let us consider the case $\Psi = 0$. Then Eqs. (40)-(41) become somewhat simpler:

$$Rm^{\text{osc}} = \frac{(Pr+1)\,\sigma^2\,(\alpha\pi^2+k^2)\,(\pi^2+k^2)^3}{4\pi k^4\,(\xi\mathcal{L}')^2\,(Pr\,\hat{\mu}+\sigma)}\,,$$

$$\omega^2 = \frac{3Pr\,(\mu_0-1)\mathcal{L}^2\,(\pi^2+k^2)^2}{Pr\,\hat{\mu}+\sigma}\,. \tag{44}$$

The function $Rm^{\text{osc}}(k)$ has a minimum $Rm_{\text{c}}^{\text{osc}}$ at the critical wave number $k_{\text{c}}$ determined by the formula

$$k_{\text{c}}^2 = (\pi/2)^2\,(1-\alpha+\sqrt{1+14\alpha+\alpha^2})\,. \tag{45}$$

Here $\alpha = \hat{\mu}/\mu \le 1$ depends on the dimensionless magnetic field $\xi$ and the initial magnetic permeability $\mu_0$. For commonly used ferrofluids (with $1 < \mu_0 < 5$) $k_{\text{c}}$ weakly depends on $\xi$. For instance, at $\mu_0 = 3$ the critical wave number lies inside the interval $2.97 < k_{\text{c}} < 3.14$ for any $\xi$, so that $k_{\text{c}}$ variations do not exceed 6%.

Substituting $k_{\text{c}}^2$ from (45) into (44) we find the critical Rayleigh number and the frequency of neutral oscillations. The values $\log Rm_{\text{c}}^{\text{osc}}$ and $f = \omega_{\text{c}}/2\pi$ as functions of $\xi$ are presented in Fig. 2. In the same figure it is shown the logarithm of $Rm_{\text{c}}^{(0)}$ determined by Eq. (43) under the replacement $k^2$ by $k_{\text{c}}^2$ from (45).

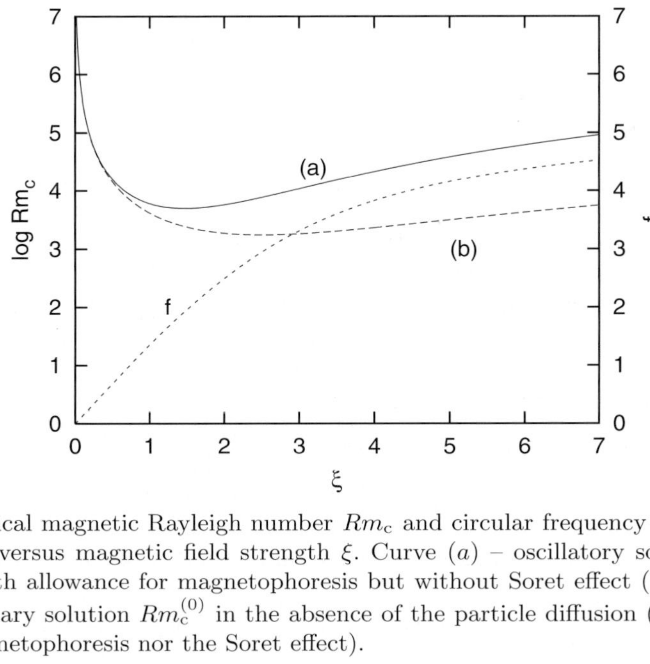


**Fig. 2.** Critical magnetic Rayleigh number $Rm_{\text{c}}$ and circular frequency of neutral oscillations $f$ versus magnetic field strength $\xi$. Curve $(a)$ – oscillatory solution $Rm_{\text{c}}^{\text{osc}}$ obtained with allowance for magnetophoresis but without Soret effect ($\Psi = 0$), curve $(b)$ – stationary solution $Rm_{\text{c}}^{(0)}$ in the absence of the particle diffusion (i.e., there are neither magnetophoresis nor the Soret effect).

We dare say that both the critical Rayleigh numbers, $Rm_c^{osc}$ and $Rm_c^{(0)}$ have a direct physical meaning. Depending on conditions of a real experiment it will set in either stationary convection above $Rm_c^{(0)}$ or oscillatory convection above $Rm_c^{osc}$. The point is that ferrofluids are characterized by two very different time scales [13]. One of two is the above-mentioned long mass diffusion time $\tau_D = d^2/\pi^2 D$: even in a thin layer ($\sim 1\,$mm) the concentration equilibrium is established for a time longer than *one hour.* Another one is a short temperature diffusion time $\tau_T = d^2/\pi^2\kappa$: in the same fluid layer the stationary temperature gradient sets in merely in *one second.* The ratio $\tau_T/\tau_D$ is equal just to the Lewis number $Le = D/\kappa \sim 10^{-4}$. The presence of such different time scales makes possible two scenarios of the onset of convection. First, the temperature difference between the layer boundaries can be increased from zero till a supercritical value during a time $\tau \ll \tau_D$. Then diffusion processes have no time to evolve, so the magnetic colloid behaves like a pure fluid. Indeed, magnetic grains do not move without diffusion relative to the surrounding liquid matrix, they are "frozen" into it. Just that very case has been studied in [3-8]. Stationary convection starts in this case at $Rm_c^{(0)}$. The latter value has a minimum at a certain $\xi$; for a water-based ferrofluid ($Pr = 7$) with $\mu_0 = 3$ this minimum, $\min[Rm_c^{(0)}] = 1742$, is reached at $\xi = 2.52$ and $k_c = 3.02$.

Second scenario can be realized in a thin enough fluid layer, where the long diffusion time, $\tau_D$, is *not too long.* Then it is possible to adjust so low heating rate, that the temperature difference across the layer, $\Delta T$, will be formed for a time $\tau > \tau_D$. Under this condition the concentration gradient induced by the temperature gradient due to magnetophoresis (the Soret effect is absent at $\Psi = 0$) will be built up undisturbed by convection. In this case, as it shown above, there arises oscillatory instability at $Rm_c^{osc}$. For a ferrofluid under consideration the critical Rayleigh number reaches the minimum, $\min[Rm_c^{osc}] = 4993$, at $\xi = 1.47$; corresponding wave number and frequency of neutral oscillations are $k_c = 3.08$ and $\omega_c = 12.13$. The frequency increases monotonously with $\xi$ approaching the saturation value $\omega_c = 34.19$ at $\xi \to \infty$.

Let us estimate the temperature difference and other parameters required to cause oscillatory convection by the thermomagnetic mechanism. Assuming the mean diameter of magnetite particles ($M_s = 480\,$G) to be $2a = 11\,$nm, we obtain the particle magnetic moment $m = (4\pi a^3/3)M_s = 3.345 \times 10^{-16}\,$erg/Oe, so the value $\mu_0 = 1 + 4\pi\phi M_s m/3k_B T = 3$ is reached at the room temperature at $\phi = 12\%$, and $\xi = mH/k_B T = 1$ at $H = 122\,$Oe. Thus the minimum in the curve of $\log Rm_c^{osc}$ in Fig. 2 takes place at $H = 180\,$Oe (to get such a field within the ferrofluid layer one must apply externally $H_e = \mu_0 H = 540\,$Oe). The minimum Rayleigh number ($\approx 5000$) is reached at $(\Delta T)_c = (1.02/d)\,$K/cm, thus the critical temperature difference is about 10 K in the layer of $d = 1$mm. For this layer the unit of time, $d^2/\kappa$, is equal to 6.9 s, so the above-pointed dimensionless frequency $\omega_c = 12.13$ of critical convective oscillations corresponds to the period of 3.6 s.

Maps of stability in the $(Rm_c, \xi)$–plane are presented in Figs. 3 and 4 for $\Psi = -0.6$ and $\Psi = +0.6$, respectively. As it seen in Fig. 1, for both these

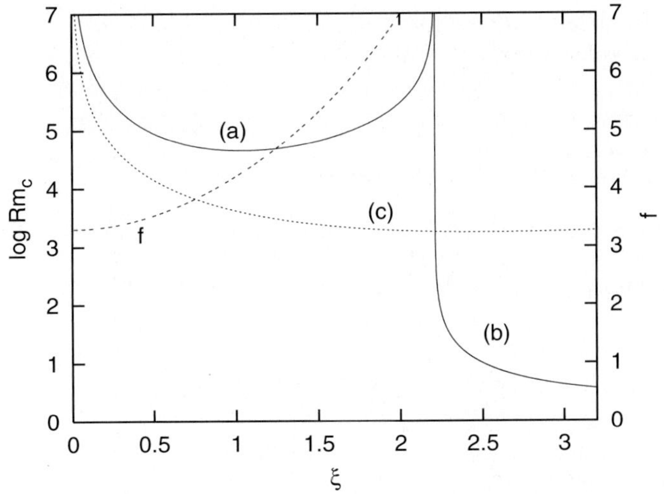


**Fig. 3.** Critical magnetic Rayleigh number and frequency of neutral oscillations versus magnetic field strength. Curve $(a)$ – oscillatory solution $Rm_c^{osc}$ and $(b)$ – stationary solution $Rm_c^{st}$ with allowance for magnetophoresis and the negative Soret effect ($\Psi = -0.6$); curve $(c)$ – stationary solution $Rm_c^{(0)}$ in the absence of the particle diffusion.

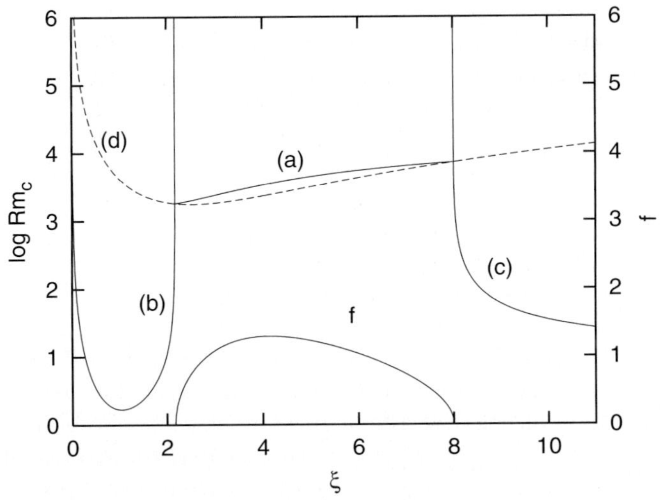


**Fig. 4.** Critical magnetic Rayleigh number and frequency of neutral oscillations versus magnetic field strength. Curve $(a)$ – oscillatory solution $Rm_c^{osc}$ and $(b)$, $(c)$ – stationary solution $Rm_c^{st}$ with allowance for magnetophoresis and the positive Soret effect ($\Psi = 0.6$); curve $(d)$ – stationary solution $Rm_c^{(0)}$ in the absence of the particle diffusion.

magnitudes of the separation ratio there are certain intervals of $\xi$ in which it can occur oscillatory instability. Out of the intervals there arises stationary Soret convection which is characterized by very small Rayleigh numbers (37) due to the smallness of the Lewis number (for a water-based ferrofluid $Le = 1.9 \times 10^{-4}$).

The critical Rayleigh number for stationary instability, $Rm_c^{st}$, is obtained from (37) under the substitution $k_c^2$ from (45) with the replacement of $\alpha$ by $\beta$.

For a fixed negative separation ratio, $0 > \Psi > -1$, oscillatory instability takes place within an interval of the field $0 < \xi < \xi_0$ (see Figs. 1 and 3), where $\xi_0$ satisfies the equation

$$\Psi + \xi\mathcal{L}'(\xi)/\xi = 0 . \tag{46}$$

On the borders of the interval, the critical Rayleigh number (40) turns into infinity as $\xi^{-2}$ for $\xi \to 0$ and as $(\xi - \xi_0)^{-1}$ for $\xi \to \xi_0$ (see Fig. 3). The frequency of neutral oscillations, $\omega_c(\xi)$, commences from a finite magnitude at $\xi = 0$,

$$\omega_c(0) = 2\pi^2 \sqrt{\frac{Pr\,|\Psi|}{1 + Pr\,(1 - |\Psi|)}} ,$$

and grows monotonously with $\xi$ until it reaches $\xi_0$.

The upper border of the interval (38) is described by

$$\Psi(\xi) = \frac{\hat{\mu} - 1}{\hat{\mu}}\, \xi\mathcal{L}(\xi) , \tag{47}$$

where $\hat{\mu} = 1 + 3(\mu_0 - 1)\mathcal{L}'(\xi)$. The function (47) has a maximum $\Psi_m$ at a certain value $\xi_m$ dependent on the initial magnetic permeability $\mu_0 = 1 + 4\pi\chi_0$. For the fluid under consideration ($\mu_0 = 3$) we find $\xi_m = 3.90$, $\Psi_m = 0.806$. For any positive $\Psi < \Psi_m$ there exists an interval $\xi_1 < \xi < \xi_2$ of oscillatory instability – see Figs. 1 and 4. The frequency of oscillations reaches a maximum within the interval and turns into zero on its borders. Oscillatory branch ($a$) of $Rm_c(\xi)$ links in Fig. 4 two codimension-2 points, $Rm_c(\xi_1)$ and $Rm_c(\xi_2)$, in which stationary branches ($b$ and $c$) bifurcate.

A diagram of stability in the $(Rm_c,\ \Psi)$–plane at the fixed magnitude of the field, $\xi = 2$, is shown in Fig. 5. This diagram looks more or less traditional for binary mixtures [14,15] with exception of two things. Firstly, the codimension-2 point (i.e., the intersection point of $a$ and $c$ branches) is usually located in the close vicinity of $\Psi = 0$. In Fig. 5 this point has essentially shifted towards positive $\Psi$ *due to magnetophoresis.* Secondly, the stationary branch $b$ as a general rule is located in the region of negative Rayleigh numbers, i.e., the branch determines the onset of convection in the binary mixture heated from *above*. Thermomagnetic mechanism of convection knows, however, neither the top nor the bottom: the magnetic Rayleigh number (25a) is proportional to $(\Delta T)^2$, so it is always positive.

Thus, convective instability of magnetized ferrofluids is strongly effected by the magnetophoretic and thermophoretic (the Soret effect) transfer of magnetic grains. The most interesting consequence of the transfer is an opportunity to observe oscillatory convective instability. However, as we have already commented, to observe predicted oscillations one needs to satisfy certain experimental conditions. Namely, the temperature difference must not be increased faster than the limit imposed by concentration diffusion. In the opposite case, a ferrofluid behaves like a pure fluid what results in stationary instability at the critical

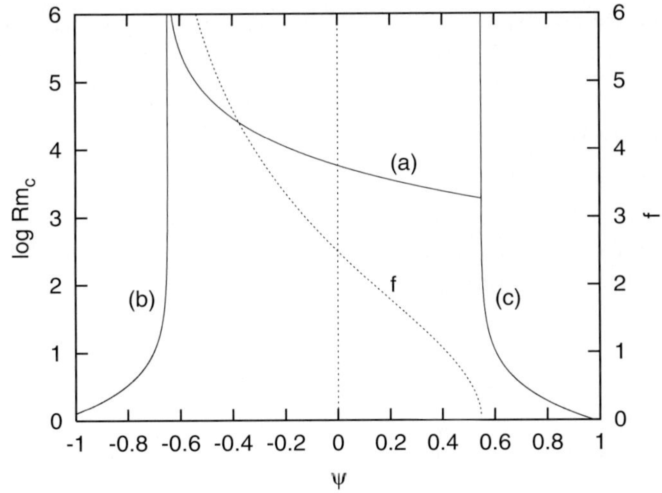


**Fig. 5.** Critical magnetic Rayleigh number and frequency of neutral oscillations versus separation ratio at the fixed magnetic field strength $\xi = 2$. Curve $(a)$ – oscillatory instability, curves $(b)$ and $(c)$ – stationary instability.

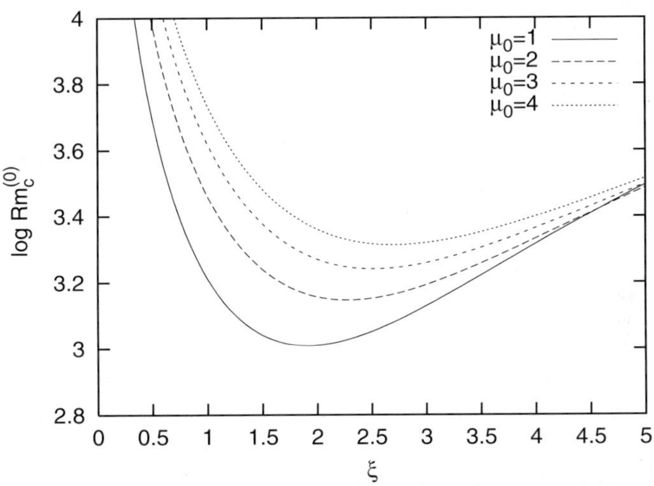


**Fig. 6.** Stationary instability curves for different $\mu_0$ in the case when the particle diffusion is not operative.

Rayleigh number $Rm_c^{(0)}$; its dependence on $\xi$ for some magnitudes of $\mu_0$ is shown in Fig. 6. In conclusion, it is worth to remind that all our results are relevant to thin ferrofluid layers ($\sim$ 1mm) where magnetic mechanism of convection predominates over the buoyancy mechanism, and the characteristic diffusion time, $\tau_D$, is not too long. To compare above-stated theoretical predictions with future experimental results one should perform a more precise analysis, which would satisfy realistic boundary conditions (27), (28) and (34) instead of idealized boundary conditions (35) used in the paper.

**Acknowledgements**

I am grateful to the Alexander von Humboldt Foundation for a Meitner-Humboldt research award and acknowledge the hospitality at the Bayreuth and Saarlandes Universities. This work has been also supported by a Grant 336/00–15.3 from the Israel Science Foundation.

## References

1. M.I. Shliomis: Sov. Phys.-Uspekhi **17**, 153 (1974)
2. R.E. Rosensweig: *Ferrohydrodynamics* (Cambridge University Press, Cambridge 1985)
3. V.M. Zaĭtsev, M.I. Shliomis: J. Appl. Mech. Techn. Phys. **9**, 24 (1968)
4. B.A. Finlayson: J. Fluid Mech. **40**, 753 (1970)
5. M.I. Shliomis: Fluid Dynamics **6**, 957 (1973)
6. P.J. Stiles, M. Kagan: J. Magn. Magn. Mater. **85**, 196 (1990)
7. C.L. Russell, P.J. Blennerhassett, P.J. Stiles: J. Magn. Magn. Mater. **149**, 119 (1995)
8. A. Recktenwald, M. Lücke: J. Magn. Magn. Mater. **188**, 326 (1998)
9. J.-C. Bacri: private communication
10. M.I. Shliomis: Sov. Phys.-JETP **34**, 1291 (1972)
11. B.V. Derjaguin, S.S. Dukhin, A.A. Korotkova: Kolloidn. Zh. **23**, 53 (1961)
12. L.D. Landau, E.M. Lifshitz: *Electrodynamics of Continuous Media* (Pergamon Press, New York 1984)
13. M.I. Shliomis, M. Souhar: Europhys. Lett. **49**, 55 (2000)
14. D.T.J. Hurle, E. Jakeman: J. Fluid. Mech. **47**, 667 (1971)
15. E. Knobloch, D.R. Moore: Phys. Rev. A **37**, 860 (1988)

## Supplementary glossary

$a$ particle radius
$\mathbf{B}$ magnetic induction: $\mathbf{B} = \mathbf{H} + 4\pi\mathbf{M}$
$f$ frequency of neutral oscillations: $f = \omega/2\pi$
$\mathbf{F}$ magnetic force (volume density)
$\mathbf{H}$ magnetic field
$k$ wave number
$Le$ Lewis number: $Le = \tau_{\mathrm{T}}/\tau_{\mathrm{D}}$
$\mathbf{M}$ magnetization
$Pr$ Prandtl number
$Ra$ Rayleigh number
$Rm$ magnetic Rayleigh number
$\mathbf{v}$ ferrofluid velocity
$W$ vertical component of velocity

| | |
|---|---|
| $\eta$ | fluid viscosity |
| $\kappa$ | thermal diffusivity |
| $\lambda$ | increment of a normal mode |
| $\mu$ | magnetic permeability: $\mu = 1 + 4\pi(M/H)$ |
| $\mu_0$ | initial magnetic permeability: $\mu_0 = 1 + 4\pi\chi_0$ |
| $\chi_0$ | initial magnetic susceptibility |
| $\chi$ | magnetic susceptibility: $\chi = M/H$ |
| $\Psi$ | separation ratio for magnetic convection: $\Psi = (T/\phi)S_\mathrm{T}$ |
| $\phi$ | concentration of magnetic grains (volume fraction) |
| $\tau_\mathrm{D}$ | characteristic diffusion time |
| $\tau_\mathrm{T}$ | characteristic temperature diffusion time |
| $\xi$ | Langevin's parameter (dimensionless magnetic field) |
| $\mathcal{L}(\xi)$ | Langevin's function: $\mathcal{L}(\xi) = \coth\xi - \xi^{-1}$ |
| $\omega$ | angular frequency of neutral oscillations |
| $\Omega$ | matter flux potential |

# On the Soret-Driven Thermosolutal Convection in a Vibrational Field of Arbitrary Frequency

Boris L. Smorodin[1], Bela I. Myznikova[2], and Igor O. Keller[3]

[1] Perm State University, Department of Theoretical Physics, Bukirev str., 15, Perm, 614600, Russia
[2] Institute of Continuous Media Mechanics, UB of Russian Academy of Science, Korolev Str., 1, Perm, 614013, Russia
[3] Intel Israel(74) Ltd. P.O.Box 1659, Haifa, 31015 Israel

**Abstract.** The instability of a plane horizontal layer of an incompressible binary mixture with Soret effect is investigated in the presence of transversal vibrations of arbitrary frequency. The boundaries of the layer are assumed to be rigid, impermeable and isothermal. A linear instability analysis for normal modes is carried out using the Floquet theory. The boundaries of instability and the characteristics of the critical disturbances are determined. In addition to the gravity field modulation, the instability can be associated with quasi-periodic disturbances. It is shown that, depending on the amplitude and frequency of modulation, the vibrations can stabilize an unstable base state or destabilize the equilibrium of the liquid. The limit case of low-frequency modulation of the gravity field is studied.

## 1 Introduction

Convection in a binary mixture is more complicated phenomenon than that in a single-component fluid. The additional buoyancy force appears in the gravity field due to inhomogeneity of concentration as well as complementary dissipative mechanism caused by diffusion and (thermodiffusion). The influence of the Soret effect on the convective instability was studied in detail in [1,2].

Thermosolutal convection can produce undesirable effects in solidification of binary alloys and solutions. Convective instabilities modify the distribution of solute concentration and macrosegregation in the solid can occur. This phenomena take place in processing materials on board of the orbital laboratories, where low level of the gravity field cannot exclude buoyancy-induced convection. Gravity field under the orbital conditions is not constant, it randomly fluctuates in direction and magnitude (g-jitter). Time-dependent forces acting on a fluid can modify convective thresholds and provide an effective way to control the convection in various situations.

The influence of time-dependent gravitational acceleration or vibration on the onset of Rayleigh-Benard convection was studied in [3,4]. The linear stability problem for the single component system was reduced to the damped Mathieu equation. It was shown that high-frequency vibrations can make the equilibrium absolutely stable, whereas low-frequency vibrations can destabilize the equilibrium in the layer with stable stratification at any moment of time due to the

W. Köhler and S. Wiegand (Eds.): LNP 584, pp. 372–388, 2002.

resonance phenomena. Two types of critical disturbances were found: the first varies synchronously with the external modulation, another one oscillates with the half frequency.

In [5,6] the problem with time-periodic modulation of temperature difference in the layer was studied. The influence of vibration on convection excitation in differently heated infinite vertical cylinders was studied in [7] for a wide range of frequencies. Using asymptotical method of small parameter authors of [5–7] studied the effect of small amplitude of modulations on stability characteristics. Convective instability in a horizontal layer of incompressible fluid subjected to uniform steady gravity field and finite frequency longitudinal vibrations was investigated in [8]. Only synchronous response to vibrations was detected. The stability analysis was carried out by using the Floquet method applied to the linearized Boussinesq convection equations.

The influence of vibration on convection in binary mixture was studied in [9–12]. The effect of harmonic time-dependent acceleration on the onset of thermosolutal convection without cross-diffusion in an infinite horizontal layer with stress-free boundaries was investigated in [9]. Besides synchronous and subharmonic responses to variations of the external gravity field, the instability in the binary-mixture layer can be associated with two–frequency, quasi-periodic disturbances. It was found that for diffusive-like configurations the regions of resonant instability are strongly coupled with unmodulated oscillatory frequency.

The stability of double diffusive system with the Dufour and the Soret effects was studied in [10]. A strong destabilizing effect of modulation on stability of the equilibrium and resonant frequencies was found for the anomalous Soret parameter. The stress-free and rigid boundaries were used in linear analysis. It was assumed that concentration disturbances vanish on the rigid boundaries. This assumption may not be admitted as reasonable. The boundary condition for concentration corresponds to the case of rigid, but permeable boundary plates. Impermeability of the rigid boundaries implies that there is no concentration transport through the plates. It is necessary to note, that these boundary conditions reflect real experimental situations [2].

The influence of longitudinal or transversal vibrations on convection excitation in a binary mixture plane layer with Soret effect for real (NSI, no slip impermeable) boundary conditions was investigated in a high-frequency limit [11,12]. In this case the period of vibration is much less than any of the characteristic convective time-scales. The amplitude and frequency of vibration are not independent parameters and all resonance phenomena are excluded from consideration. So an effective averaging method can be used. Instability is stipulated by stratification of binary mixture in a vibrating field.

In the present paper we study the instability of a horizontal binary mixture layer with the Soret effect under action of a cross difference of temperature and transversal vibration of finite amplitude and arbitrary frequency. The linear stability with respect to two-dimensional disturbances of normal type is studied. The instability may be caused by different mechanisms: vibrational, gravitational or thermosolutal. Unstable disturbances can exhibit two different characteristic frequencies. One of them is reflected the frequency of the external modulated gravity field and the other one corresponds to the fundamental frequency of neutral disturbances for a binary mixture in steady gravity field.

## 2 Statement of a problem

Let us consider an infinite plane horizontal layer of binary mixture with the Soret effect. The layer is confined between two rigid impermeable and isothermal plates located at $z = \pm h$ ($h$ is the layer half-thickness). Constant different temperatures $T = \mp \Theta$ are maintained on these boundaries. Temperature of the upper boundary $\Theta < 0$ corresponds to heating from above, the case $\Theta > 0$ represents heating from below. Both boundaries are impermeable, and there is no external difference of concentration. The only reason for concentration inhomogeneity is the Soret effect. Non-homogeneous distribution of temperature in a mixture leads to temperature and concentration (because of thermodiffusion effect) non-homogenity of density.

$$\rho = \bar{\rho}(1 - \alpha T + \beta c). \tag{1}$$

Here $\bar{\rho}$ is density of a mixture at average values of temperature and concentration, $T$ and $c$ are deviations from average values; $\alpha$; $\beta$ are coefficient of thermal expansion and concentration density coefficient, respectively. Let us $c$ be concentration of weight fraction.

The whole layer (together with the boundaries) linearly harmonically oscillates in a direction of the cross axis $z$ with angular frequency $\Omega$ and amplitude of modulation $b$. If oscillation period is much more then acoustic time $h/c_{\rm sd}$ ($c_{\rm sd}$ is the sound velocity), the situation is quite analogues to the modulation of gravity field. To take into account the vibrational acceleration we use the non inertial coordinate system, connected with the vibrating layer, in which modified acceleration vector has background constant component and harmonically modulated one:

$$\boldsymbol{g} = \boldsymbol{g}_0 + \boldsymbol{n}\, b\Omega^2 \sin \Omega t, \tag{2}$$

where $\boldsymbol{n}$ is the unit vector along an axis of vibration and is directed vertically upwards.

The variables describing the convection of binary mixture in modulated gravity field are as follows: velocity $\boldsymbol{u}$, pressure $p$, temperature $T$ and concentration $c$. The following equations describe the possible thermosolutal Soret-driven convection:

$$\begin{aligned}
\frac{\partial \boldsymbol{u}}{\partial t} + (\boldsymbol{u}\nabla)\boldsymbol{u} &= -\frac{1}{\bar{\rho}}\nabla p + \nu \triangle \boldsymbol{u} + (g_0 + b\Omega^2 \sin \Omega t)(\alpha T - \beta c)\cdot \boldsymbol{n} \\
\frac{\partial T}{\partial t} + \boldsymbol{u}\nabla T &= \chi \triangle T \\
\frac{\partial c}{\partial t} + \boldsymbol{u}\nabla c &= D \triangle (c + S_{\rm T} T) \\
div\, \boldsymbol{u} &= 0
\end{aligned} \tag{3}$$

Here $\nu$ is the kinematic viscosity of the fluid, $\chi$ is the heat diffusivity coefficient, $D$ is the coefficient of diffusion and $S_{\mathrm{T}}$ is the Soret coefficient. Hereafter we shall assume that characteristic differences of temperature and concentration are not large, and coefficients $D$ and $S_{\mathrm{T}}$ are independent on temperature and concentration.

Different variants of boundary conditions are used for investigation of convection in binary mixture. But the case of rigid isothermal and impermeable planes corresponds to real situation, which is realized in experiments. The concentration current on the boundary vanishes.

$$z = \pm 1: \qquad \boldsymbol{u} = 0, \qquad T = \mp \Theta, \qquad \frac{\partial c}{\partial z} + S_{\mathrm{T}} \frac{\partial T}{\partial z} = 0. \tag{4}$$

We introduce dimensionless variables with the help of the following scales: $h$ for distance, $h^2/\nu$ for time, $\nu/h$ for velocity, $\Theta$ for temperature, $\alpha\Theta/\beta$ for concentration, $\rho\nu^2/h^2$ for pressure and obtain the system of governing equations in the following form:

$$\begin{aligned} \frac{\partial \boldsymbol{u}}{\partial t} + (\boldsymbol{u}\nabla)\boldsymbol{u} &= -\nabla p + \triangle \boldsymbol{u} + (Gr + Gr_{\mathrm{v}} \sin \omega t)(T - c) \cdot \boldsymbol{n} \\ \frac{\partial T}{\partial t} + \boldsymbol{u}\nabla T &= \frac{1}{Pr} \triangle T \\ \frac{\partial c}{\partial t} + \boldsymbol{u}\nabla c &= \frac{1}{Sc} \triangle (c + \psi_1 T) \\ div \boldsymbol{u} &= 0 \end{aligned} \tag{5}$$

The dimensionless form of boundary conditions is

$$z = \pm 1: \qquad \boldsymbol{u} = 0, \qquad T = \mp 1, \qquad \frac{\partial c}{\partial z} + \psi_1 \frac{\partial T}{\partial z} = 0. \tag{6}$$

The formulated problem involves the following dimensionless parameters: the Grashof number $Gr = g_0 \alpha \Theta h^3/\nu^2$, the vibrational analogue of the Grashof $Gr_{\mathrm{v}} = b\Omega^2 \alpha \Theta h^3/\nu^2$, The Prandtl and Schmidt numbers $Pr = \nu/\chi$, $Sc = \nu/D$, the frequency $\omega = \Omega h^2/\nu$, the separation ratio, which is characterized of the Soret effect $\psi_1 = \beta S_{\mathrm{T}}/\alpha$. The separation ratio $\psi_1$ is positive in the case of normal thermodiffusion effect (the heaviest component migrates to the cold side) and negative in the case of anomalous effect (the heaviest component moves to the hot side).

The temperature distribution creates a vertical gradient of temperature, providing an opportunity of equilibrium of a liquid layer. The equilibrium solution has the following form:

$$\boldsymbol{u}_0 = 0, \quad T_0 = -z, \quad c_0 = \psi_1 z. \tag{7}$$

To investigate the basic state (7) stability we shall consider its small disturbances $\boldsymbol{v}$, $T'$, $c'$, $p'$. We shall enter auxiliary variable $H = -c - \psi_1 T$. After linearization one can obtain the following system of equations for small disturbances:

$$\frac{\partial \boldsymbol{u}}{\partial t} = -\nabla p + \triangle \boldsymbol{u} + (Gr + Gr_{\mathrm{v}} \sin \omega t)(T'(1+\psi_1) + H')$$

$$\begin{aligned} \frac{\partial T'}{\partial t} + \boldsymbol{u}\nabla T_0 &= \frac{1}{Pr}\triangle T' \\ \frac{\partial H'}{\partial t} &= \frac{1}{Sc}\triangle H' - \frac{\psi_1}{Pr}\triangle T' \\ div \boldsymbol{u} &= 0, \end{aligned} \tag{8}$$

with homogeneous boundary conditions:

$$\boldsymbol{u} = 0, \quad T' = 0, \quad \frac{\partial H'}{\partial z} = 0. \tag{9}$$

After elimination of the horizontal velocity components and pressure we write the perturbations of vertical velocity $u_{\mathrm{z}}$, the temperature $T'$ and function $H'$ as

$$\begin{pmatrix} u_{\mathrm{z}} \\ T' \\ H' \end{pmatrix} = \begin{pmatrix} w(z,t) \\ \theta(z,t) \\ \xi(z,t) \end{pmatrix} \exp\left[iqx\right], \tag{10}$$

here $q$ is the wave number, $w$, $\theta$, $\xi$ are amplitudes. (The vectors $\boldsymbol{g}$, $\boldsymbol{n}$ and $\nabla T_0$ are directed normally to boundaries, therefore the problem is isotropic in the plane of the layer, and it is possible to direct the $x$-axis along the wave vector $\boldsymbol{q}$).

Upon introduction of the disturbed fields in (8,9) we obtain the set of equations for the amplitudes:

$$\begin{aligned} \frac{\partial Dw}{\partial t} &= D^2 w - q^2 (Gr + Gr_{\mathrm{v}} \sin \omega t)(\theta(1+\psi_1) + \xi) \\ \frac{\partial \theta}{\partial t} - w &= \frac{1}{Pr} D\theta \\ \frac{\partial \xi}{\partial t} &= \frac{1}{Sc} D\xi - \frac{\psi_1}{Pr}\theta, \qquad D = \partial^2/\partial z^2 - q^2 \\ at \qquad z = \pm 1: &\qquad w = 0, \qquad \frac{\partial w}{\partial z} = 0, \qquad \theta = 0, \quad \frac{\partial \xi}{\partial z} = 0. \end{aligned} \tag{11}$$

The amplitude problem (11) defines behaviour of "normal" disturbances. The Floquet theory allows one to determine the conditions when periodic solution of the amplitude problem exists and to provide the instability boundary.

## 3 Numerical method

The solution of the amplitude problem can be found as a superposition of spatial functions with time-dependent coefficients,

$$w = \sum_{n=1}^{N} a_{\mathrm{n}}(t) w_{\mathrm{n}}, \quad \theta = \sum_{m=1}^{M} b_{\mathrm{m}}(t)\theta_{\mathrm{m}}, \quad \xi = \sum_{l=1}^{L} c_{\mathrm{l}}(t)\xi_{\mathrm{l}} \tag{12}$$

The basis $w_n$ (12) can be formed using the normalized hydrodynamic eigenfunctions of the amplitude problem for a quiescent layer. The eigenfunctions satisfy the fourth-order eigenvalue problem [13]:

$$D^2 w_n = -\mu_n D w_n; \qquad w_n(\pm 1) = w_n'(\pm 1) = 0. \tag{13}$$

The hydrodynamic functions $w_n$ form a complete orthonormalized system:

$$\int_{-1}^{1} w_i D w_j dz = -\delta_{ij}. \tag{14}$$

The bases for approximation of temperature and concentration amplitudes were constructed with the help of eigenfunctions for the following problems:

$$\begin{aligned} &\frac{1}{Pr} D\theta_m = -\nu_m \theta_m; \quad \theta_m(\pm 1) = 0 \\ &\frac{1}{Sc} D\xi_l = -\varrho_l \xi_l; \quad \xi_l'(\pm 1) = 0, \end{aligned} \tag{15}$$

The orthogonalization procedure of Galerkin method yields a set of $K = M + N + L$ ordinary differential equations for the coefficients $a_n$, $b_m$ and $c_l$

$$\frac{\partial a_r}{\partial t} = -\mu_r a_r + q^2 (Gr + Gr_v \sin \omega t) \left[ \sum_{m=1}^{M} (1 + \psi_1) C_{rm} b_m + \sum_{l=1}^{L} D_{rl} c_l \right]$$

$$(r = 1, 2, ... N)$$

$$\frac{\partial b_s}{\partial t} = \sum_{n=1}^{N} C_{ns} a_n - \nu_s b_s \quad (s = 1, 2, ... M)$$

$$\frac{\partial c_t}{\partial t} = \psi_1 \nu_s \sum_{s=1}^{L} F_{st} b_s - \varrho_t c_t \quad (t = 1, 2, ... L)$$

$$C_{ns} = \int_{-1}^{1} w_n \theta_s dz, \quad D_{rl} = \int_{-1}^{1} w_r \xi_l dz, \quad F_{st} = \int_{-1}^{1} \theta_s \xi_t dz \tag{16}$$

Applying the Floquet theory one can write any solution of system (16) in the following form [14]

$$F(t) = \begin{pmatrix} a_m \\ b_n \\ c_l \end{pmatrix}, \qquad F(t) = e^{\Lambda t} \cdot F_0(t) = \gamma \cdot F_0(t), \tag{17}$$

with $F(t)$ and $F_0(t)$ vector functions, $F_0(t)$ is periodic in time with period $T = 2\pi/\omega$, $\Lambda = \Lambda_r + i\Lambda_i$ is the complex growth rate whose real part determines stability, and $\gamma$ is the Floquet multiplier. For convenient initial conditions we find $K$ linearly independent solutions for the time-dependent coefficients at the end of one external period $T$. These fundamental solutions make up the $K$ columns of the $K \times K$ monodromy matrix.

The system (16) was integrated using the 4-th order Runge-Kutta method. The Floquet multipliers $\gamma_K$ are the eigenvalues of the monodromy matrix. The eigenvalues $\gamma_K$ were found by means of the QR- algorithm.

The basic rest state is stable, when the modulus of any multiplier $|\gamma_K|$ does not exceed unity. If multipliers are ordered, then $1 \geq |\gamma_1|, \geq |\gamma_2|, \cdots \geq |\gamma_K|$. The condition $|\gamma_1| = 1$ [15] determines the boundary of neutral manifold in the space of parameters $Gr$, $Gr_v$, $Pr$, $Sc$, $\psi_1$, $\omega$ and $q$. The critical values of parameters $Gr$ or $Gr_v$ depend on all other parameters of system:

$$Gr_* = Gr_*(Gr_v, \omega, Pr, Sc, \psi_1, q), \quad Gr_{v*} = Gr_{v*}(Gr, \omega, Pr, Sc, \psi_1, q).$$

If $\gamma_1 = 1$ ($\Lambda_r = 0$, $\Lambda_i = \omega$), we have synchronous response to the external field. Solutions with $\gamma_1 = -1$ ($\Lambda_r = 0$, $\Lambda_i = \omega/2$) correspond to the subharmonic response. If instability comes with complex conjugate pairs of eigenvalues with unit modulus, $|\gamma_1| = 1$ ($\Lambda_r = 0$, $\Lambda_i \neq 0$) then neutral disturbances are quasiperiodic. They have two different characteristic frequencies not rationally related.

Most of the solutions computed here use 15 trial functions (M=N=L=5). In the case of 21 basis functions (M=N=L=7) in expansions (12), the threshold of convection changes by less than 1%. All calculations of the present work have been made for $P = 0.75$ and $Sc = 1.5$, typical for a gaseous mixture.

## 4 Low frequencies of modulation

The smallness of the frequency, $\omega$, means that the period of the gravity field modulation $T$ is much greater than characteristic hydrodynamic, heat and diffusive time scales of the liquid system:

$$T \gg \max\left[\frac{h^2}{\nu}, \frac{h^2}{\chi}, \frac{h^2}{D}\right] \tag{18}$$

or in dimensionless form

$$\omega \ll \left[1, \quad Pr^{-1}, \quad Sc^{-1}\right] \tag{19}$$

But it is very "expensive" to apply the numerical procedure, described in the previous section in this case: the computation time increases proportionally with period $T$. Therefore, we perform asymptotical analysis using the Wentzel–Kramers–Brillouin (WKB) method [16] with $\omega$ as the small parameter. Let us rewrite the system of amplitude equations (11) in matrix form:

$$\frac{\partial}{\partial t} AU = NU \tag{20}$$

Here $U$ is the vector - function $\begin{pmatrix} w \\ \theta \\ \xi \end{pmatrix}$, $A$ and $N$ are the operators with matrices

$$A = \begin{pmatrix} D & 0 & 0 \\ 0 & 1 & 0 \\ 0 & 0 & 1 \end{pmatrix}, \; N = \begin{pmatrix} D^2 & -q^2(1+\psi_1)f(t) & -q^2 f(t) \\ 1 & D/Pr & 0 \\ 0 & -\psi_1 D/Pr & D/Sc \end{pmatrix}$$

and $f(t) = Gr + Gr_{\mathrm{v}} \sin \omega t$. Introducing the new variable $\tau = \omega t$, the (20) takes the form:

$$\omega \frac{\partial}{\partial \tau} AU = NU, \tag{22}$$

where $N$ is now the $2\pi$ - periodic operator with respect to $\tau$.

At small $\omega$ it is possible to expand the amplitude of modulation $Gr_v$ in the power series of $\omega$:

$$Gr_{\mathrm{v}} = Gr_{\mathrm{v}0} + \omega Gr_{\mathrm{v}1} + \omega^2 Gr_{\mathrm{v}2} + ... \tag{23}$$

Then we write the solution of (22) in the following form

$$U = \exp \left[ \frac{-1}{\omega} \int_0^{\tau} \Lambda(\tau') \, d\tau' \right] \cdot (U_0 + \omega U_1 + \omega^2 U_2 + ...)$$

$$\Lambda = \Lambda_0 + \omega \Lambda_1 + \omega^2 \Lambda_2 + ... \tag{24}$$

where $\Lambda$ is a $2\pi$-periodic function of $\tau$, meaning the characteristic decrement of disturbances for slow variation of the gravity field.

The stability boundaries in the zeroth-order approximation are given by the integration condition:

$$\int_0^{2\pi} \Lambda_{0\mathrm{r}}(\tilde{Gr}) \, d\tau = 0; \quad \tilde{Gr} = Gr + Gr_{\mathrm{v}} \cdot \sin \tau. \tag{25}$$

where $\Lambda_{0\mathrm{r}}$ is the real part of $\Lambda_0$. Thus, the convection thresholds in low-frequency approximation are determined by a real part of the lower level of the decrement spectrum in the absence of vibrations. The integration (25) is carried out in a region $Gr - Gr_{\mathrm{v}} \div Gr + Gr_{\mathrm{v}}$. To solve the eigenvalue problem $\Lambda_0$ we use the Galerkin method with basis, determined by (13),(15) . The basis up to 21 trial functions was used in calculations.

The dependencies of critical amplitude of modulation $Gr_{\mathrm{v}*}$ on a heating $Gr$ in low-frequency approximation are presented in Fig. 1 at various values of the Soret parameter $\psi_1$. The stability regions are situated between axes of coordinates

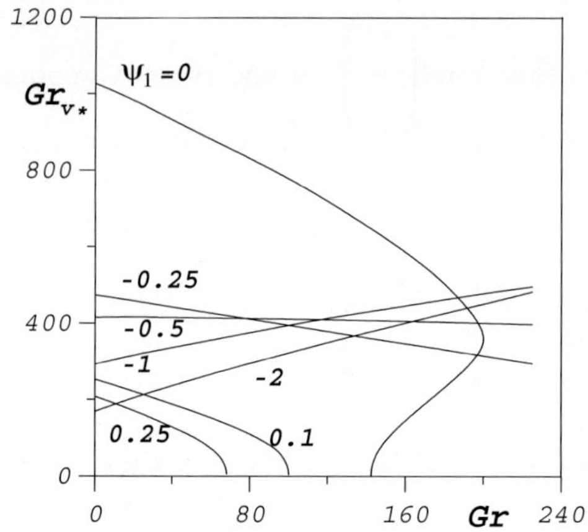


**Fig. 1.** Low frequency limit. The critical values of $Gr_{v*}$ versus $Gr$ for several values of the separation ratio $\psi_1$.

and curves. The points on abscissa axis $Gr_{v*} = 0$ correspond to threshold values of Grashof number in the absence of vibrations. In the case of weightlessness $Gr = 0$ and the absence of the Soret effect ($\psi_1 = 0$), the amplitude of critical modulation is the greatest $Gr_{v*} = 1026$. At $\psi_1 = -0.54$ critical amplitude of vibration practically does not vary in the interval $0 < Gr < 240$.

## 5 Arbitrary frequency of vibrations

The thresholds of the instability $Gr_*$ are presented in Fig. 2 as functions of separation ratio $\psi_1$ for the cases of absence of vibrations and high-frequency vibrations (the minimization is made with respect to the wave number $q$). The solid

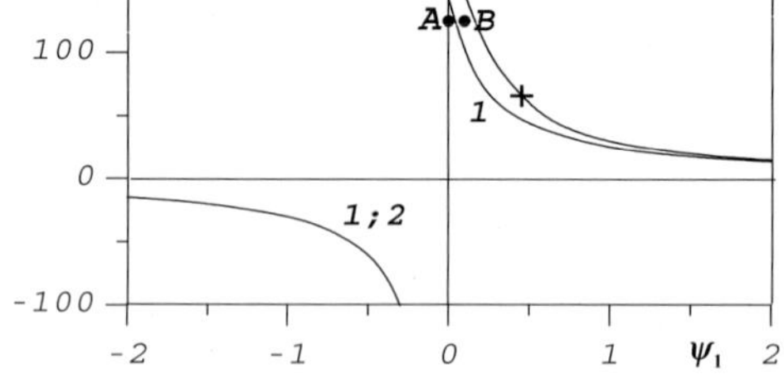


**Fig. 2.** Critical values of $Gr_*$ versus $\psi_1$: 1- for case without vibrations; curve 2 corresponds to case of high frequency vibrations; $Ra_v = 1000$.

lines correspond to the monotonic critical mode, the dashed lines correspond to the boundaries of oscillatory instability. The Rayleigh-Benard instability boundary for a single-component fluid ($\psi_1 = 0$) in static gravity field ($Gr_v = 0$) is $Ra = g\alpha\vartheta L^3 / \nu\chi = 1708$ [3]. Note that $Ra$ is determined with the help of total layer thickness $L = 2h$ and total temperature difference $\vartheta = 2\Theta$). This value of the convective threshold corresponds at $Pr = 0.75$ to our critical Grashof number $Gr_* = 142.37$ ($Ra = 16Gr_*Pr$). In the case of the normal Soret effect only monotonic instability exists and thermodiffusion plays a destabilizing role. This is cellular instability (the minimum of wave number is not zero, $q_* \neq 0$). In the range of the anomalous Soret effect $\psi_1 < 0$ monotonic cellular, monotonic long wave ($q_* = 0$) and oscillatory cellular instability modes are possible. The long wave monotonous disturbances are more dangerous when the layer is heated from above $Gr_* < 0$ .

Vibrations change the position of instability boundaries. In the limit of high frequencies the boundaries of stability coincide with results obtained by the averaging technique [12]. In this case only one parameter describes external influence. It is the vibration Rayleigh number $Ra_v = (\alpha b\Omega\vartheta L)^2 / 2\nu\chi$. The curve 2 in Fig. 2 corresponds to $Ra_v = 1000$, which is determined on the total thickness of a layer $L$ and total difference of temperatures $\vartheta$. Note that we use different dimensionless parameters than [12]. $Ra_v$ connects with our amplitude $Gr_v$ and frequency of vibrations $\omega$ in the following way $Ra_v = 16{Gr_v}^2 Pr/(2\omega^2)$. The presence of high-frequency vibrations brings to long wave instability in the region of normal Soret effect $\psi_1 > \psi_{1*}(Gr_v)$. The transition point on the instability boundary is marked in Fig. 2 as follows +.

At finite frequencies the amplitude and the frequency of vibrations remain independent parameters. The influence of vibrations on convection thresholds $Gr_*(\psi_1)$ is illustrated in Fig. 3 at fixed frequency $\omega = 4\pi$ and for a few values of the amplitude $Gr_v$.

Here curve 1 is the boundary of instability in the case of absence of vibrations (the same, as in Fig. 2); curves 2, 3, 4 are boundaries of instability for various $Gr_v = 162.23; 229.43; 324.46$ (in terms of [12] they correspond to values $Ra_v = 1000; 2000; 4000$, respectively). Here and hereinafter the solid lines present boundaries of synchronous response, dashed lines are boundaries of quasiperiodic ones, dash-and-dot lines are boundaries of growth subharmonic disturbances. Instability boundary with respect to long wave synchronous disturbances exists when heating from above. This type of the instability does not practically depend on amplitude and frequency of modulation and connects with difference of diffusive and thermal characteristic time scales (the so-called double-diffusive mechanism of instability).

Convective thresholds for synchronous and quasi-periodic disturbances weakly grow with $Gr_v$. Modulations stabilize an unstable base solution. The long wave monotonic instability appears in regions of the normal Soret effect $\psi_1 > \psi_{1*}$. But in contrary to the high-frequency limiting case [12] the critical separation ratio $\psi_{1*}$ depends on amplitude $Gr_v$ and frequency $\omega$ of modulation. The transition points to long wave instability are presented in Table 1 for $\omega = 4\pi$.

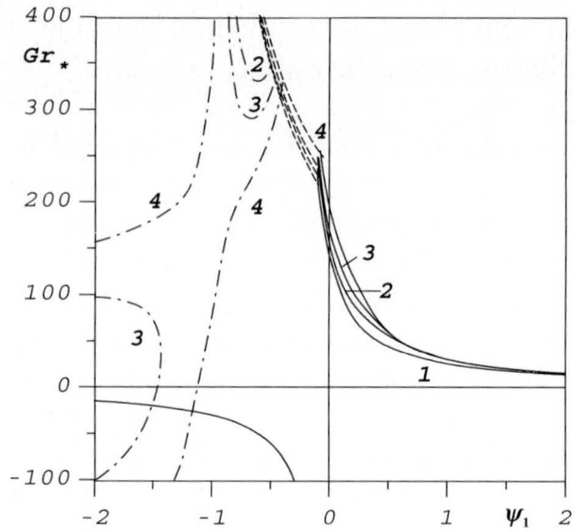


**Fig. 3.** Critical values of $Gr_*$ versus $\psi_1$: 1- for case without vibrations; curves 2, 3, 4 corresponds to frequency $\omega = 4\pi$ and $Gr_\text{v} = 162.23; 229.43; 324.46$ ($Ra_\text{v} = 1000, 2000, 4000$ in [12] notation), respectively.

**Table 1.** Transition to long wave disturbances

| $\psi_1$ | $Gr$ | $Gr_\text{v}$ |
|---|---|---|
| 0.724 | 162.23 | 497.73 |
| 0.582 | 229.43 | 619.17 |
| 0.387 | 324.46 | 931.95 |

The regions of synchronous response appear under the anomalous Soret effect independently on direction of temperature gradient. Further growth of $Gr_\text{v}$ leads to the confluence of these regions.

Dependencies of critical amplitude of modulation $Gr_{\text{v}*}$ (minimized with respect to the wave number $q$) on inverse frequency at fixed $Gr$ and various values of parameter $\psi_1$ are presented in Fig. 4–6.

The equilibrium of the single-component liquid $\psi_1 = 0$ without vibrations $Gr_\text{v} = 0$ is stable at $Gr = 125$ (the A point in Fig. 2 , $Gr < Gr_* = 142.73$). The growth of modulation amplitude $Gr_\text{v}$ leads to instability. The instability domains are the regions located above the curves $\psi_1 = 0$. The first resonant band for the homogeneous liquid $\psi_1 = 0$ corresponds to subharmonic disturbances and its minimum is located at $\omega = 8.69$; $q_* = 1.608$, $Gr_{v*} = 484.033$. Domains of synchronous and subharmonic responses alternate with decrease of the frequency.

For binary mixture with the normal Soret effect $\psi_1 = 0.1$ in the absence of vibrations the value $Gr = 125$ corresponds to the unstable region (the B point in Fig. 2). It is seen that vibrations suppress instability in wide region of frequencies. The stability region, marked by the symbol "st", is inside the border $\psi_1 = 0.1$. The vibrations cannot stabilize equilibrium at frequencies $\omega < \omega_* = 4.07$. This result corresponds to low-frequency limit (Fig. 1) where the value of

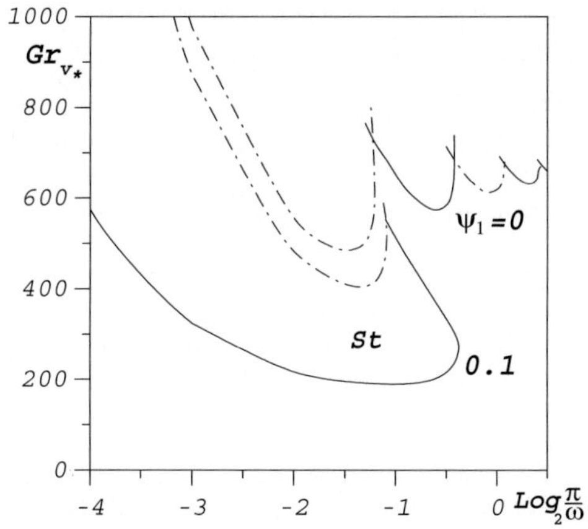


**Fig. 4.** Dependence of critical relative amplitude of modulation $Gr_{\text{v}*}$ on the inverse frequency $1/\omega$ in the cases of homogeneous liquid $\psi_1 = 0$ and normal Soret effect $\psi_1 = 0.1$ $(Gr = 125)$. "St" are stability areas.

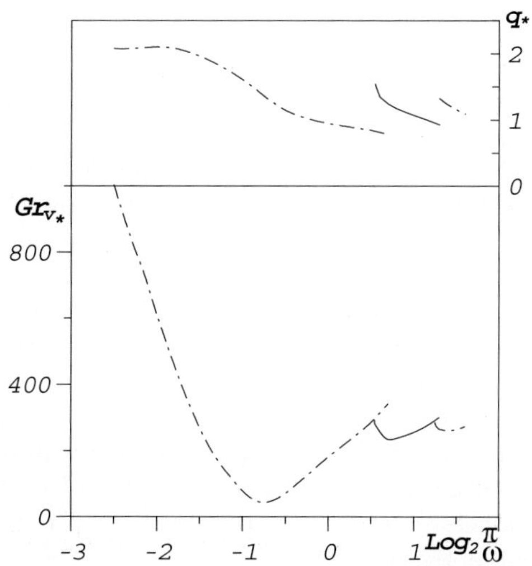


**Fig. 5.** Dependence of critical wave number $q_*$ and critical relative amplitude of modulation $Gr_{\text{v}*}$ on the inverse frequency $1/\omega$. The anomalous Soret effect $(\psi_1 = -0.3)$. Without modulation the system is stable $(Gr = 250)$.

Grashof number $Gr = 125$ belongs to the instability domain at any values of the modulation amplitude.

Note that too large values of amplitude $Gr_{\text{v}}$ are necessary in order to excite the convection $\psi_1 = 0$ or to dump it in the case $\psi_1 > 0$. There is no wonder, because for $\psi_1 \geq 0$ the liquid system has not its fundamental oscillatory frequency in the absence of vibration and only monotonic instability exists (Fig. 2).

The influence of vibrations on stability of equilibrium relatively to oscillatory mode with fundamental frequency $\omega_0$ is presented in Figs. 5–6 in the case of the

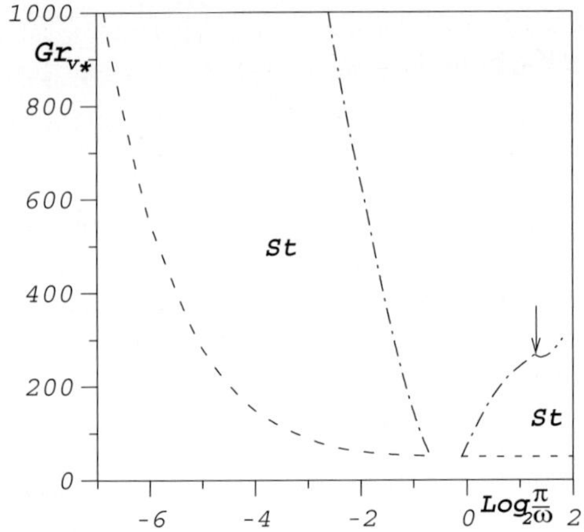


**Fig. 6.** Dependence of critical relative amplitude of modulation $Gr_{v*}$ on the inverse frequency $1/\omega$. The anomalous Soret effect $\psi_1 = -0.22$). Without modulation the system is unstable ($Gr = 250$).

anomalous Soret effect ($\psi_1 < 0$). When there is no modulation the convective instability at $\psi_1 = -0.3$ appears at critical Grashof number $Gr_* = 271.27$ and oscillation frequency $\omega_0 = 2.661$. At subcritical heating conditions $Gr = 250 < Gr_*$ (point C in Fig. 2) the growth of amplitude modulation $Gr_v$ can lead to destabilization of the equilibrium. The effect of parametric resonance destabilization is most significant at $\omega = 5.276 \simeq 2\omega_0$ and presents subharmonic response on external action. The critical value of amplitude modulation is $Gr_{v*} = 42.519$, the critical wave number $q_* = 1.355$. The other minima on boundaries of synchronous and subharmonic resonant bands correspond to frequency values $2\omega/m$, where $m$ is integer. The destabilizing effect for these bands is rather weak. Inside the regions of the resonance frequencies the crisis of the equilibrium comes at relatively small values of the amplitude. The dependence of critical wave number on inverse frequency is shown in the upper part of the Fig. 5. It is composed of synchronous and subharmonic curves. The critical wave numbers monotonically decrease in each resonant band.

The stabilizing action of modulations on the unstable equilibrium is shown in Fig. 6. Without modulations at $\psi_1 = -0.22$ the equilibrium is unstable respectively to oscillatory mode ($Gr = 250 > Gr_* = 248.46$, point D in Fig. 2). The vibrations can stabilize equilibrium. The stability domains are marked in Fig. 6 by "St". They are symmetrically located with respect to resonant frequency $\omega = 4.079 \simeq 2\omega_0$. The bottom boundaries of the stability bands represent boundary of quasi-periodic mode; the top ones are boundaries of the subharmonic disturbances. The break on the top boundary of stability domain (arrow in Fig. 6) corresponds to transition between subharmonic disturbances with various spatial period. As it can be seen, at all values of $\omega$ vibrations stabilize basic state, but they excite parametric instability $\omega \simeq 2\omega_0$. The region of resonance frequencies becomes wider as the amplitude of vibrations increases.

In order to separate thermal and vibration action on the Soret convection we rewrite alternating coefficient $Gr + Gr_v \cdot \sin \omega t$ in the following form $Gr(1 + \eta \cdot \sin \omega t)$ where $\eta = Gr_v/Gr = b\omega^2/g$. The dependencies $Gr_*(\psi_1)$ for several values $\eta$ and $\omega$ are shown in Figs. 7–8. The boundaries of instability are presented for two amplitudes $\eta = 1; 2$.

When heating from above vibrations do not practically influence on location of the boundary of critical long wave synchronous disturbances. As the heating from below the cellular disturbances are more dangerous. For $\omega = 4\pi; \eta = 1$ (Fig. 7) the boundary of instability consists of three parts: boundary of subharmonic response $-1.092 < \psi_1 < -0.4$; boundary of quasi-periodic disturbances $-0.4 < \psi_1 < -0.08$; boundary of the synchronous response on external action

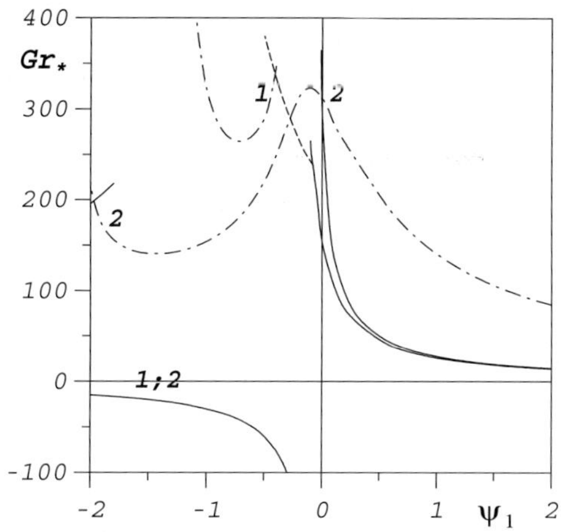


**Fig. 7.** Maps of stability on a plane $(\psi_1, Gr_*)$ in a modulated field, for various amplitudes of external modulation: $\eta = 1$, $\eta = 2$, $\omega = 4\pi$.

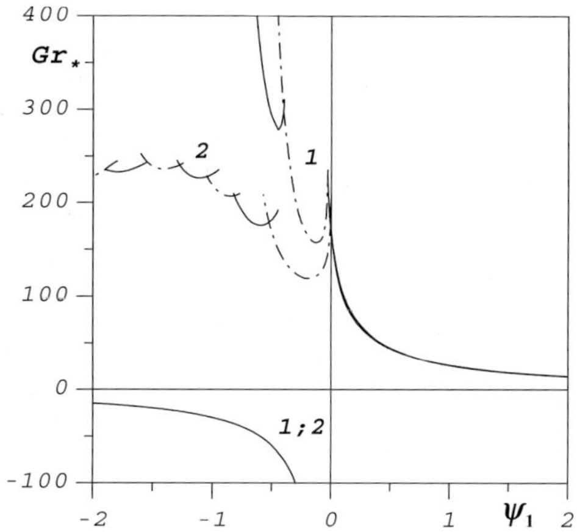


**Fig. 8.** Maps of stability on a plane $(\psi_1, Gr_*)$ in a modulated field, for various amplitudes of external modulation: $\eta = 1$, $\eta = 2$, $\omega = \pi$.

$-0.08 < \psi_1 < 2$. Growth of amplitude ($\eta = 2$) leads to decrease of the threshold of the subharmonic mode and increase of the threshold of synchronous one. The boundary of instability in this case consists of two parts: the subharmonic disturbances increase in region $-2 < \psi_1 < 0$, synchronous ones at $\psi_1 \geq 0$. The destabilization can be explained by the resonant interaction between the oscillatory disturbances with frequency $\omega_0$ and oscillation of the layer with modulated gravity field. The maximum of destabilization takes place at $\omega \simeq 2\omega_0$.

Decrease of the modulation frequency $\omega = 4\pi; \pi$ (Figs. 7,8) at fixed amplitude (for instance $\eta = 1$ or $\eta = 2$) shifts the parametric resonance bands into region of smaller oscillatory frequency $\omega_0$, which corresponds to smaller absolute value of the separation ratio. The new regions of the instability appear. They are connected with resonances at frequencies $2\omega_0/m$. The disturbances in neighboring regions have different oscillation period. They present either synchronous or subharmonic response on external modulation.

## 6 Conclusion

Thus, the problem of the convection onset in a plane horizontal layer of binary mixture with thermodiffusion effect in the presence of vibration of arbitrary frequency has been investigated. The boundaries of the layer are assumed to be rigid, isothermal and impermeable to both mixture components. The transverse orientation of the vibration axis is considered. There is not external difference of concentration, the concentration inhomogeneity is produced by the Soret effect. The Floquet theory is applied to find the instability thresholds under the linear approximation. Let us draw attention to the merits of the work. The influence of the time-dependent gravity field modulation on binary mixture rest state is studied in cases of either normal or anomalous Soret effect. Domains of resonant parametric excitation are found. The dependencies of the critical values of the vibrational analog of the Grashof number on the inverse frequency are obtained. The influence of the separation ratio and the Grashof number on the stability thresholds is also examined. It has been shown that in addition to synchronous and subharmonic responses to variation of the external gravity field, the instability can be associated with quasiperiodic disturbances. In the case of the normal Soret effect synchronous or subharmonic responses on external action are more dangerous for liquid equilibrium. In the case of the anomalous Soret effect all types of disturbances compete. The threshold of instability due to the parametric resonance mechanism of the instability strongly depends on the separation ratio. The dependencies of the Grashof number on the separation ratio are found at various amplitudes and frequencies of modulations. It is shown that, depending on the amplitude and frequency of modulation, the vibrations can stabilize the unstable base state or destabilize the equilibrium of the binary mixture layer. The asymptotic behavior of the critical parameters in the limiting case of low frequency modulation is studied with the help of the Wentzel–Kramers–Brillouin method.

## Acknowledgements

Participation of one of the authors (Smorodin B.L.) in the 'Fourth International Meeting on Thermodiffusion (IMT4) was supported by the 'Deutsche Forschungsgemeinschaft' (DFG), grant 436114/169/00.

## References

1. J.C. Legros, J.K. Platten, P.G. Poty: Phys. Fluids **15**, 1383 (1972)
2. J.K. Platten, J.C. Legros: *Convection in Fluids* (Springer-Verlag, Berlin 1984)
3. G.Z. Gershuni, E.M. Zhukhovitsky: *Convective Stability of Incompressible Fluids.* (Keter, Jerusalem 1976).
4. P.M. Gresho, R.L. Sani: J. Fluid Mech. **40**, 783 (1970)
5. G. Venezian: J. Fluid Mech. **35**, 243 (1969).
6. D. Rosenblat and G.A. Tanaka: Phys. Fluids **14**, 1319 (1971).
7. M. Wadih, B. Roux: J. Fluid Mech. **193**, 391 (1988).
8. G.Z. Gershuni, I. O. Keller, B. L. Smorodin: Physics-Dokl. **41**, 211 (1996)
9. B.V. Saunders, B.T. Murray, G.B. McFadden, S.R. Coriell, A.A Wheeler: Phys. Fluids A **4**(6), 1176 (1992)
10. G. Terrones, C.F. Chen: J. Fluid Mech. **255**, 301 (1993)
11. G.Z. Gershuni, A.K. Kolesnikov, J.-C. Legros, B.I. Myznikova: J. Fluid Mech. **330**, 251 (1997)
12. G.Z. Gershuni, A.K. Kolesnikov, J.-C. Legros, B.I. Myznikova: Int. J. of Heat Mass Transfer **42**, 547 (1999)
13. G.I. Petrov: Prikl. Mat. Mekh. **4**, 3 (1940)
14. E.A. Coddington, N. Levinson: *Theory of ordinary differential equations* (McGraw-Hill, New York 1955)
15. A.H. Nayfeh: *Introduction to Perturbations Techniques* (John Wiley, New York 1981)
16. L.D. Landau, E.M. Lifshitz: *Theoretical Physics. Quantum Mechanics* (Pergamon Press, Oxford 1981)

## Supplementary glossary

$h$ layer half-thickness
$\Theta$ upper plate temperature
$\boldsymbol{g}_0$ gravity acceleration
$\boldsymbol{n}$ unit vector along an axis of vibration
$c_{\mathrm{sd}}$ sound velocity
$b$ amplitude of vibration
$\Omega$ angular frequency of vibration
$p$ pressure
$\nu$ kinematic viscosity
$\chi$ heat diffusivity coefficient
$Gr$ Grashof number: $Gr = g_0 \alpha \Theta h^3/\nu^2$
$Gr_{\mathrm{v}}$ vibrational analogue of the Grashof number: $Gr_{\mathrm{v}} = b\Omega^2 \alpha \Theta h^3/\nu^2$
$\omega$ dimensionless frequency of vibration: $\omega = \Omega h^2/\nu$

$\psi_1$ separation ratio: $\psi_1 = \frac{\beta}{\alpha} S_{\mathrm{T}}$
$\Lambda$ complex growth rate whose real part determines stability: $\Lambda = \Lambda_{\mathrm{r}} + i\Lambda_{\mathrm{i}}$
$\gamma$ Floquet multiplier
$Ra_{\mathrm{v}}$ vibrational analogue of the Rayleigh number : $Ra_{\mathrm{v}}{=}(\alpha b\Omega 2\Theta 2h)^2/2\nu\chi$; $Ra_{\mathrm{v}}{=}16{Gr_{\mathrm{v}}}^2 Pr/(2\omega^2)$
$\eta$ relative amplitude of the modulated component of the acceleration: $\eta = Gr_{\mathrm{v}}/Gr = b\omega^2/g$

# Thermodiffusion in Porous Media and Its Consequences

Pierre Costesèque[1], Daniel Fargue[2], and Philippe Jamet[3]

[1] Université Paul Sabatier, 39, allées Jules Guesde, 31000 Toulouse, France
[2] Ecole des Mines de Paris, 60, bd Saint Michel, 75272 Paris cedex 06, France
[3] Ecole des Mines de Paris, 35, rue Saint-Honoré, 77305 Fontainebleau cedex, France

**Abstract.** This contribution presents a synthesis of experimental, theoretical and numerical studies performed on thermodiffusion and thermodiffusion-convection transport in porous media.

The study of thermodiffusion in heterogeneous domains is fostered by the potential implications of this coupled transport phenomenon in the *in situ* differentiation of geochemical fluids (concentration of natural oils, promotion of ore genesis, fractionation of isotopes). The technological simplifications brought by the introduction of synthetic porous media in the thermodiffusion columns of laboratory or industrial use is another reason for investigating thermodiffusion in discrete systems.

A review of published and original laboratory data on the thermodiffusive behaviour of selected inorganic and organic binary and of multicomponent mixtures is exposed here. A phenomenological scheme using the formalism of Irreversible Thermodynamics and the macroscopic description of the porous system is proposed and non-linear developments are suggested to account for the dispersive feature of thermodiffusion transport. Data and original equations are evaluated in a numerical model with reasonable success and compared to other existing analytical and numerical approaches.

## 1 An overview of thermodiffusion in porous media

Thermodiffusion is such an intricate and subtle second-order transport process that one may be confused by the fact that it has attracted so much attention since the founding experiments of Ludwig ([1], gaseous mixtures), and Soret ([2], ionic solutions). One could have expected this strange physical phenomenon which induces but few differentiation in the chemical composition of homogeneous systems to remain nothing more than an experimental curiosity.

There are maybe two reasons why thermodiffusion deserved a better fate. First, it is one of the various cross effects relevant to irreversible thermodynamics ([3–5]) which is easily accessible to experimental measurements. Second, the potential interest of thermodiffusion in chemical engineering has been early recognized and has been fueling theoretical, experimental and analytical research for a long time. Thermodiffusion as an industrial separation process now raises much less interest, but it has not always been so.

### 1.1 The engineering foundations of thermodiffusion

The best way to grasp the evolution from early research in free media to the present focus on porous media is to look back at history, particularly to the

W. Köhler and S. Wiegand (Eds.): LNP 584, pp. 389–427, 2002.

various attempts to design thermodiffusion into efficient chemical engineering devices.

In static, initially homogeneous fluid systems, thermodiffusion alone cannot induce more than a few percents separation. This was clearly a too limited efficiency that prevented any application to chemical engineering unless some way of coupling it with first order transport processes was imagined. This matter was cleared up in 1938, at least on a technological point of view, by the device of Clusius and Dickel ([6,7]), which used the so-called thermogravitational coupling to separate chlorine isotopes in a chlorhydric acid gas phase. They used a 36-meter long column with axial heating provided by a hot wire. This setting creates a radial temperature gradient and its accompanying colinear thermodiffusion flux - *generally* directed to the cold side - together with a global fluid instability in the gas. The coupling that takes place between the radial differential mass flux on the one hand, the bulk gas convection flux on the second hand, result in a net mass transport of the heavier compound ($H^{37}Cl$ in the Clusius and Dickel experiments) to the bottom part of the cell. As the mixture separation grows, it is progressively counterbalanced by ordinary diffusion, so that in a closed system a stationary state is reached where the ratio between bottom cell and top cell mass fractions of the heavier compound may reach several orders of magnitude (a factor of ca. 4000 in the mentioned experiment).

As shown in the analytical derivations of Debye [8], Furry, Jones and Onsager [9] and Jones and Furry [10], the efficiency of coupling in free thermogravitational cells, for a given mixture (i.e. given specific mass and kinematic viscosity) and a fixed temperature difference between the axis and the outer wall of the cell,

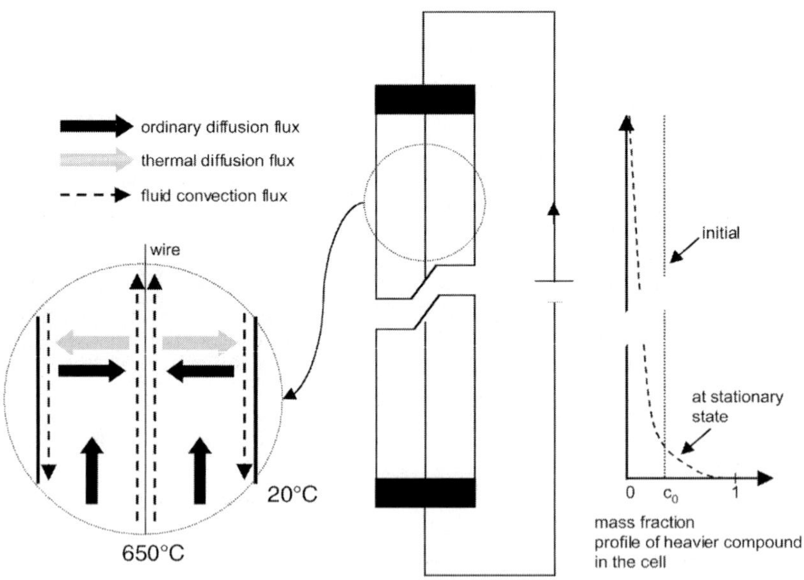


**Fig. 1.** The Clusius and Dickel thermodiffusion colum: principles

depends only from the cell's dimensions, especially width, which is the main factor acting on the fluid velocity. This property can be easily managed when working on gases, but causes important technological constraints when it comes to condensed fluids. In that case, optimal convection velocity is obtained for very narrow interspaces between the isothermal boundaries of the system ($\simeq 0.02$cm). The cost of such cells, added to the complexity in operating them cannot be tolerated in industrial conditions.

It was suggested ([11,12]) that an efficient solution for controlling fluid velocity was to introduce a packing into the columns. The packing permeability is then the only transport parameter to adjust. Changing the packing into the column is indeed easier than modifying the geometry of the separation device. Besides being convenient to control the convection regime, the packing also offers the great advantage of allowing the thermogravitational coupling to be operated in wider cells. Typically, the width of the cell can be increased by a factor 10 or 100.

The analytical extension of the Furry, Jones and Onsager theory to Packed Thermodiffusion Columns operated under horizontal temperature gradients (which will be further referred to as the horizontal gradient PTC device) was proposed by Lorenz and Emery [13] together with technological and economic evaluations of the PTC performance to refine crude oils. One important result was that, all things being equal, the separation efficiency, measured by the magnitude of the relative separation of the mixture at both ends of the cell, presented a maximum for a particular value of the packing pemeability.

In the 70's Sanchez [14–16] performed a complete assessment of the separation of chlorinated organic solvents by PTC, and investigated numerically the serial system of open columns [17] which presents a much better production performance than the closed "batch" system of Lorenz and Emery.

### 1.2 Geochemical implications of thermodiffusion

Though the case about PTC as a routine chemical engineering process was progressively settled in the 60s, mainly because of the associated energy cost, the research focus on thermodiffusion in porous media had meanwhile shifted to the field of geochemical studies. Already in 1946, Wahl [18] suggested that thermodiffusion could be an interesting candidate process to explain mass fractionation in molten rocks. The geosphere and upper mantle sometimes host important temperature gradients and geological fluids (of which many saturate porous materials) often evolve into open systems. In these conditions, mass fractionations are not blocked at moderate levels but are rather sustained, because the concentrated parts of the fluids migrate to other areas where they may further develop to higher degrees of concentration.

Schott [19], Carrigan and Cygan [20] followed the original idea of Wahl [18] on magmatic differentiation. A lot of experimental studies on ionic solutions was performed at the Paul Sabatier University, Toulouse, France, ([21–23]) mainly using horizontal gradient PTC devices, but also under vertical downwards thermal gradients. This research was motivated by the potential role of thermodiffusion and thermogravitation in the field of mineral concentration and crystallization in deep aquifers, such as the Salton Sea geotherm [23,24]. This laboratory also produced a large amount of experimental material on the thermogravitational separation of crude oil ([25–30]), as thermodiffusion is also believed to play some role in the *in situ* evolution of oil reservoirs composition ([31]).

As in laboratory experiments, *in situ* differentiation of geological fluids is likely to remain limited in magnitude were it be no thermogravitational-like coupling. This supposes that convective flow may be sustained within the system. However, in most cases, the background thermal gradients in the geosphere are vertical, i.e. not destabilizing *per se* conversely to the horizontal case. In that context, a crucial question is to assess how thermodiffusion may contribute to fluid instability. It has been shown ([32–34]) that its impact in free media is potentially tremendous in lowering the value of the effective critical Rayleigh number of the marginal stability case in the Bénard configuration. The same property is also valid in the case of porous media [35].

### 1.3 Revisiting the industrial interest of thermodiffusion

The question of using the PTC device for industrial chemical applications has not been reconsidered for near 40 years. However, the conditions are now different. For sure, the production fluxes and the high energy consumption required by the PTC separation are not on a level with any of the existing chemical engineering processes. However, thermodiffusion could regain interest for very specific, high value added applications where the present separation technologies reach their limits. It is worth remembering that thermodiffusion can separate very special mixtures such as isomers, isotopes, eutectics. As regards the energy question, there is no specification for the absolute mean temperature to operate PTCs: the problem is to apply a *gradient*. Hence, PTCs can be operated with waste energy, the so-called "low enthalpy" fluids produced and discharged into the environment by any industrial process (e.g. power stations) [36].

## 2 Experimental investigations on thermodiffusion phenomena in porous media

In this section, we will review experimental material addressing thermodiffusion in porous media, mainly using the packed thermal diffusion column device (PTC) taking advantage of thermodiffusion-convection coupling. There are but few systematic studies about thermal diffusion in porous media. This explains why the majority of bibliographic references, either as reviewed papers or as non-published literature (PhDs, etc.) originate from the research performed at the Paul Sabatier University, Toulouse, France, where such experiments have been performed since the early 70s.

The focus on PTC experiments is justified for three reasons. First, it is a good first-order analogue for the characterization of thermal diffusion transport in geological systems , not only because we work on discontinuous systems (pores and solid), but also because we observe fluxes associated with a fluid flow, which is a typical situation in geologic media. Second, PTC remains a good candidate process for industrial separations of high value added chemical substances. Third - this aspect will be mainly developed here - it is potentially a reliable and reproducible experiment for the metrology of the Soret coefficient and the attached phenomenological investigations.

At first sight, one could have expected static (no flow, pure Soret effect) experiments to be more relevant for an accurate measurement of the coefficient. This would certainly be the case, could uncontrollable thermal instabilities of the fluid always be avoided. Comparisons between static experiments involving the same binary mixture in similar experimental conditions (solute mass fraction, mean temperature) show that large discrepancies may be expected [37] due to the perturbating effect of gravity. To avoid such perturbation, static measurements of thermal diffusion coefficients might be performed under microgravity conditions [38], but at very high experimental costs.

On Earth, rather than being constantly under threat from random fluid instabilities, a good, low-cost strategy is to use a controlled instability as an "experimental partner". One may use for example the method [39] which consists in measuring the frequency of a particular oscillatory instability. The thermogravitational experiment is also of great interest [40], because of the obtained magnification of separation which increases the accuracy of the measurement.

## 2.1 Typical phenomena and experimental devices

We will first present an inventory of the various experimental apparatus that are currently used, and relate each of them to typical obervable transfer phenomena. All devices use porous packings made of corundum or zirconium oxide balls of selected granulometric brackets to yield specific packing porosity and permeability.

**Pure thermodiffusion in porous media.** Though static experiments require great care, it is however interesting to study the value of the Soret coefficient in a porous medium when the saturating fluid is at rest.

The cell represented in figure 2 has been used to perform such measurements for three selected electrolyte solutions. The characteristic time to achieve the pure thermodiffusion stationary state is given by the relation: $\theta = (a^2)/(\pi^2 D^*)$ where $a$ is the size of the cell along the thermal gradient and $D^\star$ is the effective diffusion coefficient in porous media defined by $D^\star = D/\tau^2$ where $\tau^2$ is the packing tortuosity. The order of magnitude for $\theta$ is about 5 days. A minimum notice of 20 days was respected before sampling the solutions at both ends. For dilute solutions, the analytical expression for the Soret coefficient is given by [41]:

$$s_{\mathrm{T}}^\star = \frac{D_{\mathrm{T}}^*}{D^*} = \frac{\Delta n}{n_0 \Delta T}$$

Examples of values of the Soret coefficient obtained with this experimental method [42] are presented in table 1 and compared with available literature data.

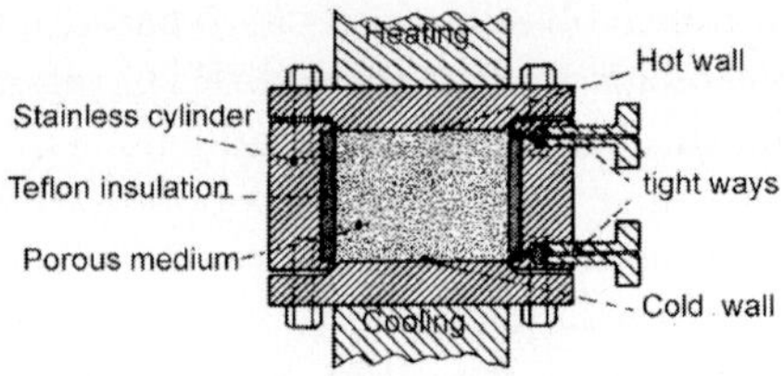


**Fig. 2.** Principle of the vertical thermal gradient cell used for the measurement of Soret coefficient in porous media under static conditions [42]

**Table 1.** Measurements of thermodiffusion coefficients in porous medium. Comparison with values in free media: ($^{(1)}$: $\bar{T} = 25°C, n_0 = 0.01$ [43]; $^{(2)}$: $\bar{T} = 60°C, n_0 = 0.05$) [44].

| compound | $n_0$ mol.l$^{-1}$ | $\bar{T}$ in °C | $s_T^\star$ in K$^{-1}$ | $s_T$ (lit.) |
|---|---|---|---|---|
| $BaCl_2$ | 0.05 | 85 | $8.4 \times 10^{-3}$ | - |
| " | " | 75 | $7.2 \times 10^{-3}$ | - |
| " | " | 66 | $5.1 \times 10^{-3}$ | $4.5\, 10^{-3}$ (1) |
| " | " | " | $5.3 \times 10^{-3}$ | - |
| LiI | 0.1 | 60 | $3.1 \times 10^{-3}$ | $4.2\, 10^{-4}$ (2) |
| " | " | 95 | $6.0 \times 10^{-3}$ | - |
| $UO_2NO_3$ | 0.017 | 74 | $13.9 \times 10^{-3}$ | - |

However, it appears difficult to go deeper into the investigation of the relationship between values in free and in porous media because the experimental conditions are not the same and because the Soret coefficient may vary significantly even in a narrow range of mean temperature and initial composition.

**Thermodiffusion-convection coupling.** The main interest of porous media for the measurement of thermodiffusion coefficients is of course represented by the thermodiffusion-convection coupling, also called "thermogravitational effect", as described in section 1. The packed columns which are used are of two types: parallelepipedic or cylindrical. The porous packing is enclosed between two isothermal and two adiabatic walls (plates or cylinders). The temperature is maintained by the circulation of water (hot or cold) or by means of an electric wire when the hot temperature exceeds 100°C.

The geometrical characteristics of the column and porous packing which govern the efficiency of the separations are: (i) the length of the column ($L$), the isothermal plates spacing (or porous packing width, $a$), (iii) the intrinsic permeability of the packing ($k$). Once the former characteristics are fixed, the experimental conditions which also impact on the magnitude and the kinetics of separations are: (i) the applied temperature difference ($\Delta T$), (ii) the direction of thermal gradient (horizontal or vertical), (iii) the physical and chemical characteristics of the solutions (firstly the thermo- and ordinary diffusion transport

**Table 2.** Typical experimental parameters for the horizontal and vertical PTC.

| parameter | horizontal PTC | vertical PTC |
|---|---|---|
| $L$ | 0.4 to 1.2 m | 0.05 to 0.2 m |
| $a$ | $4 \times 10^{-3}$m | 0.03 m |
| $\Delta T$ | 25°C | 25°C |
| $\bar{T}$ | 45°C | 45°C |

coefficients). Typical experimental parameters of horizontal- and vertical gradient PTCs are presented in table 2.

*Horizontal gradient columns.* The most frequent configuration for observing thermodiffusion-convection coupling is the PTC column operated under an horizontal thermal gradient. The layer of porous medium and the separating solution are enclosed between two coaxial isothermal cylinders. The value of $a$, though considerably larger than in the case of free thermodiffusion columns ($\simeq$ 0.02cm), remains quite small ($\simeq$ 1cm). This induces practical problems in assembling the cell but no major difficulties when it comes to use the macroscopic formalism for transport in porous media (see section 3).

The principle of the horizontal PTC is quite simple. Thermodiffusion induces an horizontal separation into the solution which is recycled by the fluid convection roll (always existing because the configuration is always unstable). When the thermodiffusion coefficient of the heavier compound (taken as the solute) is positive, the net mass transport is downwards and the stationary state corresponds to a vertical and stable mass fraction gradient. When this coefficient is negative there is a net accumulation of the solute on top of the column, a fact which may result in an additional instability which would make it difficult to interpret the experiments. Hence, the experiment can be used with no restriction for $s_{\mathrm{T}}^{\star} > 0$ but restricted to limited separations for $s_{\mathrm{T}}^{\star} < 0$. In what follows, we will refer by default to experiments involving compounds of positive thermodiffusion coefficient.

Drawings of the usual horizontal gradient PTCs are shown on figure 3.

The typical duration of PTC experiments is about 10 to 50 days, and consequently they require many precautions and laboratory skills to avoid leaks or uncontrolled phenomena such as heterogenity of the porous filling. Moreover, as the porous packing must always remain saturated, sampling during the course of the experiments is limited to very small quantities of fluids, either by means of end reservoirs or intermediary access.

For a given compound, the horizontal gradient PTC experiment may yield two (non-exclusive) experimental conclusions: (i) in transient conditions, (ii) at stationary state.

In the transient conditions case, the separation process is followed at both ends of the column from the initial state where the mass fraction of the solute is homogeneous up to a sufficient amount of time until the stationary state is

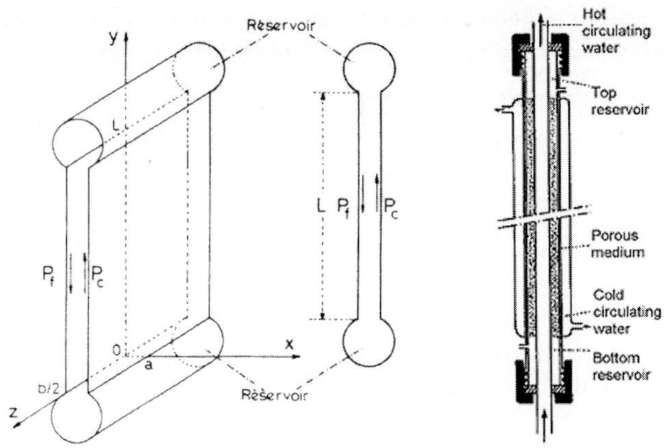


**Fig. 3.** Design of horizontal gradient PTCs.

reached. As shown on figure 4, a simultaneous evolution of the reservoirs average mass fractions is observed, with an increase (decrease) of the bottom (top) mass fraction until asymptotic values are obtained. For a given column and applied temperature difference, and a fixed packing permeability, the magnitude of each end mass fraction variation or of the separation difference is a unique function of $s_{\mathrm{T}}^{\star}$. However the kinetics of separation is a function of the ordinary diffusion coefficient $D^{\star}$. As a consequence, two transport parameters, viz. $D^{\star}$ and $D_{\mathrm{T}}^{\star}$ are needed to fit the observed $c^{\mathrm{top}}(t)$ and $c^{\mathrm{bottom}}(t)$ curves.

A good way to use stationary state results is to vary the packing permeability while keeping the other experimental parameters fixed. The separation

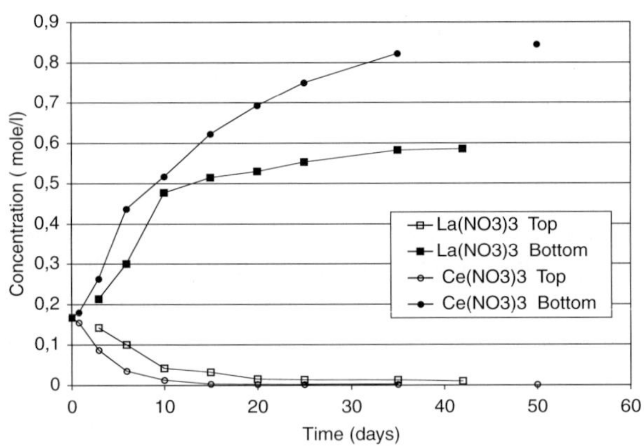


**Fig. 4.** Typical evolution of top and bottom mass fractions of the solute (rare earth salts) in a packed thermodiffusion column [42].

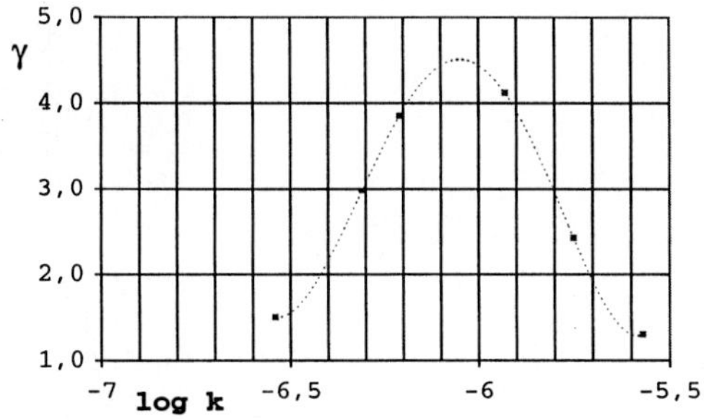


**Fig. 5.** Typical curve showing the separation magnitude in a horizontal gradient PTC as a function of packing permeability ($FeCl_3$, [45])

magnitude at stationary state $\left(c_{\text{stat}}^{\text{bottom}} - c_{\text{stat}}^{\text{top}}\right)$ or the separation ratio

$$\gamma_{\text{stat}} = \frac{c_{\text{stat}}^{\text{bottom}}}{c_{\text{stat}}^{\text{top}}} \times \frac{\left(1 - c_{\text{stat}}^{\text{top}}\right)}{\left(1 - c_{\text{stat}}^{\text{bottom}}\right)}$$

can be plotted against $k$ to yield the characteristic curve of figure 5. It is easy to understand from figure 5 how the thermodiffusion-convection coupling works. All things being equal, the convection velocity varies with the packing permeability. When small, the separation inside the cell is mainly an horizontal Soret effect. When high, the convective flux tends to annihilate the separation by mixing the fluid. In both cases the vertical separation ratio vanishes, i.e. $\gamma = 1$. In the intermediary domain, convective and thermodiffusive transport are of similar magnitude, and the coupling is much more efficient, leading to $\gamma \gg 1$.

As will be shown in section 4, the position and magnitude of the maximum of the separation ratio as a function of packing permeability only depends from the effective transport parameters $D^\star$ and $D_{\text{T}}^\star$. This curve determines the value of the thermodiffusion coefficient. This method is maybe more reliable than interpreting a unique transient experiment because it involves a cross validation of several PTC experiments, however it requires much more time and laboratory equipment.

*Vertical gradient columns.* Packed columns can also be designed to investigate thermodiffusion-convection coupling under a vertical destabilizing thermal gradient. In the case of a binary mixture with $s_{\text{T}}^\star > 0$, the applied gradient is downwards. The porous medium is contained into a vertical cylinder of stainless steel or glass whose horizontal walls are maintained at constant temperature. The vertical wall is kept adiabatic by adding a second coaxial cylinder of larger diameter separated from the former by a primary vacuum. The cell used at the Toulouse laboratory is presented in figure 6.

Using vertical gradient PTCs presents several advantages. First, moderate temperature gradients are applied, using about the same temperature difference as in the horizontal gradient column, but along the length of the cell. This may be important, particularly for potential chemical engineering applications, because

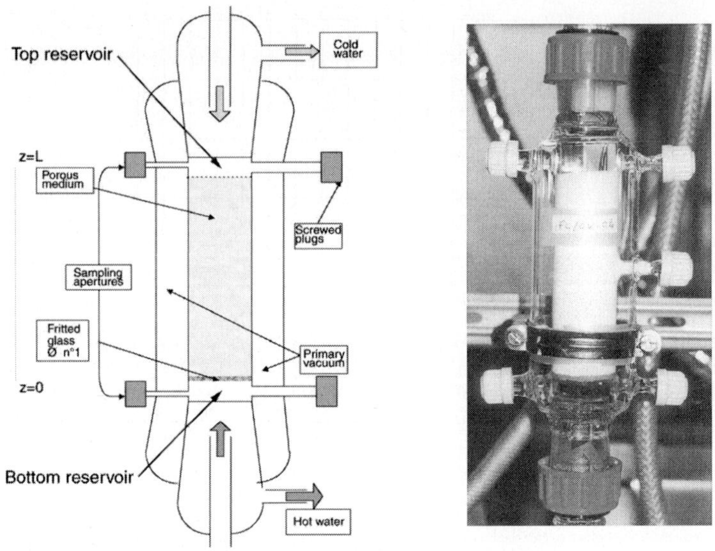


**Fig. 6.** Vertical gradient PTCs.

exergetic losses would be substantially smaller in that case. Second, the volume of the involved porous medium is larger. Third, the experimental configuration is closer to the natural case of vertical thermal flux, a situation that is more representative to evaluate potential segregation phenomena in geologic systems.

Although separation of similar magnitude as in the horizontal gradient case is observed, there is still no convincing physical scenario to explain what happens. How the coupling between the convection flux (of undetermined geometry) and the thermodiffusion flux (of undetermined direction) is achieved is the major unknown about this experiment. While waiting for the phenomenological modelling to give an appropriate characterisation of the transport phenomena inside

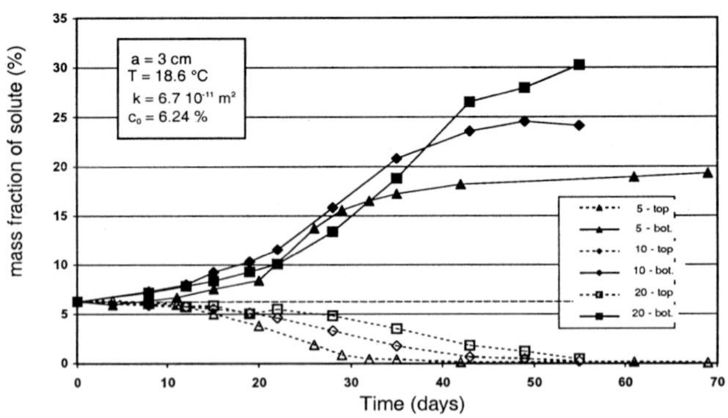


**Fig. 7.** Differenciation of a CuSO4 solution in vertical gradient PTCs.

the column, vertical PTC experiments are, until now, only of qualitative and exploratory interest.

Figure 7 shows a set of new vertical gradient PTC experiments involving a solution of copper sulfate. The three experiments correspond to three column length of 5, 10 and 20 centimeters. The separation phenomenon is qualitatively the same as for the horizontal gradient (see figure 5). However, from these curves, there are some evidences that the coupling is probably not of similar nature. For example, the characteristic separation time does not seem to vary as the square of the cells length ratio but rather to be proportional to the length. The vertical gradient PTC case will be further discussed in section 4.

## 2.2 More experimental results on inorganic and organic mixtures

The horizontal gradient PTC has been used on various binary, ternary and complex mixtures of inorganic and organic compounds. The experimental results listed below are not exhaustive but, while they will give further insight into the dynamics of the PTC experiment, they are also meant to focus on the following aspects: (i) comparison of PTC experiments and static experiments in porous media, (ii) the selectivity of thermodiffusive separation (behaviour of trace element and isotopes), (iii) multicomponent solutions vs. binary solutions, (iv) geochemical and industrial consequences of thermodiffusion-convection coupling. At this stage, only raw results are presented. Here, we do not mention the values of the Soret or thermodiffusion coefficients proposed by the authors, for they always used the analytical solution of PTC separation provided by Lorenz and Emery [13], based on the determination of the separation maximum (figure 5). As will be shown in section 4, this solution is not entirely satisfying, mainly because it does not take into account the particular nature of the effective transport coefficients in a macroscopic approach of the porous medium.

**Examples of binary electrolyte solutions.** As shown by the static measurement reported in table 1, the Soret coefficient of uranyl nitrate is particularly high. This compound has been investigated using the horizontal PTC device which confirmed its high reactivity to thermodiffusive transport. Figure 8 presents the results of four experiments on $UO_2NO_3$[42,46]. Two of the experiments led to saturated solutions at the bottom of the cell.

Cristallization of the solute was also obtained in another experiment using the vertical gradient PTC device on a $CuSO_4$ solution (initial concentration: 2 N) [42].

**Examples of binary mixture of organics.** A lot of experimental results were obtained on various binary mixtures of organic solvents and alcohols, especially $CCl_4$-$C_2H_5OH$ and $CCl_4$-$C_2HCl_3$ [14–16]. The main purpose was to investigate the efficiency and the selectivity of the batch or open PTC device for potential chemical engineering applications. This was a short-lived tentative from the industrial point of view, but which provided a considerable amount of data and a

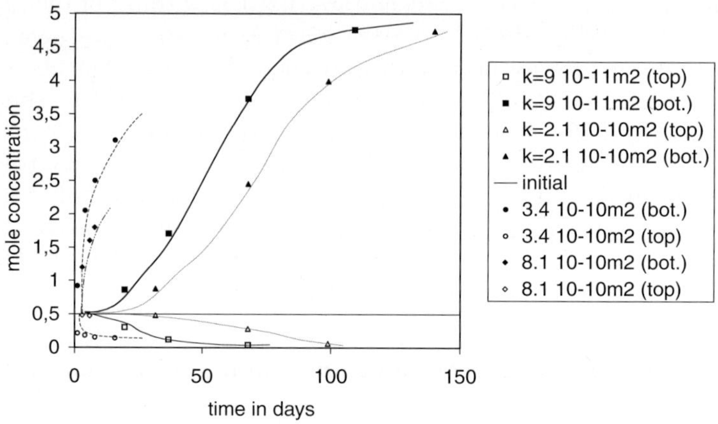


**Fig. 8.** Separation of uranyl nitrate in horizontal gradient PTCs: transient mass fraction curves [42,46].

thorough investigation of every experimental parameter governing the PTC separation dynamics. The following illustrations show part of the results obtained.

In figure 9a, the impact of varying the thickness of the porous layer (hence the temperature gradient) on the separation curve of figure 5 is illustrated. Although only one branch of the curve is displayed, it is obvious that, all other parameters being equal, both the separation maximum and the corresponding value of the packing permeability decrease with increasing values of $a$, somewhat linearly for $k$ and rather geometrically or exponentially for $\Delta n$ (which is qualitatively similar to $\gamma$ for measuring the magnitude of separation).

In figure 9b, the second geometrical factor of the cell, i.e. $L$, is varied. It is clear that the separation grows with $L$ and that closer to the separation maximum (the smaller $k$ value) the growth is not linear.

From these results, we infer that the maximum separation is a non linear function of $L/a$.

As the thermodiffusion coefficient depends on the state variables of the system, mainly $T$ and $n$ (or $c$), it may happen that $D_T^*$ changes sign locally inside the column. The experiments suggest that this may be the case for ethanol mixtures, as shown on figure 10 where all experimental parameters except the initial composition are fixed.

**Examples of isotopes.** Apart from its interest for geochemistry, the study of isotope separation by PTC is an opportunity for studying a binary system where the perturbation of the state variables induced by the presence of the solute can be neglected because it is generally a trace element.

Experiments using the horizontal gradient PTC have been performed on two binary mixtures of isotopes, namely $HDO/H_2O$ and $H_2^{16}O/H_2^{18}O$ and are reported in table 3 [42]. These results suggest that the intensity of separation in

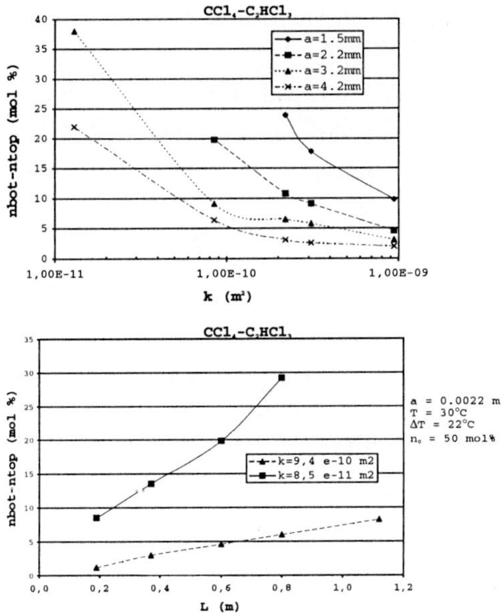


**Fig. 9.** a. Separation of binary organic mixtures. Effect of PTC width. Magnitude of separation obtained in four families of horizontal gradient PTCs and for different values of the packing permeability. b. Separation of binary organic mixtures. Effect of PTC length. Magnitude of separation obtained in two families of horizontal gradient PTCs and for different values of the packing permeability [15,16].

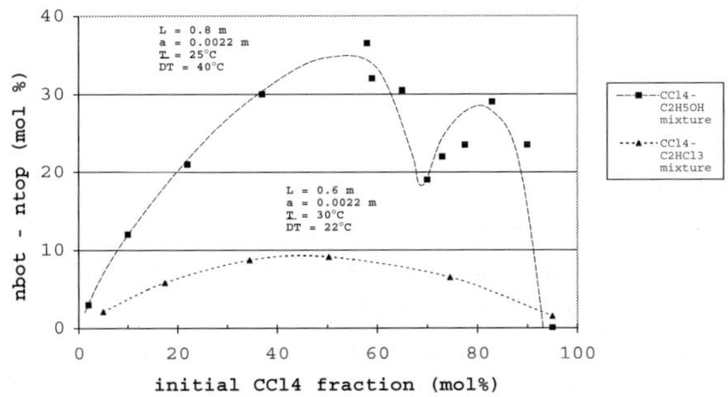


**Fig. 10.** Separation of binary organic mixtures. Comparison of $CCl_4$-$C_2H_5OH$ and $CCl_4$-$C_2HCl_3$ mixtures as regards the impact of changing the initial composition inside the cell [16]

the columns, all things being equal, increases with the mean temperature with a perceptible maximum around $\bar{T} \approx 50°C$. These Soret coefficients measurements can be compared with available data performed in static conditions, see [48].

**Table 3.** Experimental results on the segregation of two binary isotope mixtures $HDO/H_2O$ and $H_2\,^{16}O/H_2\,^{18}O$ [42].

$HDO/H_2O$ relative concentration

| mean T (°C) | $\Delta T$ (°C) | $k$ $10^{-10}$ m$^2$ | $n_0 \times (10^{-6})$ Initial | $n_T \times (10^{-6})$ Top | $n_B \times (10^{-6})$ Bottom | $\frac{\Delta n}{n_0}$ $10^{-2}$ | $\gamma = \frac{n_B}{n_T}$ |
|---|---|---|---|---|---|---|---|
| 9,8 | 20,5 | 0,81 | 291,2 | 289,6 | 295,7 | 2,09 | 1,021 |
| 9,8 | 20,3 | 1,27 | 291,7 | 289,2 | 296,3 | 2,43 | 1,024 |
| 47,5 | 19 | 0,21 | 291,8 | 291,3 | 298,9 | 2,59 | 1,026 |
| 47,5 | 19 | 0,81 | 291,4 | 285,1 | 300,9 | 5,44 | 1,056 |
| 47,5 | 19 | 1,27 | 290,7 | 283,3 | 297,8 | 4,99 | 1,051 |
| 65,5 | 17 | 0,81 | 292,8 | 283,6 | 300,2 | 5,64 | 1,058 |
| 65,5 | 17 | 1,27 | 292,8 | 286,9 | 298,9 | 4,11 | 1,042 |
| 67,5 | 19 | 0,81 | 291,7 | 285,1 | 297,7 | 4,31 | 1,044 |
| 67,5 | 19 | 3,21 | 292,2 | 290,1 | 294,7 | 1,55 | 1,016 |

$H_2\,^{16}O/H_2\,^{18}O$ relative concentration

| mean T (°C) | $\Delta T$ (°C) | $k$ $10^{-10}$ m$^2$ | $n_0 \times (10^{-6})$ Initial | $n_T \times (10^{-6})$ Top | $n_B \times (10^{-6})$ Bottom | $\frac{\Delta n}{n_0}$ $10^{-2}$ | $\gamma = \frac{n_B}{n_T}$ |
|---|---|---|---|---|---|---|---|
| 9,8 | 20,5 | 0,81 | 1981,7 | 1968,6 | 2003,0 | 1,74 | 1,017 |
| 9,8 | 20,3 | 1,27 | 1982,1 | 1967,4 | 2006,8 | 1,99 | 1,020 |
| 47,5 | 19 | 0,21 | 1982,0 | 1969,5 | 2005,9 | 1,84 | 1,018 |
| 47,5 | 19 | 0,81 | 1981,8 | 1946,0 | 2024,9 | 3,98 | 1,041 |
| 47,5 | 19 | 1,27 | 1981,2 | 1943,5 | 2015,3 | 3,62 | 1,037 |
| 65,5 | 17 | 0,81 | 1982,9 | 1947,7 | 2016,8 | 3,48 | 1,035 |
| 65,5 | 17 | 1,27 | 1982,9 | 1954,3 | 2013,3 | 2,98 | 1,030 |
| 67,5 | 19 | 0,81 | 1981,6 | 1971,4 | 1992,1 | 1,04 | 1,011 |
| 67,5 | 19 | 3,21 | 1981,6 | 1959,7 | 2011,7 | 2,62 | 1,027 |

**Examples of ternary mixtures.** When there are more than two constituents into the column, it is no more possible to characterise thermodiffusion by a single parameter, the Soret coefficient of the solute. The mass balance inside the column results from the relative thermodiffusion mobility of each compound versus the others and it may happen that specific compounds, while affected by diffusion processes in the barycentric reference frame remain motionless relatively to the cell.

Experiments have been performed with an equimolar mixture of xylene isomers [49] (meta-, ortho-, and paraxylene). Figure 11 displays the profile at stationary state of the three components inside an horizontal gradient PTC. We observe three well differentiated profiles: (i) two antagonistic, quasi-binary profiles for ortho- and para-xylene, with respectively positive and negative apparent Soret coefficients, (ii) a flat profile for meta-xylene, with an apparent Soret coefficient close to zero. It is important to emphasise that when these three constituents are involved together into more complex mixtures, their relative

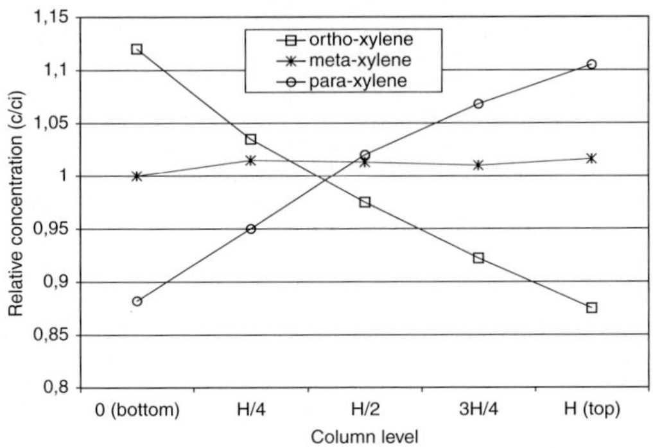


**Fig. 11.** Xylene isomers. Relative sensibility to thermodiffusion [49].

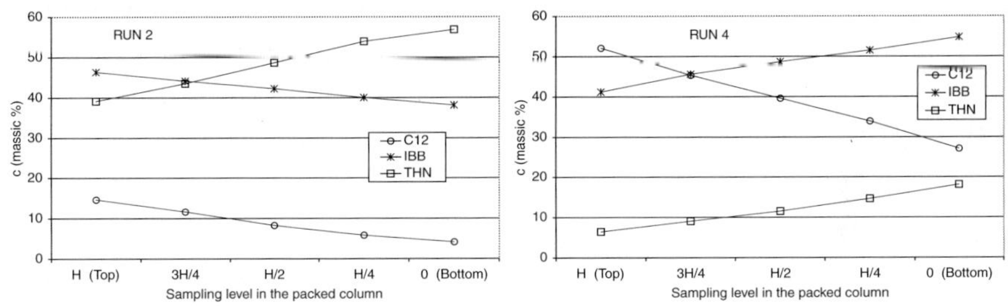


**Fig. 12.** Ternary mixtures. Stationary state mass fraction profiles in an horizontal gradient PTC for two different initial conditions and same packing permeability [50].

thermodiffusive mobility is conserved. So, experiments on crude oil (as reported below) have shown that all three isomers migrate towards the top of the cell (apparent negative Soret coefficients) but in the same order as in the ternary mixture experiment.

The behaviour of a ternary mixture of three hydrocarbons representative of a "synthetic model" of crude oil has also been investigated. The three hydrocarbons are: n-dodecan (C12), iso-butyl-benzene (IBB) and tetrahydronaphtalene (THN). The experimental conditions have been varied to check the effect of changing the components' proportion or the packing permeability.

Figures 12a and 12b show the experimental results for two different initial conditions. As it might have been expected, the relative thermodiffusive mobility is not conserved when changing the proportion in the mixture. In the displayed case, the profile of IBB is not stable when the proportions of THN and C12 are exchanged.

Also interesting to observe is the effect of changes in the packing permeability [50]. Should each constituent react as in the binary case, changing the permeability would decrease or increase the average slope of the mass fraction profiles. This is not so clear on figures 13a and 13b, but a change in the patterns of the curves and a displacement of the neutral point (where the composition of the

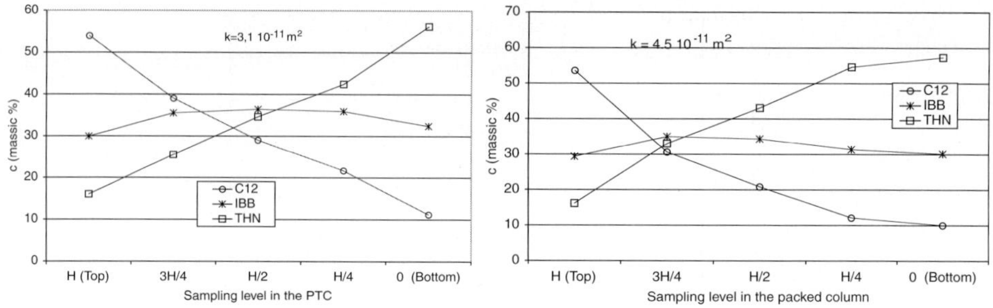


**Fig. 13.** Ternary mixtures. Stationary state mass fraction profiles in an horizontal gradient PTC for two different packing permeabilities and same initial conditions.

mixture is close to the initial one) to the upper part of the cell are associated with an increase in the permeability.

**Complex mixtures.** As will be seen in section 4, modelling binary PTC experiments to determine values of the Soret coefficient is not straightforward and still more difficult for ternary mixtures. One may expect to face an intractable problem when dealing with experiments involving complex mixtures.

There are however two main reasons why such experiments must complement more elementary investigations of binary and ternary systems. First, as the number of constituents present in the mixture becomes larger, a given compound may be considered as a solute while the others constitute a background solvent phase, so that increasing the complexity of the system may paradoxically decrease the phenomenological intricacy. Second, the objective of gathering a sufficiently large number of elementary experiments for the reconstitution of the thermodiffusive behaviour of complex mixtures seems out of reach considering the present phenomenological and experimental capacities. On the contrary, the global experimental and phenomenological study of the complex mixtures as such, even if it requires adaptations inside the physical model to better fit the macroscopic approach of the system, seems more reasonable to yield workable experimental facts.

Recent studies now take into account experiments on complex mixtures to improve thermodynamic models by including thermodiffusion processes. The goal is to achieve a better evaluation of *in situ* segregations in geological reservoirs [51,52,50,53–58]

*An experiment on crude oils.* This intuition gained ground from the experimental study of two natural oil sampled in the North Sea oil fields of Alwyn and Oseberg [27–30,51,45,59,49]. Figure 14 shows the composition of the crude oil presented by family of hydrocarbon constituants, and compare them to the transformed composition of the Alwyn oil by thermodiffusion-convection separation inside a PTC (see figure 3). The striking similarity between PTC-enriched Alwyn oil and the crude Oseberg oil suggests that thermodiffusion could induce oil segregation processes at reservoir scale.

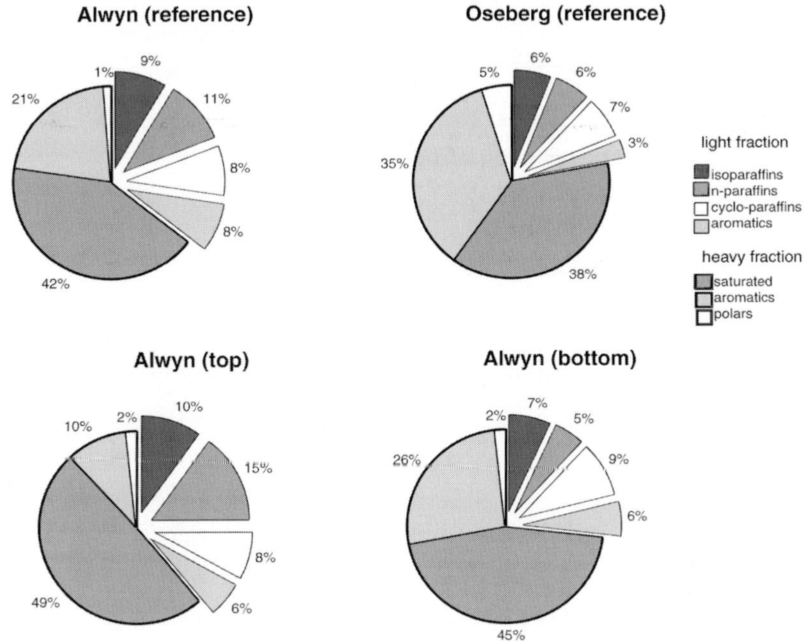


**Fig. 14.** Compositions of the Alwyn and Oseberg crude oils. Comparison with the PTC-transformed Alwyn oil composition [28].

*Impact of PTC separation on typical oil compounds families.* On this basis, a global survey was performed on several oils from North-Sea fields, using horizontal gradient PTCs to further investigate the role of thermodiffusion on the separation of well defined categories of hydrocarbons. It was first observed that the number of carbon atoms present in a given compound was a primary indicator of the sign of its apparent Soret coefficient in the mixture. Figure 15 shows that according to the C-scale, the compounds are divided into two families: compounds with less than 13 C-atoms (light fraction) generally migrate to the top of the PTC cell (negative apparent Soret coefficient), whereas compounds with more than 13 C-atoms (heavy fraction) migrate towards the bottom (positive apparent Soret coefficient). C-13 compounds appear to be indifferent, with an apparent Soret coefficient close to zero.

Inside each family the role of the structural type was investigated. Figures 16a and 16b show that there are important differences inside each fraction. Linear carbon chains n-paraffins present the largest mobility compared to the ramified chains iso-paraffins. Cyclic compounds (with one benzenic or naphtenic cycle) rather concentrate to the bottom or the centre of the PTC.

Therefore the molar mass appear to be determinant in the PTC separation for constituents of simple structure (e.g. linear or weakly branched) and of limited number of carbon atoms. When the structure of the molecule becomes more complex and the number of carbon atoms increases, molecular mass is less pre-

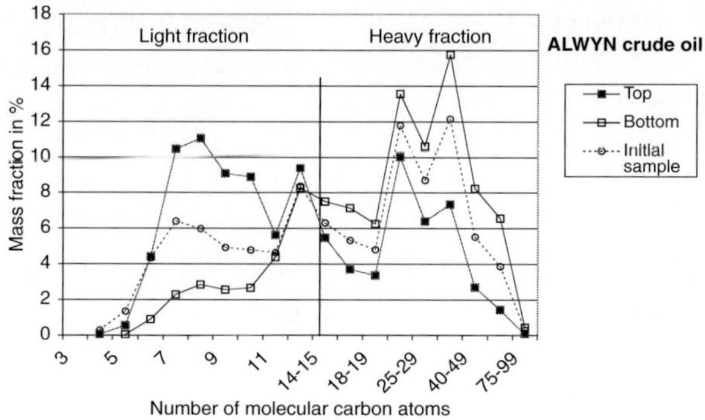


**Fig. 15.** Partition of hydrocarbons in the PTC as a function of the carbon atoms number in the molecule [28].

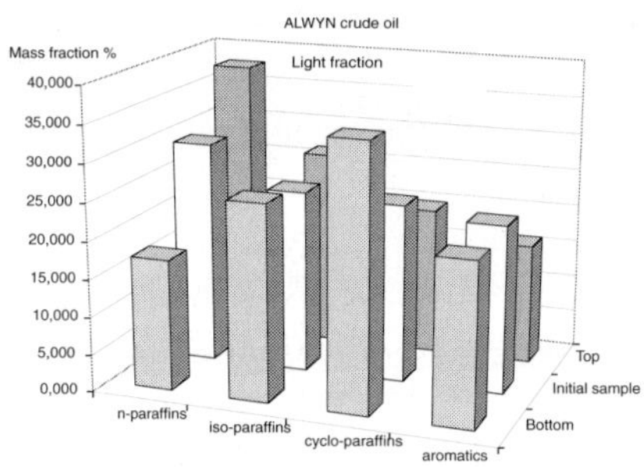


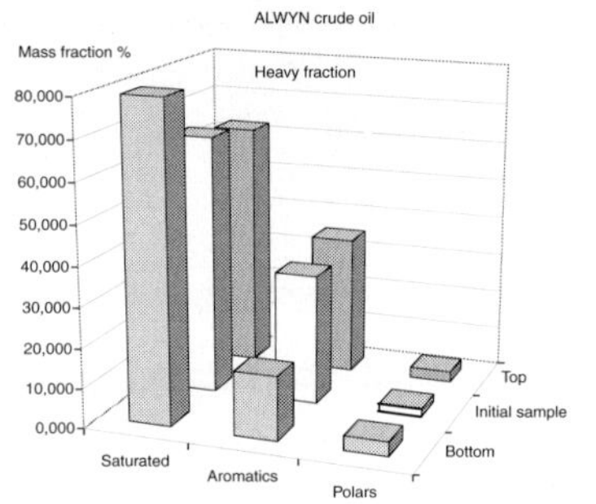


**Fig. 16.** PTC partition of structural families inside the crude oil's light and heavy fractions.

ponderant. In that case, the value of the thermodiffusion coefficient is rather determined by structural factors.

## 2.3 Remaining uncertainties in experiments

Sections 3 and 4 will expose how some critical aspects in the interpretation of the PTC experiments and the derivation of effective values for the Soret coefficient were overcome in the case of binary solutions subject to horizontal gradient separation. However, there remains some difficulties in understanding the phenomena governing thermodiffusion-convection separation.

A first category of uncertainties is linked to the PTC device itself. For example, the end reservoirs which are necessary to sample the solutions without perturbing the convection flux, are suspected to impact on the kinetics of separation as shown on figure 4. The curves show a separation time lag at the beginning and an inflexion point that are unpredictable by the current PTC theory.

A second category of difficulties is induced by the still incomplete phenomenology of PTC separation which makes it hazardous to transform the separation data into reliable Soret coefficients values. This is especially true for the vertical gradient PTC, for which there is still no satisfying description, even qualitative, of the observed segregation.

This explains why PTC experiments must still be supported and complemented by phenomenological advances and numerical simulations which we will now present in the two coming sections.

# 3 The equations of thermodiffusion in porous media

## 3.1 The scaling problem: different approaches in modelling porous media

In constrast with the case of free fluids, establishing the phenomenology of thermal diffusion and related transport phenomena in porous media, and checking it against experimental data is not completely achieved. As there is an abundant literature on porous media (mainly centered on mechanics see [60]), we will just give here some typical references, very far from being exhaustive.

As the system considered is heterogeneous, being made of (i) pores filled with fluid, (ii) a solid matrix, (iii) interfaces, the direct application of continous media equations is, in principle, no longer relevant. To overcome this phenomenological barrier, we have two basic options. First we may start from the smaller scale, where continuous equations are valid and upscale them by averaging methods which introduce the relevant macroscopic quantities [61–63]. But this leads to a closure problem, the number of new macroscopic quantities thus introduced being larger than the number of equations. It can be solved either by neglecting some correlations, or with the help of methods borrowed from macroscopic thermodynamics. For instance the so-called Representative Elementary Volume method, a very classical approach of fluid mechanics in porous media (see e.g. [64,63,65,66]) establishes the balance equations by averaging methods, but calls upon the methods of non equilibrium thermodynamics to derive the phenomenological equations needed to close the system.

As this microscale approach is not entirely self-sufficient, we may also start from the beginning at the larger scale, approximating the heterogeneous medium by an equivalent fictive continuous system, and postulate that methods valid for continuous media can still be used (see [67] for equilibrium, [68–70] for transfer laws). We'll use here this latter approach, which has proved, as will be shown below, to be a good compromise between complexity on the one hand, and numerical efficiency on the second hand.

This approach not only adds thermodiffusion in the set of macroscopic equations describing flow, heat and mass transport equations in porous media, but also shows that numerous other physical aspects are currently discarded. As thermal diffusion is a second-order transport process, other second-order phenomena should perhaps be considered (thermo-osmosis, ultrafiltration ...) in order to give a better account of the experiments, which are not up to now completely understood.

In addition to the problem of establishing valid transport equations, porous media poses a second problem: what does the value of thermodiffusion (or Soret) coefficients of compounds become in porous media? It is well-known in the case of the ordinary diffusion coefficient that some Darcy velocity-dependent corrections are required. This is the hydrodynamic dispersion theory [60,71]. We will show in the following that something similar probably occurs in the case of the thermodiffusion coefficient.

### 3.2 The non-equilibrium thermodynamics approach

In this approach, the porous medium is considered as made up of two interacting continuous media, the solid and the liquid solution, which both fill the *same* space domain $\mathcal{D}$. For the class of phenomena which we intend to investigate, it is sufficient to assume the solid at rest, that is with a constant specific mass and a zero velocity. For the sake of brevity, we'll only consider here a binary liquid.

**Mass conservation.** Proper specific masses, i.e. relative to solid or liquid are $\rho_{\mathrm{s}}, \rho_{\mathrm{f}}, \rho_1, \rho_2$, respectively for the solid, the fluid, and its two constituents. The mass fraction of the solute is $c_1 = \rho_1/\rho_{\mathrm{f}}$ etc.

Relative to the total volume (representative elementary volume), they are $\bar\rho_{\mathrm{s}}, \bar\rho_{\mathrm{f}}, \bar\rho_1, \bar\rho_2$. The liquid specific mass is $\bar\rho_{\mathrm{f}} = \bar\rho_1 + \bar\rho_2$, and the total specific mass $\bar\rho = \bar\rho_{\mathrm{s}} + \bar\rho_{\mathrm{f}}$. If the medium porosity (ratio of liquid volume to total volume) is $\varepsilon$ we have:

$$\bar\rho_{\mathrm{s}} = (1-\varepsilon)\rho_{\mathrm{s}} \ , \ \bar\rho_1 = \varepsilon\rho_1 \ , \ \bar\rho_2 = \varepsilon\rho_2 \ , \ \bar\rho_{\mathrm{f}} = \varepsilon\rho_{\mathrm{f}} \ , \ \bar\rho = (1-\varepsilon)\rho_{\mathrm{s}} + \varepsilon\rho_{\mathrm{f}}$$

Relative to the laboratory, the velocities are $\boldsymbol{w}_{\mathrm{s}}(=0), \boldsymbol{w}_1, \boldsymbol{w}_2$, such that the flow rate of constituent $\alpha$ through a surface element $\boldsymbol{n}\, dA$ ($\boldsymbol{n}$ outer normal, $dA$ area) of the porous medium is $\bar\rho_\alpha \boldsymbol{w}_\alpha.\boldsymbol{n}\, dA$. As usual, the barycentric velocity of the liquid is $\boldsymbol{w}_{\mathrm{f}}$ defined by

$$\bar\rho_{\mathrm{f}}\boldsymbol{w}_{\mathrm{f}} = \bar\rho_1\boldsymbol{w}_1 + \bar\rho_2\boldsymbol{w}_2 .$$

The so-called Darcy velocity, or filtration velocity is $\boldsymbol{U} = \varepsilon\boldsymbol{w}_{\mathrm{f}}$ such that the liquid flow rate through $dA$ is $\rho_{\mathrm{f}}\boldsymbol{w}_{\mathrm{f}}.\boldsymbol{n}\,\varepsilon dA = \rho_{\mathrm{f}}\boldsymbol{U}.\boldsymbol{n}\, dA$

In the absence of chemical reactions, the mass conservation equations read:

$$\bar\rho_{\mathrm{s}} = \text{const.} \ , \ \partial_{\mathrm{t}}\bar\rho_\alpha + \nabla.(\bar\rho_\alpha\boldsymbol{w}_\alpha) = 0 \ , \ \alpha = 1,2 \quad \Rightarrow \quad \partial_{\mathrm{t}}(\varepsilon\rho_{\mathrm{f}}) + \nabla.(\rho_{\mathrm{f}}\boldsymbol{U}) = 0 \quad (1)$$

If the (barycentric) diffusion flows in the liquid are $\boldsymbol{J}_{\mathrm{f}1}, \boldsymbol{J}_{\mathrm{f}2}$:

$$\boldsymbol{J}_{\mathrm{f}1} = \bar\rho_1(\boldsymbol{w}_1 - \boldsymbol{w}_{\mathrm{f}}) \ , \ \boldsymbol{J}_{\mathrm{f}2} = \bar\rho_2(\boldsymbol{w}_2 - \boldsymbol{w}_{\mathrm{f}})$$

the mass conservation equations may also be written ($c_\alpha = \bar\rho_\alpha/\bar\rho_{\mathrm{f}} = \rho_\alpha/\rho_{\mathrm{f}}$) :

$$\bar\rho_{\mathrm{f}}\frac{dc_\alpha}{dt} + \nabla.\boldsymbol{J}_{\mathrm{f}\alpha} = 0 \ \text{ with } \ \frac{d}{dt} = \frac{\partial}{\partial t} + \boldsymbol{w}_{\mathrm{f}}.\nabla \quad (2)$$

**Equations of dynamics.** The manner in which the laws of dynamics are written form the critical hypotheses in building a model of a porous medium, as they depart from the usual ones in irreversible thermodynamics of continuous media. Instead of writing a single equation for the global medium, one writes two equations, one for the solid, the other for the liquid, linked by a coupling force expressing their mutual interaction. This choice is adapted to the very different behaviour of the two media, in opposition to the similarity of the different solutes present in a solution.

In the simple case considered here, the forces are:

- $\boldsymbol{F}_\alpha$ density of force per unit of mass acting on constituent $\alpha$, for instance gravity $\boldsymbol{g}$.
- $\bar{\sigma}_\alpha^{\mathrm{ij}}, (\alpha = s, \ell)$ "apparent" stress tensor acting on solid and on liquid, such that the force applied on an element of total surface $\boldsymbol{n}\, dA$ is $\bar{\sigma}_\alpha . \boldsymbol{n}\, dA$. In the liquid it can be split in the usual way, pressure and viscosity:
$$\bar{\sigma}_{\mathrm{f}}^{\mathrm{ij}} = -\bar{p}\delta^{\mathrm{ij}} + \bar{\mathcal{V}}^{\mathrm{ij}}$$
The fluid pressure inside the pores is thus $p = \bar{p}/\varepsilon$ and the viscosity $\bar{\mathcal{V}}$ is an internal one.
- $\boldsymbol{\Phi}_\alpha, (\alpha = s, \ell)$ force per unit of total volume expressing the viscous interaction between liquid and solid along the pore walls. The action-reaction principle implies $\boldsymbol{\Phi}_{\mathrm{f}} + \boldsymbol{\Phi}_{\mathrm{s}} = 0$. Introducing such a specific quantity for an interfacial phenomenon is not unusual (cf for instance the specific area used in emulsions) and comes from the very idea of averaging over the representative elementary volume.

With these assumptions, the equations of dynamics read

$$0 = \bar{\rho}_{\mathrm{s}} \boldsymbol{F}_{\mathrm{s}} + \nabla . \bar{\sigma}_{\mathrm{s}} + \boldsymbol{\Phi}_{\mathrm{s}} \quad \text{(solid at rest)} \tag{3}$$
$$\bar{\rho}_{\mathrm{f}} \frac{d\boldsymbol{w}_{\mathrm{f}}}{dt} = \sum_{1,2} (\bar{\rho}_\alpha \boldsymbol{F}_\alpha) - \nabla \bar{p} + \nabla . \bar{\mathcal{V}} + \boldsymbol{\Phi}_{\mathrm{f}} \tag{4}$$

The equation for the solid is the same as for equilibrium, and will not be used in the following, in agreement with the simple case considered here.

**Principles of energy and entropy balances.** As we have already said, our aim is to justify the set of equations governing the thermodiffusion phenomena occuring in the experiments presented in the previous section. As in every mass transfer problem, the equations of dynamics are insufficient for this purpose, and we need the first principle to determine the temperature field, and the phenomenological laws governing diffusion processes.

Following the general method of irreversible thermodynamics, we will first write the first law, then the second law, which will give us the expression for the entropy production. In this latter expression we will identify the relevant fluxes and thermodynamics forces, and write the phenomenological laws coupling these fluxes and forces. Although there are some differences, the calculations are analogous to those given by Benet and Jouanna [68].

The medium will be described by the following state variables, depending on space and time: the specific masses defined above and the unique temperature $T$ (the same for solid and fluid). As the motion of fluid and solid are so different, the idea of total barycentric velocity, which forms the basis of the extension of thermodynamics to open systems may be questioned. Although this direct approach leads also to interesting results [70], and allows the use of standard free energy methods [5], we will avoid it here. But keeping two equations of dynamics instead of one implies an extension of the two principles, as the porous medium is essentially an open system. we will use the two following postulates:

- The internal energy and entropy of an arbitrary domain $\mathcal{D}$ of the porous medium are respectively the sum of the corresponding quantities in the solid and the fluid

$$U = \int_{\mathcal{D}} \bar{\rho}_{\mathrm{s}} u_{\mathrm{s}} + \bar{\rho}_{\mathrm{f}} u_{\mathrm{f}} \, d\varpi \quad , \quad S = \int_{\mathcal{D}} \bar{\rho}_{\mathrm{s}} s_{\mathrm{s}} + \bar{\rho}_{\mathrm{f}} s_{\mathrm{f}} \, d\varpi$$

  In these expressions, the specific state functions $u$, $s$ have the same properties as those of the separate media. This is clearly an approximation, neglecting interfacial phenomena.
- The first law reads for an arbitrary domain $\mathcal{D}$

$$\frac{\partial}{\partial t} \int_{\mathcal{D}} \bar{\rho}_{\mathrm{s}}(u_{\mathrm{s}} + \frac{\boldsymbol{w}_{\mathrm{s}}^2}{2}) + \bar{\rho}_{\mathrm{f}}(u_{\mathrm{f}} + \frac{\boldsymbol{w}_{\mathrm{f}}^2}{2}) \, d\varpi = \\ - \int_{\partial\mathcal{D}} \left[\bar{\rho}_{\mathrm{s}}(u_{\mathrm{s}} + \frac{\boldsymbol{w}_{\mathrm{s}}^2}{2})\boldsymbol{w}_{\mathrm{s}}.\boldsymbol{n} + \bar{\rho}_{\mathrm{f}}(u_{\mathrm{f}} + \frac{\boldsymbol{w}_{\mathrm{f}}^2}{2})\boldsymbol{w}_{\mathrm{f}}.\boldsymbol{n}\right] dA + \dot{W}_{\mathrm{ext}} + \dot{Q}_{\mathrm{ext}}$$

  Although unusual, this seems in agreement with the views expressed by Prigogine ([3] p. 90). $\dot{W}_{\mathrm{ext}}$ and $\dot{Q}_{\mathrm{ext}}$ are respectively the work and the heat supplied per unit of time by external bodies.

**First law.** Starting from the above principles, a simple calculation leads to the equation describing locally the conservation of energy in our case:

$$\bar{\rho}_{\mathrm{s}} \frac{\partial u_{\mathrm{s}}}{\partial t} + \bar{\rho}_{\mathrm{f}} \frac{du_{\mathrm{f}}}{dt} = \sum_{1,2} \boldsymbol{F}_{\alpha}.\boldsymbol{J}_{\mathrm{f}\alpha} - \bar{p}\nabla.\boldsymbol{w}_{\mathrm{f}} + \bar{\mathcal{V}} : \nabla\boldsymbol{w}_{\mathrm{f}} - \boldsymbol{\Phi}_{\mathrm{f}}.\boldsymbol{w}_{\mathrm{f}} - \nabla.\psi \tag{5}$$

In this equation, we have introduced a (global) heat flow vector $\psi$ assuming that heat transfer is due to conduction only, and we have used the equations of dynamics to remove the kinetic energy term.

**Second law.** As we have assumed that specific state functions have the same properties as usual, Gibbs identity is still valid for the two media. We can then calculate the change in entropy of an arbitrary fixed domain $\mathcal{D}$ using the above expression of the first law and following the standard steps (cf [4]):

$$\begin{aligned}\frac{\partial}{\partial t}\int_{\mathcal{D}}(\bar{\rho}_{\mathrm{s}}s_{\mathrm{s}}+\bar{\rho}_{\mathrm{f}}s_{\mathrm{f}})\,d\varpi &= -\int_{\partial\mathcal{D}}\Big\{\bar{\rho}_{\mathrm{f}}s_{\mathrm{f}}\boldsymbol{w}_{\mathrm{f}}+\frac{\boldsymbol{\psi}-\sum\mu_\alpha\boldsymbol{J}_\alpha}{T}\Big\}.\boldsymbol{n}\,dA \\ &+\int_{\mathcal{D}}\Big\{\boldsymbol{\psi}.\nabla\big(\frac{1}{T}\big)+\frac{\bar{\mathcal{V}}:\nabla\boldsymbol{w}_{\mathrm{f}}}{T}-\frac{\boldsymbol{\Phi}_{\mathrm{f}}}{T}.\boldsymbol{w}_{\mathrm{f}}+\sum_{1,2}\big[\frac{\boldsymbol{F}_\alpha}{T}-\nabla\big(\frac{\mu_\alpha}{T}\big)\big]\Big\}\,d\varpi\end{aligned} \tag{6}$$

In agreement with the general expression of the second law, the first line represents the entropy supplied to the domain $\mathcal{D}$ by its surroundings and the second the entropy production due to irreversible processes.

If we group the terms involving $\nabla T$, the entropy production (per unit of volume) reads

$$\begin{aligned}\sigma = &\big[\boldsymbol{\psi}-\boldsymbol{J}_{\mathrm{f1}}(h_1-h_2)\big].\nabla\big(\frac{1}{T}\big)+\boldsymbol{J}_{\mathrm{f1}}.\big(\frac{\boldsymbol{F}_1-\boldsymbol{F}_2-\nabla_{\mathrm{T}}(\mu_1-\mu_2)}{T}\big) \\ &-\frac{\boldsymbol{\Phi}_{\mathrm{f}}}{T}.\boldsymbol{w}_{\mathrm{f}}+\frac{\bar{\mathcal{V}}:\nabla\boldsymbol{w}_{\mathrm{f}}}{T}\end{aligned} \tag{7}$$

In this expression, $\nabla_{\mathrm{T}}$ is the constant temperature gradient; chemical potentials $\mu_\alpha$ and partial specific enthalpies $h_\alpha$ in the liquid have their usual values.

The above entropy production is the sum of 4 terms corresponding to 4 types of irreversibility: heat transfer, mass diffusion, Darcy flow, bulk viscosity. Each term is the product of two quantities, a flux (the first one) and a thermodynamic force (the second one).

### 3.3 Heat and mass transfer equations

The equations describing the time evolution of continuous systems include not only the balance equations given above, but also the laws governing transfer phenomena such as heat conduction, diffusion and so on. Non equilibrium thermodynamics states that they may be deduced from the general principle that the fluxes appearing in the entropy production are functions of all the thermodynamic forces, vanishing with these forces.

**Linear phenomenological laws.** In order to avoid unnecessary complications, we will make two further assumptions: the only force per unit mass is gravity, i.e. $\boldsymbol{F}_\alpha=\boldsymbol{g}$ and bulk viscosity is negligible, i.e. $\bar{\mathcal{V}}=0$ (valid for porosity not too high as in the above experiments). We also assume that inertial effects are negligible (the so-called hypothesis of mechanical equilibrium [3,73]), meaning that $\boldsymbol{\Phi}_{\mathrm{f}}$ can be deduced from the equation of motion, i.e. $\boldsymbol{\Phi}_{\mathrm{f}}=\nabla\bar{p}-\bar{\rho}_{\mathrm{f}}\boldsymbol{g}$. The entropy production becomes

$$\sigma=\big[\boldsymbol{\psi}-\boldsymbol{J}_{\mathrm{f1}}(h_1-h_2)\big].\nabla\big(\frac{1}{T}\big)-\boldsymbol{J}_{\mathrm{f1}}.\big(\frac{\nabla_{\mathrm{T}}(\mu_1-\mu_2)}{T}\big)-\frac{\nabla\bar{p}-\bar{\rho}_{\mathrm{f}}\boldsymbol{g}}{T}.\boldsymbol{w}_{\mathrm{f}} \tag{8}$$

Other equivalent expressions may also be used, but lead *in fine* to equivalent results.

Transfer phenomena can thus be expressed by the linear phenomenological laws (valid for an isotropic medium)

$$\begin{aligned} \boldsymbol{J}_{\mathrm{f1}} = \bar{\rho}_1(\boldsymbol{w}_1 - \boldsymbol{w}_{\mathrm{f}}) &= -L_{1\mathrm{q}}\nabla T - L_{11}\big(\nabla_{\mathrm{T}}(\mu_1 - \mu_2)\big) - L_{1\mathrm{w}}(\nabla\bar{p} - \bar{\rho}_{\mathrm{f}}\boldsymbol{g}) \\ \boldsymbol{w}_{\mathrm{f}} &= -L_{\mathrm{wq}}\nabla T - L_{\mathrm{w1}}\big(\nabla_{\mathrm{T}}(\mu_1 - \mu_2)\big) - L_{\mathrm{ww}}(\nabla\bar{p} - \bar{\rho}_{\mathrm{f}}\boldsymbol{g}) \\ \boldsymbol{J}_{\mathrm{q}} = \boldsymbol{\psi} - \boldsymbol{J}_{\mathrm{f1}}(h_1 - h_2) &= -L_{\mathrm{qq}}\nabla T - L_{\mathrm{q1}}\big(\nabla_{\mathrm{T}}(\mu_1 - \mu_2)\big) - L_{\mathrm{qw}}(\nabla\bar{p} - \bar{\rho}_{\mathrm{f}}\boldsymbol{g}) \end{aligned} \tag{9}$$

where the coefficients $L_{\mathrm{ij}}$ are functions of the state variables.

The first equation includes thermodiffusion, Fickian diffusion, and chemical osmosis, the second thermoosmosis, filtration phenomena and Darcy law, the last one concerns heat transfer, generalized Fourier's law, and Dufour effect. The main terms are the "diagonal" ones, coupling a flux with its associated force in the entropy production; they correspond to Fick, Darcy and Fourier laws.

Many of these phenomena seem, from the few known experiments, unimportant, and equations (9) will be further simplified in the effective set of equations used in the practical model described below. But it is not impossible that they may play some part in natural phenomena, or even in laboratory experiments.

**Dispersion phenomena and non linear laws.** As we have already said, modelling the experiments of thermodiffusion in porous media through the known phenomenological laws leads to severe difficulties. It is well-known that linear relations are insufficient to give full account of diffusion phenomena in porous media, and the "hydrodynamic dispersion theory" has been devised to address this problem. In porous media, the diffusion term depends on the flow velocity through the expression $-\bar{\rho}_{\mathrm{f}}(D + \alpha_{\mathrm{d}} w_{\mathrm{f}})\nabla c_1$. Unfortunately, this is insufficient to remove the discrepancies we have observed.

We have thus suggested to extend this same idea to thermodiffusion and introduce a term which could be called "dispersive thermodiffusion" [74]. Following a macroscopic approach already successful in dispersion theory [60,71] and assuming the medium to be isotropic, the generalized law of diffusion becomes (taking into account the fact that Darcy law allows the interchange of $\boldsymbol{\Phi}_{\mathrm{f}}$ and $\boldsymbol{w}_{\mathrm{f}}$)

$$\begin{aligned} \boldsymbol{J}_{\mathrm{f1}} = &-L_{11}\nabla_{\mathrm{T}}(\mu_1 - \mu_2) - L_{1\mathrm{q}}\nabla T - L_{1\mathrm{w}}\boldsymbol{w}_{\mathrm{f}} \\ &-L^{(1)}_{11\mathrm{ww}}\nabla_{\mathrm{T}}(\mu_1 - \mu_2)\|\boldsymbol{w}_{\mathrm{f}}\|^2 + L^{(2)}_{11\mathrm{ww}}\boldsymbol{w}_{\mathrm{f}}\big(\boldsymbol{w}_{\mathrm{f}}.\nabla_{\mathrm{T}}(\mu_1 - \mu_2)\big) \\ &+L^{(1)}_{1\mathrm{qww}}\nabla T\|\boldsymbol{w}_{\mathrm{f}}\|^2 + L^{(2)}_{1\mathrm{qww}}\boldsymbol{w}_{\mathrm{f}}(\boldsymbol{w}_{\mathrm{f}}.\nabla T) \quad . \end{aligned} \tag{10}$$

In the experiments $\boldsymbol{w}_{\mathrm{f}}$ is mainly orthogonal to $\nabla T$, and it seems sufficient to keep only the term $L^{(1)}_{1\mathrm{qww}}\nabla T\|\boldsymbol{w}_{\mathrm{f}}\|^2$.

### 3.4 Practical set of equations

If we consider only terms of significant value, equations (9) and (10) can be rewritten in the form

$$\begin{aligned} \boldsymbol{J}_{\mathrm{f1}} &= -\varepsilon\rho_{\mathrm{f}}(D^* + \alpha_{\mathrm{d}}\|\boldsymbol{U}\|)\nabla c_1 - \varepsilon\rho_{\mathrm{f}}(D_{\mathrm{T}}^* + \alpha'_{\mathrm{d}}\|\boldsymbol{U}\|)c_1(1-c_1)\nabla T \\ \boldsymbol{U} &= -\frac{k}{\eta}(\nabla p - \rho_{\mathrm{f}}\boldsymbol{g}) \quad \text{and} \quad \boldsymbol{\psi} = -\lambda_{\mathrm{T}}\nabla T \end{aligned} \tag{11}$$

In the first equation the diffusion and thermodiffusion coefficients are the effective ones, valid in porous media and often written as $D^* = D/\tau^2$ where $\tau$ is called the tortuosity of the medium.

Combining equations (11) with the two mass conservation equations (1), (2) and the first law (5), we obtain a set of 12 equations for the 12 unknowns $\rho_{\mathrm{f}}, c_1, T, \boldsymbol{U}, \boldsymbol{w}_1, \boldsymbol{\psi}$, as $p, u_{\mathrm{s}}, u_{\mathrm{f}}$ are known state functions.

The next step consists in making the so-called Boussinesq approximation, stating that the fluid is quasi-incompressible, i.e. $\rho_{\mathrm{f}}$ = const. except in the equation of dynamics, which is here Darcy's law. This leads to the following practical set

$$\begin{aligned} &\nabla.\boldsymbol{U} = 0 \quad , \quad \boldsymbol{U} = -\frac{k}{\eta}(\nabla p - \rho_{\mathrm{f}}\boldsymbol{g}) \\ &\varepsilon\frac{\partial c_1}{\partial t} + \boldsymbol{U}.\nabla c_1 - \varepsilon\nabla.(D^* + \alpha_{\mathrm{d}}\|\boldsymbol{U}\|)\nabla c_1 \\ &\qquad - \varepsilon\nabla.(D_{\mathrm{T}}^* + \alpha'_{\mathrm{d}}\|\boldsymbol{U}\|)c_1(1-c_1)\nabla T = 0 \\ &\left[(1-\varepsilon)\rho_{\mathrm{s}}c_{\mathrm{s}} + \varepsilon\rho_{\mathrm{f}}c_{\mathrm{f}}\right]\frac{\partial T}{\partial t} + \rho_{\mathrm{f}}c_{\mathrm{f}}\boldsymbol{U}.\nabla T = \lambda_{\mathrm{T}}\nabla^2 T \end{aligned} \tag{12}$$

which will be used in the next section.

## 4 Quantitative studies

This section will be devoted to the presentation of various analytical and numerical methods for integrating the equations of mass and energy transport in porous media in the particular context of thermodiffusive systems. The major interest of this quantitative approach is to derive reliable values of the effective Soret (or thermodiffusion) coefficient in porous media from laboratory PTC experiments. However, the major difficulty is that the transport coefficients are not the only unknowns. There also exist several possible physical descriptions that have to be confronted to the experiments.

### 4.1 Analytical derivations

A first category of quantitative models is represented by analytical solutions. The equations of thermodiffusive transport in porous media can be solved for simple cases such as the PTC experiment and the Bénard configuration of hydrodynamic instability.

**Separation of binary mixtures in horizontal gradient PTC: the Lorenz and Emery solution.** The Lorenz and Emery solution [13] is an extension of the Furry, Jones and Onsager theory of thermodiffusion-convection in free columns. The PTC is described in a two-dimensional fixed referential $(\hat{x}, \hat{z})$ colinear with the dimensions $(a, L)$ of the column. Like these authors, Lorenz and Emery decouple the fluid flux and mass transport problem and assume (i) that the thermal gradient is not perturbed, (ii) that $L \gg a$. They combine a second order, one-dimensional equation of dynamics, which reads:

$$\frac{\partial^2 U(x)}{\partial x^2} - \frac{\varepsilon}{k} U(x) + \frac{\rho g \alpha \Delta T}{a \eta} c = 0$$

with an approximate expression of the vertical mass flux:

$$m_{\mathrm{z}} = -\rho a D^{\star} \frac{\partial c}{\partial z} - \rho \int_0^a G(x) \frac{\partial c}{\partial x} dx \text{ where: } G(x) = \int_0^x U(x) dx$$

After some additional transformations, they obtain the approached expression of $m_{\mathrm{z}}$:

$$m_{\mathrm{z}} = -Hc - K \frac{\partial c}{\partial z}$$

where $H$ and $K$ are two transport constants equal to:

$$H = -\rho s_{\mathrm{T}} \frac{\Delta T}{a} \int_0^a G(x) dx \text{ and } K = \frac{\rho}{D^{\star}} \int_0^a G^2(x) dx + \rho D^{\star} a$$

The stationary state equation is obtained by equating $m_{\mathrm{z}}$ to zero. The concentration factor of the PTC can be approximated by:

$$\ln \gamma = \frac{H}{K} L = \frac{\gamma_1 k}{\gamma_2 + \gamma_3 k^2} L$$

which means that the $\ln \gamma = f(k)$ curve presents a maximum $(k_M, \ln \gamma_M) = \left(\sqrt{\frac{\gamma_2}{\gamma_3}}, \frac{1}{2}\gamma_1 \sqrt{\frac{1}{\gamma_2 \gamma_3}} L\right)$, and is symmetrical on a logarithmic scale in $k$ around $\ln k = \ln k_M$. The expressions given by Lorenz and Emery for the maximum values are:

$$k_M = \frac{\sqrt{120} \varepsilon D^{\star} \eta}{\rho a g \alpha \Delta T} \text{ and } \ln \gamma_M = \frac{s_{\mathrm{T}} \Delta T L \sqrt{30}}{12a}$$

In principle, the Lorenz and Emery solution enables the determination of the solute's Soret coefficient by plotting the $\gamma = f(\ln k)$ curve and determining its maximum value at stationary state. As shown in figure 5, experiments qualitatively confirm Lorenz and Emery's predictions. However, it was early concluded [42] that the analytical curve was located about one order of magnitude lower than the experimental curve on the $\log k$ scale (an example is shown on figure 17). Though the above equations suggest that the value of $\gamma$ at the maximum suffices to determine $s_{\mathrm{T}}^*$, this systematic discrepancy raised serious doubt about the validity of this analytical solution [75]. On the other hand, FJO theory for free media is generally considered to agree with experimental data.

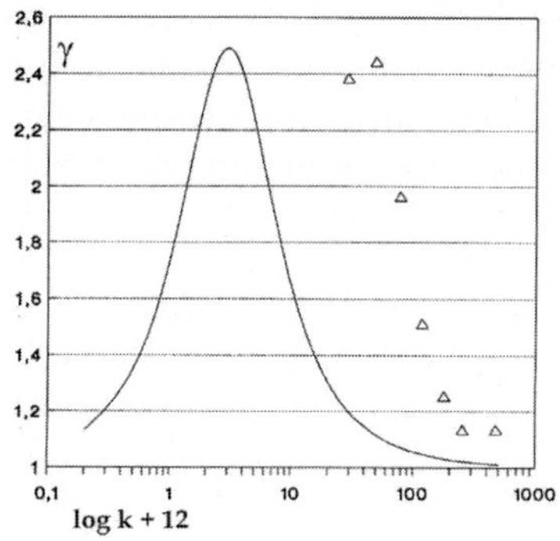


**Fig. 17.** Typical comparison of the Lorenz and Emery solution to experimental data obtained with horizontal gradient PTCs [45].

**Separation of ternary mixtures in horizontal gradient PTC.** The approach of Lorenz and Emery can be extended to ternary mixtures without major difficulties ([76,50]). The following expressions are obtained for the mass fractions $c_1$ and $c_2$:

$$\frac{\partial c_1}{\partial z} = I\frac{A_1 + D_T^1 J}{B + CJ + J^2} \text{ and } \frac{\partial c_2}{\partial z} = I\frac{A_2 + D_T^2 J}{B + CJ + J^2}$$

where:

$$\begin{aligned} A_1 &= D_T^1(D_{22}^2 + D_{12}D_{21}) - D_T^2(D_{11}D_{12} + D_{12}D_{22}) \\ A_2 &= D_T^2(D_{11}^2 + D_{12}D_{21}) - D_T^1(D_{22}D_{21} + D_{21}D_{11}) \\ B &= \left(D_{11}D_{22} - D_{12}D_{21})\right)^2 \\ C &= D_{11}^2 + D_{22}^2 - 2D_{12}D_{21} \\ I &= \tfrac{\Delta T}{a}\int_0^a G(x)dx = \tfrac{1}{12}\varepsilon\Delta T^2 k \\ J &= \tfrac{1}{a}\int_0^a G^2(x)dx = \tfrac{1}{120}(\varepsilon\Delta T a k)^2 \end{aligned}$$

As in the binary mixture case, the separation of each constituent, $\gamma_i$, $i = 1, 2$ can be expressed by a simple expression:

$$\gamma_i = \frac{k(\gamma_1 + \gamma_2 a^2 k^2)}{\gamma_3 + \gamma_4 a^2 k^2 + a^4 k^4}$$

this solution is qualitatively in agreement with the experimental data about ternary mixtures (see section 2).

**Separation in vertical gradient PTC.** The similarity between horizontal gradient PTC and vertical gradient PTC separation have suggested to some authors [6] the existence of an analogous analytical solution. It was empirically proposed to replace the dimensions of the horizontal gradient PTC inside the Lorenz and Emery solution by the dimensions of the vertical gradient PTC. However, though they appear to agree qualitatively with the experiments, these transformations are not founded on a phenomenological point of view.

**The marginal stability problem in thermodiffusive systems.** When dealing with the horizontal gradient PTC, the solution of the thermal problem is never mentioned. There are two reasons: (i) the horizontal PTC system is always unstable, (ii) the thermal field is assumed to be purely conductive, leading to a constant temperature gradient. This is not the case for the vertical PTC and, more generally in media subject to vertical thermal gradients (e.g. geothermal configuration). In that case, temperature and solute mass fraction fields compete to determine the fluid instability, and the sign and magnitude of the Soret coefficient is decisive. This competition is entirely different from the so-called thermosolutal case, where temperature and mass fraction are independent. The originality of the Soret-driven instability is that both fields are strongly and explicitly coupled.

Several authors [32–34] have shown that thermodiffusion has a considerable impact on the determination of the fluid stability conditions. An indicator to measure this impact is the value of the critical Rayleigh number.

A similar evaluation can be performed in porous media, for example in the typical situation of figure 18 which corresponds to a semi-infinite horizontal, isotropic porous layer of thickness $d$, filled with a binary mixture and enclosed between two rigid, impervious, non-diffusive, and isothermal boundaries.

The stability problem can be solved easily by a classical linear stability analysis method. Many authors [77,78] obtained a simple, explicit solution by expanding the perturbations according to a base of circular functions. They obtain the following expression for the critical Rayleigh number (the * superscript refers to parameters evaluated in the porous medium):

$$\mathbf{Ra}_{\mathrm{T}}^{\star} = \frac{\mathbf{Ra}^{\star}}{1 - \mathcal{S}^{\star}\,(\mathbf{Le}^{\star}\,+1)} \quad \text{where } \mathcal{S}^{\star} = s_{\mathrm{T}}^{*} c_0 \frac{\alpha\prime}{\alpha}$$

where $\mathbf{Ra}^{\star}$ is the critical Rayleigh number for a pure fluid ($4\pi^2$ for the configuration of figure 18), and $\mathbf{Le}^{\star}$ is the Lewis number.

However, this solution is only approximate, because the circular expansion does not exactly obey the boundary conditions of figure 18. A second, more relevant solution is obtained with a base of hyperbolic components [79], leading to the following equation:

$$\begin{aligned}\det(\mathbf{S}) = M_2 \mathbf{Le}^{\star}\,\mathcal{S}^{\star}\,\sin\tfrac{M_2}{2} - M_1 \mathbf{Le}^{\star}\,\mathcal{S}^{\star}\,\cos\tfrac{M_2}{2}\tanh\tfrac{M_1}{2} \\ +2b(1-\mathcal{S}^{\star}\,)\tanh\tfrac{b}{2}\cos\tfrac{M_2}{2} = 0\end{aligned}$$

where:

$$\begin{aligned}M_1 &= \sqrt{b}\sqrt{b+\sqrt{\mathbf{Ra}^{\star\prime}}} \quad \text{where } \mathbf{Ra}^{\star\prime} = (1-\mathcal{S}^{\star}\,)\mathbf{Ra}^{\star}\,+\mathbf{Le}^{\star}\,\mathbf{S}^{\star} \\ M_2 &= \sqrt{b}\sqrt{\sqrt{\mathbf{Ra}^{\star\prime}}\,-b}\end{aligned}$$

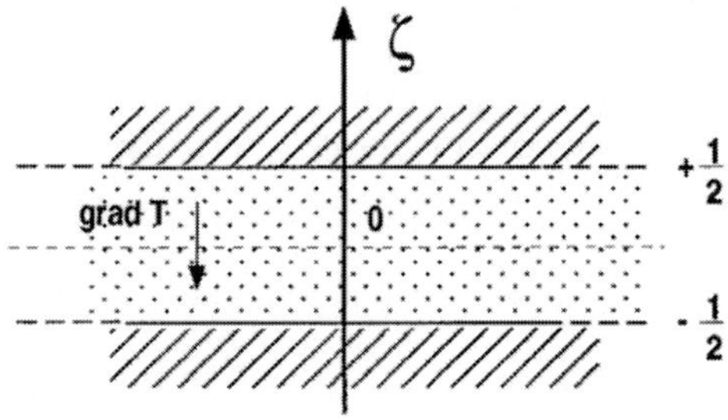


**Fig. 18.** Typical configuration of fluid instability in porous media.

and $h$ is the perturbation's wave number.

The two above analytical solutions are displayed on figure 23 for a typical case-study. We see that they are separated by a factor two. However, they give the same qualitative answer to the problem: even with a small Soret coefficient, the impact of thermodiffusion flux on lowering the instability frontier is very important.

### 4.2 Operational equations and numerical models

Except in simple configurations, the set of transport equations in porous media cannot be solved by analytical methods, and the use of numerical solutions is required. Another reason for using models is also that they represent a simple way to test different phenomenological scenarios. Finally, numerical models require fewer assumptions, for example in describing the dependency of transport and dynamic parameters on state variables.

**Equations and governing parameters.** In section 3 we have presented the operational equations that are generally considered to describe thermodiffusive transport in porous media following a macroscopic approach, at least for the linear part of them. In that case, fitting the models is achieved through the adjustment of the governing physical parameters. Some of those are known explicitely, like $\rho$ and $\eta$, but the major problem is to find appropriate values for the effective transport parameters $D^*_{\mathrm{th}}$, $D^*$ and $D^*_{\mathrm{T}}$. This question is made even more complex by dispersion non-linearities, as described in section 3.

The macroscopic approach can be applied using different descriptions of the fluid phase. Some authors consider one equation of dynamics per constituent instead of considering one unique flow equation (Darcy law) and diffusion equations [80].

Finally, the porous medium can also be described as a discrete medium in which pure fluid phase transport equations are used [61,81,82].

**Semi-analytical models.** Numerical modelling generally involves the discretization of the domain (e.g. finite differences, finite elements/volumes). For simple geometries however, one may use spectral expansions of the transport equations which are an efficient way of solving partial derivatives systems in simple geometries. Such a method has been used to solve the thermodiffusion-convection problem in an annular PTC [83]. Its result is consistent with the classical analytical solution of the problem. However, no comparison with experimental data is provided.

**Stochastic approach of thermodiffusion-convection in PTC.** The porous medium can be approached by the generation of random obstacles inside a continuous fluid medium. In that case, the governing equations are taken from fluid mechanics and the role of the porous media is to induce additional, complex boundary conditions.

This method has been formalised by many authors (e.g. [61]) and numerical applications have been provided. A finite volume code coupled to a Monte-Carlo generation of random obstacles is used by Chavepeyer and Platten [82] to describe the horizontal gradient PTC separation. 10 realisations of the porous medium suffice to achieve stable transport properties. The comparison with the analytical solution is good and leads to the conclusion that introducing obstacles does not change thermodiffusion characteristics. However, one may stress the fact that until now, the density of obstacles taken into account is far from representing a real porous medium.

**Equivalent continuous medium approach: the METSOR model.** As the equivalent continuous medium description of the porous medium is consistent with irreversible thermodynamics, this representation was used throughout the quantitative studies performed by the authors since 1987. One might have raised some doubts about the relevance of this approach for porous domains of very small extensions such as the ones that are used in the PTCs. However, as will be shown, this choice is justified *a posteriori* by the very good agreement with experiments.

We implemented a finite model of flow, heat and solute transport in porous media (METSOR code [75]) that was adapted from an existing transport code by introducing non-linear, dispersive diffusion and thermodiffusion. This model is able to cope with complex geometries, although we have used it mainly on very simple configurations like the PTCs, while simpler models would have been technically sufficient. Our intention is indeed to apply a validated code on natural systems which require generalizable numerical methods. Moreover, introducing explicit dependencies of the transport parameters on the state variables is trivial in such a code.

As has been already mentioned, the linear Lorenz and Emery solution presents a major discrepancy with experiments, which is not explained by further refining the theory [81].

## 4.3 Numerical modelling of the horizontal gradient PTC

Fitting the METSOR model on complete datasets described in §2.2, we evidenced ([52,74]) non-linear, dispersion-like behaviour of the effective thermodiffusion coefficient in porous media. Figure 19 shows the value of $D_{\mathrm{T}}^*$ increasing with the Péclet number in the cell, for a same $CCl_4$-$C_2HCl_3$ mixture. On figure 20 an example of the improvement brought by the non-linear extension of thermodiffusive transport is displayed. Under this assumption, the model is able to fit closely the data, while the Lorenz and Emery solution differs from one order of magnitude in $k$ from the experimental curve.

The METSOR code was also used for more detailed phenomenological studies in the transient separation curve (figure 21) and the dependency of the thermodiffusion coefficient on the initial mass fraction (figure 22). With this model, we

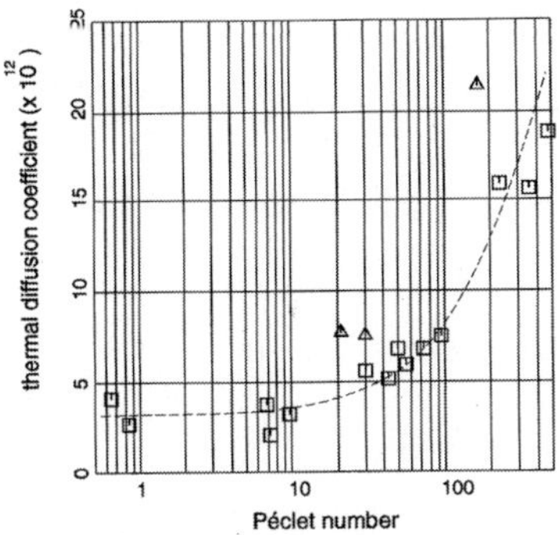


**Fig. 19.** An example of the dependency of the thermodiffusion coefficient on the flow velocity in the horizontal gradient PTC. Experimental data on $CCl_4$-$C_2HCl_3$ mixture [14,47,15,17,16,52].

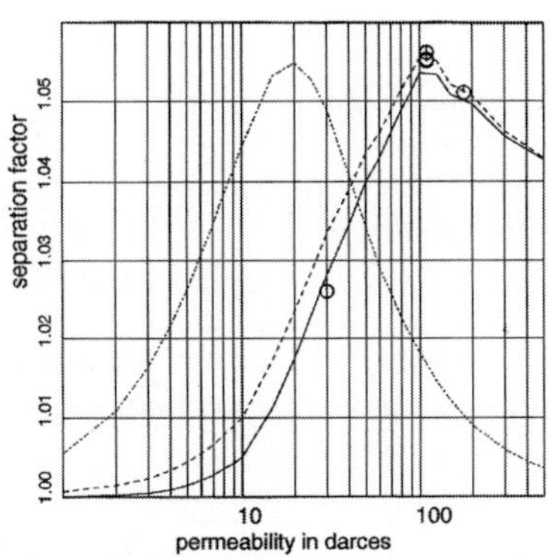


**Fig. 20.** Comparison of the linear and non-linear approaches of thermodiffusion transport. Circles: experimental data on the $HDO/H_2O$ mixture [42]; dotted curve: linear model or Lorenz and Emery analytical solution; solid curve: purely dispersive thermodiffusion coefficient; dashed curve: combination of a constant Soret coefficient of $10^{-5}\mathrm{K}^{-1}$ and of a non linear dispersive component [74]. (1 darce= $10^{-12}\mathrm{m}^2$)

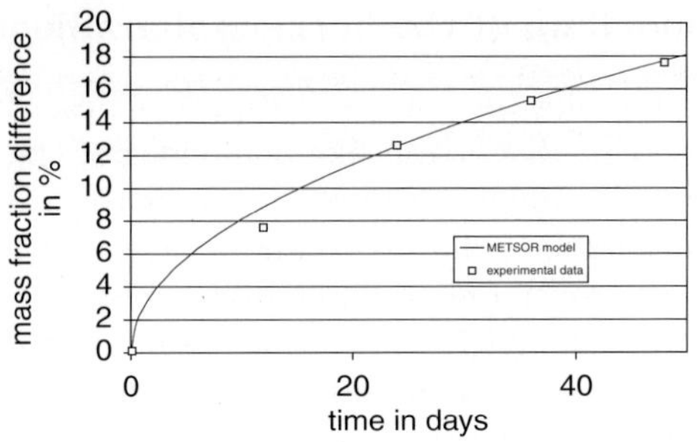


**Fig. 21.** Example of fitting the METSOR model on experimental transient separation data (squares, $C_{24}H_{50}$-$C_{12}H_{26}$ mixture, [45]. From [52].

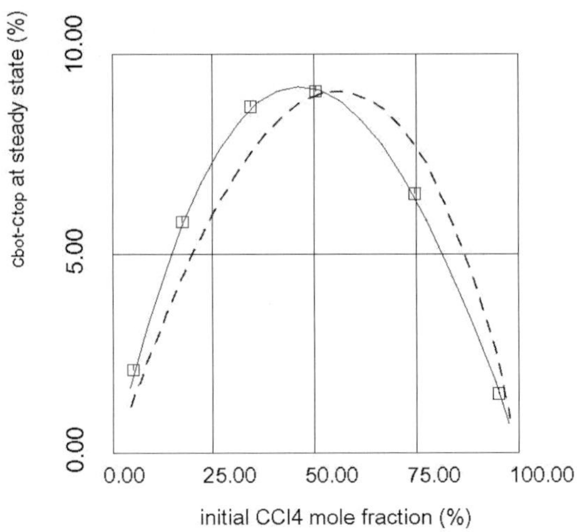


**Fig. 22.** Model-assisted phenomenological investigation about the dependency of the thermodiffusion coefficient on the initial mass fraction. Experimental data (squares) on the $CCl_4$-$C_2HCl_3$ mixture [14,47,15,17,16]. Dashed line: model output with constant thermodiffusion coefficient; solid line: model output with linear dependency of $s_T^*$ on $c$.

have hence an operational tool for the simultaneous determination of the effective thermodiffusion and ordinary diffusion coefficient from the fitting of the $\gamma = f(\log k)$ curve.

### 4.4 Modelling Soret-driven hydrodynamic instabilities

The successfull confrontation of numerical models on the horizontal gradient PTC is only a partial validation as the flow problem, except the weak coupling through the dependency of $\rho$ and $\eta$ upon $c$, is largely separated from the solute mass flux problem. A second validation step is to address the vertical gradient configuration in conditions where the thermal Rayleigh number is subcritical. In this case the convection flow in the thermodiffusion-convection coupling depends on the magnitude of thermodiffusive separation.

**Properties.** The Soret-driven instability problem presents some specific and characteristic features that are exposed in more details in other contributions of this book. It is interesting to explore these features with models in order to make further progress into the validation process.

The relative signs of the thermodiffusion coefficient and of the thermal gradient result in a large diversity of instability phenomena. Two typical properties are particularly suitable to benchmark numerical models: (i) in the steady convection domain, a positive contribution (relative to thermal gradient) of thermodiffusion results in decreasing the effective value of the critical Rayleigh number and of the perturbation wave number (i.e. convection rolls become larger in size), (ii) for a negative contribution, oscillatory regimes can be observed [84].

**Semi-analytical models.** The stability and bifurcation diagrams can be obtained using spectral expansions (Galerkin method) for the perturbations [76,85]. When the geometry of the system is simple, it is by far the most efficient way to have a global view of the possible states of the system (stability diagram, stream functions, concentration fields).

**Modelling case-studies.** Building stability diagrams by means of meshed models such as the METSOR code is a harder process, because each simulation leads to the description of a unique state of the system. Moreover, the problem is highly non-linear, a fact that generates careful and painful adjustments of numeric parameters such as the time discretisation. However, such a deterministic model allows to go deeper into the understanding of the physics, and the initial investments accepted to make it run properly are largely paid back, for the full model, conversely to the semi-analytical resolutions, can be used in any geometries.

Below we present some preliminary results obtained with the METSOR model while investigating the Soret-driven instability problem. The model is operated under initial conditions corresponding to the applied thermal gradient and an homogenous composition. We observed that numeric noise sufficed to start convection. Figure 23 compares the model outputs to the two analytical solutions. For fixed values of the thermal Rayleigh number, we performed several runs with increasing Soret coefficients to cross the instability frontier. The response of the model was close to the exact implicit analytical solution.

Figure 24 shows that the model gives a good account of the qualitative behaviour with respect to the perturbation wave number. In the pure thermal case, it is easy to show that the convection rolls are squares. For the dimensions of the domain considered here, the number of rolls is $L/d \approx 15$.

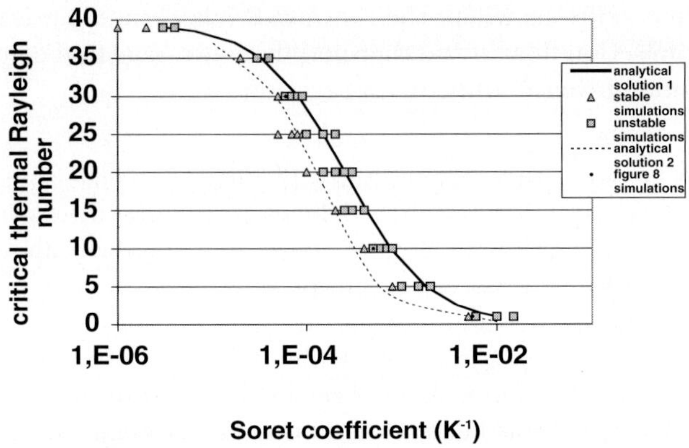


**Fig. 23.** Modelling the Soret-driven steady instability problem. Comparison between (i) the approximate (explicit, solid line) analytical solution, (ii) the exact (implicit, dashed line) analytical solution, (iii) METSOR model outputs (squares and triangles).

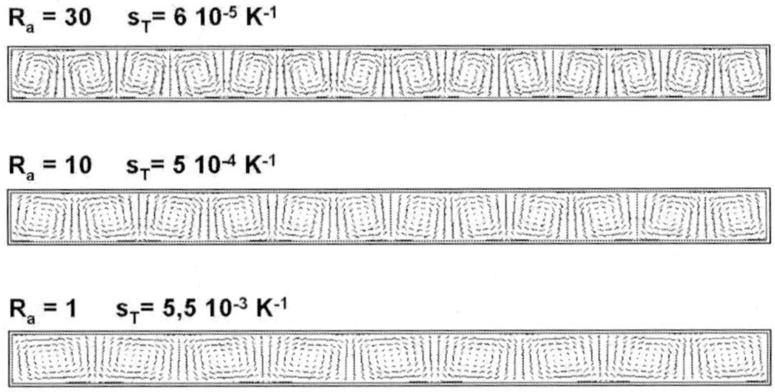


**Fig. 24.** Two-dimensional evolution of the convection roll geometry with an increasing contribution of the Soret effect on the instability. We see that 14 cells remain in the $(30; 6 \times 10^{-5})$ case, 12 in the $(10; 5 \times 10^{-4})$ case, and 9 only in the $(1; 5.5 \times 10^{-3})$ case.

Figure 25 is a modelling output of an overstability situation corresponding to $\mathbf{Ra}^\star = 42$ and $s_{\mathrm{T}}^* = -10^{-3}\mathrm{K}^{-1}$, characterised by a growing oscillatory instability. This result is also qualitatively consistent with results observed in the laboratory in free mixtures [84] and predicted in porous media by semi-analytical solutions [85].

### 4.5 Further investigations

The comparison of the METSOR model with characteristic features of the Soret-driven instability problems led to the conclusion that this model was representative of the physics of the system, at least on a qualitative level. Further modelling has to be performed on systems of more complex geometries. Another important aspect is to investigate the coupling between the flow resulting from the instability and the thermodiffusive transport. The vertical gradient PTC experiments

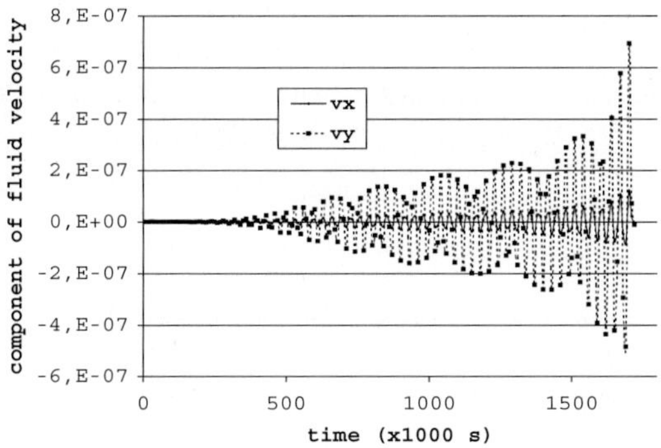


**Fig. 25.** Modelling an oscillatory convection regime in the Hopf bifurcation domain. The figure displays the time evolution of the horizontal ($v_x$) and vertical ($v_y$) fluid velocity at a fixed location in a 2-D porous domain.

will provide data to achieve a comprehensive phenomenological model for this coupling which, until now, is not understood.

# 5 Conclusions

In the past ten years, research about thermodiffusion has developed and now covers a wide range of topics, from material sciences to fluid mechanics, from geosciences to chemical engineering. This research effort will probably grow further, as a lot of aspects remain to be understood, measured and modelled. At this stage, one of the interest of such a research is to involve multidisciplinary teams of physicists, experimentalists and numericians, as illustrated by this contribution.

As regards porous media, thermodiffusion still offers large research perspectives, beginning with a "new frontier" which is to decipher the coupling dynamics occurring in vertical gradient PTCs, i.e. the thermodiffusion and Soret-driven instability coupling. Once this phenomenological and numerical difficulty is overcome, this will open the path to the modelling of any complex geological system hosting thermodiffusion transport. At the laboratory level, a lot of work remain to be done for the description of the behaviour of multicomponent mixtures in order to yield the appropriate effective values of thermodiffusion, diffusion and cross-diffusion coefficients using the same model-assisted methodology as for the study of binary mixtures in horizontal gradient PTCs.

**Acknowledgements**

The authors thank the French Atomic Energy Commission Control Authority (CEA/IPSN), the French Agency for the Management of Nuclear Waste (ANDRA), the French Agency for Environment and Energy Conservation (ADEME) and the Elf-Aquitaine Company for having supported their research.

## References


1. C. Ludwig: Akad. Wiss. Wien, Math. Naturwiss., **20**, 539 1856
2. Ch. Soret: Arch. Sci. Phys. Nat. Genève, **2**, 48 (1879)
3. I. Prigogine: *Etude Thermodynamique des Phénomènes Irréversibles.* (Dunod et Desoer 1947)
4. S.R. De Groot, P. Mazur: *Non-equilibrium Thermodynamics.* (North-Holland, Amsterdam 1961)
5. F. Fer: *Thermodynamique Macroscopique : Systèmes Ouverts.* (Gordon and Breach, London 1971)
6. K. Clusius, G. Dickel: Naturwiss., **6**, 546 (1938)
7. K. Clusius, G. Dickel: Naturwiss., **27**, 148 (1939).
8. P. Debye: Ann. Phys., **36**, 284 (1939)
9. W.H. Furry, R.C. Jones, L. Onsager: Phys. Rev., **55**, 1083 (1939)
10. R.C. Jones, W.H. Furry: Rev. Mod. Phys., **18**, 151 (1946)
11. P. Debye: Method of an apparatus for effecting thermodiffusion, U.S. Patent No 567 (1951)
12. L.J. Sullivan, T.C. Ruppel,C.B. Willingham: Ind. Eng. Chem., **49**, 110 (1957)
13. M. Lorenz, A.H. Emery: Chem. Eng. Sci., **11**, 16 (1959)
14. V. Sanchez, P. Brugel, M. Sagnes: J. Chim. Phys., **69**, 931 (1972)
15. V. Sanchez, M. Sagnes, P. Brugel: Chem. Eng. Science, **28**, 759 (1973)
16. V. Sanchez: La diffusion thermique et son application au fractionnement des mélanges binaires. PhD Thesis no 677, (Université Paul-Sabatier, Toulouse 1975)
17. V. Sanchez, L. Estaque: Chem. Eng. Science, **29**, 2283 (1974)
18. W. Wahl: Am. J. Sci., **244**, 6, 417 (1946)
19. J. Schott: Bull. Min., **106**, 247 (1983)
20. C.R. Carrigan, R.T. Cygan, J. Geophys. Res., **91** (11), 451 (1986)
21. J. Estèbe, J. Schott: C.R. Acad. Sci. D, **272**, 905 (1971)
22. P.Costesèque, J. Schott: C.R. Acad. Sci. D, **282**, 1913 (1976)
23. P. Costesèque: Bull. Min., **108**, 305 (1985)
24. R.O. Fournier: Geothermics, **19**, 481 (1990)
25. P. Costesèque, M. Hridabba, J.Sahores: C.R. Acad. Sc., Paris, **304** II, 1069 (1987)
26. P. Costesèque, M. El Maâtaoui, J.Sahores: C.R. Acad. Sc., Paris, **305** II, 1531 (1987)
27. P. Costesèque,E. Rivière: C.R. Acad. Sc., Paris, **315** IIa, 179 (1992)
28. P. Costesèque, M. El Maâtaoui, E. Rivière: Entropie, **184-185**, 94 (1994)
29. P. Costesèque, M. El Maâtaoui: Entropie, **184-185**, 101 (1994)
30. P. Costesèque, M. El Maâtaoui, J. Plique: C.R. Acad. Sci. IIa, **320**, 265 (1995)
31. B. Faissat, F. Montel: Entropie, **32**, 198, 107 (1996)
32. D.T.J. Hurle, E. Jakeman: J. Fluid Mech., **47**, 667 (1971)
33. R.S. Schechter, I. Prigogine, J.R. Hamm: Phys. Fluids, **15**, 379 (1972)
34. J.C. Legros,P. Goemaere, J.K. Platten: Phys. Rev. A, **32**, 1903 (1985)
35. H. Brand, V. Steinberg: Physica, **119A**, 327 (1983).
36. P. Costesèque, Ph. Jamet: Entropie, **198-199**, 119 (1996)
37. J.C. Legros, P. Poty, G. Thomaes: Physica, **64**, 481 (1973)
38. S. Van Vaerenbergh, J.C. Legros, J.Ch. Dupin: Adv. Space. Res., **16**, 69 (1995)
39. O. Lhost, J.K. Platten: Phys. Rev. A, **40**, 4552 (1989)
40. M. Bou-Ali, O. Ecenarro, J.A. Madariaga, C.M. Santamaria, J.J. Valencia: J. Phys. Cond. Matter, **10** (15), 3321 (1998)
41. S.R. De Groot: J. Phys. Radium, **8**, 129 (1947)

42. P. Costesèque: *Sur la migration sélective des isotopes et des éléments par thermodiffusion dans les solutions. Application de l'effet thermogravitationnel en milieu poreux; observations expérimentales et conséquences géochimiques*, Doctoral Thesis no 1049, Université Paul Sabatier, Toulouse (1982)
43. A.D. Payton, J.C.R. Turner: Trans Faraday Soc., **58**, 55 (1962)
44. K.F. Alexander: Zs. Phys. Chem., **203** A, 181 (1954)
45. M. El Maâtaoui: Conséquences de la thermodiffusion en milieu poreux sur l'hydrolyse des solutions de chlorures ferriques et sur les migrations d'hydrocarbures dans des mélanges de n-alcanes et dans un pétrole brut; implications géochimiques. Thèse de 3e cycle, (Université Paul Sabatier, Toulouse 1986)
46. J. Estèbe: Etude de l'effet thermogravitationnel en milieu poreux. PhD Thesis, no 240, Paul Sabatier University, Toulouse (1973)
47. V. Sanchez, C. Guiglion, J. Mahenc: Chem. Eng. Science, **28**, 751 (1973)
48. I. Prigogine, L. de Brouckere, R. Buess, , Physica, **18**, 915 (1952)
49. E. Rivière: Sur la migration des composés hydrocarbonés dans les mélanges complexes (huiles naturelles) par diffusion thermogravitationnelle en milieu poreux. Etude expérimentale et état des modélisations. PhD Thesis, (Université Paul Sabatier, Toulouse 1991)
50. P. Costesèque, J.C. Loubet: Entropie, **214**, 53 (1998)
51. P. Costesèque, M. El Maâtaoui, J.C. Garrigues: Entropie, **198-199**, 15 (1996)
52. Ph. Jamet, P. Costesèque, D. Fargue: Chemical Engineering Science, **51**, 4463 (1996)
53. L. J. Kempers: J. Chem. Phys., **90**, pp. 6541 (1989)
54. A. Firoozabadi, B. Dindoruk, E. Chang: Entropie, **198/199**, 109 (1996)
55. A.A.Shapiro, E.H. Stenby: Entropie, **217**, 55 (1999)
56. K. Ghorayeb, A. Firoozabadi: A. I. Ch. E. Journal, **46**, 883 (2000)
57. K. Ghorayeb, A. Firoozabadi: S. P. E. Journal, **5**, 158 (2000)
58. A. Firoozabadi, K. Ghorayeb, K. Shulka: A. I. Ch. Journal, **46**, 892 (2000)
59. M. El Maâtaoui: *Migration sélective des hydrocarbures dans les fluides pétroliers par diffusion thermogravitationnelle en milieu poreux. Conséquences possibles sur la stratification des espèces dans les réservoirs naturels*, PhD Thesis, (Université Paul Sabatier, Toulouse 1995)
60. J. Bear: *Dynamics of fluids in porous media*, (American Elsevier, N-Y 1972).
61. P. Adler: *Porous media: geometry and transport*, (Butterworths-Heinemann, 1992)
62. P.H. Groenevelt, G.H. Bolt: J. of Hydrology, **7**, 358-388 (1969)
63. O.A. Plumb, S. Whitaker: Water resources res. **24**, 913 and 927 (1988)
64. C.L. Carnahan, J. Hydrol., **31** 125 (1976)
65. C.M. Marle: Int. J. Eng. Sc. **20**, 643 (1982); Ann. Mines, may 1984, 51 (1984)
66. S.M. Hassanizadeh: Adv. Wat. Res., **9**, 196, **24**, 321 (1986)
67. M.A. Biot: J. Appl. Phys., **33** (4), pp. 1482 (1962)
68. J.C. Benet, P. Jouanna: Int. J. Heat Mass Transfer, **26** (11), 1585 (1983)
69. P. Jamet: Sur certains aspects du couplage en milieu poreux entre les champs de température et de concentration. Ph. D. Thesis, (University of Paris-6 1991)
70. D. Fargue, Ph. Jamet: Entropie, **184-185**, 17 (1994)
71. M. Poreh: J. Geophys. Res., **70**, 3909 (1965)
72. I. Prigogine: *Introduction to thermodynamics of irreversible processes*, (Wiley, N-Y 1962)
73. J. Meixner, H.G. Reik - Thermodynamik der irreversiblen Prozesse, in *Handbuch der Physik*, vol.III/2, (Springer, Heidelberg 1959)
74. D. Fargue, Ph. Jamet, P. Costesèque: Transport in Porous Media, **30**, 323 (1998)

75. Ph. Jamet, D. Fargue, P. Costesèque, G. de Marsily, A. Cernes: Transport Porous Med., **9**, 223 (1992)
76. M. Marcoux, M.C. Charrier-Mojtabi, A. Bergeon: , Entropie, **214**, 31 (1998)
77. P.G. Bedrikovetsky, D.G. Polonsky, A.A. Shapiro: Entropie, **184-185**, 73 (1994)
78. D.A. Nield, A. Bejan: , *Convection in porous media*, (Springer, 2nd Edition 1999)
79. Ph. Jamet: *Influence des processus de couplage non diagonaux sur la dynamique du transfert de masse et de chaleur dans le milieu géologique*, Research report, ANDRA (1999)
80. J.F. Figué, J.-P. Caltagirone, F. Montel: Entropie, **198-199**, 102 (1996)
81. J.-F. Dutrieux, G. Chavepeyer, J.-K. Platten, E. Istasse: Entropie, **218**, 18 (1999)
82. G. Chavepeyer, J.-K. Platten: *Entropie*, **214**, 27 (1998)
83. M. Marcoux, M.C. Charrier-Mojtabi, M. Azaiez: , Int. J. Heat Mass Transf., **42**, 2313 (1999)
84. H. Touiri, J.K. Platten, G. Chavepeyer: Eur. J. Mech., B/Fluids, **15**, 241 (1996)
85. Sovran, Charrier-Mojtabi, Azaiez, Mojtabi: Poster to IMT 4 conference (2000)

## Supplementary glossary

$\mathcal{D}$ space domain
$a$ width of the porous layer between isothermal walls
$b$ perturbation wave number
$c_0$ initial solute mass fraction
$D^*$ apparent ordinary diffusion coefficient in porous medium
$D^*_{\mathrm{T}}$ apparent thermodiffusion coefficient in porous medium
$D^*_{\mathrm{th}}$ apparent thermal diffusion coefficient in porous medium
$dA$ elementary surface
$d\varpi$ elementary volume
$d/dt$ transport derivative operator in the fluid phase
$f$ free energy per mass
$\boldsymbol{F}_\alpha$ force per unit of mass
$\boldsymbol{g}$ gravity vector
$\boldsymbol{J}_{\mathrm{f}\alpha}$ barycentric diffusion flow in the fluid phase
$K$ kinetic energy
$k$ intrinsic permeability of the porous packing
$L$ length of PTC
$\mathbf{Le}^\star$ effective Lewis number in a porous medium
$L_{\mathrm{ij}}$ Onsager's phenomenological coefficients
$n$ specific mole number
$n_0$ initial specific mole number
$\boldsymbol{n}$ external normal vector
$\bar{p}$ apparent pressure (relative to representative elementary volume)
$p$ liquide pore pressure
$\mathbf{Ra}^\star$ critical Rayleigh number in a porous medium
$s_{\mathrm{T}}$ soret coefficient in free medium
$s^*_{\mathrm{T}}$ soret coefficient in porous medium
$\bar{T}$ mean temperature in the PTC

| | |
|---|---|
| $\boldsymbol{U}$ | filtration velocity ($= \varepsilon \boldsymbol{w}_{\mathrm{f}}$) |
| $\bar{\mathcal{V}}^{ij}$ | viscosity tensor |
| $\boldsymbol{w}_\alpha$ | true velocity of component $\alpha$ |
| $\boldsymbol{w}_{\mathrm{f}}$ | fluid barycentric velocity |
| $\alpha_{\mathrm{d}}$ | dispersivity of the porous packing |
| $\alpha'_{\mathrm{d}}$ | thermodispersivity of the porous packing |
| $\gamma$ | separation factor in the PTC |
| $\Delta n$ | mole fraction difference between bottom end and top end of the PTC |
| $\Delta T$ | temperature difference between isothermal walls in the PTC |
| $\eta$ | dynamic viscosity |
| $\varepsilon$ | porosity |
| $\rho_\alpha$ | specific mass of constituent $\alpha$ relative to its own phase |
| $\bar{\rho}_\alpha$ | specific mass of constituent $\alpha$ relative to total volume |
| $\sigma$ | specific entropy production |
| $\sigma^{ij}_\alpha$ | apparent stress tensor in the $\alpha$ phase |
| $\tau$ | tortuosity |
| $\boldsymbol{\Phi}_\alpha$ | volumic interphase interaction forces |
| $\boldsymbol{\psi}$ | heat flux |

# Soret Effect and Mixed Convection in Porous Media

Mohamed Najib Ouarzazi[1], Annabelle Joulin[1], Pierre-Antoine Bois[1], and Jean K. Platten[2]

[1] Université de Lille, LML, 59655 Villeneuve d'Ascq cedex, France
[2] Université de Mons-Hainaut, General Chemistry, Avenue du Champ de Mars, 24, B-7000 Mons, Belgium

**Abstract.** We study the pattern formation of a binary mixture in a porous medium heated from below in the presence of an horizontal flow. The asymptotic response of the system to a localized perturbation in the both directions of the horizontal plane is evaluated in two different situations. When the system exhibits a supercritical bifurcation, a linear stability analysis is performed. Presence of through flow breaks the rotational symmetry and the system at the absolute instability threshold, selects transversal rolls among an infinity of unstable modes. We show that in binary mixtures with negative separation ratio $\psi$, contrary to the case of positive $\psi$, the through flow shrinks the region of convective instabilities, and even may suppress the convective/absolute transition. In the case of subcritical bifurcation, a quintic Ginzburg-Landau equation is derived. A threshold of nonlinear absolute instability is found below the linear one when nonlinear front propagation dominates the dynamics. The velocity and wavenumber of these fronts are determined. An exists in nonlinear absolute instability region and its extent is ruled by the through flow rate. Moreover, special emphasis on the determination of $\psi$ is given.

## 1 Introduction

The necessety of the distinction between convective and absolute nature of the instabilities in open flows is now widely recognized. In the convectively unstable regime a spatially localized perturbation is convected downstream and eventually is blown out of the system. On the contrary, in the absolutely unstable regime, such perturbations expands in the downstream as well as in the upstream direction. The behaviour of the system is thus qualitatively very different in both regimes. In the convectively unstable regime, noise is spatially amplified and gives rise to noise-sustained structures. While, in absolutely unstable regime, structures are intrinsically sustained by the deterministic dynamics, which provides the relevant selection and stability criteria. The problem of convection in pure fluid layer heated from below and subjected to an horizontal though flow constitutes a good candidate to investigate the nature of the instabilities in open flows. Recent studies on this problem in fluid media [1–4] as well as in porous media [5] show notably that the through flow creates necessarily a region of convective instabilities. Convection in binary mixtures can reveal additionnal pattern formation phenomena that do not occur in pure fluids. In this system, there are two contributions to the density gradient. One of them comes

W. Köhler and S. Wiegand (Eds.): LNP 584, pp. 428–447, 2002.

from the thermal expansion of the fluid and the other is due to the concentration gradient. The nature of the bifurcation depends on the separation ratio $\psi$, which measures the on the stability of the system. When $\psi > \psi_{\mathrm{CT}}$, the system undergoes as in the case of pure fluids ($\psi = 0$) a pitchfork bifurcation at the onset. On the other hand, when $\psi < \psi_{\mathrm{CT}}$, a Hopf bifurcation may set up leading to oscillations of traveling waves. The problem of bidimensional mixed convection of binary mixtures in a fluid medium has been studied in [6] using linear stability criteria. In this work, we extand the study to the tridimensional case and assume that the duct is a porous medium of infinite extension. We perform a spatio-temporal instability analysis by evaluating the asymptotic response of the system to localized 3D perturbations. As far as a supercritical bifurcation is concerned, linear criteria [7] is valid to locate the absolute/convective transition curves of the conductive state. However, in the case of subcritical bifurcation, the leading nonlinearities destabilize the system and we can not apply the linear criteria any more. Chomaz [8] has defined a concept of nonlinear convective and absolute instabilities required for such situation. This concept is based on the asymptotic response of the system to a localized saturated wave packet of finite size, instead of a localized infinitesimal impulse. The aim of this paper is to study both situations, namely the supercritical case as well as the subcritical one. The paper is organized as follows. In section 2, we recall the governing equations of mixed convection of binary mixtures in porous media and their basic solution. Section 3 deals with linear stability analysis performed as an initial value problem to obtain both thresholds for the onset of convective and absolute instabilities. When the system exhibits a subcritical bifurcation, a nonlinear analysis is performed in section 4. The results of section 3 and 4 are proposed as criteria to determine the separation ratio $\psi$ of binary mixtures used in experiments.

## 2 Problem formulation and basic state

In this section we first recall the basic equations of motion of a fluid mixture in a porous medium. We consider that the entire medium is constituted of a solid phase with small granulometry and a fluid phase which is a mixture of two miscible fluids. The isotropic and homogeneous porous layer is assumed of infinite extent in the horizontal ($\boldsymbol{e}_{\mathbf{x}}, \boldsymbol{e}_{\mathbf{y}}$) plane with uniform thickness H and heated from below. Furthermore, we assume that a through-flow is driven by a pressure gradient in the x-direction. The filtration velocity $\boldsymbol{u}$ which is the barycentric velocity of the fluid phase is supposed not too large and satisfies the Darcy law. We denote by $\rho$ the density of the fluid and by C the concentration of one of the fluid constituents. The Boussinesq approximation is applied indicating that the density variations are neglected everywhere except for the gravity in the Darcy equation. Subject to these conditions the following set of governing equations is obtained [9]:

$$
\begin{aligned}
&\nabla\cdot\boldsymbol{u} = 0\\
&\frac{1}{\rho_0}\nabla P + (\alpha(T-T_0) - \beta(C-C_0))\boldsymbol{g} + \frac{\nu}{K}\boldsymbol{u} = 0\\
&\frac{\partial T}{\partial t} + \boldsymbol{u}\cdot\nabla T = D_{\mathrm{th}}\nabla^2 T\\
&\frac{\partial C}{\partial t} + \boldsymbol{u}\cdot\nabla C = D(\nabla^2 C + S_{\mathrm{T}}C_0(1-C_0)\nabla^2 T)
\end{aligned}
\tag{1}
$$

where P and T are the deviation from the hydrostatic pressure corresponding to a uniform density $\rho_0$ and the temperature field, respectively. The coefficients $\alpha$ (thermal expansion coefficient), $\beta$ (concentration expansion coefficient), $\nu$ (kinematic viscosity), K (permeability of the medium), $D_{\mathrm{th}}$ (thermal diffusivity), D (concentration diffusivity) and $S_{\mathrm{T}}$ (Soret coefficient) are assumed constants. The boundary conditions at the impermeable horizontal walls are:

$$
\begin{aligned}
&T = T_0 \text{ at } z = 0\\
&T = T_1 \text{ at } z = H\\
&\boldsymbol{u}\cdot\boldsymbol{e_z} = 0 \text{ and } \frac{\partial C}{\partial z} = -S_{\mathrm{T}}C_0(1-C_0)\frac{\partial T}{\partial z} \text{ at } z = 0 \text{ and } z = H
\end{aligned}
\tag{2}
$$

and the rate of the driven horizontal through-flow is:

$$
\int_0^H \boldsymbol{u}\cdot\boldsymbol{e_x}dz = Q \tag{3}
$$

A basic solution of the problem (1,2,3) is a combination of vertical thermal and concentration stratifications and a homogeneous flow in the $\boldsymbol{e_x}$ direction

$$
\begin{aligned}
T_{\mathrm{b}} &= T_0 - \frac{T_0 - T_1}{H}z\\
\boldsymbol{u_{\mathrm{b}}} &= \frac{Q}{H}\boldsymbol{e_x}\\
C_{\mathrm{b}} &= C_0 + S_{\mathrm{T}}C_0(1-C_0)\frac{T_0-T_1}{H}z\\
P(x,z) &= -\frac{\nu\rho_0}{KH}Qx - \rho_0 g\frac{T_0-T_1}{2H}(\alpha - \beta S_{\mathrm{T}}C_0(1-C_0))z^2
\end{aligned}
\tag{4}
$$

# 3 Linear convective and absolute instability

## 3.1 Response to 3D localized perturbations in the case of supercritical bifurcation

The linear stability analysis is performed by linearizing (1) around the basic solution (4). If we choose H, $\frac{D_{\mathrm{th}}}{H}$, $\frac{H^2}{D_{\mathrm{th}}}$, $\frac{\nu D_{\mathrm{th}}}{\alpha g K H}$ and $\frac{-\nu D_{\mathrm{th}} S_{\mathrm{T}} C_0(1-C_0)}{\alpha g K H}$ as references for length, filtration velocity, time, temperature deviation $\theta$ and concentration deviation C, we obtain a set of nondimensional equations for the vertical component W of the filtration velocity, $\theta$ and the field $\xi = C - \theta$:

$$\begin{aligned} &\Delta W - (1+\psi)\Delta_{\rm h}\theta - \psi\Delta_{\rm h}\xi = \delta(x)\delta(y)\delta(z-z_0)\delta(t) \\ &\frac{\partial\theta}{\partial t} + Q\frac{\partial\theta}{\partial x} - \Delta\theta - RaW = \delta(x)\delta(y)\delta(z-z_0)\delta(t) \\ &\frac{\partial\xi}{\partial t} + Q\frac{\partial\xi}{\partial x} - Le\Delta\xi + \Delta\theta = \delta(x)\delta(y)\delta(z-z_0)\delta(t) \end{aligned} \tag{5}$$

with boundary conditions: $W = \theta = \frac{\partial\xi}{\partial z} = 0$ at $z = 0$ and $z = 1$.

The system is characterized by the following dimensionless parameters: Q (the through-flow rate), $Ra = \frac{\alpha(T_0 - T_1)gKH}{\nu D_{\rm th}}$ (the filtration Rayleigh number), $Le = \frac{D}{D_{\rm th}}$ (the Lewis number) and $\psi = \frac{\beta}{\alpha}S_{\rm T}C_0(1-C_0)$ (the separation ratio) which can be stabilizing or destabilizing depending on the particular mixture that is used.

The forcing terms added to the right-hand side of (5) represent a localized initial impulse in $(x, y, z)$ space for W, $\theta$ and $\xi$ ; $\delta$ being the Dirac function. Problem (5) can be solved using Fourier transforms in space $(x, y)$ and Laplace transform in time, defined by:

$$\widehat{G}(k_{\rm x}, k_{\rm y}, z, \omega) = \int_{-\infty}^{\infty}\int_{-\infty}^{\infty}\int_{0}^{\infty} G(x, y, z, t) exp(i(\omega t - k_{\rm x}x - k_{\rm y}y))dx\, dy\, dt \tag{6}$$

where $G = (W, \theta, \xi)^T$ and $\boldsymbol{q} = (k_{\rm x}, k_{\rm y})$ is the wave vector in the $(\boldsymbol{e_{\rm x}}, \boldsymbol{e_{\rm y}})$ plane. Proceeding as in [2] by solving the problem in Fourier space and performing the Laplace transform by the Residue Theorem, the solution $G(x, y, z, t)$ may be written as:

$$G(x, y, z, t) = \frac{1}{4\pi^2}\sum_{\rm n=1}^{2}\int_{-\infty}^{\infty}\int_{-\infty}^{\infty} Z^{\rm n}(k_{\rm x}, k_{\rm y}, z_0)\widehat{G_{\rm n}}(k_{\rm x}, k_{\rm y}, z)e^{(-{\rm i}(\omega_{\rm n}(k_{\rm x}, k_{\rm y})t - k_{\rm x}x - k_{\rm y}y))}dk_{\rm x}dk_{\rm y} \tag{7}$$

with $\omega_{\rm n}(n = 1, 2)$ are the two complex frequency modes obtained by solving the homogeneous boundary value problem (5), and

$$Z^n(k_{\rm x}, k_{\rm y}, z_0) = \left[\widehat{W}^*_{\rm n}(k_{\rm x}, k_{\rm y}, z_0) + \widehat{\theta}^*_{\rm n}(k_{\rm x}, k_{\rm y}, z_0) + \widehat{\xi}^*_{\rm n}(k_{\rm x}, k_{\rm y}, z_0)\right],$$

where (*) denotes complex conjugate.

Let us recall some ideas on convective and absolute instabilities. The system is stable when $|G(x + Ut, y + Vt, z, t)| \to 0$ as $t \to \infty$ for any arbitrary $(x, y, z)$ position and any velocity $(U, V)$. The condition to have stability is $(\omega_{\rm n})_{\rm i} < 0$ for any $(k_{\rm x}, k_{\rm y}) \in \mathbb{R}^2$, because it yields the integral (7) decaying with time. When $(\omega_{\rm n})_{\rm i} > 0$, for some wave vector $(k_{\rm x}, k_{\rm y})$, then we have to distinguish between two types of instabilities. We can either have that $|G(x, y, z, t)| \to \infty$ for any arbitrary position $(x, y, z)$ (absolutely unstable) or that $|G(x+Ut, y+Vt, z, t)| \to \infty$ for some velocity $(U, V)$ and any arbitrary position $(x, y, z)$ (convectively unstable). According to the definition given above, the propagation velocity $(U, V)$ of a wave packet has an important role in the unstable region, i.e. $(\omega_\mathrm{n})_\mathrm{i} > 0$. The instability is convective when in the $(U, V)$ plane, the origin $(U, V) = (0, 0)$ does not belong to the unstable wave packet. The instability becomes absolute when the wave packet spreads but that at least one spatial mode has a vanishing velocity $(U, V)$. Thus, in the absolutely unstable regime, the packet expands in the laboratory frame into the direction of packet motion as well as opposite to it [6]. The asymptotic behaviour of the integral (7) can be obtained by applying the method of the steepest descent. It consists, in particular, in deforming the real wave vector $k_\mathrm{x}$ and $k_\mathrm{y}$ contours of integration into the two complex planes $k_\mathrm{x}$ and $k_\mathrm{y}$, provided that the integrand has no singularities in the area bounded by the original and the new contours. The dominant part of the integrand arises in the region of the complex saddle point $(k_\mathrm{x}^s, k_\mathrm{y}^s)$ defined by:

$$\frac{\partial \omega_\mathrm{n}}{\partial k_\mathrm{x}} = U \qquad \text{and} \qquad \frac{\partial \omega_\mathrm{n}}{\partial k_\mathrm{y}} = V \tag{8}$$

The necessary, but not sufficient, condition for absolute instability is the existence of a saddle point $(k_\mathrm{x}^0, k_\mathrm{y}^0)$ in the two complex $k_\mathrm{x}$ and $k_\mathrm{y}$ planes for some complex $\omega^{(0)}$ with $\omega_\mathrm{i}^{(0)} > 0$ and $U = V = 0$. However, for sufficiency, it is required that the saddle point emerges from the pinching of spatial branches issuing from the upper and lower parts of both the complex $k_\mathrm{x}$ and $k_\mathrm{y}$ planes. This is the so-called pinching condition [10].

These two points are investigated in the following. First, we look for the necessary condition by solving (8) with $U = V = 0$. We notice that the presence of the through-flow ($Q \neq 0$) has no influence on the spatial dependance of the solutions of the set equations (5) without forcing. Indeed, the use of the Darcy model allows the homogeneous system (5) to remain invariant under a Galilean transformation. This invariance breaks down if we substitute the Brinkman's equation for the Darcy's law. The Brinkman's model introduces an additional viscous term analogous to the Laplacian term that appears in the Navier-Stockes equations. Therefore, it constitutes a useful model if one wishes to compare mixed convection of binary mixtures in a porous media with that in clear fluids [6]. These two problems are recovered respectively in the limit of small and large permeability. Although this strategy is interesting, our purpose remains to describe mixed convection in common porous materials where the values of the permeability vary widely, i.e. $10^{-7}m^2 - 10^{-20}m^2$ [11].

In the framework of the Darcy equation, the dependence on the through-flow rate Q is trivial and amounts to a single frequency shift:

$$\omega^{\pm}(k_\mathrm{x}, q^2, Q \neq 0) = k_\mathrm{x} Q + \Omega^{\pm}(q^2, Q = 0) \tag{9}$$

where $\Omega^{\pm}$ are the complex frequencies of the system in the absence of through-flow. In the temporal instability analysis, the two components $k_\mathrm{x}$ and $k_\mathrm{y}$ of the horizontal wave number are assumed to be real and $\omega^{\pm}$ are complex. It is then straightforward to deduce from (9) that the temporal growth rate of the instability is $\omega_{\mathrm{i}}^{\pm} = \Omega_{\mathrm{i}}^{\pm}(q^2, Q = 0)$. Consequently, both the instability threshold $Ra^c$ and the wave number $k_{\mathrm{c}}$ do not depend on the through-flow rate. The basic state (4) is linearly unstable with respect to all 3D extended perturbations whose wave numbers $\boldsymbol{k_c} = (k_{\mathrm{x}}^c, k_{\mathrm{y}}^c)$ belong to a circle centered on $\boldsymbol{q} = 0$, with a radius $R = ||\boldsymbol{k_c}||$. In particular, for $\psi < \psi_{\mathrm{CT}}$, two convective structures $S-$ and $S+$ with axes perpendicular to the direction of the through-flow (transversal rolls) may simultaneously appear beyond the threshold with frequencies:

$$\omega_{\mathrm{c}}^{\pm} = k_{\mathrm{c}} Q \pm \Omega_{\mathrm{c}}(\psi, Le, Q = 0) \tag{10}$$

Note that the presence of the through flow, which breaks the reflexion symmetry $((x \to -x))$, compels the phase velocities $V_{\phi}^{\pm} = \dfrac{\omega_{\mathrm{c}}^{\pm}}{k_{\mathrm{c}}}$ to take different values for each mode $S+$ and $S-$. Unless $Q = 0$, the system can not exhibit standing waves in the form of transversal rolls.

We now proceed to determine the transition curves $Ra^A(\psi, Q)$ separating the regions of absolute and convective instability. By applying the necessary criteria (8) to the frequencies (9), we obtain:

$$\begin{aligned} \frac{\partial \omega^{\pm}}{\partial k_{\mathrm{x}}} &= Q + 2k_{\mathrm{x}}^{(0)} \frac{\partial \Omega^{\pm}}{\partial q^2} = 0 \\ \frac{\partial \omega^{\pm}}{\partial k_{\mathrm{y}}} &= 2k_{\mathrm{y}}^{(0)} \frac{\partial \Omega^{\pm}}{\partial q^2} = 0 \end{aligned} \tag{11}$$

For a non vanishing through flow rate, the system (11) has a solution only if $k_{\mathrm{y}}^{(0)} = 0$. This means that the system selects the travelling waves propagating in the through flow direction among an infinity of unstable spatio-temporal modes. Thus traveling waves with two different frequencies $\omega\pm^{(0)}$ and wave vectors $(k_{\mathrm{x}}^{(0)}, 0)$ are selected at the different thresholds of absolute instability. In order to determine these thresholds, we need the expressions of $\Omega^{\pm}$ given by the dispersion relation of the linear problem.

### 3.2 Dispersion relation and results

We will use a Galerkin method [12] to expand the z-dependence of $\widehat{G}_{\mathrm{n}}$. We then set:

$$\begin{aligned} \widehat{W}_{\mathrm{n}}(k_{\mathrm{x}}, k_{\mathrm{y}}, z) &= \sum_{\mathrm{m}=0}^{\mathrm{m}=\mathrm{M}} W_{\mathrm{nm}}(k_{\mathrm{x}}, k_{\mathrm{y}}) f_{\mathrm{W,m}}(z) \\ \widehat{\theta}_{\mathrm{n}}(k_{\mathrm{x}}, k_{\mathrm{y}}, z) &= \sum_{\mathrm{m}=0}^{\mathrm{m}=\mathrm{M}} \theta_{\mathrm{nm}}(k_{\mathrm{x}}, k_{\mathrm{y}}) f_{\theta,\mathrm{m}}(z) \\ \widehat{\xi}_{\mathrm{n}}(k_{\mathrm{x}}, k_{\mathrm{y}}, z) &= \sum_{\mathrm{m}=0}^{\mathrm{m}=\mathrm{M}} \xi_{\mathrm{nm}}(k_{\mathrm{x}}, k_{\mathrm{y}}) f_{\xi,\mathrm{m}}(z) \end{aligned} \tag{12}$$

with:

$$f_{\mathrm{W,m}}(z) = f_{\theta,\mathrm{m}}(z) = \begin{pmatrix} \cos(m\pi z) & \text{m odd} \\ \sin(m\pi z) & \text{m even} \end{pmatrix}$$

and:

$$f_{\xi,\mathrm{m}}(z) = \begin{pmatrix} 1 \ \ \text{m} = 0 \\ \sin(m\pi z) \ \ \text{m odd} \\ \cos(m\pi z) \ \ \text{m} \neq 0 \text{ even} \end{pmatrix}$$

In the sequel, we restrict the analysis to 1-mode truncation of the expansions (12), i.e. we will consider only $f_{\mathrm{W},1}$, $f_{\theta,1}$ and $f_{\xi,0}$, which represents the approximate dispersion relation obtained in [13] for clear fluids in the absence of the through flow and for free and impermeable boundary conditions.

A straightforward calculation leads to the approximate dispersion relation:

$$\begin{aligned} D(\omega, k_{\mathrm{x}}, k_{\mathrm{y}}) &= (\Omega^{\pm})^2 + ia_1\Omega^{\pm} + a_0 = 0 \\ a_1 &= (Le+1)q^2 + \pi^2 - \frac{(1+\psi)q^2 Ra}{q^2+\pi^2} \\ a_0 = Leq^2(q^2+\pi^2) - \frac{8Ra\psi q^2}{\pi^2} &- \frac{Le(1+\psi)q^4 Ra}{q^2+\pi^2} \end{aligned} \tag{13}$$

The absolute thresholds of the instabilities is determined by numerical integration of (11) and (13) with $k_{\mathrm{y}}^{(0)} = 0$ and $\omega_{\mathrm{i}}^{\pm} = 0$ using a Newton-Raphson method. The results are shown on figure (1.a) for $\psi < \psi_{\mathrm{CT}}$ and on figure (1.b) for $\psi > \psi_{\mathrm{CT}}$ for selected parameters and for $Le = 10^{-2}$.

As we think that mixed convection of binary mixtures constitutes an alternative way to determine the separation ratio $\psi$, the result of the integration $Ra^A$ versus $\psi$ for a fixed value of Q, is depicted in figure (2.a) and (2.b) and it will be discussed in connection with the results of section 3.3.

The second step is to check that the selected frequency at the threshold of absolute instability corresponds to a pinching point in the complex $k_{\mathrm{x}}$ plane, i.e.

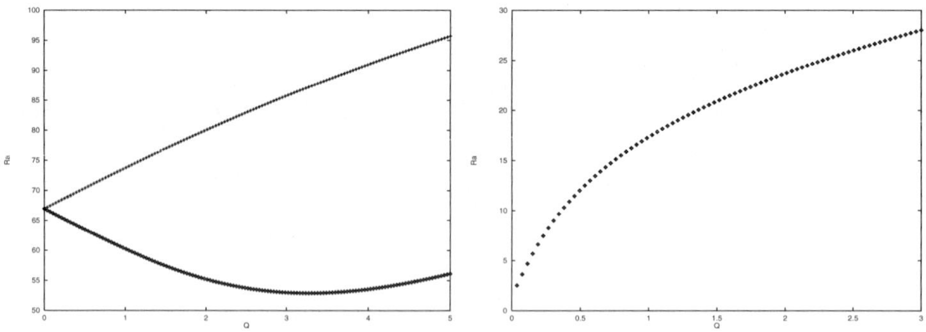


**Fig. 1.** On the left (1.a) absolute threshold $Ra = f(Q)$ with $\psi = -0.25$, ($\diamond$) represents $S-$ (lower curve) and $+$ represents ($S+$) (upper curve); on the right (1.b) absolute threshold $Ra = f(Q)$ with $\psi = 0.1$

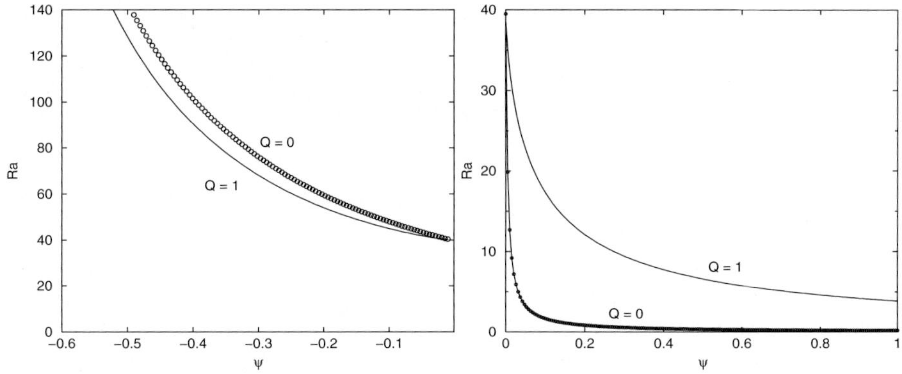


**Fig. 2.** Absolute threshold $Ra = f(\psi)$ with $Q = 0$ (circles) and $Q = 1$ (solid line) (Fig 2.a) $\psi > \psi_{\mathrm{CT}}$ and (Fig 2.b) $\psi < \psi_{\mathrm{CT}}$

the sufficient condition. To do this, we set $k_{\mathrm{y}} = 0$ in (13) and, by expanding it, obtain a sixth degree polynomial of $k_{\mathrm{x}}$. We then numerically solve it in the complex $k_{\mathrm{x}}$ plane for different values of the filtration Rayleigh number in the vicinity of the saddle point for fixed values of Q. We then obtain the spatial branches defined by: $\Gamma_{\omega_{\mathrm{i}}=0} = \{k_{\mathrm{x}} \in \mathbb{C}/D(k_{\mathrm{x}}, \omega_{\mathrm{r}}, Ra, Q) = 0/\omega_{\mathrm{r}} \in \mathbb{R}\}$.

As we observe in Figure 3., when Ra reaches the absolute threshold $Ra^A$, two branches emerging from the lower and from the upper half complex $k_{\mathrm{x}}$ plane pinch at the saddle point $k_{\mathrm{x}}^{(0)}$ with a selected frequency $\omega^0$.

For a fixed value of Q, the selected frequencies at the threshold of absolute instability are plotted in figure 4. versus $\psi$.

The main conclusion to be drawn from the results of this section may be summarized as follows:

i) For values of $\psi < \psi_{\mathrm{CT}}$ and in the absence of through flow, the system undergoes a Hopf bifurcation to symmetry degenerate left (mode $S^-$) and right (mode $S^+$) traveling wave patterns and the instability is convective (fig 1.a). We notice that, for $Q = 0$ the absolute thresholds $S^+$ and $S^-$ are equal. When $Q \neq 0$, the growth rate of the mode $S^-$ is bigger than the one of the mode $S^+$.
The presence of the through flow breaks this symmetry and alters the absolute thresholds for both modes $S^-$ and $S^+$. Figure 1 shows that the mode $S^-$ becomes absolutely unstable first regardless of the through flow rate Q. Thus at onset of the absolute instability, the system selects a traveling wave propagating in the through flow direction with a frequency $\omega_-^0$ and a wave number $q^{(0)}(\omega_-^0)$. In addition, the through flow may reduce the region of convective instabilities and there exists a through flow rate value Q for which both absolute and convective instability thresholds coincide. At this value, the conductive state becomes directly absolutely unstable with no transition through a convective instability region as is observed for other values of Q. This phenomenon occurs for a certain relation between $\psi$ and Q, obtained

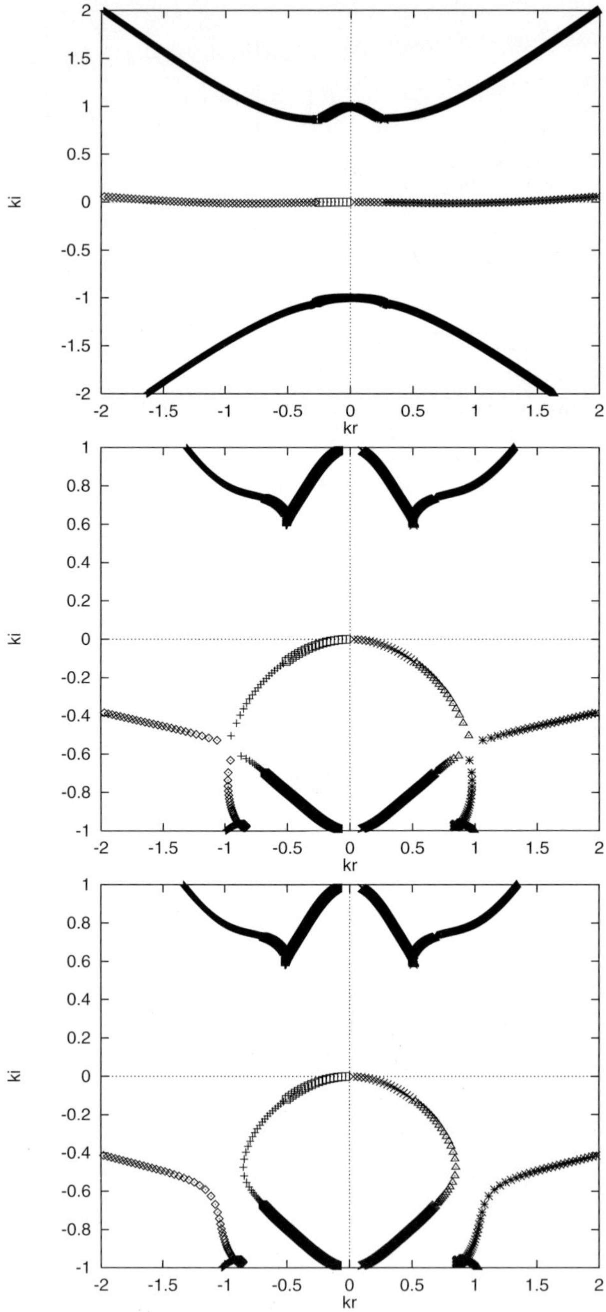


**Fig. 3.** Spatial branches of the dispersion equation for $\psi = 0.1$ and $Q = 1$, on the top $Ra = 3.7$, on the middle $Ra^A = 17.27$ and on the bottom $Ra = 18$

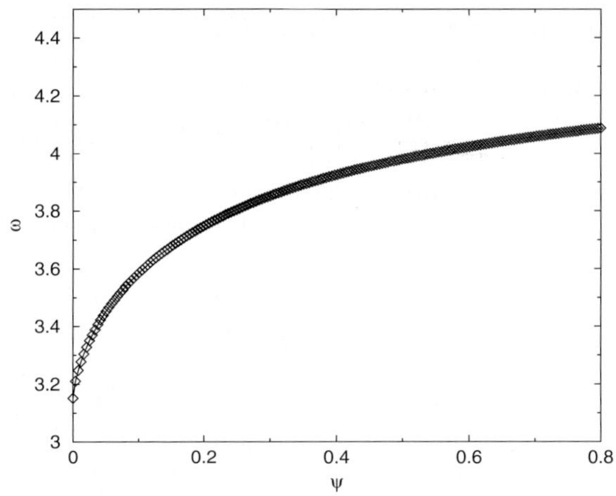


**Fig. 4.** $\omega = f(\psi)$ for $\psi > 0$ with $Q = 1$

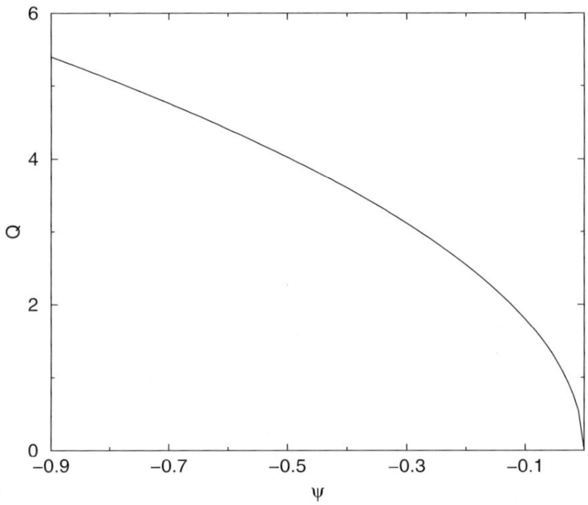


**Fig. 5.** $Q = f(\psi)$ and $V_g = 0$

by stating that the group velocity of the most unstable mode vanishes, i.e. $\dfrac{\partial\omega-}{\partial q}|_{q=k_c} = 0$. The result is depicted in fig 5.

ii) For $\psi > \psi_{CT}$, fig (1.b), the through flow creates necessarily a region of convective instability as in the case of pure fluids ($\psi = 0$). As for $\psi < \psi_{CT}$ at the absolute instability thresholds, the system selects a traveling wave propagating in the through flow direction.
The link between our findings and the determination of the separation ratio $\psi$ will be discussed in the next section.

### 3.3 Scaling laws and determination of $\psi$

**Comparison with experiments for $\psi = 0$.** First of all and to eliminate any possible confusion, we recall that all the results stated above were obtained within the framework of the Darcy's law, valid for moderate through flow rate. Second, in our theoretical model, we have assumed that the medium is infinite in both $\boldsymbol{e}_\mathbf{x}$ and $\boldsymbol{e}_\mathbf{y}$ directions. In a laboratory experiment, this later assumption means that each of two aspect ratios is large enough. Therefore, we have to take into account these two remarks if we try to make comparison with experiments. For lack of experimental investigations dealing with mixed convection of binary mixtures, we restrict ourselves to a qualitative comparison with results issuing from experiments of mixed convection of one component fluid ($\psi = 0$) in both porous and clear fluids media. The experiments of [14, 15] were performed in large aspect ratios in porous media for water and oil alternatively. When Ra exceeds a critical value, they observed that the conductive state loses its stability with respect to transversal rolls oscillating with a global frequency provided that the through flow rate Q is less than 5. This experimental result confirms our analytical prediction. However, for larger values of Q, a transition to longitudinal rolls was observed. Similar phenomena occur in experiments with clear fluids [3, 4]. In particular, transition from Poiseuille flow to longitudinal rolls takes place for $Re > 0.3$ in the convectively unstable region despite the fact that no external perturbations are applied. We think that these observed patterns are probably sustained and amplified by unavoidable and intrinsic noise in the convectively unstable system. This explanation is based on the results of similar open flows problems, namely the Couette-Taylor system with an axial flow [16]. In their experimental and theoretical study, emphasis is given to the amplification of noise observed when the base flow is convectively unstable. These noise sustained patterns are characterized by a broadened frequency spectrum, contrary to self sustained patterns in the absolutely unstable region where the frequency spectrum is a thin peak.

We now turn to a discussion of how in the absolutely unstable region, a pattern grows spatially from zero at the inlet, i.e. $x = 0$ to a fully developed transverse traveling rolls. A characteristic length, or healing length $\Delta x$ of this spatial growth can be defined as the distance from $x = 0$ required for the pattern to reach half of its saturated amplitude. Using recent theoretical studies obtained by [17] within the Complex Ginzburg-Landau approximation, an analytical scaling law is derived which links the characteristic length $\Delta x$ and the dimensionless parameters Ra, Q and $\psi$.

**Scaling laws.** In order to obtain the scaling laws for the healing length and the global frequency when the Rayleigh number exceeds the threshold of absolute instability, we follow the approach of [17]. The wave packet centered on the unstable mode $\omega_c^-$ can be characterized as a plane wave $A(x,t)e^{i(k_c x - \omega_c^- t)}$ with complex amplitude $A(x,t)$ which may be slowly modulated in space and time. The weakly nonlinear dynamics of this wave can be described by a complex

Ginzburg-Landau equation for the amplitude $A(x,t)$:

$$\frac{\partial A}{\partial t} + U_0 \frac{\partial A}{\partial x} = (1 + ic_1)\frac{\partial^2 A}{\partial x^2} + \mu A - (1 - ic_3)A^3 \tag{14}$$

subject to the homogeneous boundary conditions $A(x=0)=0$ at the inlet since the through flow is enforced there. The different coefficients $c_1$, $c_3$ were obtained in [18] and $\mu$ is proportional to $Ra - Ra^c$.

The problem to resolve is to look for time-periodic solution of (14) which links the solution $A=0$ at the inlet and a finite amplitude state at infinity represented by a traveling wave. This solution is termed a nonlinear global mode. In the sequel, we briefly recall some results of practical applications on nonlinear global mode and the reader should consult [17] for more details of analysis. A nonlinear global mode may be written in the form:

$$A(x,t) = a(x)e^{\mathrm{i}\int^{\mathrm{x}} \mathrm{q}'(\mathrm{x}')\mathrm{dx}'} e^{-\mathrm{i}\omega t} \tag{15}$$

with $a(x), q'(x), \omega$ real. $A(x)$ is enforced to satisfy the boundary condition at the inlet $A(0)=0$. At infinity, $A(x)$ is asymptotic to a traveling wave solution of (14) which is of the form: $A = a_{\mathrm{N}} e^{\mathrm{i}q'_{\mathrm{N}}\mathrm{x} - \mathrm{i}\omega t}$ where $a_{\mathrm{N}}, q'_{\mathrm{N}}$ and $\omega$ are real and defined by: $a_{\mathrm{N}}^2 = \mu - q_{\mathrm{N}}'^2$ and $\omega = U_0 q_{\mathrm{N}} + c_1 q_{\mathrm{N}}'^2 - c_3 a_{\mathrm{N}}^2$

The frequency $\omega$ is considered as an eigenvalue and is evaluated near the absolute instability threshold $\mu_{\mathrm{A}}$ by a matched asymptotic expansion method. If we set $\mu = \mu_{\mathrm{A}} + \varepsilon$ and $\omega = \omega_{\mathrm{A}} + \varepsilon\omega_1$, with $\varepsilon \ll 1$, this method consists of looking for a solution by using an asymptotic expansion with respect to $\varepsilon$ (the outer solution). This solution breaks down in a region of magnitude $O(\varepsilon^{1/2})$ near the inlet $x=0$, and an inner solution is obtained by a local study. The matching of the outer solution and the inner solution allows one to determine both the shift $\omega_1$ of the global frequency and the healing length $\Delta x$. $\omega_1$ is then the solution of the following equation [17]:

$$\alpha_1\beta_1 = -\sinh^2\left(\frac{\alpha_1\pi}{\beta_1}\right)\sum_{\mathrm{j}=1}^{\infty} j\rho_{\mathrm{j}} a_{\mathrm{N}}^{\mathrm{j}-1} \sum_{\mathrm{j}=1}^{\infty}(\lambda_{\mathrm{j}}^0 + \omega_1\lambda_{\mathrm{j}}^1)a_{\mathrm{N}}^j \tag{16}$$

with $\alpha_1 = \sqrt{\dfrac{|\xi_1| - \kappa_1}{2}}$, $\beta_1 = \sqrt{\dfrac{|\xi_1| + \kappa_1}{2}}$, $|\xi_1| = \sqrt{c_0(1+\omega_1^2)}$, $\kappa_1 = c_0(1 + c_1\omega_1)$ and $c_0 = \dfrac{1}{1+c_1^2}$. and $\rho_{\mathrm{j}}$, $\lambda_{\mathrm{j}}^0$ and $\lambda_{\mathrm{j}}^1$ appear in the expression of the outer solution and are computed numerically.

The healing length of the nonlinear global mode takes the form:

$$\Delta x = \frac{\pi}{\beta_1\sqrt{\varepsilon}} \tag{17}$$

We note from this scaling law that when the Rayleigh number approaches the absolute instability threshold $Ra^A$ (i.e. $\varepsilon \to 0$), the characteristic length $\Delta x$ diverges. As an illustrative example, we evaluate numerically the frequency shift $\omega_1$ for $\psi = -0.1$ and $U_0 = 1$ as the solution of (16) and we obtain $\omega_1 = -1.3933$. Therefore, $\beta_1$ is determined and the characteristic length $\Delta_{\mathrm{x}}$is depicted in log-log plot of Figure 6 with respect to $\varepsilon$.

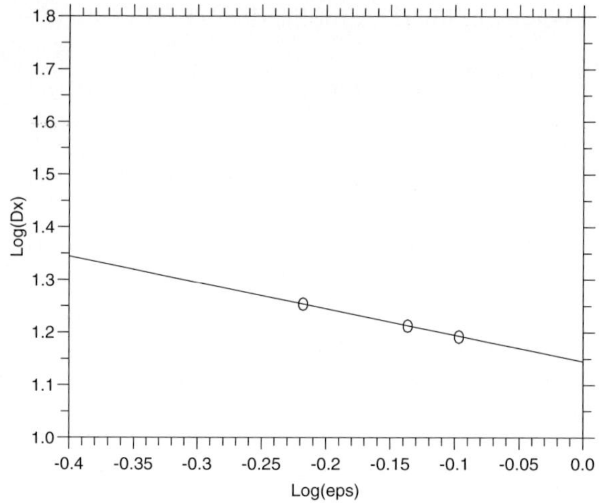


**Fig. 6.** Growth length of the nonlinear global mode for $\psi = -0.1$ and $U_0 = 1$

**Determination of $\psi$.** Now, we address the question of the determination of $\psi$ by using the theoretical results in connection with experiments. Suppose that two laboratory experiments are performed with the same mixture and at a fixed through flow rate Q with two different Rayleigh numbers ($Ra^{(1)} \neq Ra^{(2)}$). We also suppose that for each value $Ra^{(1)}$ and $Ra^{(2)}$ of the Rayleigh number, a global nonlinear mode appears in the experimental cell with a characteristic length $\Delta x^{(1)}$ and $\Delta x^{(2)}$ respectively. According to the scaling law (17), $Ra^{(1)}$, $Ra^{(2)}$, $\Delta x^{(1)}$ and $\Delta x^{(2)}$ satisfy the equalities:

$$\frac{\pi}{\beta_1} = [Ra^{(1)} - Ra^A]^{1/2} \Delta x^{(1)} = [Ra^{(2)} - Ra^A]^{1/2} \Delta x^{(2)} \tag{18}$$

$\beta_1$ and $Ra^A$ are numerical constants since they depend on $\psi$ and Q which are constants in the two performed experiments. We deduce from (18) that the measured absolute threshold $Ra^{\mathrm{A(m)}}$ is:

$$Ra^{\mathrm{A(m)}} = \frac{Ra^{(1)}(\Delta x^{(1)})^2 - Ra^{(2)}(\Delta x^{(2)})^2}{(\Delta x^{(1)})^2 - (\Delta x^{(2)})^2}$$

In order to obtain a good accuracy, we can perform $n$ experiments and obtain an average value of $Ra^{\mathrm{A(m)}}$. Therefore, we can use the theoretical curve transition to absolute instability for the same through flow rate Q used in experiments (Fig 2.a and Fig 2.b) to deduce $\psi$.

This strategy holds for negative as well as for positive $\psi$ provided that the system exhibits a supercritical bifurcation.

The numerical value of third-order nonlinearities is always positive for $\psi < \psi_{\mathrm{CT}}$ indicating that the bifurcation is supercritical contrary to the clear fluids case. For $\psi > \psi_{\mathrm{CT}}$, the nonlinearities may be stabilizing or destabilizing and the system may exhibit a supercritical as well as a subcritical bifurcation. Figure 7.

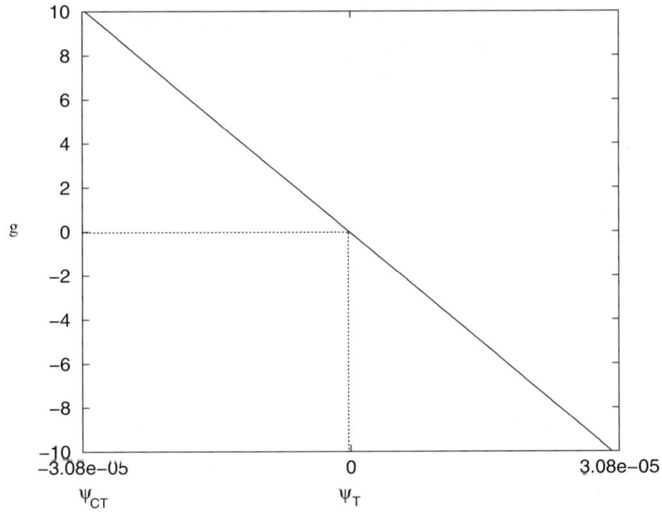


**Fig. 7.** Third order nonlinearities $g$ versus $\psi$, $\psi_t = -1.5 \times -10^{-7}$

shows that the subcritical zone takes place in the parameter range: $\psi_{CT} < \psi < \psi_t$ with $\psi_t = -1.5 \times 10^{-7}$.

# 4 Nonlinear absolute and convective instability

## 4.1 Subcritical Ginzburg-Landau equation for $\psi > \psi CT$

In order to determine the absolute instability threshold and to analyse the transition from convective to absolute instability, linear criteria are appropriate in the case of supercritical bifurcations. However, in the case of subcritical bifurcations, the nonlinearities are destabilizing and the linear instability criteria fail. This section is devoted to describe some properties of the nonlinear global mode when the system exhibits a subcritical bifurcation. To describe the dynamics of the system, we follow the approach of [19], based on the real fifth-order Ginzburg-Landau equation.

We start our analysis by deriving an equation describing the nonlinear behaviour near threshold of the complex amplitude $A(x,t)$ of the most unstable mode. The derivation is obtained by a classical multiple scale method and will be published elsewhere. This equation takes the form:

$$\frac{\partial A}{\partial t} = a\frac{\partial^2 A}{\partial x^2} + \mu A - Q\frac{\partial A}{\partial x} + gA^3 - hA^5 \tag{19}$$

We set : $F = \left(\dfrac{\Lambda k_c^2\left[8\psi(k_c^2+\pi^2) + \Lambda(1+\psi)k_c^2\pi^2\right]}{8\psi((\Lambda+1)k_c^2+\pi^2)}\right)$ with $\mu = \dfrac{Ra - R_c}{R_c}F$

$$a = \frac{1}{2}\left[\frac{4(3k_c^2+\pi^2)}{(k_c^2+\pi^2)^2} - \frac{(16\psi/\pi^2) + 2\Lambda(1+\psi)}{(8\psi/\pi^2)(k_c^2+\pi^2) + \Lambda(1+\psi)k_c^2}\right]F$$

$$g = \left[\frac{1}{(k_c^2+\pi^2)} + \frac{16\psi R_c(4k_c^2\Lambda + 3(k_c^2+\pi^2))}{3\pi^2\Lambda^3 k_c^2(k_c^2+\pi^2)^2}\right] F > 0 \text{ for subcritical condition}$$

$$h = \left[\frac{2}{\Lambda^2 k_c^2(k_c^2+\pi^2)}\right] F$$

$$k_c^2 = \frac{\pi^2(\Lambda + \Lambda\psi - \frac{8\psi}{\pi^2})}{(\Lambda + \Lambda\psi + \frac{8\psi}{\pi^2})}, \; R_c = \frac{4\pi^2\Lambda^2(1+\psi)}{(\Lambda + \Lambda\psi + \frac{8\psi}{\pi^2})^2}.$$

Homogeneous solutions of (19) are $A = 0$ and the non trivial solution $A_\pm^2 = \dfrac{g \pm \sqrt{g^2 + 4\mu h}}{2h}$

In the absence of through flow, the upper branch $A_+$ exists and is stable for $\mu > \mu_1$, with the middle branch $A_-$ exists and is unstable for $\mu_1 < \mu < 0$ with $\mu_1 = -g^2/4h$ which corresponds to the saddle-node bifurcation.

### 4.2 Front propagation, hysteresis and determination of $\psi$

The nonlinear stability analysis of the zero state relies on its response to perturbations of finite extent and amplitude. In the absence of through flow, it is sufficient to consider a front solution connecting the zero state to the $A_+$ state as discussed in [8].

In the framework of (19) with $Q = 0$, the front velocities $v_\mathrm{f}$ and wave numbers $k$ are:

$$v_\mathrm{f} = v_\mathrm{f}^{\mathrm{NL}} = \frac{g\sqrt{a}}{\sqrt{3h}}\left[-1 + 2(1 + 4\mu\frac{h}{g^2})^{1/2}\right] \text{ for } \mu_1 < \mu < \frac{3g^2}{4h} \tag{20}$$

$$v_\mathrm{f} = v_\mathrm{f}^{\mathrm{L}} = 2\sqrt{a\mu} \text{ for } \mu > \frac{3g^2}{4h} \tag{21}$$

$$k_\mathrm{i}^L = -\sqrt{\frac{\mu}{a}} \tag{22}$$

$$k_\mathrm{i}^{\mathrm{NL}} = -\sqrt{\frac{h}{3a}A_+^2} \tag{23}$$

If the front velocity $v_\mathrm{f}$ is positive (i.e. $\mu > \mu_\mathrm{M}$), the zero state is nonlinearly unstable, otherwise it is stable. $\mu_\mathrm{M} = \dfrac{-3g^2}{16h}$ is the Maxwell point where the two stable states $A = 0$ and $A_+$ have equal potential.

When $Q \neq 0$, we define a front velocity $V_\mathrm{f}$ in the laboratory frame by setting:

$$V_\mathrm{f} = v_\mathrm{f} - Q \tag{24}$$

If $V_\mathrm{f} < 0$, the instability is advected out of the medium and the instability is termed nonlinearly convective (NLC). On the contrary, when $V_\mathrm{f} > 0$, the instability is nonlinearly absolute (NLA) and the up branch $A_+$ invades the medium.

Therefore the criterion for zero front velocity ($v_f = Q$) gives the threshold for nonlinear and linear absolute instability:

$$Q < g\sqrt{3a/h} \qquad \mu_A = \mu^{NL} = \frac{3Q^2}{16a} + \frac{3}{16}\left(\frac{2gQ}{\sqrt{3ah}} - \frac{g^2}{h}\right) \tag{25}$$

$$Q > g\sqrt{3a/h} \qquad \mu_A = \mu^{L} = \frac{Q^2}{4a} \tag{26}$$

We can state from this result that when the through flow effects dominate over nonlinear ones ($Q > g\sqrt{3a/h}$), the absolute instability threshold is the linear one. Conversely, the absolute instability threshold is the nonlinear one when nonlinearities dominate over the through flow effects.

These thresholds coincide for $Q = Q_c = g\sqrt{3a/h}$. We have plotted in Figure 8. the thresholds corresponding to different instabilities.

Let us now discuss the influence of absolute instabilities on the hysteresis cycle observed when the Rayleigh number is varied (Fig 9). The hysteresis cycle has a lower limit at $\mu = \mu^{NL}$ and an upper one at $\mu = 0$, and these two limits depend on the through flow rate. Consequently, the hysteresis cycle vanishes for $Q \geq Q_L = \frac{g\sqrt{a}}{\sqrt{3h}}$.

For the determination of $\psi$, we have to distinguish two different situations:

i) For $Q \geq Q_c$, the scaling law $\Delta x = \frac{\beta_1}{(\mu - \mu^L)^{1/2}}$ holds. We can then proceed as for the supercritical bifurcation to determine $\psi$.
ii) For $Q < Q_c$, the nonlinear effects cause the transition to absolute instability. Analysis close above the convective/absolute transition [19] predicts that $\Delta x$

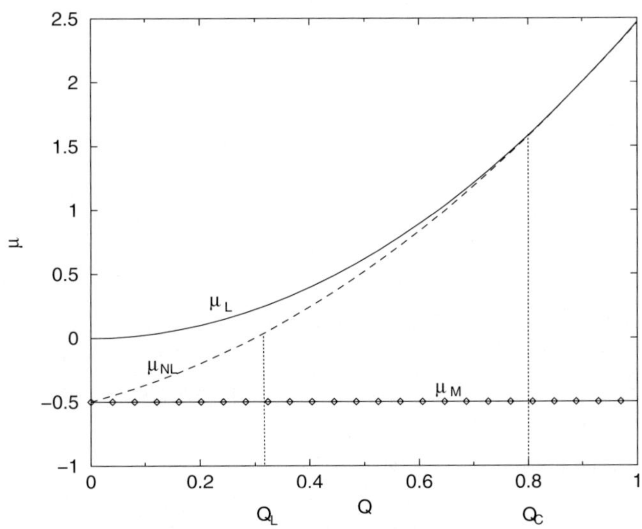


**Fig. 8.** Nonlinear and linear absolute threshold versus Q for $\psi = -5.5 \times 10^{-5}$

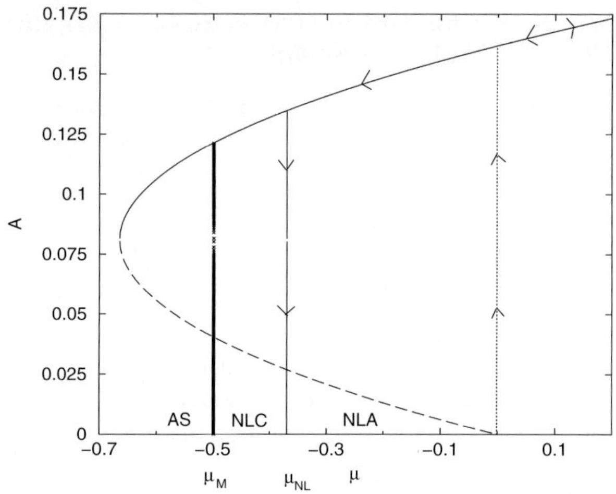


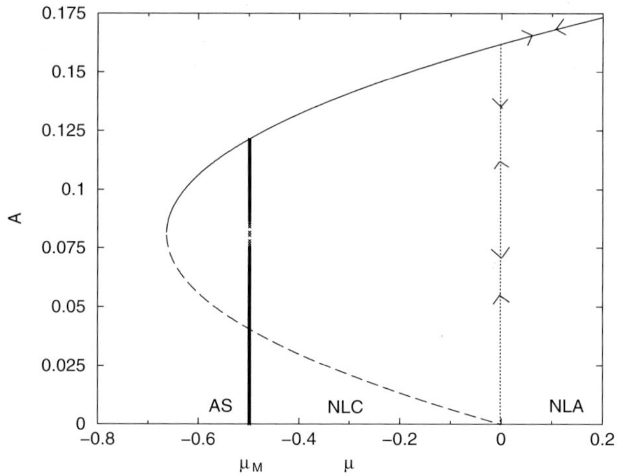


**Fig. 9.** Above: hysteresis cycle for $Q = 0.1 < Q_{\mathrm{L}}$ and $\psi = -5.5 \times 10^{-5}$, and below: disappearance of the hysteresis cycle $Q = 0.5 > Q_{\mathrm{L}}$ and $\psi = -5.5 \times 10^{-5}$

scales as Log $\dfrac{1}{(\mu - \mu^{\mathrm{NL}})}$. This scaling law indicates that $\mu^{\mathrm{NL}} = \mu - e^{-\Delta x}$ which means that the nonlinear global mode appears almost at $\mu^{\mathrm{NL}}$ near the outlet of an experimental cell of aspect ratio $L \gg 1$.

Consequently, the appearance of this global mode allows the measure of $\mu^{\mathrm{NL}}$ directly and by comparison to its theoretical value (Fig 10.), one can deduce the value of $\psi$ corresponding to the mixture used in the experiments.

## 5 Conclusion

In the present paper we investigate the stability of binary mixture in porous media subject simultaneously to an horizontal through flow and to a vertical

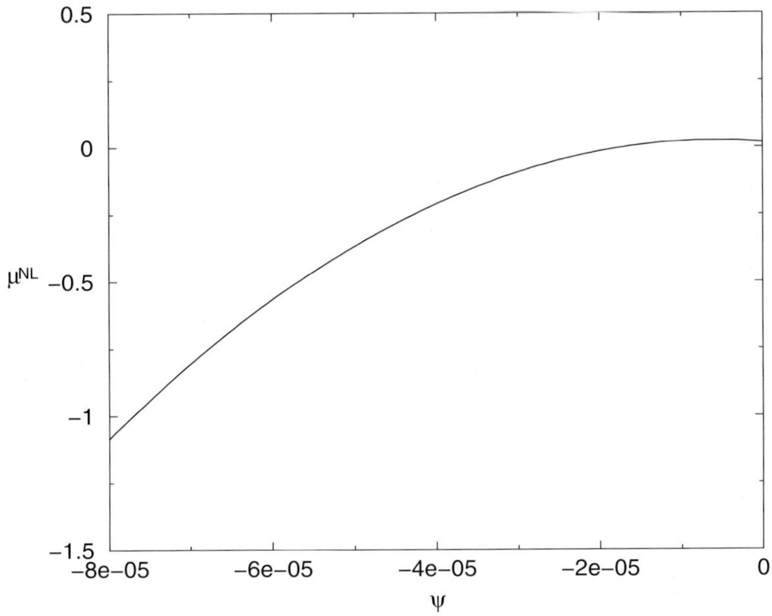


**Fig. 10.** Nonlinear ($\mu^{\rm NL}$) absolute instability thresholds versus $\psi$ for $Q = 0.1$

temperature gradient. In the framework of Darcy's law, we show that tridimensional spatially extended wave perturbations become unstable for the same values of the Rayleigh number as in the absence of through flow. The dependence on the through flow rate is trivial and amounts to a simple frequency shift. This picture changes drastically if, instead of considering extended perturbations, localized tridimensional perturbations must be regarded as it is often the case of open flow problems. The strategy followed to study the response of the system to these localized perturbations depends on the bifurcation nature. In the case of binary mixtures systems exhibiting supercritical bifurcations, linear criteria are appropriate to analyse the transition from convective to absolute instability. By using the linear criteria, we show that, at the absolute instability threshold, only traveling waves in the form of transversal rolls are amplified locally in the medium. The other tridimensional patterns are convected downstream and leave the experimental cell eventually. The transition curve from convective to absolute instability of transversal rolls indicates that the through flow may reduce and even may suppress the region of convective instability for $\psi < \psi_{\rm CT}$. On the contrary, for $\psi > \psi_{\rm CT}$, we observe a raise of this region induced by increasing the through flow rate. The interesting problem of determining the global frequency and the healing length of these transversal rolls beyond the absolute instability threshold is resolved by using recent theoretical advances [17, 19].

When the system undergoes a subcritical bifurcation, the linear stability criteria fail. This happens for some mixtures in a certain range of values of $\psi > \psi_{\rm CT}$. In order to describe the dynamics of these mixtures, we first derived a quintic

Ginzburg-Landau equation near threshold of subcritical bifurcation. After that, we carried out the analysis of nonlinear absolute and convective instabilities, i.e. instabilities with respect to finite amplitude perturbations. We showed that the front connecting the pure conductive state to the transversal rolls may be ruled either by the linear stability or by nonlinear properties. Which of these dominates depends mainly on the relative velocities of linear and nonlinear fronts. We have obtained analytically the characteristics of the switching fronts, namely their velocity, wavenumber and the corresponding thresholds. We also showed that an hysteresis cycle exists whose width is ruled by the through flow rate.

The last but not the least point of this conclusion is that we argue in this paper that an alternative way to determine the separation ratio $\psi$ is possible by combining our theoretical predictions and highly needed and desirable laboratory experiments dealing with mixed convection of binary mixtures.

## References


1. Nicolas X., Mojtabi A., Platten J.K.: Phys. Fluids, **9**, 337 (1997)
2. Carriere P., Monkewitz P.A.: J. Fluid Mech, **384**, 243 (1999)
3. Ouazzani M.T., Platten J.K., Müller H.W., Lücke M: Int. J. Heat and Mass Transfer, **38**, 875 (1995)
4. Ouazzani M.T., Caltagirone J.P., Meyer G., Mojtabi A.: Int. J. Heat and Mass Transfer, **32**, 261 (1989)
5. Dufour F., Néel M.C.: Phys. Fluids, **10**, 2198 (1998)
6. Jung.Ch., Lücke M., Büchel P.: Phys. Rev. E, **54**, 1510 (1996)
7. Joulin A., Ouarzazi M.N.: Comptes Rendus de l'Académie des Sciences, **328**, Série IIb, 311 (2000)
8. Chomaz J.M.: P.R.L., **69**, 1931 (1992)
9. Ouarzazi M.N.,Bois P.A.,: Eur. J. Mech., B/Fluids, **13**, 423 (1994)
10. Brevdo L.: ZAMP, **42**, 911 (1991)
11. Nield D.A., Bejan A.: Convection in porous media, (Springer)
12. Huke B., Lücke M., Büchel P.,Jung.Ch.: J. Fluid Mech, **408**, 121 (2000)
13. Linz S.J.,Lücke M.,: Phys. Rev. A, **35**, 3997 (1987)
14. Combarnous M.: Convection naturelle et convection mixte en milieu poreux. PhD Thesis, University of Sciences of Paris, 1970
15. Le Cotillec M: Instabilités dans une couche plane poreuse horizontale en convection mixte. PhD Thesis, University Pierre et Marie Curie, Paris, 1983
16. Babcock K.L., Ahlers G., Cannell D.S.: Phys. Rev. E, **50**, 3388 (1994)
17. Couairon A., Chomaz J.M.: Phys D, **132**, 428 (1999)
18. Brand H., Steinberg V.: Phys. Letters, **93A**, 333 (1983)
19. Couairon A., Chomaz J.M.: Phys. D, **108**, 236 (1997)


## Supplementary glossary

$A(x,t)$ complex amplitude
$C_0$ initial concentration
$D_{\mathrm{th}}$ thermal diffusivity
$D(\omega, k_{\mathrm{x}}, k_{\mathrm{y}})$ dispersion relation

| | |
|---|---|
| $(\boldsymbol{e_x}, \boldsymbol{e_y})$ | horizontal plane |
| $H$ | vertical thickness |
| $(k_x, k_y)$ | horizontal components of the wave vector q |
| $K$ | permeability |
| $Le$ | Lewis number |
| $P$ | pressure |
| $Q$ | through flow |
| $Ra$ | filtration Rayleigh number |
| $Re$ | Reynolds number |
| $S_{\mathrm{T}}$ | Soret coefficient |
| $T_0$ | temperature of the lower wall |
| $T_1$ | temperature of the upper wall |
| $\boldsymbol{u}$ | filtration velocity |
| $(U, V)$ | propagation velocity of a wave packet |
| $v_{\mathrm{f}}$ | front velocity |
| $V_{\mathrm{f}}$ | front velocity in the laboratory frame |
| $\Delta x$ | healing length |
| $\alpha$ | thermal expansion coefficient |
| $\beta$ | concentration expansion coefficient |
| $\mu_{\mathrm{M}}$ | Maxwell point |
| $\nu$ | kinematic viscosity |
| $\psi$ | separation ratio |
| $\psi_{\mathrm{CT}}$ | codimension two-point |
| $\rho$ | fluid density |
| $\omega_{\mathrm{n}}$ | complex frequency |

# Soret Effect in Multicomponent Flow Through Porous Media: Local Study and Upscaling Process

Bruno Lacabanne[1], Serge Blancher[2], René Creff[2], and François Montel[3]

[1] Laboratoire de Mathématiques Appliquées, Université de Pau, 64000 Pau, France
[2] Laboratoire de Transferts Thermiques, Université de Pau, France
[3] Elf Exploration Production, Centre Scientifique Jean Feger, Pau, France

**Abstract.** An efficient model for Soret effect in porous media is determined. After having described a local effect that can be generated at a microscopic scale, we present an homogenization technique that will allow us to determine precisely the macroscopic Soret coefficient and other caracteristic lengths such as thermal diffusivity and mass diffusive coefficient. Numerical simulations have been made for several geometrical configurations and corresponding porous media influence is discussed.

## 1 Introduction

Characterization and understanding of porous media is necessary in many situations such as thermal diffusion in petroleum reservoirs (the hydrocarbons trapped in porous reservoirs are submitted to the geothermal gradient which is partly responsible for species segregation by Soret effect, *cf.* [6], [15]). In a thermogravitational porous column, the coupling between slow convection and radial Soret effect generates extremely large separations that current models are unable to predict accurately (*cf.* [7]). In unsteady conditions, a small concentration gradient induced by Soret effect will relax much slower thant the initial thermal perturbation. This kinetic aspect for example deeply impacts the hydrodynamic stability of a mixture heated from below ([16]) which leads to complex spatial and temporal flow structures ([14]). The uncoordinated flow in a complex network inside a porous stone, mixes the species carried by the fluid, but it is well known that external force fields can also, on the contrary, separate species. In such media, the presence of numerous grains increases the effects of diffusion by the well-known dispersion phenomena. In recent works, we showed that it was possible to accumulate matter locally by thermal diffusion in certain configuration of velocity and thermal gradient ([9]). If, in most cases, the component distribution may not be changed, the introduction of a porous matrix could locally alter it. Numerous attemps in the field of thermal diffusion in porous media have been made (*cf.* [8]), by using Soret coefficient determined in fluid phases, but these simulations did not take into account local effects, that can generate the introduction of a porous matrix such as thermal gradient deviation (due to conductivity differences between solid and fluid media).

Fluid flow simulation on a geometrically complicated medium poses some mathematical important problems. When geometry and microscopic system are

W. Köhler and S. Wiegand (Eds.): LNP 584, pp. 448–465, 2002.

such hopelessly complicated, one has to pass to an homogenized macroscopic model system. In a first part, we describe the modelling of Soret effect in free medium and particularly the local coupling that can be made between Soret effect and convection. Then we present a technique of homogenization, we will adapt in order to upscale the energy and Soret equations. In a last part, the macroscopic coefficients are computed, and the results are discussed.

## 2 A local approach: the pore scale

We consider a $n$ components fluid flowing in a domain $\Omega$.

If $\boldsymbol{f}_i$ (resp. $\boldsymbol{J}_i$) denote the forces (resp. the fluxes), we dispose of the Onsager generalized linear laws

$$\boldsymbol{J}_i = \sum_j L_{ij} \boldsymbol{f}_j .$$

The Onsager phenomenologic coefficients matrix $L = (L_{ij})$ owns then a symmetry property, due to the "ergodicity" assumption. The entropy generation of the system is then expressed by the quadratic form

$$\frac{dS}{dt} = \sum_i \boldsymbol{J}_i . \boldsymbol{f}_i$$

$$\frac{dS}{dt} = \sum_{i,j} L_{ij} \boldsymbol{f}_i . \boldsymbol{f}_j$$

which, according to the $2^{nd}$ thermodynamic principle, must be non negative. Prigogine's theorem allows then to define the steady state of the system as being the one where entropy production is minimized. Hence a species distribution is created and we can write the heat and matter fluxes under the form

$$\boldsymbol{J}_q = -L_{qq} \frac{\nabla T}{T^2} - \sum_{k=1}^{n-1} \frac{L_{qk}}{T} [\nabla_{\mathrm{T}} (\mu_k - \mu_n)]$$

$$\boldsymbol{J}_i = -L_{iq} \frac{\nabla T}{T^2} - \sum_{k=1}^{n-1} \frac{L_{ik}}{T} [\nabla_{\mathrm{T}} (\mu_k - \mu_n)]$$

where we recognize the following terms

- diagonal effects:
  - Fourier's law $\boldsymbol{J}_q^{\mathrm{T}} = -\lambda \dfrac{\nabla T}{T^2}$
  - Fick's law: $\boldsymbol{J}_i^{\mu} = -\sum_{k=1}^{n-1} \dfrac{L_{ik}}{T} [\nabla_T (\mu_k - \mu_n)]$
- cross effects:
  - Dufour effect: $\boldsymbol{J}_q^{\mu} = -\sum_{k=1}^{n-1} \dfrac{L_{qk}}{T} [\nabla_T (\mu_k - \mu_n)]$

– Soret effect: $\boldsymbol{J}_i^{\mathrm{T}} = -L_{iq}\dfrac{\nabla T}{T^2}$.

By reducing the isothermal gradient of chemical potential $\nabla_T \mu_i$ to mass fraction gradient $\nabla c_i$ (isobare medium, buoyancy effects are neglected)[1], we obtain, for each component $i$, its flux expression

$$\begin{aligned}\boldsymbol{J}_i &= \boldsymbol{J}_i^{\boldsymbol{u}} + \boldsymbol{J}_i^{\mathrm{c}} + \boldsymbol{J}_i^{\mathrm{T}} \\ &= \rho c_i \boldsymbol{u} - \rho \sum_{j=1}^{n} D_{ij} \nabla c_j - \rho D_{\mathrm{T}}^{i} c_i (1 - c_i) \nabla T\end{aligned} \tag{1}$$

$\boldsymbol{J}_i^{\boldsymbol{u}}$, $\boldsymbol{J}_i^{c}$, $\boldsymbol{J}_i^{\mathrm{T}}$ being respectively the convective, diffusive and thermal diffusive fluxes of the $i^{th}$ component. In a same way, we can write the heat flux under the form

$$\begin{aligned}\boldsymbol{J}_q &= \boldsymbol{J}_q^{\boldsymbol{u}} + \boldsymbol{J}_q^{\mathrm{T}} + \boldsymbol{J}_q^{\mathrm{c}} \\ &= \rho C_p T \boldsymbol{u} - \lambda_{\mathrm{T}} \nabla T - \sum_{j=1}^{n} D_{\mathrm{c}}^{j} \nabla c_j\end{aligned}$$

$\boldsymbol{J}_q^{\boldsymbol{u}}$, $\boldsymbol{J}_q^{\mathrm{T}}$, $J_q^c$ being the convective, diffusive and Dufour fluxes . Energy and mass conservation can then be rewritten

$$\begin{cases} \rho C_p \partial_t T + \mathrm{div}(\boldsymbol{J}_{\mathrm{q}}) = 0 \\ \rho \partial_t c_i + \mathrm{div}(\boldsymbol{J}_{\mathrm{i}}) = 0 \end{cases}$$

or again, as $\rho C_p$ is taken constant,

$$\boxed{\begin{aligned} &\partial_t T + \mathrm{div}(\mathrm{T}\boldsymbol{u} - \mathrm{D_{th}}\nabla \mathrm{T} - \sum_{\mathrm{j=1}}^{\mathrm{n}} \mathrm{D_c^j} \nabla \mathrm{c_j}) = 0 \\ &\partial_t c_i + \mathrm{div}(\mathrm{c_i}\boldsymbol{u} - \sum_{\mathrm{j=1}}^{\mathrm{n}} \mathrm{D_{ij}} \nabla \mathrm{c_j} - \mathrm{D_T^i} \mathrm{c_i}(1 - \mathrm{c_i}) \nabla \mathrm{T}) = 0 \end{aligned}} \tag{2}$$

Soret coefficient of each component is then defined by the relation

$$S_{\mathrm{T}}^{i} = \frac{D_{\mathrm{T}}^{i}}{D_{ii}}.$$

For this study, we neglect the Dufour effect (which is justified in liquids) and restrain our interest in the Soret effect. The strong coupling of the two equations could also be treated, but one needs to introduce new tools which are not described here.

In the following part, we present a local configuration of the velocity and thermal field that allows the accumulation of matter at a microscopic scale.

[1] The validity of this assumption is discussed in section 6

## 3 Local coupling between velocity and Soret effect

We present shortly[2] here a situation that shows how variations of velocity profile coupled with Soret effect can influence locally the segregation of species in a multicomponent mixture. We modelize the pore with a tube of lentgh $L$ through which a binary mixture flows slowly ($Re \ll 100$). The velocity profile is given by the Poiseuille's flow profile of the dimensionless form :

$$u(x,r) = u(r) = u_{\max}(1 - r^2) \tag{3}$$

where $x$ and $r$ denote the dimensionless cylindrical coordinates. Governing equations for heat and mass fraction given by the system (2) can be rewritten (in case $n = 2$) in the following form

$$\begin{cases} \partial_t T + (1 - r^2)\dfrac{\partial T}{\partial x} = \dfrac{1}{Pe_{\mathrm{T}}}\Delta T \\ \partial_t c + (1 - r^2)\nabla c = \dfrac{1}{Pe}\Delta c + \dfrac{S_{\mathrm{T}}^*}{Pe}\Delta T \end{cases} \tag{4}$$

where $S_{\mathrm{T}}^*$ denotes the dimensionless quantity $S_{\mathrm{T}}^* = \Delta T \dfrac{D_{\mathrm{T}}}{D}(1 - c(x = 0))$, and $Pe = \dfrac{u_{\max} L}{D}$ the mass Peclet number . We apply on the walls of the tube a thermal profile (illustrated in figure 2) of the form

$$T_{\mathrm{wall}}(x) = \begin{cases} T_0 \text{ if } x < x_1 \\ T_0 + \dfrac{\Delta T(x - x_1)}{x_2 - x_1} \text{ if } x_1 \leq x \leq x_2 \\ T_0 + \Delta T \text{ otherwise} \end{cases} \tag{5}$$

where $T_0 < T_1$, $0 < x_1 < x_2 < L$ and $\Delta T = T_1 - T_0$. Numerical resolution of system (4) has been made. The simulations made showed that it was possible to accumulate matter locally (as depicted in figure 1). In a first time Soret effect is taking place in the capillary at the place where the thermal gradient is not equal to zero (step (1) in figure 1). The shape of the peak is steeper on the capillary wall as convection does not occur at this place. Then convection coupled with Soret effect and diffusion removes the "negative" part of the separation (step (2) and (3) in figure 1). At steady state, it remains an accumulation of matter (step (4) in figure 1). One has to be carefull with the fact that this accumulation is a dynamic one, this peak is continuously fed by convection. With the help of an integration in the 1-D case, the quantity of matter accumulated with this process has been determined analytically: at steady state, the total amount of matter trapped in the capillary is of the order of $\dfrac{S_{\mathrm{T}}^*}{Pe\bar{u}}$ where $\bar{u}$ is the dimensionless mean value of the velocity in the tube. It follows that the Soret coefficient value can be determined by measuring this quantity. A multistage process has even been envisaged, in

[2] This part will be developped in a future publication [9] and was presented at IMT3 in a poster session.

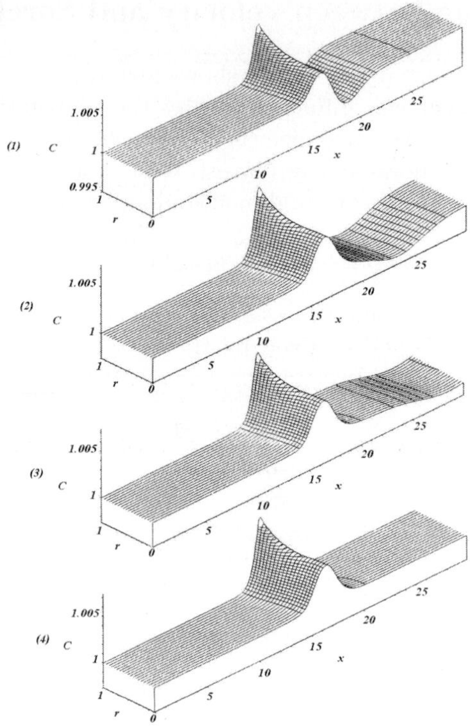

**Fig. 1.** Relative mass fraction $\frac{c(.)}{c(x=0)}$ in the capillary for $Pe = 10$, $S_{\mathrm{T}}^{*} = 0.0625$

order to increase quantity of matter accumulated in the capillary and so to determine more precisely the real value of $S_{\mathrm{T}}$. This process will be described in [9]. All this demonstrates that variations of velocity and thermal gradient can create local perturbations of species segregations in a porous medium. The following part is devoted to the process that allows us to obtain semi analytic expressions of macroscopic coefficients, giving the "free media" ones.

## 4 From microscopic to macroscopic scale

There are several way to obtain macroscopic laws for a porous medium : direct numerical techniques (such as the one used by E. Arquis and J.P. Caltagirone in [3]), statistical methods or analytical ways. In a general way, the last example is the most commonly used for the derivation of microscopic laws. It can be divided in two techniques: the REV method (representative elementary volume method) (used by M. Quintard in [17]) and the homogenization . U Hornung recalls in [10] that the basic difference between these two methods is that REV methods used averaging formulas such as

$$\langle T \rangle (x) = \int_{\Omega_x} T(y) dy \tag{6}$$

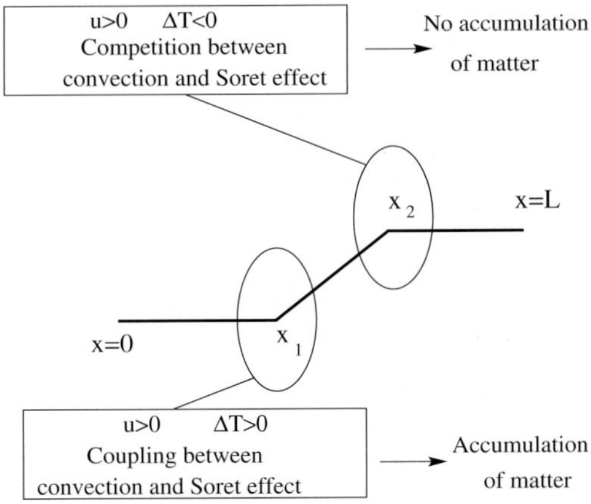


**Fig. 2.** Thermal profile along the capillary and corresponding effect on concentrations

whereas homogenization allows the derivation of microscopic laws by letting the microscale tend to zero. This technique has already been used a lot and some examples of applications can be found in [5]. This technique is applied to both thermal diffusivity and diffusion coefficient. Homogenization is a powerfull tool used to derive global models from microscopic processes, which supports the modelling process whereever spatial upscaling is necessary. We base this study on the assumption of the porous medium periodicity . The medium is assumed to be the periodical repetition of an elementary cell, denoted by $Y_\varepsilon$ ($\varepsilon$ being the characteristic length of the elementary cell which is destined to tend to 0). We assume that we have a single phase flow (in order to understand basic mechanism of all effects studied here). The homogenization process consists in searching the limit state of the corresponding equations, when $\varepsilon \longrightarrow 0$.

### 4.1 Specific notations

In the following section we use special notations in order to differenciate local and global scale, free and porous media, or fluid and solid phase. This differenciation is very important in this part because homogenization gives a lot of cases where the two notions can be mixed (in particular in case of nonlinear effect).

As a classical notation, the subscript $\varepsilon$ in a variable, will indicate that this variable is taken at a local scale. When the variable is noted without subscript, it indicates that we consider it's macroscopic state. In a same way, the notation $x_f$ will denote a length considered in fluid phase, on the opposite $x_s$ will represent this same length in the solid phase.

Our aim here is not to calculate the global solution of thermal equation or Soret equation, but rather to obtain an efficient model of Soret effect at a reservoir scale.

We have to introduce the notion of multiple scale equations. In order to use homogenization device, we must differenciate mathematically the two scales that

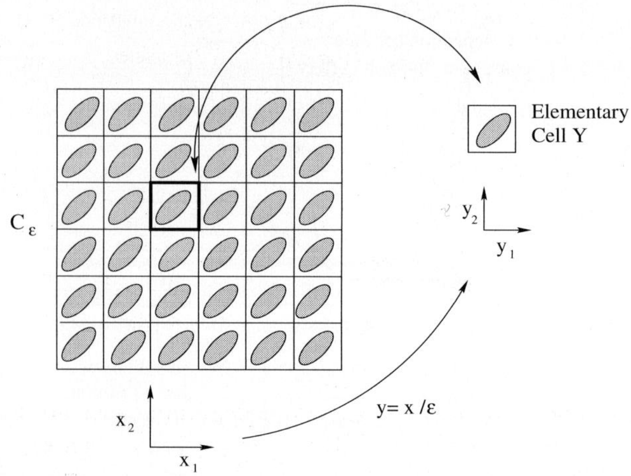


**Fig. 3.** Heterogeneous periodic porous medium: multiple scale notion

concern us: the macroscopic and the microscopic one. If the spatial coordinates for the first one are classically noted $x = (x_1, x_2, x_3)$, we endow the microscopic system with the new local coordinates

$$y = \begin{pmatrix} y_1 \\ y_2 \\ y_3 \end{pmatrix} = \frac{1}{\varepsilon} \begin{pmatrix} x_1 \\ x_2 \\ x_3 \end{pmatrix} \tag{7}$$

as depicted in figure 3. One has to be cared with the fact that this two scales notion introduces new derivation operator. Indeed, the usual operators must be rewritten under the form

$$\nabla = \nabla_x + \frac{1}{\varepsilon} \nabla_y \tag{8a}$$

$$\mathrm{div} = \mathrm{div}_\mathrm{x} + \frac{1}{\varepsilon} \mathrm{div}_\mathrm{y} \tag{8b}$$

$$\Delta = \Delta_x + \frac{2}{\varepsilon} \Delta_{xy} + \frac{1}{\varepsilon^2} \Delta_y \tag{8c}$$

where we have introduced the notation $\Delta_{xy} = \sum_{i=1}^{3} \frac{\partial^2}{\partial x_i \partial y_i}$.

The domain $\Omega$ will be separated in two parts $\Omega_\mathrm{f}$ et $\Omega_\mathrm{s}$ as for the unit cell $Y$ (separated in its fluid part $Y_\mathrm{f}$ and its solid part $Y_\mathrm{s}$).

### 4.2 Thermal diffusivity

Homogenization of thermal transport has been studied a lot and numerous results have been given for a macroscopic model for Energy equation ([10] or [17]). We describe in this part the process that will allow us to obtain the macroscopic

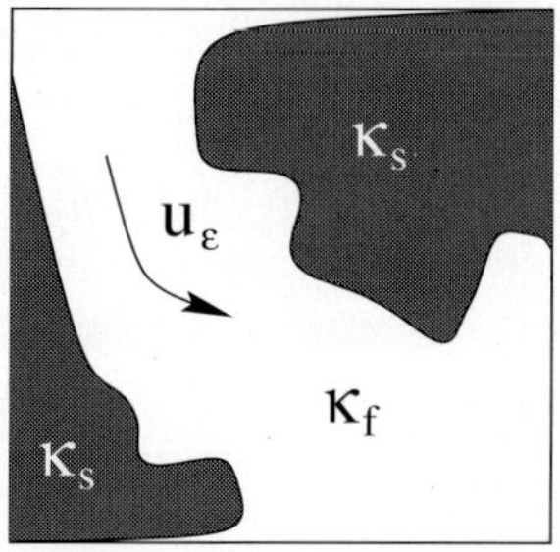


**Fig. 4.** Difference of thermal diffusivity between solid and fluid phase at a local scale

model which is of great importance for Soret effect modelisation. Porous materials are saturated with oils which, under a local pressure gradient, migrates and transports energy through the media. The energy equation is given by

$$\mathrm{div}(-\mathrm{D_{th}}(.)\nabla \mathrm{T}_\varepsilon + \chi(.)\mathrm{T}_\varepsilon \boldsymbol{u}_\varepsilon) = 0 \tag{9}$$

$D_{\text{th}}$ being the function defined by

$$D_{\text{th}} = \begin{cases} D^{\text{f}}_{\text{th}} \text{ in fluid medium } Y_{\text{f}} \\ D^{\text{s}}_{\text{th}} \text{ in solid medium } Y_{\text{s}} \end{cases}$$

(where $D^{\text{f}}_{\text{th}}$ et $D^{\text{s}}_{\text{th}}$ denote the thermal diffusivity of fluid and solid phase), and $\chi$ is the flow characteristic function such as,

$$\chi = \begin{cases} 1 \text{ in fluid medium} \\ 0 \text{ anywhere else.} \end{cases}$$

By this way we own a unique equation, modelling thermal transfer in porous media at a local scale. This equation is associated to thermal fluxes condition at the fluid/solid interface.

**Derivation of microscopic law.** We start our study by the assumption that the temperature $T_\varepsilon$ defined on a part $\Omega_\varepsilon$ admits one asymptotic developement with respect to $\varepsilon$, *i.e.* that $T_\varepsilon$ can be rewritten under the form

$$T_\varepsilon(x) = T_0(x,y) + \varepsilon T_1(x,y) + \varepsilon^2 T_2(x,y) + \varepsilon^3 T_3(x,y)... \tag{10}$$

with $y = \dfrac{x}{\varepsilon}$ and $T_i(x,y)$ being periodic functions of variable $y$ (these developement have been already introduced by **Sc**$^\star$ A. Bensoussan and al. in [4] and later by **Sc**$^\star$ U. Hornung in [10] and [11]). In the following part, we assume that macroscopic state of velocity $\boldsymbol{u}$ is known (Darcean flow), and we do not study the homogenization of velocity profile (the derivation of Navier-Stokes equations has been already made in numerous studies, such as [1]). The proof of this transition is based on the same process as the following one.

By reintroducing (10) in (9) we get, at the order 1 a polynomial expression of $\varepsilon$ of the form

$$A_{-2}(x,y)\varepsilon^{-2} + A_{-1}(x,y)\varepsilon^{-1} + A_0(x,y)\varepsilon^0 + A_1(x,y)\varepsilon^1 = 0 \tag{11}$$

where

$$\begin{aligned} A_{-2}(x,y) &= \mathrm{div_y}(\mathrm{D_{th}}(.)\nabla_{\mathrm{y}}\mathrm{T_0}) &\text{(12a)}\\ A_{-1}(x,y) &= \mathrm{div_x}(\mathrm{D_{th}}(.)\nabla_{\mathrm{y}}\mathrm{T_0}) + \mathrm{div_y}(\mathrm{D_{th}}(.)\nabla_{\mathrm{x}}\mathrm{T_0}) + \mathrm{div_y}(\mathrm{D_{th}}(.)\nabla_{\mathrm{y}}\mathrm{T_1}) \\ A_0(x,y) &= \mathrm{div_x}(\mathrm{D_{th}}(.)\nabla_{\mathrm{x}}\mathrm{T_0}) + \mathrm{div_x}(\mathrm{D_{th}}(.)\nabla_{\mathrm{y}}\mathrm{T_1}) + \mathrm{div_y}(\mathrm{D_{th}}(.)\nabla_{\mathrm{x}}\mathrm{T_1}) \\ &\quad + \mathrm{div_x}(\mathrm{T_0}\boldsymbol{u}_0(\mathrm{x})) &\text{(12b)}\\ A_1(x,y) &= \mathrm{div_x}(\mathrm{D_{th}}(.)\nabla_{\mathrm{x}}\mathrm{T_1}). &\text{(12c)} \end{aligned}$$

We now identify the different terms in $\varepsilon$ powers in the equation (11). In this way, the equality $A_{-2}(x,y) = 0$ gives

$$\mathrm{div_y}(\mathrm{D_{th}}(.)\nabla_{\mathrm{y}}\mathrm{T_0}) = 0 \ \text{ for } \mathrm{y} \in \mathrm{Y}. \tag{13}$$

which proves that $T_0(x,y) = T_0(x)$ as $T_0(x,y)$ is an $Y$-periodic function. In a same way, using this result and the identification of terms in $\varepsilon^{-1}$, we get

$$\mathrm{div_y}(\mathrm{D_{th}}(.)\nabla_{\mathrm{y}}\mathrm{T_1}) = -\mathrm{div_y}(\mathrm{D_{th}}(.)\nabla_{\mathrm{x}}\mathrm{T_0})$$

that can be rewritten

$$\mathrm{div_y}(\mathrm{D_{th}}(.)\nabla_{\mathrm{y}}\mathrm{T_1}) = -\sum_{\mathrm{j}=1}^{3} \partial_{\mathrm{y_j}}(\mathrm{D_{th}}(.)\partial_{\mathrm{x_j}}\mathrm{T_0}). \tag{14}$$

By the consideration of the following unit-cell problem,

$$\begin{aligned} \mathrm{div_y}(\mathrm{D^f_{th}}(\nabla_{\mathrm{y}}\sigma_{\mathrm{j}}(\mathrm{y}) + \boldsymbol{e}_{\mathrm{j}}) &= 0 \ \text{ for } y \in Y_f &\text{(15a)}\\ \mathrm{div_y}(\mathrm{D^s_{th}}\nabla_{\mathrm{y}}\sigma_{\mathrm{j}}(\mathrm{y}) + \boldsymbol{e}_{\mathrm{j}}) &= 0 \ \text{ for } y \in Y_s &\text{(15b)} \end{aligned}$$

where $(\sigma_j)$ are $Y$-periodical solutions, we can express $T_1$ as

$$T_1(x,y) = \sum_{j=1}^{3} \sigma_j(y)\partial_{x_j} T_0(x) + T_1(x) \tag{16}$$

$T_1(x)$ being independent of $y$. We integrate the equality $A_0(x,y) = 0$ on $Y$ and get

$$\begin{aligned} &-\int_{\mathrm{Y}} \mathrm{div_x}(\mathrm{D_{th}}(.)\nabla_{\mathrm{x}}\mathrm{T_0})\mathrm{dy} - \int_{\mathrm{Y}} \mathrm{div_x}(\mathrm{D_{th}}(.)\nabla_{\mathrm{y}}\mathrm{T_1})\mathrm{dy} \\ &-\int_{\mathrm{Y}} \mathrm{div_y}(\mathrm{D_{th}}(.)\nabla_{\mathrm{x}}\mathrm{T_1})\mathrm{dy} + \int_{\mathrm{Y}} \chi(\mathrm{y})\mathrm{dy}\,(\nabla_{\mathrm{x}}\mathrm{T_0}\boldsymbol{u}_0(\mathrm{x})) = 0 \end{aligned} \tag{17}$$

First term can be rewritten $-\int_Y D_{\mathrm{th}}(y)dy\,\Delta_x T_0$ with the help of (13).

Second term can be rewritten $-\sum_{i,j=1}^{3}\int D_{\mathrm{th}}(.)\partial_{y_i}\sigma_j(y)\partial_{x_ix_j}T_0(x)$ after the derivation of the equality (16).

The third term is removed with the help of a Green formula and the periodicity of function $T_1$.
The last term is kept, observing that

$$\int_Y \chi(y)dy = \phi.$$

where $\phi$ is the porosity of the medium. We can define now the tensor $\tilde{\Lambda} = (\Lambda_{kl})$ by the expression

$$\Lambda_{kl} = \int_Y D_{\mathrm{th}}(y)(\delta_{kl} + \partial_{y_k}\sigma_l(y))dy. \tag{18}$$

The equation (17) is then rewritten

$$\sum_{i=1}^{3}\left(-\sum_{j=1}^{3}\lambda_{ij}\partial_{x_ix_j}T_0(x) + \phi\partial_{x_i}T_0(x)\boldsymbol{u}_0(x)\right) = 0$$

which is equivalent to

$$\mathrm{div}(-\tilde{\Lambda}\nabla \mathrm{T} + \phi \mathrm{T}\boldsymbol{u}) = 0. \tag{19}$$

**Result 1**
*The macroscopic equation for heat transfer in the porous medium is*

$$\mathrm{div}(-\tilde{\Lambda}\nabla \mathrm{T} + \phi \mathrm{T}\boldsymbol{u}) = 0$$

*where $\tilde{\Lambda}$ is a symmetrical tensor defined by*

$$\begin{aligned}\forall(k,l)\in\{1..3\}^2 \quad \Lambda_{kl} &= \int_Y D_{\mathrm{th}}(y)[\boldsymbol{e}_k + \nabla_y\sigma_k(x,y)].\boldsymbol{e}_l dy \\ &= \int_{Y_\mathrm{f}} D^{\mathrm{f}}_{\mathrm{th}}(\nabla_y\sigma_k + \boldsymbol{e}_k).\boldsymbol{e}_l dy + \int_{Y_\mathrm{s}} D^{\mathrm{s}}_{\mathrm{th}}(\nabla_y\sigma_k + \boldsymbol{e}_k).\boldsymbol{e}_l dy\end{aligned}$$

*$(\sigma_k)$ being the solutions of the following unit-cell problem*

$$\begin{cases}\sigma_k\ Y-\textit{periodic}\\ \mathrm{div_y}(\mathrm{D^f_{th}}[\nabla_\mathrm{y}\sigma_\mathrm{k} + \boldsymbol{e}_\mathrm{k}]) = 0 \textit{ in } \mathrm{Y_f}\\ \mathrm{div_y}(\mathrm{D^s_{th}}[\nabla_\mathrm{y}\sigma_\mathrm{k} + \boldsymbol{e}_\mathrm{k}]) = 0 \textit{ in } \mathrm{Y_s}\\ \left[D^{\mathrm{f}}_{\mathrm{th}}(\nabla_y\sigma_k + \boldsymbol{e}_k) - D^{\mathrm{s}}_{\mathrm{th}}(\nabla_y\sigma_k + \boldsymbol{e}_k)\right].\boldsymbol{n} = 0 \textit{ on } \partial Y_\mathrm{f}\backslash\partial Y.\end{cases} \tag{20}$$

*Remark 1 A time dependant problem would have a similar macroscopic formulation. Indeed, as it is shown in next part, the homogenized problem would have a time derivative term, similar to the microscopic one, but weighted wih the porosity of the medium $\phi$.*

The macroscopic energy equation has been determined. One can notice that macroscopic heat transfer are traduced by equations of same type that at a local scale. The porosity which is present in front of convective term is used to ponderate effects of fluid phase convective motion relatively to diffusion effects. A study has been made on comparison for this two terms with homogenization (H. Hene, in [10]), so that it will be not discussed here. The determination of macroscopic energy equation in porous medium will not be complete until the problem (20) (and by the same way the tensor $\tilde{\Lambda}$) is computed. This is done in part 5.

### 4.3 Global diffusion and Soret coefficients

We consider in this part the treatement of mass conservation equations for each component in the fluid phase. We attempt to apply the method described in the last part to the diffusive coefficients. P. V. Jamet assumed in [12] that macroscopic equations of Soret effect would be of the same form than the microscopic ones, differing from a multiplicative constant. This is what we demonstrate here, and moreover we give the "analytical" expression of the macroscopic coefficients.

We consider the following convective-diffusive equations with Soret effect on the domain $\Omega_{\mathrm{f}}$
$\forall i \in \{1..n\}$,

$$\partial_t c_{i,\varepsilon} + \boldsymbol{u}_\varepsilon . \nabla c_{i,\varepsilon} - \sum_j D_{ij} \Delta c_{j,\varepsilon} - D_{\mathrm{T}}^i \mathrm{div}(\mathrm{c}_{\mathrm{i},\varepsilon}(1 - \mathrm{c}_{\mathrm{i},\varepsilon}) \nabla \mathrm{T}_\varepsilon) = 0 \tag{21}$$

associated with initial and boundary conditions [3] on the fluid/solid interface

$$\begin{cases} \displaystyle\sum_j D_{ij} \frac{\partial c_{j,\varepsilon}}{\partial n} + D_{\mathrm{T}}^i c_{i,\varepsilon}(1 - c_{i,\varepsilon}) \frac{\partial T_\varepsilon}{\partial n} = 0 \\ c_{i,\varepsilon}(x, t = 0) = c_i^0. \end{cases}$$

The problem of homogenization for Soret effect is a little more complicated than the previous one and the determination of macroscopic equations requires mathematical tools that we will not develop here. We need to define particularly the notion of asymptotic developement for time dependent functions. Hence we need to assume that the functions $c_{i,\varepsilon}$, admit the following asymptotic developement

$$c_{i,\varepsilon}(x, t) = c_i^0(x, y, \tau) + \varepsilon c_i^1(x, y, \tau) + \varepsilon^2 c_i^2(x, y, \tau) + \varepsilon^3 c_i^3(x, y, \tau)... \tag{22}$$

[3] The viability of these conditions are discussed in part 6

where $\tau = \varepsilon^2 t$ is the microscopic time variable. Furthermore, the unknowns $c_{i,\varepsilon}$ must be defined on the whole domain. That's why we have to introduce the notation

$$\begin{cases} \bar{c}_{i,\varepsilon} = c_{i,\varepsilon} \text{ sur } \Omega_{\mathrm{f}} \\ \bar{c}_{i,\varepsilon} = 0 \text{ sur } \Omega \backslash \Omega_{\mathrm{f}} \end{cases} \tag{23}$$

Homogenization of thermal diffusive term is not detailed in the following part and may be obtained using analogous arguments.

**Result 2**

*The homogenized equations governing mass conservation (macroscopic convecto-diffusive-thermal diffusive equations) are the following ones:*

$$\phi \partial_t c_i + \mathrm{div}\left( \phi \mathrm{c_i} \boldsymbol{u} - \sum_{\mathrm{j}} \mathrm{D_{ij}} \tilde{\Upsilon} \nabla \mathrm{c_j} - \mathrm{c_i}(1 - \mathrm{c_i}) \tilde{\Sigma}_{\mathrm{i}} \nabla \mathrm{T} \right) = 0 \tag{24}$$

*where we have defined the symmetrical tensors $\tilde{\Sigma}_i$ and $\tilde{\Upsilon}$ by*

$$\begin{aligned} \forall (k,l) \in \{1..3\}^2 \quad (\tilde{\Sigma}_i)_{kl} &= D^i_{\mathrm{T}} (\tilde{\Upsilon})_{kl} \\ &= \frac{D^i_{\mathrm{T}}}{\mathrm{mes(Y)}} \int_{Y_{\mathrm{f}}} (\nabla_y \omega_k + \boldsymbol{e_k})(\nabla_y \omega_l + \boldsymbol{e_l}) dy \end{aligned}$$

*$\omega_k$ being the solutions of the unit-cell problem $Y$*

$$\begin{cases} \omega_k \; Y - periodic \\ -\mathrm{div_y}(\nabla_{\mathrm{y}} \omega_{\mathrm{k}} + \boldsymbol{e}_{\mathrm{k}}]) = 0 \;\; in \, \mathrm{Y_f} \\ (\nabla_y \omega_k + \boldsymbol{e}_k).\boldsymbol{n} = 0 \;\; on \, \partial Y_{\mathrm{f}} \backslash \partial Y. \end{cases} \tag{25}$$

*Remark 2* *The tensor $\tilde{\Upsilon} = (\Upsilon_{kl})$ can be rewritten, with the help of equations (25) in the following form:*

$$\begin{aligned} \Upsilon_{kl} &= \frac{1}{\mathrm{mes(Y)}} \left\{ \int_{Y_{\mathrm{f}}} \frac{\partial \omega_k}{\partial y_l} dy + \mathrm{mes(Y_f)} \delta_{\mathrm{kl}} \right\} \\ &= \phi \delta_{kl} + \frac{1}{\mathrm{mes(Y)}} \int_{Y_{\mathrm{f}}} \frac{\partial \omega_k}{\partial y_l} dy \end{aligned} \tag{26}$$

*We find a formulation similar to the one of $\tilde{\Lambda}$ in result 1, having on the diagonal, the porosity (modulo the diffusion coefficients). The symmetry of tensor $\tilde{\Upsilon}$ is no more obvious, but implicitely given by elementary problems.*

It's interesting to notice here that, even if diffusion and thermal diffusion are very different physical phenomenons, the tensor introduced by homogenization in their expression is the same (modulo a coefficient). That comes from the fact that mathematical treatment does not make any difference between these two effects: both are provoked by fluxes, and in this sense they are similar. Of course these likenesses would disappear if the microscopic model of Soret effect would take into account local variations of Peclet number or variations of Soret coefficient versus temperature. The obtained model would then be more complicated, and has not been computed here.

## 5 Numerical results

Numerical resolutions of "unit-cell systems" (20) and (25) have been made. A finite volume method has been used on an unstructured triangular mesh. We present in the following parts the results of these computations and the values of the corresponding macroscopic tensors.

### 5.1 Geometrical considerations

We based our study on the assumption that the porous medium was defined as the periodic repetition of an elementary $\varepsilon$-size cell partially filled with a solid. One can be convinced that the geometrical configuration chosen for unit-cell is of greater importance for the computation of equivalent coefficients. Several assumptions must be made on this geometrical configuration. These assumptions have been already defined, by example in [1]. The fluid phase $Y_f$ must be connected, and we consider a porous medium without any fluid inclusion. In our case, the porous medium is modelized by grain stacks, in a situation which is closer to a sand than to a rough rock. This type of modelling has already been used, particularly in [2] and [3]. Each medium considered (solid and fluid phase) are supposed to be isotropic, homogeneous.

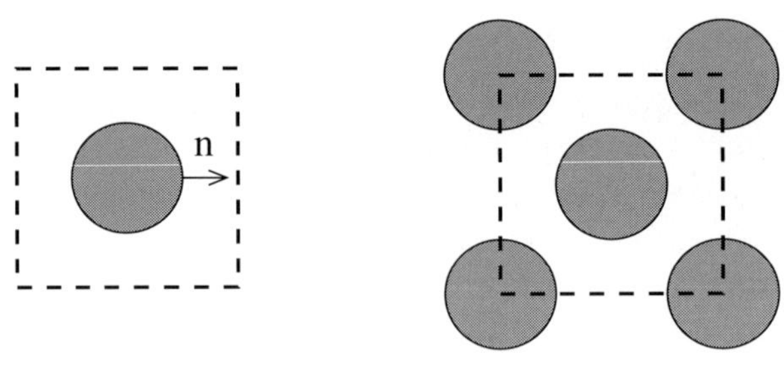


(a) Cell 1, $\phi = 0.72$ (b) Cell 2, $\phi = 0.43$

**Fig. 5.** 2 examples of elementary representative volums

One can notice that in these 2 examples of geometry, effects of the fluid-solid interface are equivalent, because of the "periodicity of the medium" assumption. The only difference between these two cases comes from the value of the porosity which is taken into account for the determination of the effective coefficients.

### 5.2 Physical parameters

We give in table 1 the data used to the computation of effective thermal diffusivity and effective diffusive coefficients. These parameters have been chosen for an oil in deposits conditions and for an unsatured mean sandstone. These data may vary depending on the saturation degree of the medium, the temperature and the pressure inside the domain. However, we consider a situation far from the critical point, where the variations against these two length are relatively slight. Soret and diffusion coefficients values are the ones experimentally determined in [13] and [18].

**Table 1.** Local physical parameters of each phase

| Physical parameters | Fluid phase ($Y_\mathrm{f}$) | Solid phase ($Y_\mathrm{s}$) |
|---|---|---|
| $\lambda$ ($\mathrm{Wm^{-1}K^{-1}}$) | $2 \times 10^{-1}$ | $9 \times 10^{-1}$ |
| $C_p$ ($\mathrm{Jkg^{-1}K^{-1}}$) | $2 \times 10^{3}$ | $7.6 \times 10^{2}$ |
| $\rho$ ($\mathrm{kgm^{-3}}$) | $7.5 \times 10^{2}$ | $2.1 \times 10^{3}$ |
| $D_\mathrm{th}$ ($\mathrm{m^2s^{-1}}$) | $1.3 \times 10^{-7}$ | $5.6 \times 10^{-7}$ |
| $D$ ($\mathrm{m^2s^{-1}}$) | $2.75 \times 10^{-7}$ | - |
| $S_\mathrm{T}$ ($\mathrm{K^{-1}}$) | $4.92 \times 10^{-3}$ | - |

### 5.3 Comments

In the following part, we present the results of numerical resolution of the elementary problems. Numerical solutions are depicted in figure 6 for energy equation and in figures 7 and 8 for the macroscopic coefficients of the mass conservation equations. One can notice the symmetry of the solutions and the effects of the discontinuity of function $D_\mathrm{th}(.)$ at the fluid-solid interface. Moreover, the solution $\omega$ depicted in figure 8 is no longer symmetrical, even if the medium is an isotropic one. It can be explained by the fact that the solution of the elementary problems are defined "modulo" an additive constant (this characteristic is not really important for the computation of macroscopic coefficients as we just consider "energy" expressions). Moreover, one knows that if we consider a symmetrical unit-cell and if each medium is isotropic, the macroscopic tensors are isotropic too. We can see that it is not the case for some of the coefficients given in table 2, but this difference can be explained by the unstructured mesh used.

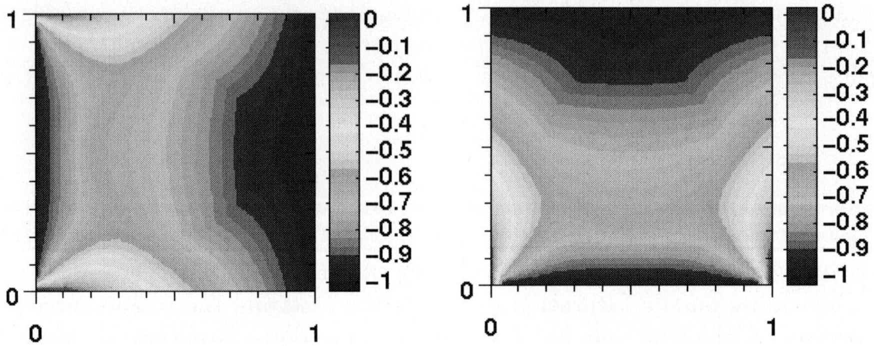


(a) Solution $\sigma_1$ (b) Solution $\sigma_2$

**Fig. 6.** Solutions $(\sigma_k)_{k=1,2}$ of problem (20) for unit cell 2

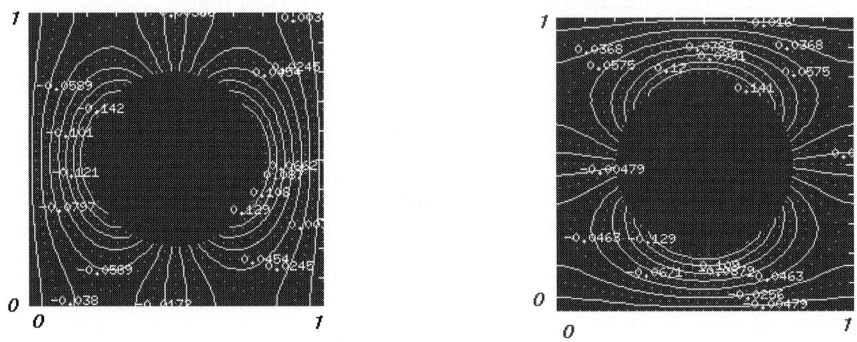

(a) Solution $\omega_1$ (b) Solution $\omega_2$

**Fig. 7.** Solutions $(\omega_k)_{k=1,2}$ of problem (25) for unit cell 1

# 6 On the "well-posedness" of local equations

Consideration of equations (Energy equation and Soret effect equation) in the form that one uses, remains available in the condition that one does not consider a multiphase system. In the porous media case, the consideration of these two equations associated with the boundary condition of type "zero flux" brings necessarily a contradiction. Indeed, if we look the "closed" equation given by the condition

$$\boldsymbol{J}_i \boldsymbol{n} = 0 \text{ on } \partial\Omega_{\mathrm{f}}$$

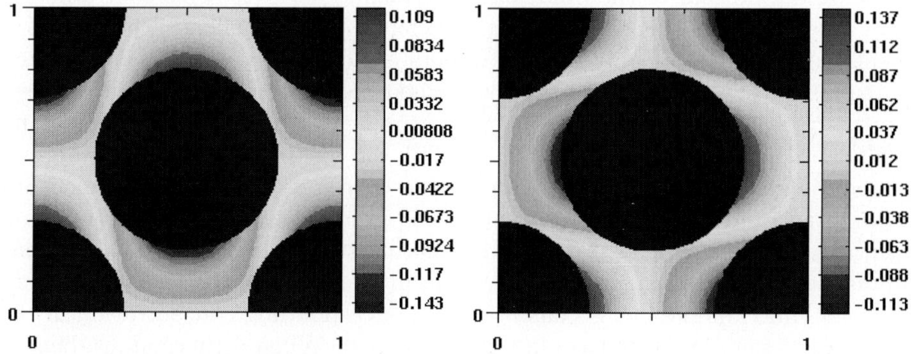


(a) Solution $\omega_1$ (b) Solution $\omega_2$

**Fig. 8.** Solutions $(\omega_k)_{k=1,2}$ of problem (25) for unit cell 2

**Table 2.** Tensors obtained for several geometrical configurations

| Tensor \ Cell | Cell 1 | Cell 2 |
|---|---|---|
| $\tilde{\Lambda}$ | $\begin{pmatrix} 3.49 \times 10^{-7} & 0 \\ 0 & 3.48 \times 10^{-7} \end{pmatrix}$ | $\begin{pmatrix} 2.67 \times 10^{-7} & 0 \\ 0 & 2.66 \times 10^{-7} \end{pmatrix}$ |
| $\tilde{\Sigma}$ | $\begin{pmatrix} 2.77 \times 10^{-3} & 0 \\ 0 & 2.77 \times 10^{-3} \end{pmatrix}$ | $\begin{pmatrix} 1.313 \times 10^{-3} & 0 \\ 0 & 1.314 \times 10^{-3} \end{pmatrix}$ |
| $\tilde{\Upsilon}$ | $\begin{pmatrix} 5.6303 \times 10^{-1} & 0 \\ 0 & 5.6304 \times 10^{-1} \end{pmatrix}$ | $\begin{pmatrix} 2.67 \times 10^{-1} & 0 \\ 0 & 2.671 \times 10^{-1} \end{pmatrix}$ |

$\Omega_f$ being the fluid part of the global domain; by splitting up $J_i$ as in identity (1), the consideration of boundary conditions

$$\boldsymbol{J}_i^{\mathrm{c}}\boldsymbol{n} = \boldsymbol{J}_i^{\mathbf{u}}\boldsymbol{n} = 0 \text{ on } \partial\Omega_{\mathrm{f}}$$

and thus

$$\boldsymbol{J}_i^{\mathrm{T}}\boldsymbol{n} = 0 \text{ on } \partial\omega_{\mathrm{f}}$$

brings us to a contradiction, as soon as one has noticed that

$$\frac{\partial T}{\partial n_{f,s}} \neq 0 \text{ on } \partial\Omega_{\mathrm{f}}$$

$n_{f,s}$ being the normal to the fluid-solid interface $\Gamma_{\mathrm{f,s}} = \partial\Omega_{\mathrm{f}}$. This contradiction comes directly from the non diagonal property of Soret effect. The $i^{th}$ component flux (which, by definition, represents a monophasic length) is provoked and expressed by a heat flux (which is defined on both fluid and solid phases). More precisely, it's the thermodynamic formalism of Soret effect which is no more adapted as soon as one considers a diphasic medium. As for diffusion in a multiphasic system, it is well known that the thermodynamic forces are chemical potential gradients and not concentration gradients (Fick's law). The main advantage of the chemical potential is that it is well defined and continuous everywhere in the system. In a close system in equilibrium, the chemical potential of any fluid component is constant. The corresponding concentration in the solid phase is very small but could still be defined. When a thermal gradient is applied to the system, the chemical potential changes slightly across the system and a concentration gradient appears in both fluid and solid phases, even if the concentrations are very small. This flux should compensate the heat flux on $\partial\Omega_{\mathrm{f}}$ and thus, equality

$$\boldsymbol{J}_i^{\mathrm{c}}\boldsymbol{n} + \boldsymbol{J}_i^{\mathrm{T}}\boldsymbol{n} = 0 \text{ on } \partial\Omega_{\mathrm{f}} \tag{27}$$

would be established.

## 7 Conclusions

A macroscopic model for Soret effect and heat transfer in porous medium has been determined. This model seems relevant as it keeps the same form as the microscopic one. Of course, the question which model is best can only be answered by experimental tests, either in the laboratory or in the field. It would be interesting to compare our model with the results of a stochastic or a direct numerical approach. More geometry configurations will be tested. We are particularly interested in non-isotropic media simulations which would give more interesting results for the Soret effect (by favouring the "main" direction of the thermal field). The most sensible conclusion we can come to, is that a general model has been given by this study. It must be tested with several configurations. It would appear, then, that the introduction of the local variations of the parameters in the local equations would give a more efficient and stronger model.

### Acknowledgements

This work was financially supported by french petroleum society Total-Fina-Elf.

## References

1. G. Allaire: Thesis, Université de Paris VI (1989)
2. E. Arquis: Transferts en milieu poreux et à l'interface: de l'échelle microscopique à l'échelle macroscopique. State thesis, Université de Bordeaux (1994)
3. E. Arquis, J.P. Caltagirone: In: Entropie **198/199**, 39 (1996)

4. A. Bensoussan, J.L. Lions, G. Papanicolaou: *Asymptotic analysis for periodic structures*, Interdisciplinary Applied Mathematics (North Holland, Amsterdam 1978)
5. A. Dmitrievsky, M. Panfilov: *Porous media: Physics, Models, Simulation*, proceedings of the conference of the same name (World Scientific, 1997)
6. B. Faissat, K. Knudsen, E.H. Stenby and F. Montel : Fluid Phase Equilibria **100**, 209 (1994)
7. D. Fargue, Ph. Jamet, P. Costesèque: Transport in Porous Media **30**, 323 (1999)
8. A. Galka, J.J. Telega, R. Wojnar: Arch. Mech. **46**, 267 (1994)
9. P. Georis, B. Lacabanne: 'Coupling between Soret effect and convection in a Poiseuille flow', In preparation
10. U. Hornung: *Homogenization and Porous Media*, Interdisciplinary Applied Mathematics (Springer, New York 1996)
11. U. Hornung, W. Jäger: J. Diff. Equ. **92**, 199 (1991)
12. P.V. Jamet: Thesis, Ecole des Mines de Paris (1991)
13. W. Köhler, B. Müller: J. Chem. Phys. **103**, 4367 (1995)
14. P. Kolodner, A. Passner, C.M. Surko and R.W. Walden: Phys. Rev. Lett. **56**, 2621 (1989)
15. F. Montel: Entropie **184/185**, 86 (1994)
16. J.K. Platten, J.C. Legros: *Convection in Liquids*, Springer Verlag., 679 p., 1990
17. M. Quintard, M. Kaviany, S. Whitaker: Advances in Water resources **20**, 77 (1997)
18. K.J. Zhang, M.E. Briggs, R.W. Gammon, J.V. Sengers: J. Chem. Phys. **104**, 6881 (1996)


## Supplementary glossary

| | |
|---|---|
| $C_p$ | specific heat |
| $D_c$ | Dufour coefficient |
| $(\boldsymbol{e}_k)_{1\leq k\leq 3}$ | space base vectors |
| $\boldsymbol{f}$ | force |
| $\phi$ | porosity |
| $\boldsymbol{J}$ | fluxes |
| $\tilde{\Lambda}$ | homogenized thermal diffusivity |
| $\mu$ | fluid phase dynamic viscosity |
| $n$ | number of component |
| $\tilde{\Upsilon}$ | homogenized diffusive coefficient |
| $Pe$ | Peclet number, $Pe = Re \times Sc = \dfrac{u_{\max} L}{D}$ |
| $Pe_{\mathrm{T}}$ | thermal Peclet number, $Pe = \dfrac{u_{\max} L}{D_{\mathrm{th}}}$ |
| $\tilde{\Sigma}$ | homogenized Soret coefficient |
| $S$ | entropy |
| $S_{\mathrm{T}}^*$ | dimensionless Soret coefficient |
| $u_{\max}$ | maximal value of the velocity |
| $Y$ | elementary cell of the medium |
| $Y_{\mathrm{f}}$ | fluid part of Y |
| $Y_{\mathrm{s}}$ | solid part of Y |
| $\partial\Omega_{\mathrm{f}}$ | fluid-solid interface |

# Index

Druck: Strauss Offsetdruck, Mörlenbach
Verarbeitung: Schäffer, Grünstadt